Physics of Fluid Flow and Transport in Unconventional Reservoir Rocks

Physics of Fluid Flow and Transport in Unconventional Reservoir Rocks

Edited by

Behzad Ghanbarian
Feng Liang
Hui-Hai Liu

WILEY

Published by John Wiley & Sons, Inc., Hoboken, New Jersey.
Published simultaneously in Canada.

Library of Congress Cataloging-in-Publication Data

Names: Ghanbarian, Behzad, editor. | Liang, Feng, editor. | Liu, Hui-Hai, editor.
Title: Physics of fluid flow and transport in unconventional reservoir rocks / edited by Behzad Ghanbarian, Feng Liang, Hui-Hai Liu.
Description: First edition. | Hoboken, NJ, USA : John Wiley & Sons, Inc., 2023. | Includes bibliographical references and index.
Identifiers: LCCN 2022052921 (print) | LCCN 2022052922 (ebook) | ISBN 9781119729877 (hardback) | ISBN 9781119727842 (adobe pdf) | ISBN 9781119729907 (epub)
Subjects: LCSH: Hydrocarbon reservoirs–Analysis. | Rocks–Permiability. | Petroleum–Migration. | Gas reservoir engineering. | Fluid dynamics. | Transport theory. | Petrology.
Classification: LCC TN870.56 .P54 2023 (print) | LCC TN870.56 (ebook) | DDC 622/.338–dc23/eng/20221230
LC record available at https://lccn.loc.gov/2022052921
LC ebook record available at https://lccn.loc.gov/2022052922

Cover Design: Wiley
Cover Image: © BalLi8Tic/Shutterstock

Set in 9.5/12.5pt STIXTwoText by Straive, Pondicherry, India

Our families for their countless support, endless inspiration, and more importantly unconditional love.

Contents

List of Contributors

Stacey M. Althaus
Aramco Americas: Aramco Research Center—Houston
Houston, TX, USA

Mustafa Basri
Saudi Aramco, Dhahran
Saudi Arabia

Athma R. Bhandari
Institute for Geophysics, Jackson School of Geosciences
The University of Texas at Austin
Austin, TX, USA

Mohammed Boudjatit
Saudi Aramco, Dhahran
Saudi Arabia

Huangye Chen
Aramco Americas: Aramco Research Center—Houston
Houston, TX, USA

Jin-Hong Chen
Aramco Americas: Aramco Research Center—Houston
Houston, TX, USA

Li Chen
Key Laboratory of Thermo-Fluid Science and Engineering of MOE
School of Energy and Power Engineering
Xi'an Jiaotong University
Xi'an, Shaanxi, China

Xiaona Cui
SimTech LLC
Katy, TX, USA

Derek Elsworth
Department of Energy and Mineral Engineering
EMS Energy Institute and G3 Center
Penn State University
University Park, PA
USA

Misagh Esmaeilpour
Porous Media Research Lab
Department of Geology
Kansas State University
Manhattan, KS, USA

Peter B. Flemings
Institute for Geophysics
Jackson School of Geosciences
The University of Texas at Austin
Austin, TX, USA
and
Department of Geological Sciences
Jackson School of Geosciences
The University of Texas at Austin
Austin, TX, USA

Zhiye Gao
State Key Laboratory of Petroleum Resources and Prospecting
China University of Petroleum, Beijing, China
and
Unconventional Petroleum Research Institute,
China University of Petroleum, Beijing, China

Behzad Ghanbarian
Porous Media Research Lab
Department of Geology
Kansas State University
Manhattan, KS, USA

Yanhui Han
Aramco Americas: Aramco Research Center—Houston
Houston, TX, USA

Qinhong Hu
Department of Earth and Environmental Sciences
University of Texas at Arlington
Arlington, TX, USA

Nayif Jama
Saudi Aramco, Dhahran, Saudi Arabia

Qinjun Kang
Earth and Environmental Sciences Division
Los Alamos National Laboratory
Los Alamos, NM, USA

Feng Liang
Aramco Americas: Aramco Research Center—Houston
Houston, TX, USA

Chuxi Liu
The University of Texas at Austin
Hildebrand Department of Petroleum and Geosystems Engineering
Austin, TX, USA

Hui-Hai Liu
Aramco Americas: Aramco Research Center—Houston
Houston, TX, USA

Rabah Mesdour
Saudi Aramco, Dhahran, Saudi Arabia

Jijun Miao
SimTech LLC
Katy, TX, USA

Feifei Qin
Chair of Building Physics
Department of Mechanical and Process Engineering
ETH Zürich, Zurich, Switzerland

Sebastian Ramiro-Ramirez
Institute for Geophysics, Jackson School of Geosciences
The University of Texas at Austin
Austin, TX, USA
and
Department of Geological Sciences
Jackson School of Geosciences
The University of Texas at Austin
Austin, TX, USA

Brandon Schwartz
Department of Energy and Mineral Engineering, EMS Energy Institute and G3 Center
Penn State University
University Park, PA, USA

Kamy Sepehrnoori
The University of Texas at Austin
Hildebrand Department of Petroleum and Geosystems Engineering
Austin, TX, USA

Pejman Tahmasebi
Colorado School of Mines
Golden 80401, USA

Cenk Temizel
Saudi Aramco
Dhahran, Saudi Arabia

Hari Viswanathan
Earth and Environmental Sciences Division
Los Alamos National Laboratory
Los Alamos, NM, USA

Junjian Wang
Klohn Crippen Berger Ltd
Brisbane, Queensland
Australia

Yuqi Wu
Shandong Provincial Key Laboratory of Deep Oil and Gas
China University of Petroleum (East China)
Qingdao, China
and
School of Geosciences, China University of Petroleum (East China), Qingdao, China

Hongbing Xie
SimTech LLC
Katy, TX, USA

Wei Yu
SimTech LLC
Katy, TX, USA
and
The University of Texas at Austin
Hildebrand Department of Petroleum and Geosystems Engineering
Austin, TX, USA

Jilin Zhang
Aramco Americas: Aramco Research Center—Houston
Houston, TX, USA

Jianlin Zhao
Chair of Building Physics
Department of Mechanical and Process Engineering
ETH Zürich
Zurich, Switzerland

Preface

This book provides basic concepts and recent advances of fluid flow and transport in unconventional reservoirs across different scales (from pore to core and to reservoir) for a broad range of audiences from various scientific disciplines, such as geology, geoscience, geochemistry, geophysics, rock mechanics, and petroleum engineering. In the Introduction chapter, we address recent progress and ongoing challenges related to hydrocarbon exploration and production in tight and ultra-tight formations. The first part of the book on pore-scale characterizations consists of three chapters. In the second chapter, Wang and his coworkers present an overview of recent progress on pore-scale simulations and digital rock physics to unconventional reservoir rocks. They emphasize that further small-scale experiments are still required to validate numerical models. In Chapter 3, Wu and Tahmasebi review digital rock models. They discuss that incorporating multiresolution and multiscale structures and generating large-scale digital models are still very challenging. In Chapter 4, Ghanbarian and Esmaeilpour address the effect of scale on permeability and formation factor. They present a simple scaling law and show reasonable agreement between theoretical estimations and pore-network simulations. Part II of the book on core-scale heterogeneity includes seven chapters. In Chapter 5, Ghanbarian et al. study theoretical modeling of single-phase and gas relative permeabilities in shales and tight porous rocks. They apply concepts of the effective-medium approximation and demonstrate that by including the physics of gas flow, one can estimate permeability reasonably well at the core scale. Chapter 6 by Chen and his coworkers addresses applications of nuclear magnetic resonance and its recent advances to determine total porosity and partial porosity in organic matter of unconventional reservoir rocks. In Chapter 7, recent progress on tight rock permeability measurement is addressed by Liu, Zhang, and Boudjatit. They present two newly developed laboratory methods and evaluate them using laboratory measurements. Chapter 8 by Bhandari et al. presents permeability evolution under cycling confining stress conditions. They demonstrate that micro-fractures might be closed due to confining stress leading to permeability reduction under cyclic loading. In Chapter 9, Gao and Hu provide insights into shale wettability using spontaneous imbibition experiments. Their results demonstrate the co-existence of water and oil in the pore network of shales proving their mixed-wet characteristics. In Chapter 10, Schwartz and Elsworth study permeability enhancement in shales induced by desorption. Those authors argue that the magnitude of permeability enhancement depends on the distribution of sorptive mineral components, geometry of flow path, and initial permeability. Chapter 11 by Liang, Liu, and Zhang provides insights that help improve oil and gas production in unconventional reservoirs. More specifically, their experimental evidence shows that aqueous-based fracturing fluid may have positive impacts on gas production from organic-rich carbonate source rocks. The last part of the book focuses on large-scale petrophysics of unconventional reservoirs, which has broad applications to field.

In Chapter 12, Ghanbarian proposes percolation-based effective-medium theory to model effective permeability in matrix-fracture systems. By comparing with numerical simulations, he shows that effective permeability can be accurately estimated at different fracture densities. In Chapter 13, Xie et al. apply embedded discrete fracture model to simulate fluid flow in complex fracture networks. They address the effect of natural fracture properties, such as fracture azimuth, length, and dip angle. Chapter 14 by Liu and his coauthors presents a closed-form relationship for production rate. By comparing with numerical simulations, Liu et al. validate their proposed model for different initial reservoir pressures, pressure drawdowns, and pressure sensitivity factors for permeability. In the last chapter, Mesdour et al. discuss sweet spots and their identification in shale reservoirs. They provide a state-of-art review of existing methods developed for sweet spot identification and address relevant challenges and knowledge gaps.

Many colleagues and students contributed to our understanding of fluid flow and transport in unconventional reservoirs. We are grateful to those who helped us with this book. We also acknowledge those who contributed to this book by writing different chapters on several topics. We hope this book helps geologists and petroleum engineers in industry as well as faculty and students in academia.

Behzad Ghanbarian
Feng Liang
Hui-Hai Liu
January 2023

Introduction

1

Unconventional Reservoirs

Advances and Challenges

Behzad Ghanbarian[1], Feng Liang[2], and Hui-Hai Liu[2]

[1] *Porous Media Research Lab, Department of Geology, Kansas State University, Manhattan, KS, USA*
[2] *Aramco Americas: Aramco Research Center–Houston, Houston, TX, USA*

1.1 Background

Energy is one of the most important components in the world. Primary sources of energy take various forms, such as fossil energy, nuclear energy, and renewable energy sources. Fossil energy resources (e.g. coal, oil, and natural gas) were formed when plants and animals died and were buried underground. The quality of hydrocarbon accordingly depends on organic content as well as temperature and pressure conditions. Although there are limited reserves of fossil energy resources and despite recent advances in renewable energy, global economy still depends on fossil fuels to a great extent (Figure 1.1). Statistics reported by the British Petroleum (BP) company based on data from 1994 to 2019 show that the world primary energy consumption growth in 2019 slowed to 1.3%. This is less than half the growth rate i.e. 2.8% in 2018. Three-quarters of the energy consumption increase was driven by natural gas and renewable resources in 2019.

Based on analyses reported by the BP company, oil has contributed to the share of global primary energy more than others since 1994 (Figure 1.2) with 33.1% contribution in 2019. After oil, coal and natural gas are the second and third largest contributors. Although coal lost its share to account for nearly 27%, the contribution of natural gas increased to 24% in 2019. The share of renewable resources rose to record highs of 5% in 2019, and they overtook nuclear energy with about 4% contribution. Figure 1.2 shows the share of hydroelectricity has been nearly constant and about 6%.

Unconventional reservoirs, including oil and gas shales and tight sandstones, are distributed around the world (Figure 1.3) with an estimated endowment of several thousand trillion cubic feet (Kim et al. 2000). Since shale reservoirs have been successfully explored and produced in the United States (Figure 1.3), they recently became one of the major contributors to energy supplies.

There exist three general types of unconventional reservoirs, i.e. (i) organic-rich source rocks, (ii) tight oil reservoirs, and (iii) hybrid plays in which production occurs from source rocks and conventional reservoirs (Zoback and Kohli 2019). These types of unconventional reservoirs are different in geologic formations and, therefore, should be optimally exploited using different and appropriate approaches.

Despite numerous practical applications in oil/gas exploration and production as well as recent progress, we are still far from fully understanding all mechanisms of flow and transport in shales

Physics of Fluid Flow and Transport in Unconventional Reservoir Rocks, First Edition.
Edited by Behzad Ghanbarian, Feng Liang, and Hui-Hai Liu.

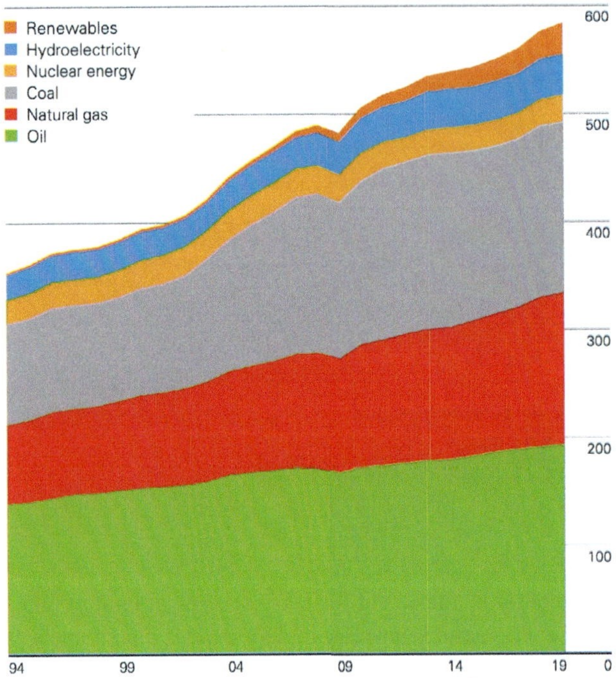

Figure 1.1 World total energy consumption between 1994 and 2019. *Source:* BP Statistical Review of World Energy (2020)/BP International Limited.

and tight sandstones across scales, *particularly from pore to reservoir*. In the following, we briefly address recent advances in unconventional reservoirs and discuss current challenges in oil and gas exploration and production.

1.2 Advances

Since 2005, the beginning of shale gas revolution in the United States, unconventional oil and gas resources as well as their developments and productions have received a remarkable amount of attention around the world (Zoback and Kohli 2019). Despite various challenges that still exist, the petroleum engineering community made tremendous progress, particularly in the past decade. In what follows, we briefly address several notable achievements. For further details and comprehensive recent advances, see e.g. Barati and Alhubail (2020), Rezaee (2021), and Moghanloo (2022).

1.2.1 Wettability

Characterizing the contact angle of fluids (e.g. water, oil, and gas) and its spatial variability within unconventional reservoirs and under in situ conditions are essential not only to understand the trapping phenomenon and enhance oil and gas recovery but also to improve greenhouse gas (e.g. carbon dioxide and hydrogen) sequestration underground. In the literature, various methods, such as contact angle measurements (Iglauer et al. 2015; Roshan et al. 2016), spontaneous imbibition (Liu et al. 2019; Siddiqui et al. 2019), and nuclear magnetic resonance (Odusina et al. 2011; Su et al. 2018) were proposed to determine wettability in unconventional reservoir rocks. Recently, Arif et al. (2021) collected

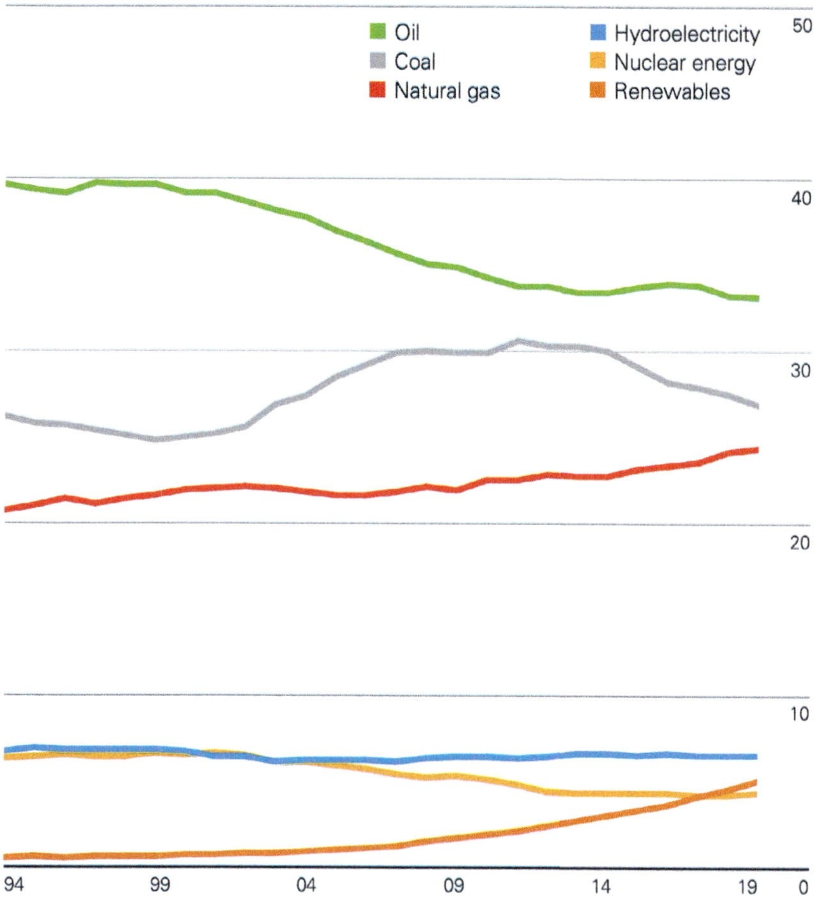

Figure 1.2 Shares of global primary energy between 1994 and 2019 (BP Statistical Review of World Energy 2020/BP International Limited).

published data on shale contact angle measurements and developed a repository. They concluded that the oil-brine mixture in shales behaved in terms of wettability over a wide range from water-wet to strongly oil-wet. Although the CO_2-brine mixture typically showed weakly water-wet to CO_2-wet behavior, the CH_4-brine mixture in shales was weakly water-wet. Arif et al. (2021) also investigated what causes high variabilities in shale wettability and found that the main factors were pressure, temperature, thermal maturity, total organic content, and mineralogy of shales.

Although our knowledge of shale wettability has improved, further investigations are still needed to study the solid–fluid and fluid–fluid contact angles under realistic reservoir conditions more comprehensively. This would help enhance oil and gas recovery and exploit unconventional reservoirs even more successfully.

1.2.2 Permeability

Liquid and gas transports in shales and tight porous rocks were widely studied, particularly at the pore and core levels. The literature on gas permeability and its modeling is indeed vast and extensive (Javadpour et al. 2021; Liu 2017; Tahmasebi et al. 2020; Zhang et al. 2019). Numerous models

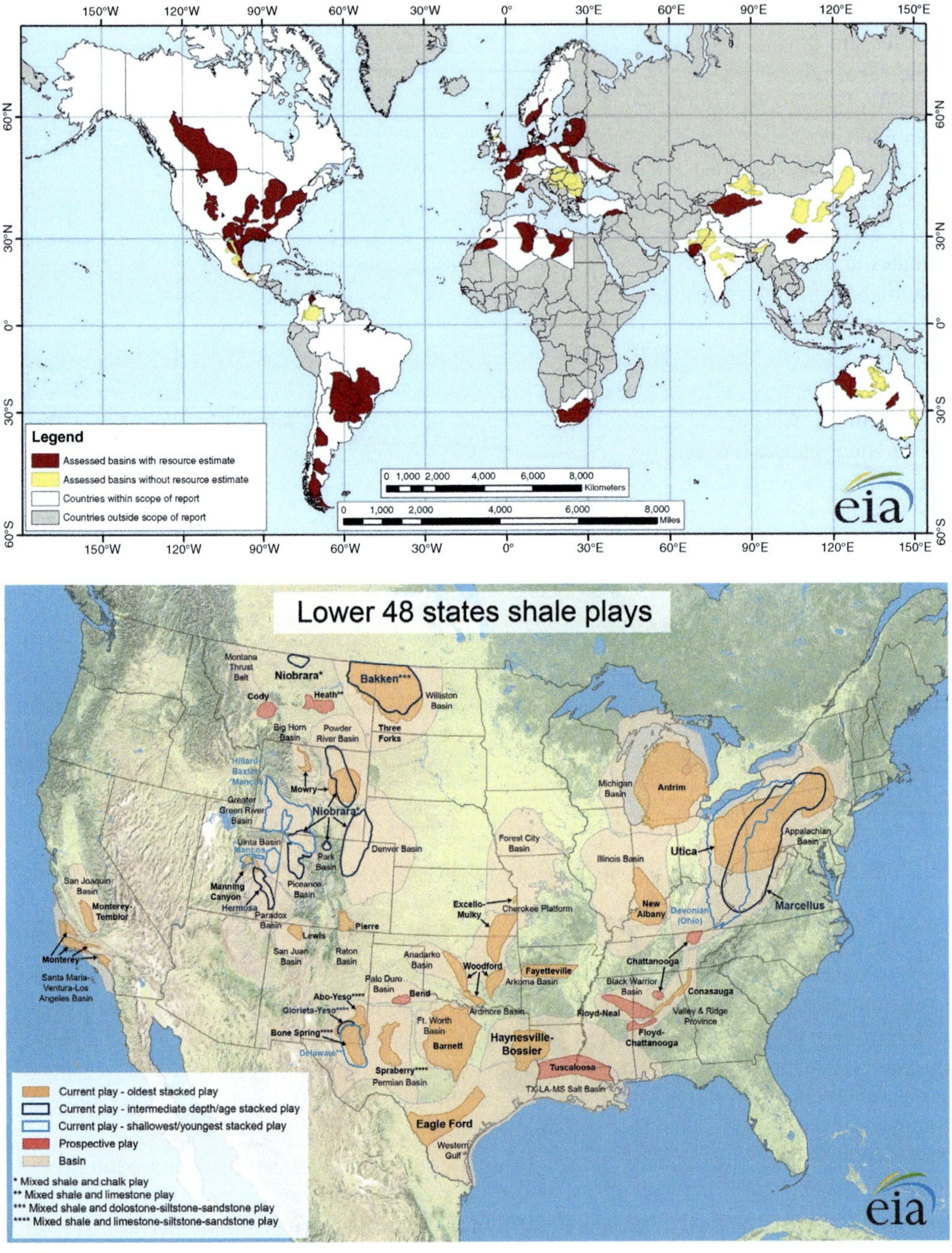

Figure 1.3 World shale gas resources (top) and shale gas and oil plays in the United States (bottom). *Source:* Both maps are from US Energy Information Administration.

were developed to address gas flow in nanostructures of shales by taking the effect of different transport mechanisms, such as slip flow, Knudsen diffusion, surface diffusion, and sorption into account. For example, Beskok and Karniadakis (1999) incorporated the effect of slip flow and modified the Poiseuille equation to describe gas flow in a cylindrical tube. Civan (2010) later applied the

Beskok–Karniadakis model to scale up gas permeability in a network of pores. Using concepts from first-order slip flow and Knudsen diffusion, Javadpour (2009) developed another model for gas transport in a nano-scale cylindrical tube.

Recently, in addition to slip flow and Knudsen diffusion, more complex mechanisms such as sorption and surface diffusion were incorporated to model gas flow in shales. For instance, Wu et al. (2016) proposed a unified theoretical model by coupling various mechanisms including slip flow, Knudsen diffusion, sorption, and surface diffusion. Those authors stated that "... surface diffusion is an important transport mechanism, and its contribution cannot be negligible and even dominates in nanopores with less than 2 nm in diameter." Jia et al. (2018) found that surface diffusion might increase gas flow significantly at low pore pressures (e.g. < 2000 psi) depending on the value of surface diffusivity.

In the literature, most theoretic models developed to study gas permeability are based on the bundle of tortuous capillary tubes approach. However, it is an oversimplifying idealization in which a porous medium with actual interconnected pores is replaced with non-interconnected tortuous tubes of equal length (Purcell 1949). An important inconsistency between the bundle of tubes model and real porous media was noted by Fatt (1956). Fatt's main objection to the parallel-tubes model was that it has no connections between the individual tubes, in contrast to the network models that he developed. An additional problem is that, in a bundle-of-tubes model, individual pores span the entire sample or problem domain, regardless of its size.

Instead of the bundle of tortuous tubes approach, one may theoretically model gas transport in shales using more appropriate upscaling techniques e.g. percolation theory, critical path analysis (CPA), and effective-medium approximation (EMA) from statistical physics (Ghanbarian et al. 2020; Hunt et al. 2014; Sahimi 2011). Such approaches take into account the effect of pore connectivity. For example, Zhang and Scherer (2012) applied the CPA approach to estimate Klinkenberg-corrected permeability in shales from mercury intrusion capillary pressure curve and formation factor. They also used the Kozeny–Carman model, compared their estimations with experiments, and found that measured permeability values matched with CPA estimations more accurately. In another study, Ghanbarian and Javadpour (2017) assumed that slip flow and Knudsen diffusion are dominant transport mechanisms and employed the EMA to estimate the pore pressure-dependent gas permeability in shales. By comparing with experiments, pore-network models, and lattice-Boltzmann simulations, they found that their proposed model estimated gas permeability within a factor of 3 of measurements and/or simulations.

1.3 Challenges

Over the past decade, productions from unconventional reservoirs were successful in the United States and Canada. However, there are several outstanding issues that critically impact hydrocarbon recovery in tight and ultra-tight formations. In the following, we discuss some notable issues briefly.

1.3.1 Multiscale Systems

Unconventional reservoirs are extremely complex multiscale formations composed of nano- and micro-scale pores as well as naturally and hydraulically induced fractures of various sizes. Morphology, pore structure, and spatial/temporal heterogeneities are important, as these factors affect fluid transport in unconventional reservoirs. Figure 1.4a and b show nano- and micro-scale pores within organic and inorganic matrices, respectively, and Figure 1.4c shows microfractures.

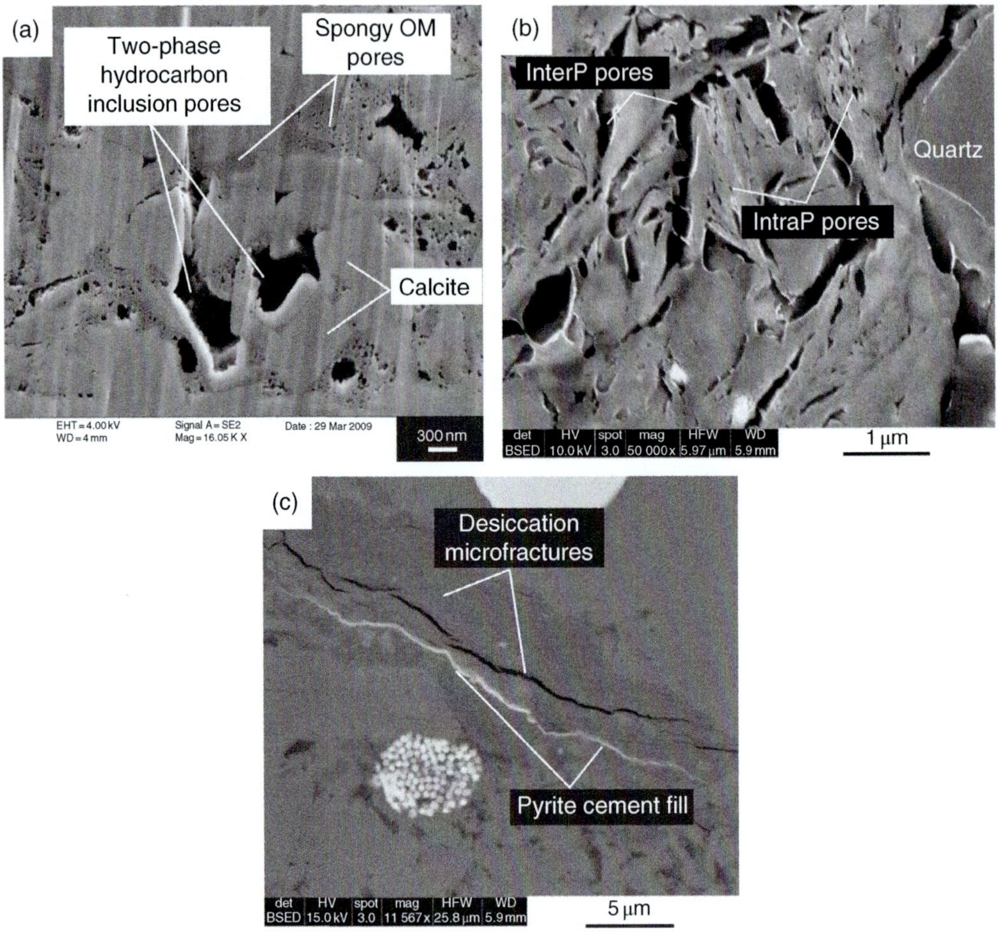

Figure 1.4 (a) SEM image of contiguous spongy-textured organic matter pore network. *Source:* Loucks and Reed (2014)/from Gulf Coast Association of Geological Societies. (b) Intra-particle pores located along cleavage planes of clay particles. *Source:* Loucks et al. (2012)/from American Association of Petroleum Geologists. (c) Irregular natural microfracture filled with pyrite *Source:* Loucks and Reed (2016)/from Gulf Coast Association of Geological Societies.

Maximizing the production of unconventional reservoirs is limited in part by the incomplete understanding of mass transport mechanisms in tight and ultra-tight reservoir rocks at the pore and even molecular scale. In the literature, various approaches, such as scan electron microscopy (Wu and Aguilera 2012), focused ion beam scanning electron microscopy (Shabro et al. 2014), atomic force microscopy (Javadpour et al. 2012; Wu and Aguilera 2012), and transmission electron microscopy (Chalmers et al. 2012; Froute and Kovscek 2020) were extensively applied to characterize the complexity and heterogeneity of pore space in unconventional reservoir rocks.

Although molecular dynamics simulations (Castez et al. 2017; Fang et al. 2017; He et al. 2020; Huang et al. 2018; Ungerer et al. 2015; Zhang et al. 2020) have been applied to gain molecular insights of transport processes, mass transport dynamics investigations in real shale rocks are still very limited. Characterization of mass transport dynamics in shales is of contemporary interest in the oil and gas industry. Shales may be permeable to some hydrocarbons and impermeable to others. Transport rates are controlled by partitioning between different phases, adsorption, and desorption from pore surfaces, and by diffusion within nano-scale pores. These molecular-level processes are dictated by chemical and electrostatic interactions. A better understanding of the molecular mechanisms of mass transport is certain to facilitate oil and gas production.

1.3.2 Hydrocarbon Production

Another issue associated with unconventional reservoirs is that production rates rapidly decrease in the first few years of production (Hakso and Zoback 2019; Sandrea and Sandrea 2014; Speight 2019).

Patzek et al. (2013) developed a simple but robust gas production model using basic concepts and geometry of the extraction process. In their model, the production rate scales with time as $t^{-0.5}$ at early times, however, declines exponentially at late times. By comparison with gas production from 8294 wells within the Barnett shale, they showed that their model provided accurate scaling of gas production. Patzek et al. (2013) also found reasonable agreement between their theory and experimental data from 2057 horizontal wells. They argued that in young wells, it would be difficult to determine the crossover between the square-root decay and exponential decay. However, their model was capable of determining lower and upper bounds.

More recently, Hakso and Zoback (2019) analyzed hydrocarbon production data from two oil (Eagle Ford and Bakken) and two gas (Marcellus and Barnett) unconventional reservoirs. The average monthly data are presented in Figure 1.5 (Hakso and Zoback 2019). To eliminate uncertainties, Hakso and Zoback (2019) removed those wells that did not produce hydrocarbon for at least two years. Each curve shown in Figure 1.5 represents the average over more than 7500 wells from each reservoir. As can be seen, monthly production rates decline sharply during the first two to three years of production, consistent with the results of Patzek et al. (2013).

1.3.3 Recovery Factor

Recovery factor is the recoverable amount of hydrocarbon initially in place whose value is relatively low for both shale gas and tight oil wells. Although it is uncertain to accurately determine the amount of total hydrocarbon in place, the recovery factor in tight gas reservoirs is around 25% after a few years of production (Zoback and Kohli 2019). For tight oil, various studies reported recovery factors about 2–10% (Hamdi et al. 2018; Pankaj et al. 2018; Wang et al. 2019). Low and extremely low recovery factors for gas and oil reservoirs indicate that a considerable amount of hydrocarbon is not produced from unconventional resources, particularly compared to conventional reservoirs. One promising enhanced oil recovery approach to improve recovery factor is gas Huff and Puff (Hoffman 2012, 2018). Various types of gas, such as carbon dioxide, methane, air, immiscible hydrocarbon, and miscible hydrocarbon, were used in the gas Huff and Puss process (Firouz and Torabi 2014; Sanchez-Rivera et al. 2015; Soh and Babadagli 2018). Particularly, the CO_2 Huff and Puff process has been applied in the field for several decades. However, there still exist challenges, such as asphaltene precipitation, corrosion, and viscous fingering, associated with field implementations (Zhou et al. 2018). One should note that the price of oil and cost of CO_2 capture and storage may also play determinative roles.

1.3.4 Unproductive Wells

Another issue in unconventional reservoirs is that a large number of drilled wells contribute minimally to hydrocarbon production (Zoback and Kohli 2019). It was reported that around 30% of shale wells were not productive (Al-Nakhli et al. 2020; Miller et al. 2011). Although unproductive

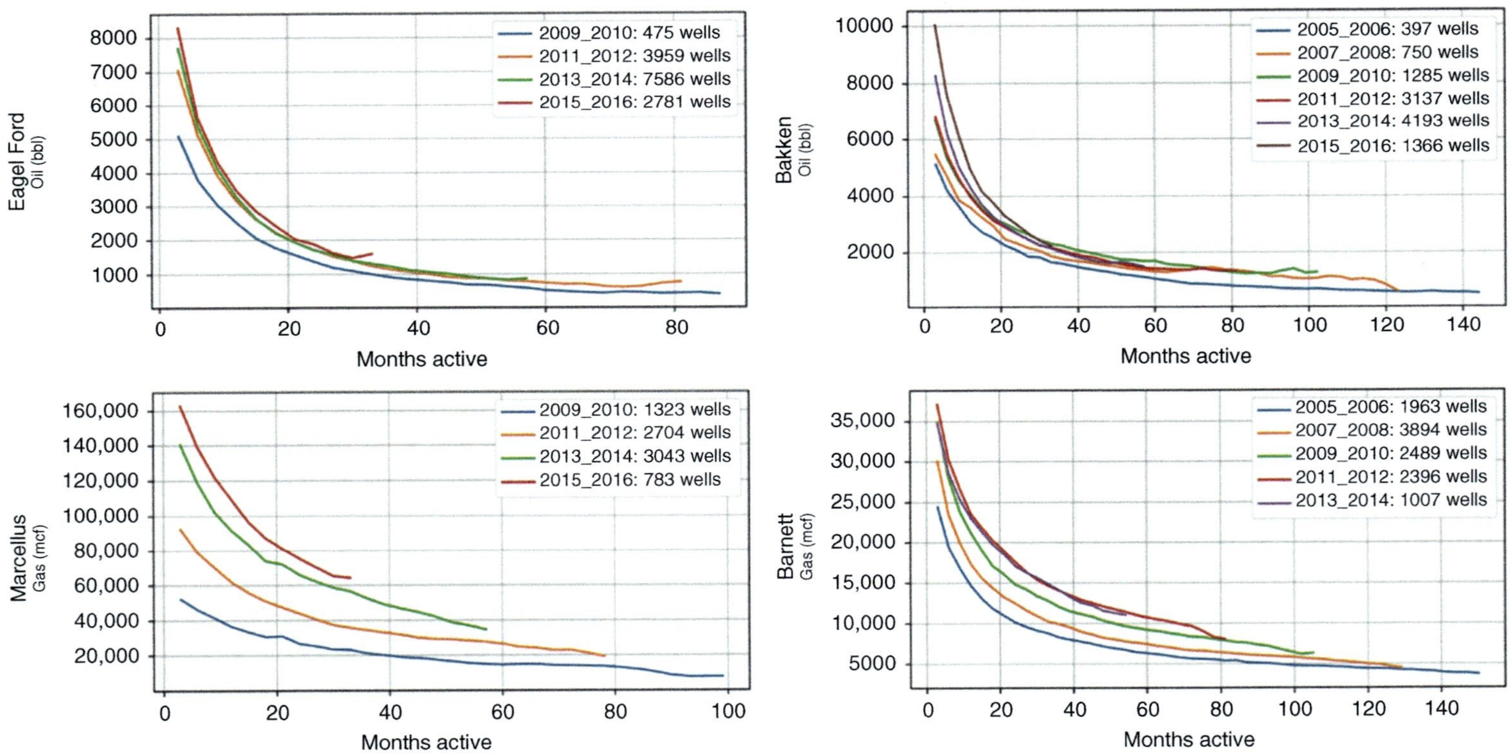

Figure 1.5 Monthly production for two oil and two gas reservoirs (*Source:* Hakso and Zoback (2019)/Society of Exploration Geophysicists).

wells may be used for CO_2 sequestration or enhanced oil recovery purposes, high priorities should be given to produce hydrocarbon more optimally from unconventional reservoirs by drilling more economic wells.

1.4 Concluding Remarks

In this chapter, we briefly reviewed several recent advances in unconventional oil and gas reservoirs. We noted that our knowledge of shale wettability as well as gas and liquid transport at the pore and core levels increased in the past decade. However, there still exist challenges, such as low hydrocarbon production and recovery factors. We also discussed that a considerable number of drilled wells in unconventional formations are unproductive. Although various technologies were developed to improve production, further investigations are still required to substantially increase production rates in tight and ultra-tight reservoirs.

The major purpose of this book is to, in a systematic way, present the most recent developments in addressing some of these technical challenges in the areas of pore-scale characterizations, core-scale heterogeneity, and large-scale petrophysics.

References

Al-Nakhli, A.R., Tariq, Z., Mahmoud, M., and Abdulraheem, A. (2020). A state-of-the-art technology to reduce fracturing pressure in tight gas formations using thermochemical pulse. *SEG Global Meeting Abstracts. Unconventional Resources Technology Conference*, pp. 177–1187 (20–22 July). Society of Exploration Geophysicists. https://doi.org/10.15530/urtec-2020-2439.

Arif, M., Zhang, Y., and Iglauer, S. (2021). Shale wettability: data sets, challenges, and outlook. *Energy & Fuels* 35: 2965–2980.

Barati, R. and Alhubail, M.M. (2020). *Unconventional Hydrocarbon Resources: Techniques for Reservoir Engineering Analysis*. Wiley.

Beskok, A. and Karniadakis, G.E. (1999). A model for flows in channels, pipes, and ducts at micro and nano scales. *Microscale Thermophysical Engineering* 3 (1): 43–77. https://doi.org/10.1080/108939599199864.

Looney, B. (2020). Statistical Review of World Energy, 2020. BP, vol. 69, pp. 1–66.

Castez, M.F., Winograd, E.A., and Sanchez, V.M. (2017). Methane flow through organic-rich nanopores: the key role of atomic-scale roughness. *The Journal of Physical Chemistry C* 121: 28527–28536.

Chalmers, G.R., Bustin, R.M., and Power, I.M. (2012). Characterization of gas shale pore systems by porosimetry, pycnometry, surfacearea, andfield emission scanning electron microscopy/transmission electron microscopy image analyses: examples from the Barnett, Woodford, Haynesville, Marcellus, and Doig units. *AAPG Bulletin* 96 (6): 1099–1119. https://doi.org/10.1306/10171111052.

Civan, F. (2010). Effective correlation of apparent gas permeability in tight porous media. *Transport in Porous Media* 82 (2): 375–384. https://doi.org/10.1007/s11242-009-9432-z.

Fang, T., Wang, M., Wang, C. et al. (2017). Oil detachment mechanism in CO_2 flooding from silica surface: molecular dynamics simulation. *Chemical Engineering Science* 164: 17–22.

Fatt, I. (1956). The network model of porous media: I. Capillary pressure characteristics. *Transaction of American Institute of Mining, Metallurgical, and Petroleum Engineers (AIME)* 207: 144–159.

Firouz, A.Q. and Torabi, F. (2014). Utilization of carbon dioxide and methane in huff-and-puff injection scheme to improve heavy oil recovery. *Fuel* 117: 966–973.

Froute, L. and Kovscek, A.R. (2020). Nano-imaging of shale using electron microscopy techniques. Paper presented at the SPE/AAPG/SEG Unconventional Resources Technology Conference, Virtual (20 July), URTEC-2020-3283-MS.

Ghanbarian, B. and Javadpour, F. (2017). Upscaling pore pressure-dependent gas permeability in shales. *Journal of Geophysical Research: Solid Earth* 122 (4): 2541–2552. https://doi.org/10.1002/2016JB013846.

Ghanbarian, B., Liang, F., and Liu, H.H. (2020). Modeling gas relative permeability in shales and tight porous rocks. *Fuel* 272: 117686.

Hakso, A. and Zoback, M. (2019). The relation between stimulated shear fractures and production in the Barnett shale: implications for unconventional oil and gas reservoirs. *Geophysics* 84: B461–B469.

Hamdi, H., Clarkson, C.R., Ghanizadeh, A. et al. (2018). Huff-n-puff gas injection performance in shale reservoirs: a case study from Duvernay Shale in Alberta, Canada. *Unconventional Resources Technology Conference*, pp. 2919–2943, Houston, Texas (23–25 July). Society of Exploration Geophysicists, American Association of Petroleum Geologists, Society of Petroleum Engineers.

He, J., Ju, Y., Lammers, L. et al. (2020). Tortuosity of kerogen pore structure to gas diffusion at molecular- and nano-scales: a molecular dynamics simulation. *Chemical Engineering Science* 215: 115460.

Hoffman, B.T. (2012). Comparison of various gases for enhanced recovery from shale oil reservoirs. Paper presented at the SPE Improved Oil Recovery Symposium, Tulsa, Oklahoma, USA (14 April), SPE-154329-MS.

Hoffman, B.T. (2018). Huff-N-Puff gas injection pilot projects in the Eagle Ford. Paper presented at the SPE Canada Unconventional Resources Conference, Calgary, Alberta, Canada (13 March), SPE-189816-MS.

Huang, L., Ning, Z., Wang, Q. et al. (2018). Molecular simulation of adsorption behaviors of methane, carbon dioxide and their mixtures on kerogen: effect of kerogen maturity and moisture content. *Fuel* 211: 159–172.

Hunt, A., Ewing, R., and Ghanbarian, B. (2014). *Percolation Theory for Flow in Porous Media*. Springer.

Iglauer, S., Al-Yaseri, A.Z., Rezaee, R., and Lebedev, M. (2015). CO_2 wettability of caprocks: implications for structural storage capacity and containment security. *Geophysical Research Letters* 42: 9279–9284.

Javadpour, F. (2009). Nanopores and apparent permeability of gas flow in mudrocks (shales and siltstone). *Journal of Canadian Petroleum Technology* 48 (8): 16–21.

Javadpour, F., Moravvej Farshi, M., and Amrein, M. (2012). Atomic-force microscopy: a new tool for gas-shale characterization. *Journal of Canadian Petroleum Technology* 51: 236–243.

Javadpour, F., Singh, H., Rabbani, A. et al. (2021). Gas flow models of shale: a review. *Energy & Fuels* 35: 2999–3010.

Jia, B., Tsau, J.S., and Barati, R. (2018). A workflow to estimate shale gas permeability variations during the production process. *Fuel* 220: 879–889.

Kim, E.M., Tinker, S.W., Fisher, W.L., and Ruppel, S.C. (2000). Benefit/cost analysis of GRI's gas supply research initiative: economics of a gas supply research trust fund to increase gas production of Federal lands: the University of Texas at Austin. *Bureau of Economic Geology*, final report prepared for Gas Research.

Liu, H. (2017). *Fluid Flow in the Subsurface*. Springer https://doi.org/10.1007/978-3-319-43449-0.

Liu, J., Sheng, J.J., Wang, X. et al. (2019). Experimental study of wettability alteration and spontaneous imbibition in Chinese shale oil reservoirs using anionic and nonionic surfactants. *Journal of Petroleum Science and Engineering* 175: 624–633.

Loucks, R.G. and Reed, R.M. (2014). Scanning-electron-microscope petrographic evidence for distinguishing organic-matter pores associated with depositional organic matter versus migrated organic matter in mudrock. *GCAGS Journal* 3: 51–60.

Loucks, R.G. and Reed, R.M. (2016). Natural microfractures in unconventional shale-oil and shale-gas systems: real, hypothetical, or wrongly defined? *GCAGS Journal* 5: 64–72.

Loucks, R.G., Reed, R.M., Ruppel, S.C., and Hammes, U. (2012). Spectrum of pore types and networks in mudrocks and a descriptive classification for matrix-related mudrock pores. *AAPG Bulletin* 96: 1071–1098.

Miller, C., Waters, G., and Rylander, E. (2011). Evaluation of production log data from horizontal wells drilled in organic shales. Paper presented at the North American Unconventional Gas Conference and Exhibition, The Woodlands, Texas, USA (14 June), SPE-144326-MS.

Moghanloo, R.G. (2022). *Unconventional Shale Gas Development: Lessons Learned*. Gulf Professional Publishing.

Odusina, E., Sondergeld, C., and Rai, C. (2011). An NMR study on shale wettability. Paper presented at the Canadian Unconventional Resources Conference, Calgary, Alberta, Canada (15 November), SPE-147371-MS.

Pankaj, P., Mukisa, H., Solovyeva, I., and Xue, H. (2018). Enhanced oil recovery in eagle ford: opportunities using huff-n-puff technique in unconventional reservoirs. Paper presented at the SPE Liquids-Rich Basins Conference, North America, Midland, Texas, USA (5 September), SPE-191780-MS.

Patzek, T.W., Male, F., and Marder, M. (2013). Gas production in the Barnett shale obeys a simple scaling theory. *Proceedings of the National Academy of Sciences* 110: 19731–19736.

Purcell, W.R. (1949). Capillary pressures – their measurement using mercury and the calculation of permeability therefrom. *Journal of Petroleum Technology* 1 (2): 39–48. https://doi.org/10.2118/949039-G.

Rezaee, R. (2021). *Development of Unconventional Reservoirs 2020*. MDPI.

Roshan, H., Al-Yaseri, A., Sarmadivaleh, M., and Iglauer, S. (2016). On wettability of shale rocks. *Journal of Colloid and Interface Science* 475: 104–111.

Sahimi, M. (2011). *Flow and Transport in Porous Media and Fractured Rock: From Classical Methods to Modern Approaches*, 2nde. Weinheim, Germany: Wiley-VCH.

Sanchez-Rivera, D., Mohanty, K., and Balhoff, M. (2015). Reservoir simulation and optimization of huff-and-puff operations in the Bakken shale. *Fuel* 147: 82–94.

Sandrea, R. and Sandrea, I. (2014). New well-productivity data provide US shale potential insights. *Oil & Gas* 112: 66–77.

Shabro, V., Kelly, S., Torres-Verdín, C. et al. (2014). Pore-scale modeling of electrical resistivity and permeability in FIB-SEM images of organic mudrock. *Geophysics* 79 (5): D289–D299. https://doi.org/10.1190/geo2014-0141.1.

Siddiqui, M.A.Q., Chen, X., Iglauer, S., and Roshan, H. (2019). A multiscale study on shale wettability: spontaneous imbibition versus contact angle. *Water Resources Research* 55: 5012–5032.

Soh, Y.J. and Babadagli, T. (2018). Use of air to improve the efficiency of foamy flow and reservoir pressurization in heavy oil recovery. *Journal of Petroleum Science and Engineering* 170: 166–176.

Speight, J.G. (2019). *Shale Oil and Gas Production Processes*. Gulf Professional Publishing.

Su, S., Jiang, Z., Shan, X. et al. (2018). The wettability of shale by NMR measurements and its controlling factors. *Journal of Petroleum Science and Engineering* 169: 309–316.

Tahmasebi, P., Javadpour, F., and Enayati, S.F. (2020). Digital rock techniques to study shale permeability: a mini-review. *Energy & Fuels* 34: 15672–15685.

Ungerer, P., Collell, J., and Yiannourakou, M. (2015). Molecular modeling of the volumetric and thermodynamic properties of kerogen: influence of organic type and maturity. *Energy Fuels* 29: 91–105.
Wang, M., Wang, L., Zhou, W., and Yu, W. (2019). Lean gas huff and puff process for eagle ford shale: methane adsorption and gas trapping effects on EOR. *Fuel* 248: 143–151.
Wu, P. and Aguilera, R. (2012). Investigation of shale gas at nanoscale using scan electron microscopy, transmission electron microscopy and atomic force microscopy, and up-scaling to a petrophysical model for water saturation evaluation in shales. Paper presented at the SPE Annual Technical Conference and Exhibition, San Antonio, Texas, USA (8 October), SPE-159887-MS.
Wu, K., Chen, Z., Li, X. et al. (2016). A model for multiple transport mechanisms through nanopores of shale gas reservoirs with real gas effect-adsorption-mechanic coupling. *International Journal of Heat and Mass Transfer* 93: 408–426. https://doi.org/10.1016/j.ijheatmasstransfer.2015.10.003.
Zhang, J. and Scherer, G.W. (2012). Permeability of shale by the beam-bending method. *International Journal of Rock Mechanics and Mining Sciences* 53: 179–191. https://doi.org/10.1016/j.ijrmms.2012.05.011.
Zhang, T., Sun, S., and Song, H. (2019). Flow mechanism and simulation approaches for shale gas reservoirs: a review. *Transport in Porous Media* 126: 655–681.
Zhang, L., Liu, C., Liu, Y. et al. (2020). Transport property of methane and ethane in K-Illite nanopores of shale: insights from molecular dynamic simulations. *Energy & Fuels* 34: 1710–1719.
Zhou, X., Yuan, Q., Peng, X. et al. (2018). A critical review of the CO_2 huff 'n'puff process for enhanced heavy oil recovery. *Fuel* 215: 813–824.
Zoback, M.D. and Kohli, A.H. (2019). *Unconventional Reservoir Geomechanics*. Cambridge University Press.

Part I

Pore-Scale Characterizations

2

Pore-Scale Simulations and Digital Rock Physics

Junjian Wang[1,*], *Feifei Qin*[2,*], *Jianlin Zhao*[2,*], *Li Chen*[3], *Hari Viswanathan*[4], *and Qinjun Kang*[4]

[1] *Klohn Crippen Berger Ltd, Brisbane, Queensland, Australia*
[2] *Chair of Building Physics, Department of Mechanical and Process Engineering, ETH Zürich, Zurich, Switzerland*
[3] *Key Laboratory of Thermo-Fluid Science and Engineering of MOE, School of Energy and Power Engineering, X'an Jiaotong University, Xi'an, Shaanxi, China*
[4] *Earth and Environmental Sciences Division, Los Alamos National Laboratory, Los Alamos, NM, USA*

2.1 Introduction

Unconventional reservoir rocks have drawn increasing attention due to the development of advanced techniques such as horizontal drilling and multi-stage hydraulic fracturing, and due to the depletion of conventional resources (Middleton et al. 2015). Unconventional reservoirs have been successfully explored in North America and are being evaluated and produced in Europe, Australia, and China (Weijermars 2014). To reduce uncertainties and increase the efficiency of production related to these systems, accurate production prediction over time is necessary (Kim et al. 2017). This is typically achieved by running large-scale fluid flow simulations on the order of meters to kilometers using constitutive equations. They, therefore, require the input of effective material properties, such as porosity and permeability, which characterize the formation where the hydrocarbon production takes place. The unconventional reservoir rocks are composed of various porous components and fractures from nano- to macro-scale. The internal micro-structure and the macroscopic behavior are indeed strongly and directly related to each other, and properties of the micro-porous system and their interaction with inherent fluid significantly impact the formation's performance. Therefore, understanding why and how the macroscopic features of the unconventional reservoirs vary over space and time requires the detailed examination of the fluid flow at the pore scale. Given the complexity and heterogeneity of these natural micro-porous media, such a deep understanding is crucial. However, experimental measurements of fluid flow in unconventional reservoirs are often difficult, expensive, and time-consuming since these reservoirs are extremely tight with the average pore space in the matrix comparable to the mean free path of the gas molecules. Therefore, analytical or semi-analytical models derived based on simplified structures (Javadpour et al. 2007) and image-based pore-scale simulation techniques (Jing et al. 2020) on the scale of the material's micro-structure have become an attractive tool to complement direct measurements and to improve our understanding of fluid flow behavior in unconventional

*The authors contributed equally to this work.

Physics of Fluid Flow and Transport in Unconventional Reservoir Rocks, First Edition.
Edited by Behzad Ghanbarian, Feng Liang, and Hui-Hai Liu.

reservoir rocks. Among the image-based pore-scale simulation tools, the lattice Boltzmann (LB) method, which has been widely used to calculate the single-phase flow and transport in the continuum flow regime, has attracted increasing attention in the recent decade (Chen et al. 2014, 2022; Wang et al. 2016a). Tremendous efforts devoted to developing its methodology or boundary treatment have extended its application to simulating fluid flow and transport phenomena in unconventional reservoir rocks.

In this chapter, the physics of fluid flow in unconventional reservoir rocks is firstly introduced, then followed by recent developments in pore-scale LB simulation methods and their applications in fluid flow simulation of unconventional reservoir rocks.

2.2 Physics of Pore-Scale Fluid Flow in Unconventional Rocks

Due to the small pore size of unconventional reservoir rocks, the inherent fluid flow falls in different regimes from those in conventional reservoirs. Additionally, the liquid flow is under the control of wettability, and high content of organic matter with strong adsorption capacity would introduce extra mass transport to the free gas flow. Therefore, a reasonable description of the physics of fluid flow in unconventional reservoir rocks in numerical models is the prerequisite of accurate estimation of hydrocarbon production. A brief introduction of fluid flow physics is described in the following subsections.

2.2.1 Physics of Gas Flow

2.2.1.1 Gas Slippage and Knudsen Layer Effect

Unlike conventional hydrocarbon reservoirs, abundant micro- and nano-pores with irregular geometric structures coexist in unconventional reservoir rocks. Classic theories under continuum assumptions fail to describe gas flow in these pores, and the deviations of gas flow in micropores from that in macropores are usually characterized by the Knudsen number (Kn), which is defined as $Kn = \lambda/L$ with λ and L representing the mean free path (MFP) and the characteristic length, respectively. Many studies have confirmed that the gas flow in micro-systems like unconventional reservoir rocks is dynamically similar to rarefied gas flow at high altitudes at the same Kn (Beskok and Karniadakis 1999), and according to the magnitude of Kn, the gas flow regimes can be classified as continuum flow ($Kn < 0.001$), slip flow ($0.001 < Kn < 0.1$), transition flow ($0.1 < Kn < 10$), and free molecular flow ($Kn > 10$). Because the pore sizes in inorganic minerals are normally less than 2 μm, and pores in kerogen are considerably smaller with sizes ranging from 2 to 50 nm, gas flow in unconventional reservoirs falls into the slip flow ($0.001 < Kn < 0.1$) and early transition flow regimes ($0.1 < Kn < 1$).

From slip to transition flow regimes, the rarefaction effect, as well as the Knudsen layer (KL), which is defined as a region of local non-equilibrium with a thickness of a few *MFP* from the wall start to take effect (Zhang et al. 2012). The rarefaction effect indicates that the non-zero slip velocity at the solid boundary is noticeable, and the classic linear stress/strain-rate behavior breaks down inside the KL. It is believed that in micro-systems, approximately 70% of the mass flow rate increase could result from the velocity slip at the wall, and 30% of this increase is attributed to the non-Newtonian structure of the KL (Lockerby et al. 2005). Therefore, it is essential to capture KL when its relative thickness cannot be ignored. The typical velocity slip profile near a solid wall is shown in Figure 2.1 (Watari 2010).

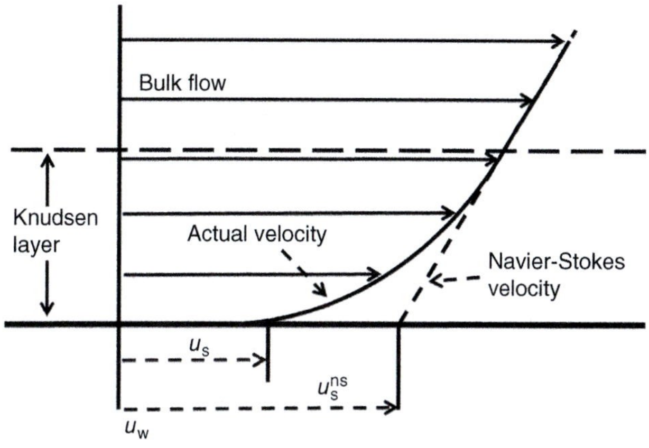

Figure 2.1 Schematic of the Knudsen Layer near a solid wall. The Navier–Stokes solution with a slip boundary condition (u_s^{ns}) departs significantly from the actual slip velocity (u_s), and u_w is the velocity of the wall itself. *Source:* Adapted from Lockerby et al. (2005), Watari (2010).

In the slip flow regime, the no-slip boundary condition fails to provide an agreement between theoretical predictions and experimental results, although continuum conservation equations can still be used to describe the bulk flow. Therefore, in this flow regime, the gas flow could be analyzed by solving the Navier–Stokes (N–S) equation with slip boundary condition, i.e. allowing non-zero fluid velocity near the solid wall. Various gas–solid interaction models have been proposed since Maxwell in 1879 (Maxwell 1879), and the first-order or second-order slip models in which the slip velocity is a function of λ or Kn have been mostly used. In the first-order slip model, the slip velocity satisfies:

$$u_s = \sigma_p \lambda \left(\frac{\partial u_s}{\partial n} \right)_s, \tag{2.1}$$

where n is the coordinate normal to the solid wall, and $\sigma_p = (2 - \sigma_v)/\sigma_v$ is the slip coefficient with σ_v being the tangential momentum accommodation coefficient (TMAC).

In the second-order slip model, the slip velocity satisfies:

$$u_s = C_1 \lambda \left(\frac{\partial u_s}{\partial n} \right)_s + C_2 \lambda^2 \left(\frac{\partial^2 u_s}{\partial n^2} \right)_s, \tag{2.2}$$

where C_1 and C_2 are the first- and second-order slip coefficients, respectively. Many methods have been proposed to determine the slip coefficients, including but not limited to Kinetic-model based methods, direct simulation Monte Carlo (DSMC) methods, experimental measurements, etc.

In the transition flow regime, the rarefaction effect dominates the flow and the thermodynamic equilibrium assumptions of the N–S equation begin to break down. The difficulty in analyzing the transition flow from a continuum assumption arises from the fact that the stress–strain relationship from the gas flow becomes nonlinear near KL. Therefore, the gas dynamic viscosity change in KL needs to be proposed, and the effective gas viscosity in KL can be written as:

$$\mu_e{}^{KL} = J \times \mu, \tag{2.3}$$

where μ is gas viscosity, J is the modification factor, and various J have been proposed (see Table 3 in Wang et al. (2016)). Among them, wall function methods or Bosanquet-type models are most

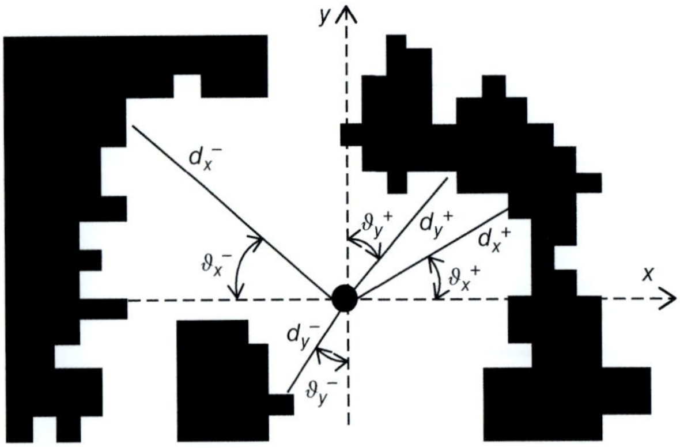

Figure 2.2 A gas molecule confined by solid walls in a porous medium. The gas molecule has an equal probability to travel in any zenith angle ϑ.

popular. Wall function approach is a phenomenological method and has been widely used to study micro-channel/tube flows. Various correction functions have been proposed. Most recently, Wang et al. (2017b) further modified the power–law (PL) based correction function proposed by Dongari et al. (2011) and extended its application to modeling gas flow in micro-porous media (as shown in Figure 2.2), in which the local mean free path is an average of the mean free path confined by surrounding solid walls:

$$\lambda_{\text{loc}} = \frac{1}{4}\sum_{m=1}^{4} \lambda_{\text{loc},m} = \frac{1}{4}\lambda \sum_{m=1}^{4} J(d_m, \lambda), \tag{2.4}$$

where d is the traveling distance, m indicates the wall confinement from + or − direction, and J-(d_m, λ) satisfies:

$$J(d_m, \lambda) = 1 - \frac{1}{48}\left[\left(1 + \frac{d_m}{\lambda}\right)^{-2} + 4\sum_{i=1}^{8}\left(1 + \frac{d_m}{\lambda\cos[(2i-1)\pi/32]}\right)^{-2}\right. \\ \left. + 2\sum_{i=1}^{7}\left(1 + \frac{d_m}{\lambda\cos[i\pi/16]}\right)^{-2}\right]. \tag{2.5}$$

It can be seen from wall function expression that ${\mu_{\text{e}}}^{\text{KL}}$ is a function of the distance from the wall. Alternatively, the Bosanquet-type effective viscosity is independent of the distance due to its value averaging over the cross-section, and

$$J = \frac{1}{1 + \alpha_{\text{k}} Kn}, \tag{2.6}$$

where α_{k} is a coefficient and its value changes with different micro-systems. It is observed that the value of α_{k} depends on the Knudsen number, and it varies slightly over the majority of the transition flow regime (Michalis et al. 2010).

2.2.1.2 Gas Adsorption/Desorption and Surface Diffusion

In addition to the free gas flow as a bulk phase in unconventional reservoir rocks, a considerable amount of adsorbed gas exists in the organic matter and clay minerals under reservoir conditions and therefore provides a significant portion of gas in place (GIP). As pressure declines under continuous reservoir depletion during production, adsorbed gas starts to desorb, leading to sustainable gas flow and affecting the ultimate recovery rate.

It is believed that the gas adsorption in unconventional reservoir rocks is under the control of physical interactions of gas/solid walls, and in a situation where the potential energy barrier to diffusion is smaller than the energy of desorption for diffusion to occur, an adsorbed particle may remain mobile in a metastable phase while still being physically bound to the surface. This phenomenon is known as the surface diffusion (Higashi et al. 1963). Besides, there is also a probability that adsorbed molecules may escape from the surface before reaching thermal equilibrium with the surface after a short time and leave prematurely at a velocity with a portion of its energy dissipated (Levesque et al. 2013). Therefore, after the incident gas molecules collide with the solid wall, in addition to being directly reflected, they would also be adsorbed on the surface and move in terms of surface diffusion or escape back into the bulk gas before reaching the thermal equilibrium with the surface.

A complete mathematical description of gas-surface interaction is under development. However, as the adsorption phase contributes to the total mass transport to a great extent in unconventional reservoir rocks, many studies have been carried out. The Langmuir model (Langmuir 1918), the Brunauer–Emmett–Teller (BET) model (Brunauer et al. 1938), or gas condensation theory are normally adopted to model adsorbing capacity of unconventional reservoir rocks. The Langmuir isotherm model is based on the assumption that adsorption is a mono-layer process. It does not account for the variation of pore size and wettability of hydrocarbon in unconventional reservoirs. However, it has been widely used in the literature because of its simplicity and acceptable accuracy in modeling adsorption in micro-pores. The Langmuir isotherm curve is illustrated in Figure 2.3.

Surface diffusion of adsorbed gas was introduced either by hopping process (Higashi et al. 1963) or random walk statistics (Theodorou and Wei 1983), to explain the mass flow in excess of that of the bulk phase. Krishna and Wesselingh (1997) proposed a Maxwell-Stefan approach combined with the Langmuir adsorption isotherm (which implicitly implies equilibrium), and yielded a reasonable expression for the description of surface diffusion, which is confirmed by experimental work and simulation studies (Marković et al. 2009; Moon et al. 2005). In their approach, the molar flux of surface diffusion in a single component system can be expressed as:

$$N_{\text{surface}} = q_{\text{sat}} \frac{D_{\text{s}}}{1-\theta_{\text{L}}} \nabla \theta_{\text{L}}, \tag{2.7}$$

where D_{s} is the Maxwell-Stefan diffusivity that can be estimated from experimental studies or molecular dynamic (*MD*) simulations, q_{sat} is the Langmuir volume, and θ_{L} is the fractional coverage of adsorbed gas which satisfies $\theta_{\text{L}} = q/q_{\text{sat}} = bP/(1+bP)$ based on the Langmuir model, where b is the Langmuir constant, P is the pressure, and q is the surface concentration. Therefore, the surface diffusion velocity u_{surface} can be written as:

$$u_{\text{surface}} = \frac{N_{\text{surface}} M_{\text{ads}}}{\rho_{\text{ads}}} = \frac{q_{\text{sat}} M_{\text{ads}}}{\rho_{\text{ads}}} \frac{bD_{\text{s}}}{1+bP} \nabla P, \tag{2.8}$$

where M_{ads} and ρ_{ads} are molecular weight of adsorbed gas and the density of adsorbed gas, respectively.

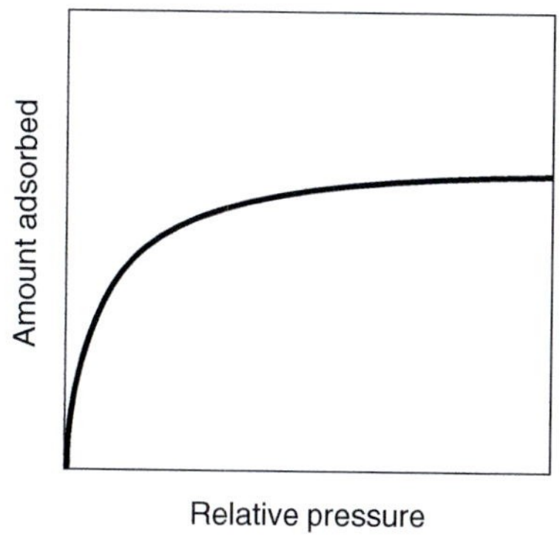

Figure 2.3 The Langmuir gas adsorption isotherm curve.

2.2.2 Physics of Water Flow

Because of the significant difference in physical properties between liquid and gas, the gas flow mechanisms do not always apply to liquid flow in unconventional reservoir rocks. Liquid flow, especially water flow, has also been widely investigated. The micro- to nano-scale effect can either increase or decrease water flux compared with that obtained by no-slip Hagen-Poiseuille equation, depending on the difference between fluid–solid interaction force and water intermolecular interaction force (Wu et al. 2017). The continuum model with slip length and effective viscosity is widely used to quantitatively describe the nano-scale effect for water flow, and the slip length and effective viscosity can be extracted by curve fitting the velocity profile obtained by molecular dynamics (MD) simulations. Different fluid–solid interaction forces can induce different slip lengths and effective viscosity. If the fluid–solid interaction force is much larger than the water intermolecular interaction force, water molecules would be stuck onto the solid surface, thus water mobility near the solid surfaces would decrease. In this situation, the no-slip boundary condition is still valid on the solid surface and the viscosity of water near the solid surface would increase (Raviv et al. 2001). As a result, the nano-scale effect decreases the flow rate. If the fluid–solid interaction force is much smaller than the water intermolecular interaction force, water molecules would slip on the solid surface and the viscosity of water near the solid surface would decrease (Barrat and Bocquet 1999). Then the nano-scale effect would increase the flow rate. The fluid–solid interaction force can be characterized by the wettability or contact angle (θ_w) measured from a water droplet forming on the walls. A small θ_w means strong fluid–solid interaction force while a large θ_w indicates weak fluid–solid interaction force. Wu et al. (2017) established a model describing water flow in nano-pores considering the fluid–solid interaction effect and nano-pore dimension. In their model, the slip length is calculated as:

$$l_s{}^w = \frac{C}{(\cos\theta_w + 1)^2}, \tag{2.9}$$

where l_s is the slip length defined as the ratio of slip velocity to the shear rate at the solid walls, C is a constant for a certain fluid which should be fitted with experimental or MD simulation data ($C = 0.41$ for water), and θ_w is the contact angle of water. This relationship matches well MD simulation results for θ_w up to 150° (figure S1 in Wu et al. (2017)).

Affected by the fluid–solid interaction force, the viscosity of water near the solid wall would change. Considering the interface region affected by the solid walls, the area-weighted-average effective viscosity $\mu_e{}^w$ can be calculated by:

$$\mu_e{}^w = \mu_i \frac{A_i}{A_t} + \mu_\infty \left(1 - \frac{A_i}{A_t}\right), \tag{2.10}$$

where μ_i and A_i are the viscosity and area of water in the interface region, respectively, μ_∞ is the bulk viscosity of water in the middle of the channel, and A_t is the total cross-sectional area. The area of the interface region is dependent on the thickness of the interface region, which is 0.7 nm for water according to experiments and MD simulations (Table S1 in Wu et al. (2017)).

The viscosity of water in the interface region is also dependent on the wettability (Wu et al. 2017):

$$\frac{\mu_i}{\mu_\infty} = -0.018\theta_w + 3.25. \tag{2.11}$$

Incorporating the slip length and effective viscosity, the modified N–S equation can be adopted to describe water flow in nano-pores. However, for water flow in short channels or tubes, because of the bending of streamlines at the entrance or exit of the channels or tubes, additional hydrodynamic resistance can be generated by the end effect, and a complete description of fluid volume flux in slit pores satisfies:

$$Q = \frac{\delta P}{\dfrac{12L\mu_{\mathrm{e}}{}^{\mathrm{w}}}{H^3 + 6H^2 l_{\mathrm{s}}} + \dfrac{32\mu_{\mathrm{e}}{}^{\mathrm{w}}}{\pi H^2}}, \tag{2.12}$$

where the first term in the denominator accounts for the flow resistance in the pore and the second term represents the flow resistance caused by the end effect. Therefore, the end effect would become important under two conditions: ratio of channel length to height, L/H is small or l_s is very large compared with the pore size (large θ_w).

2.2.3 Physics of Condensation

Phase change is a specific process of liquid–vapor multi-phase flow, which includes evaporation, boiling and condensation (Li et al. 2016). The difference between phase change and general two-phase flows lies in the transition from one phase to another governed by a non-ideal equation of state. This chapter focuses on condensation phase change during which vapor turns into liquid when over-saturated. In conventional reservoirs, condensation consists of two types, i.e. dropwise and film condensations on hydrophobic and hydrophilic surfaces. However, condensation in unconventional reservoirs may be quite different. Capillary condensation occurs in pores when vapor turns into liquid phase before reaching the vapor saturation pressure (Barsotti et al. 2016). The reason is considerable van der Waals interactions between vapor molecules and solid walls in very small pores, which is categorized as the adsorption effect of gas phase. The gas adsorption to achieve over-saturated vapor precedes its condensation on wall surface. In general, since phase change is a specific type of two-phase flow, except for the adsorption effect, it shares the physics of gas Knudsen layer effect and liquid–solid interaction force. If liquid is formed after quasi-equilibrium condensation without the influence of other gas flow (such as air) and without capillary flow in the liquid, the liquid velocity is normally small enough to be neglected. In this case, the influence of liquid–solid interaction on liquid velocity may be negligible. Hence, the gas slippage, Knudsen layer, and adsorption effects play an important role.

2.3 Theory of Pore-Scale Simulation Methods

2.3.1 The Isothermal Single-Phase Lattice Boltzmann Method

Historically, the LBM was derived from lattice gas automata (McNamara and Zanetti 1988). Later, it was shown that the LB equation can also be directly derived by discretizing the Boltzmann equation (He and Luo 1997). Shortly after its introduction, the LBM became an attractive technique to study single/multi-phase flow and reactive transport in porous media, covering groundwater flow, fabric materials, fuel cells, etc. Detailed pore structures obtained by FIB-SEM and micro-CT have made the LBM a popular alternative to direct numerical solution of the N–S equation for flows in tight rocks. Because of its kinetic nature, the LBM is, however, far more than just an N–S solver at the pore-scale. Advances in micro electrical mechanical system (*MEMS*) and nanotechnology have spurred interest in the use of the LBM for the simulation of microfluidics, and tremendous efforts have been made in developing the LBM for micro-system. One noteworthy advance is its extension to the simulation of gaseous flows in slip flow regime and/or early transition flow regime. In general, the LB equation with a body force term can be written as (Guo et al. 2002):

$$f_i(\boldsymbol{x}+ \mathbf{c}_i\delta t, t + \delta t) - f_i(\boldsymbol{x}, t) = \Omega_i(f_i) + \delta t \boldsymbol{F}_i, \tag{2.13}$$

where $\mathbf{c}_i$ indicates the finite number of discrete velocities of particles, f_i is the distribution function of particles with speed $\mathbf{c}_i$, Ω_i is the collision operator, $\boldsymbol{F}_i$ is the force term, which is defined according to the collision operator, and δt is the time between two simulation iterations. The fluid variables (density and velocity) are defined as:

$$\rho(\boldsymbol{x}, t) = \sum_i f_i(\boldsymbol{x}, t), \tag{2.14}$$

$$\rho(\boldsymbol{x}, t)\boldsymbol{u}(\boldsymbol{x}, t) = \sum_i f_i(\boldsymbol{x}, t)\mathbf{c}_i + 0.5\delta t \boldsymbol{F}, \tag{2.15}$$

The collision operator Ω_i describes the collision behavior of particles at every lattice location, which represents the variation of distribution functions caused by the collisions between particles. The requirements are that they should be physically correct and efficiently computable. In the following, collision operators commonly used in fluid flow are presented.

2.3.1.1 Bhatnagar–Gross–Krook (BGK) Collision Operator

Bhatnagar et al. (1954) proposed an efficient simplification operator to approximate the collision term as:

$$\Omega_i(f_i) = -\tau^{-1}(f_i - f_i^{\text{eq}}), \tag{2.16}$$

where τ is the relaxation time and f_i^{eq} is the Maxwell Boltzmann distribution of $\mathbf{c}_i$ for a given fluid velocity $\boldsymbol{u}$ and density ρ. The Maxwell–Boltzmann distribution describes the density distribution of a fluid in its equilibrium state to which every fluid strives. The equilibrium distribution f_i^{eq} in its discrete form can be expressed as:

$$f_i^{\text{eq}}(\boldsymbol{x}, t) = \rho(\boldsymbol{x}, t)\omega_i \left[1 + \frac{3(\mathbf{c}_i \cdot \boldsymbol{u})}{c^2} + \frac{9(\mathbf{c}_i \cdot \boldsymbol{u})^2}{2c^4} - \frac{3(\boldsymbol{u} \cdot \boldsymbol{u})}{2c^2} \right]. \tag{2.17}$$

Suga et al. (2010) summarized $\mathbf{c}_i$, the sound speed c_s and the weight parameter ω_i in accordance with each subfigure of Figure 2.4 (see Table 2.1).

The forcing term, $\boldsymbol{F}_i$ can be specified with respect to the relaxation parameter s and the body force $\boldsymbol{F}$ as:

$$\boldsymbol{F}_i = \left(1 - \frac{1}{2\tau}\right)\omega_i \left[\frac{\mathbf{c}_i - \boldsymbol{u}}{c_s^2} + \frac{\mathbf{c}_i \cdot \boldsymbol{u}}{c_s^4}\mathbf{c}_i \right] \cdot \boldsymbol{F}. \tag{2.18}$$

The dynamics viscosity of the fluid is given by the following relation:

$$\mu = \rho c_s^2 \left(\tau - \frac{1}{2}\right) \delta t. \tag{2.19}$$

2.3.1.2 The Multi-Relaxation Time (MRT)-LB Scheme

The MRT-LB model is the most general form within the theoretical framework of the LB equation and kinetic theory. The MRT collision operator relaxes the kinetic moments separately, which can

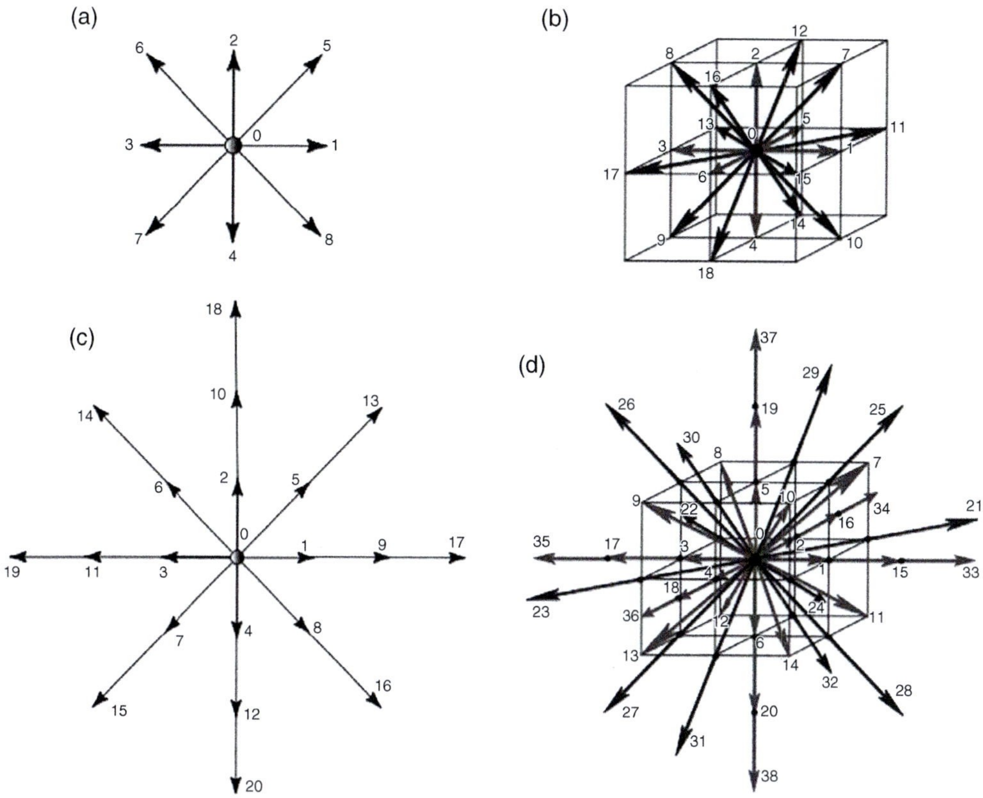

Figure 2.4 Discrete velocity models; (a) D2Q9, (b) D3Q19, (c) D2Q21, and (d) D3Q39.

be retrieved from the distribution functions. The MRT collision operator is defined as follows (Lallemand and Luo 2000):

$$\Omega_i\left(f_j\right) = -\left(\mathbf{M}^{-1}\mathbf{S}\mathbf{M}\right)_{ij}\left(f_j - f_j^{eq}\right), \tag{2.20}$$

where, $\mathbf{S} = \mathrm{diag}(\tau_\rho, \tau_e, \tau_\epsilon, \tau_j, \tau_q, \tau_j, \tau_q, \tau_s, \tau_s)^{-1}$ is a diagonal matrix, and $\mathbf{M}$ is an orthogonal transform matrix, which maps the distribution functions onto the moment space, and defined as:

$$\mathbf{M} = \begin{bmatrix} 1 & 1 & 1 & 1 & 1 & 1 & 1 & 1 & 1 \\ -4 & -1 & -1 & -1 & -1 & 2 & 2 & 2 & 2 \\ 4 & -2 & -2 & -2 & -2 & 1 & 1 & 1 & 1 \\ 0 & 1 & 0 & -1 & 0 & 1 & -1 & -1 & 1 \\ 0 & -2 & 0 & 2 & 0 & 1 & -1 & -1 & 1 \\ 0 & 0 & 1 & 0 & -1 & 1 & 1 & -1 & -1 \\ 0 & 0 & -2 & 0 & 2 & 1 & 1 & -1 & -1 \\ 0 & 1 & -1 & 1 & -1 & 0 & 0 & 0 & 0 \\ 0 & 0 & 0 & 0 & 0 & 1 & -1 & 1 & 1 \end{bmatrix} \tag{2.21}$$

Table 2.1 Parameters of the discrete velocity models are shown in Figure 2.4.

Models	c_s^2/C^2	c_i/c	w_i
D2Q9	1/3	(0, 0)	4/9 ($i = 0$)
		(±1, 0), (0, ±1)	1/9 ($i = 1 - 4$)
		(±1, ±1)	1/36 ($i = 5 - 8$)
D2Q21	2/3	(0, 0)	91/324 ($i = 0$)
		(±1, 0), (0, ±1)	1/12($i = 1 - 4$)
		(±1, ±1)	2/27 ($i = 5 - 8$)
		(±2, 0), (0, ±2)	7/360 ($i = 9 - 12$)
		(±2, ±2)	1/432 ($i = 13 - 16$)
		(±3, 0), (0, ±3)	1/1620 ($i = 17 - 20$)
D3Q19	1/3	(0, 0, 0)	12/36 ($i = 0$)
		(±1, 0, 0), (0, ±1, 0), (0, 0, ±1)	2/36 ($i = 1 - 6$)
		(±1, ±1, 0), (±1, 0, ±1), (0, ±1, ±1)	1/36 ($i = 7 - 18$)
D3Q39	2/3	(0, 0, 0)	1/12 ($i = 0$)
		(±1, 0, 0), (0, ±1, 0), (0, 0, ±1)	1/12 ($i = 1 - 6$)
		(±1, ±1, ±1)	1/27 ($i = 7 - 14$)
		(±2, 0, 0), (0, ±2, 0), (0, 0, ±2)	2/135 ($i = 15 - 20$)
		(±2, ±2, 0), (±2, 0, ±2), (0, ±2, ±2)	1/432 ($i = 21 - 32$)
		(±3, 0, 0), (0, ±3, 0), (0, 0, ±3)	1/1620 ($i = 33 - 38$)

The distribution function f_i and its equilibrium distribution function f_i^{eq} can be projected onto the moment space, and the following results can be obtained:

$$m = \mathbf{MF} = \left(\rho, e, \epsilon, j_x, q_x, j_y, q_y, p_{xx}, p_{xy}\right)^{\mathrm{T}}, \tag{2.22}$$

$$\begin{aligned} m_{\mathrm{eq}} &= \mathbf{Mf^{eq}} = \left(\rho, e^{\mathrm{eq}}, \epsilon^{\mathrm{eq}}, j_x, {q_x}^{\mathrm{eq}}, j_y, {q_y}^{\mathrm{eq}}, {p_{xx}}^{\mathrm{eq}}, {p_{xy}}^{\mathrm{eq}}\right)^{\mathrm{T}} \\ &= \rho\left(1, -2 + 3|\mathbf{u}|^2, 1 - 3|\mathbf{u}|^2, u, -v, u, -v, u^2 - v^2, uv\right)^{\mathrm{T}}, \end{aligned} \tag{2.23}$$

where e is the energy mode, ϵ is related to energy square, (j_x, j_y) are the momentum components, (q_x, q_y) correspond to energy flux, (p_{xx}, p_{xy}) are related to the diagonal and off-diagonal components of the stress tensors, and u and v are the components of the velocity $\mathbf{u}$, respectively. Therefore, the MRT-LB equation can be written as:

$$|\mathbf{f}(\boldsymbol{x} + \mathbf{c}_i\delta t, t + \delta t)\rangle - |\mathbf{f}(\boldsymbol{x}, t)\rangle = -\mathbf{M}^{-1}\mathbf{S}[|\mathbf{m}(\boldsymbol{x}, t)\rangle - |\mathbf{m}^{\mathrm{eq}}(\boldsymbol{x}, t)\rangle]. \tag{2.24}$$

2.3.1.3 The Regularization Procedure

Generally, the LBM is carried out in alternate streaming and collision steps. When this dynamic procedure is viewed as a projection of the LB equation onto the Hermite space, it introduces an error analogous to the aliasing effect because the distribution function f_i does not automatically lie entirely within the Hermite space. Such an error is usually very small and negligible. However, it can be no longer neglected when the system is far from equilibrium. Therefore, a "regularization procedure" is introduced to resolve this error and to improve the numerical stability. The procedure is implemented as follows (Suga et al. 2010):

First, the post-streaming distribution function f_i is divided into the equilibrium part f_i^{eq} and the non-equilibrium part f_i^{neq}:

$$f_i = f_i^{eq} + f_i^{neq}. \tag{2.25}$$

As f_i^{eq} already lies entirely in the subspace of Hermite space, then f_i^{neq} needs to be converted to a new distribution $\tilde{f}_i^{neq}$ which lies within the subspace spanned by the higher-order Hermite polynomials. Using the second-order Hermite polynomials, for the D2Q9 models, $\tilde{f}_i^{neq}$ is expressed as:

$$\tilde{f}_i^{neq} = \omega_i \left[\frac{1}{2c_s^2} H^{(2)} \left(\frac{\mathbf{c}_i}{c_s} \right) \sum_{j=0}^{Q-1} \tilde{f}_j^{neq} \mathbf{c}_j \mathbf{c}_j \right]. \tag{2.26}$$

where $H^{(2)}$ is the second-order Hermite polynomial. By replacing f_i^{neq} with $\tilde{f}_i^{neq}$, the BGK model can be written as:

$$f_i(\boldsymbol{x} + \mathbf{c}_i\delta t, t + \delta t) - f_i^{eq}{}_i(\boldsymbol{x}, t) = \left(1 - \tau^{-1}\right)\tilde{f}_i^{enq} + \left(1 - 0.5\tau^{-1}\right)\delta t \boldsymbol{F}_i, \tag{2.27}$$

and the MRT-LB model becomes:

$$|\mathbf{f}(\boldsymbol{x} + \mathbf{c}_i\delta t, t + \delta t)\rangle - \left|\tilde{\mathbf{f}}^{\mathbf{neq}}(\boldsymbol{x}, t)\right\rangle = -\mathbf{M}^{-1}\mathbf{S}\mathbf{M}\left|\tilde{\mathbf{f}}^{\mathbf{neq}}(\boldsymbol{x}, t)\right\rangle + \mathbf{M}^{-1}\left(\mathbf{I} - \frac{\mathbf{S}}{2}\right)\mathbf{M}|\mathbf{F}\rangle\delta t. \tag{2.28}$$

The regularization procedure filters out the higher-order non-equilibrium moments that contain strong discrete artifacts.

2.3.2 Multi-phase Lattice Boltzmann Simulation Method

To simulate multi-phase flow, interactions among different phases should be incorporated into the LB model. There are different ways to introduce interaction forces among different fluids, based on which different multi-phase LB models have been developed, such as the color-gradient model, Shan-Chen model, Free-energy model (Huang et al. 2015; Liu et al. 2016), and the former two LB models have been extended to modeling multi-phase flow in unconventional reservoir rocks.

2.3.2.1 Color-Gradient Model

The color-gradient LB model has been widely used to simulate immiscible two-phase flow in porous media as the desired interfacial tension can be specified directly without any simulation test. In addition, it can easily handle two-phase flow with large density/viscosity ratios (Liu et al. 2016). In the two-phase color-gradient LB model, two density distribution functions are used to characterize the immiscible two fluids. The multi-phase interactions are incorporated by a perturbation operator and a recoloring operator. Then the collision operator becomes:

$$\Omega_i^K = \left(\Omega_i^K\right)^{(3)} \left[\left(\Omega_i^K\right)^{(1)} + \left(\Omega_i^K\right)^{(2)} \right], \tag{2.29}$$

where $\left(\Omega_i^K\right)^{(1)}$ is the single-phase collision operator and $\left(\Omega_i^K\right)^{(2)}$ is the perturbation operator which generates an interfacial tension:

$$\left(\Omega_i^K\right)^{(2)} = A^K \omega_i \left(1 - \frac{1}{2\tau^K}\right)[3(\mathbf{c}_i - \boldsymbol{u}) + 9(\mathbf{c}_i \cdot \boldsymbol{u})] \cdot \boldsymbol{F}_{\mathrm{t}}, \tag{2.30}$$

where A^K represents the ratio of interfacial tension contributed by fluid K; $\boldsymbol{F}_{\mathrm{t}}$ is the interfacial tension calculated by:

$$\boldsymbol{F}_{\mathrm{t}} = -\frac{1}{2}\sigma K_{\mathrm{C}} \nabla \rho^N, \tag{2.31}$$

where σ is the interfacial tension coefficient, K_{C} is the local curvature of the interface, and ρ^N is the phase field function defined as:

$$\rho^N = \frac{\left(\dfrac{\rho^{\mathrm{R}}}{\rho^{\mathrm{R0}}} - \dfrac{\rho^{\mathrm{B}}}{\rho^{\mathrm{B0}}}\right)}{\left(\dfrac{\rho^{\mathrm{R}}}{\rho^{\mathrm{R0}}} + \dfrac{\rho^{\mathrm{B}}}{\rho^{\mathrm{B0}}}\right)}, \tag{2.32}$$

where R and B represent the red and blue fluids, respectively, and ρ^{R_0} and ρ^{B_0} are the densities of pure red and blue fluids.

The recoloring operator $\left(\Omega_i^k\right)^{(3)}$ is used to produce the phase segregation and maintain the phase interface:

$$\left(\Omega_i^{\mathrm{R}}\right)^{(3)}(f_i') = \frac{\rho^{\mathrm{R}}}{\rho} f_i' + \beta W_i \frac{\rho^{\mathrm{R}}\rho^{\mathrm{B}}}{\rho} \cos(\varphi_i)|\mathbf{c}_i| \tag{2.33}$$

$$\left(\Omega_i^{\mathrm{B}}\right)^{(3)}(f_i') = \frac{\rho^{\mathrm{B}}}{\rho} f_i' + \beta W_i \frac{\rho^{\mathrm{R}}\rho^{\mathrm{B}}}{\rho} \cos(\varphi_i)|\mathbf{c}_i|, \tag{2.34}$$

where f_i' is the post-perturbation total distribution function of both fluids, φ_i is the angle between the color gradient $\nabla\rho^N$ and the lattice direction $\mathbf{c}_i$, and β is a parameter related to the interface thickness. The details of the color-gradient LB model can be found in the references (Ba et al. 2016; Zhao et al. 2018a).

2.3.2.2 Shan-Chen Model

Shan and Chen (1993) developed an inter-particle potential model (also known as Shan-Chen model) through introducing microscopic interactions among nearest-neighboring particles. The mean field force is incorporated by adding an additional force term in the collision operator, and this force ensures phase separation and introduces interfacial tension. The Shan-Chen model has become very popular due to its simplicity and versatility. Different non-ideal equations of state (Yuan and Schaefer 2006) can be easily incorporated into the Shan-Chen model to simulate multi-phase flow and phase change problems. The evolution equation of MRT-LB single component multi-phase Shan-Chen model can be written as (Guo and Zheng 2008):

$$f_i(\boldsymbol{x} + \mathbf{c}_i\delta t, t + \delta t) - f_i(\boldsymbol{x}, t) = \Omega_{ij}\left(f_j\right) + F_i, \tag{2.35}$$

where $\Omega_{ij}\left(f_j\right) = -(\mathbf{M}^{-1}\mathbf{SM})_{ij}\left(f_j - f_j^{eq}\right)$ is the collision operator and $F_i = -\mathbf{M}^{-1}(\mathbf{I} - 0.5\mathbf{S})\overline{\mathbf{F}}$ is the force term in which

$$\overline{\mathbf{F}} = \left(0, 6\boldsymbol{u}\cdot\boldsymbol{G}, -6\boldsymbol{u}\cdot\boldsymbol{G}, G_x, -G_x, G_y, -G_y, 2\left(u_xG_x - u_yG_y\right), \left(u_xG_y + u_yG_x\right)\right)^{\mathrm{T}}, \tag{2.36}$$

here $\boldsymbol{G} = (G_x, G_y)$ is the external force and $\boldsymbol{u} = \frac{1}{\rho}\left(\sum_i \left(\mathbf{c}_i f_i + \frac{\delta t}{2}\mathbf{G}\right)\right)$ is the two-phase fluid velocity where $\rho = \sum f_i$ is the fluid density.

In the Shan-Chen model, the external force $\boldsymbol{G}$ is modeled with the fluid–fluid interaction $\mathbf{G_c}$ to consider phase separation and fluid–solid adhesion forces and $\mathbf{G_w}$ to realize different surface wettability, which satisfy

$$\mathbf{G_c} = -g\psi(\boldsymbol{x})\sum_i \omega\left(|\mathbf{c}_i|^2\right)\psi(\boldsymbol{x} + \mathbf{c}_i)\mathbf{c}_i, \tag{2.37}$$

and

$$\mathbf{G_w} = -g\psi(\boldsymbol{x})\sum_i \omega\left(|\mathbf{c}_i|^2\right)S(\boldsymbol{x} + \mathbf{c}_i)\psi(\rho_{vs})\mathbf{c}_i, \tag{2.38}$$

where g is the interaction strength and $w(|\mathbf{c}_i|^2)$ is the force weight, $S(\boldsymbol{x} + \mathbf{c}_i)$ is a Boolean variable equal to 1 for solid and 0 for fluid, ρ_{vs} is the virtual solid density to realize different contact angles, and $\psi(\boldsymbol{x}) = \sqrt{2\left(P_{\mathrm{EoS}} - \rho c_s^2\right)/(gc^2)}$ is the effective mass related to non-ideal equation of state P_{EoS}. The Carnahan-Starling equation of state is applied for phase change problem,

$$P_{\mathrm{EoS}} = \rho RT\frac{1 + b_v\rho + \left(\frac{b_v\rho}{4}\right)^2 - \left(\frac{b_v\rho}{4}\right)^3}{\left(1 - \frac{b_v\rho}{4}\right)^3} - a_e\rho^2, \tag{2.39}$$

where a_e is a parameter describing the attractive energy between fluid particles, b_v is the parameter representing the volume of fluid particle, and R is the gas constant. In simulations $a_e = 1$, $b_v = 4$, $R = 1$ are always used.

2.3.3 Capture Fluid Slippage at the Solid Boundary

In traditional fluid mechanics, the no-slip condition is used at the solid boundary and breaks down on micro- and nano-scales. Generally, the interfacial slip is generated by hydrophobicity in liquid flow and by rarefaction effect for gas flow. To capture the fluid slippage at the solid boundary, various slip boundary conditions have been adopted to implicitly consider the solid–fluid interaction. Among them, the three mostly adopted combined boundary forms are:
the combined specular with diffusive reflection boundary condition (MR) (Guo et al. 2006)

$$f_i(\mathbf{x}) = (1 - C_{\mathrm{mr}})f_i^{\mathrm{sr}}(\mathbf{x}) + C_{\mathrm{mr}}f_i^{\mathrm{dr}}(\mathbf{x}), \tag{2.40}$$

the combined bounce-back with specular reflection boundary condition (BR) (Guo et al. 2006)

$$f_i(\mathbf{x}) = (1 - C_{\mathrm{br}})f_i^{\mathrm{sr}}(\mathbf{x}) + C_{\mathrm{br}}f_i^{\mathrm{bb}}(\mathbf{x}), \tag{2.41}$$

and the combined bounce-back with diffusive reflection boundary condition (MB) (Verhaeghe et al. 2009)

$$f_i(\mathbf{x}) = (1 - C_{\text{mb}})f_i^{\text{bb}}(\mathbf{x}) + C_{\text{mb}}f_i^{\text{dr}}(\mathbf{x}). \tag{2.42}$$

In these three forms, $f_i^{\text{bb}}(\mathbf{x})$, $f_i^{\text{sr}}(\mathbf{x})$, and $f_i^{\text{dr}}(\mathbf{x})$ represent bounce back, specular reflection, and Maxwellian diffuse reflection distribution functions, respectively. C_{mr}, C_{br}, and C_{mb} are combination coefficients in corresponding boundary conditions. Historically, there is no consensus on how to choose combination coefficients. Recent studies indicate that the combination coefficients can be obtained by matching the slip velocity of LB model and the analytical solution of N–S equation with Maxwellian slip boundary condition. Therefore, the combination coefficients are functions of the TMAC for gas slip flow and contact angle θ_{w} for water flow, which would be inferred from experimental results or molecular simulations.

2.3.4 Capture the Knudsen Layer/Effective Viscosity

To implement water-effective viscosity in the LB model is quite straightforward. By simply replacing μ with $\mu_{\text{e}}^{\text{W}}$ to calculate relaxation times, the change of viscosity could be taken into account. To capture KL for gas flow with an effective viscosity model, a reasonable relationship between λ and viscosity needs to be applied. According to gas kinetic theory, the hard-sphere model can be adopted to calculate the μ and λ relationship:

$$\frac{\mu}{\rho} = \lambda\sqrt{\frac{2R_{\text{g}}T}{\pi}}, \tag{2.43}$$

where R_{g} is the specific gas constant and T is temperature. For dilute gas, the above equation can be written as:

$$\frac{\mu}{\rho} = \frac{5\lambda}{16}\sqrt{\frac{2\pi kT}{m_{\text{g}}}}, \tag{2.44}$$

where k is the Boltzmann constant and m_{g} is the gas molecular mass. With these functions, relaxation times could be easily estimated at different λ or Kn, and KL effect can be reflected in the LB model by replacing μ with $\mu_{\text{e}}^{\text{KL}}$ using the wall function or Bosanquet-type approach as shown in Section 2.2.1.

2.3.5 Capture the Adsorption/Desorption and Surface Diffusion Effects

2.3.5.1 Modeling of Adsorption in LBM

The adsorption in unconventional reservoir rocks is normally considered as a kind of physical adsorption caused by the strong interaction forces between solids and gas. Therefore, an interaction force between solid and gas particles could be introduced in LB models to model adsorption (Martys and Chen 1996):

$$\mathbf{F}_{\text{s}}(\boldsymbol{x}, t) = -G_{\text{s}}\psi(\boldsymbol{x}, t)\sum_{i=1}^{8} S_i\omega_i\psi_{\text{s}}(\boldsymbol{x} + \mathbf{c}_i\delta t, t)\mathbf{c}_i, \tag{2.45}$$

where $\mathbf{F}_{\text{s}}$ is the interaction force between solids and gas particles, G_{s} is a parameter that controls the interaction strength between solids and gas particles, S_i is a Boolean variable, which equals 1 if the node in a direction is solid node and 0 otherwise, and ψ_{s} is the potential function of solids and depends on ρ_{S}. So, $\mathbf{F}_{\text{s}}$ can be changed by changing G_{s} or ρ_{S}. It should be mentioned that ρ_{s} is not the real solid's density but an adsorption parameter reflecting the solid's potential. For a given system, the key issue of adopting LBM to simulate gas adsorption accurately is to obtain the accurate $\mathbf{F}_s$. Zhao et al. (2016) proposed a novel process to estimate $\mathbf{F}_s$ by matching the simulation results of LBM and molecular simulations, and the general process is implemented as follows:

1) Applying molecular simulations e.g. the grand canonical Monte Carlo (GCMC) to simulate gas adsorption in a certain porous medium and obtain the adsorption curve.
2) Applying LBM to do the same simulation under the same conditions. By matching the simulation results of LBM and the adsorption curve obtained by molecular simulation, the adsorption parameter ρ_s in the LB model can be obtained under different conditions.
3) The regression curve between the adsorption parameter ρ_s and bulk density can be obtained, and this curve is introduced into the micro-scale LB model.

By doing so the adsorption parameters in LBM are determined by molecular simulation. It is worth mentioning that Lubbers et al. (2020) have built a Machine Learning (ML) surrogate model that captures adsorption effects across a wide range of parameter space and bridges the MD and LBM scales using a relatively small number of MD calculations. The surrogate model is seven orders of magnitude faster than MD simulations.

2.3.5.2 Modeling of Surface Diffusion Via LBM

In addition to adsorption/desorption of adsorbed gas, to capture the surface diffusion, inspired by the work of Levesque et al. and their co-workers (Vanson et al. 2015), Wang et al. (2017a) proposed a similar propagation scheme by introducing a surface diffusion to the slip boundary condition (as shown in Figure 2.5).

In this boundary condition, the surface diffusion of the adsorbed gas follows the Maxwell-Stefan approach, and the streamwise surface diffusion velocity can be calculated with u_{surface} shown in Section 2.2.1.2, and the distribution function of fluid boundary nodes is defined as:

$$f_i^{\text{BC}}(\mathbf{x}, \rho, \mathbf{u}) = \theta_{\text{ads}} f_i^{\text{des}}(\mathbf{x}, \rho, \mathbf{u}_{\text{surface}}) + (1 - \theta_{\text{ads}}) f_i^{\text{fr}}(\mathbf{x}, \rho, \mathbf{u}_{\text{w}}), \tag{2.46}$$

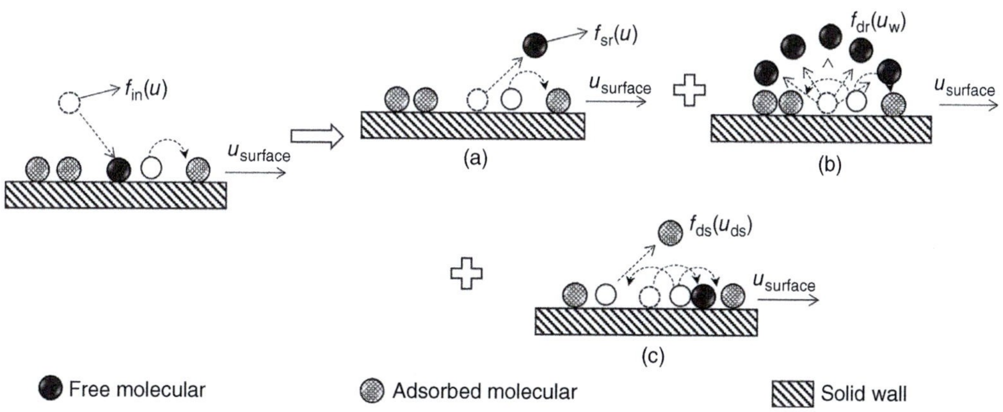

Figure 2.5 Schematic of gas–solid interactions considering surface diffusion and adsorption/desorption based on Maxwell slip theory (Maxwell 1879) and inter-facial physics with adsorption field (Higashi et al. 1963; Levesque et al. 2013). (a) Denotes the specular reflected gas molecules, (b) denotes the diffusive reflected gas molecules, and (c) denotes the adsorption of incident gas molecules and the premature desorption of adsorbed gas molecules with a velocity u_{des}. *Source:* Adapted from Maxwell (1879), Higashi et al. (1963), Levesque et al. (2013).

where θ_{ads} is the fractional coverage of adsorbed gas which satisfies $\theta_{ads} = \frac{q}{q_{sat}} = bP(1+bP)$, f_i^{des} is the "post-desorption distribution function" and is a function of the surface diffusion velocity $\mathbf{u}_{surface}$, f_i^{fr} is the reflected distribution function contributed from the free gas and can be modeled with combined forms in modeling gas slippage near the solid wall. For the calculation of f_i^{des}, the boundary condition proposed by Inamuro et al. (1995) can be used, in which f_i^{des} is assumed to be the equilibrium distribution function with a counter slip velocity to ensure the gas at the fluid boundary has the velocity $\mathbf{u}_{surface}$ and the density ρ.

2.4 Applications

In the above sections, the fundamentals of fluid flow physics and LB method for fluid flow have been briefly summarized. In this section, some representative applications of the LB methods in modeling fluid flow in unconventional reservoir rocks are presented.

2.4.1 Simulation of Gas Flow in Unconventional Reservoir Rocks

2.4.1.1 Gas Slippage

In spite of the tremendous efforts, modeling of gas flow in micro-porous media on pore-scale still remains a challenge. To clarify effects of rarefaction and pore configuration on gas flow in unconventional reservoir rocks (Wang et al. 2017b) proposed a regularized multiple-relaxation-time (MRT) lattice Boltzmann (LB) model to analyze gas flow in 2-dimensional (2D) micro-porous medium reconstructed by quartet structure generation set (QSGS) on pore-scale. A 4-CCL algorithm is also applied to remove the dead pores generated by QSGS, and the structure (100 × 400 lattices) used for analysis is illustrated in Figure 2.6.

In this study, MR boundary condition is applied for modeling gas slippage, and to match the second-order slip boundary condition, the following combination factor C_{mr} and relaxation time τ_q are applied:

$$C_{mr} = 2\left(1 + \sqrt{\frac{\pi}{6}}\frac{(2-\sigma_v)}{\sigma_v}(1-0.1817\sigma_v)\right)^{-1}, \tag{2.47}$$

$$\tau_q = \frac{1}{2} + \frac{\pi(2\tau_s-1)^2}{8(2\tau_s-1)}\left(\frac{2}{\pi} + \frac{1}{2}\frac{(2-\sigma_v)}{\sigma_v}(1-0.1817\sigma_v)\right) + \frac{3}{8(2\tau_s-1)}. \tag{2.48}$$

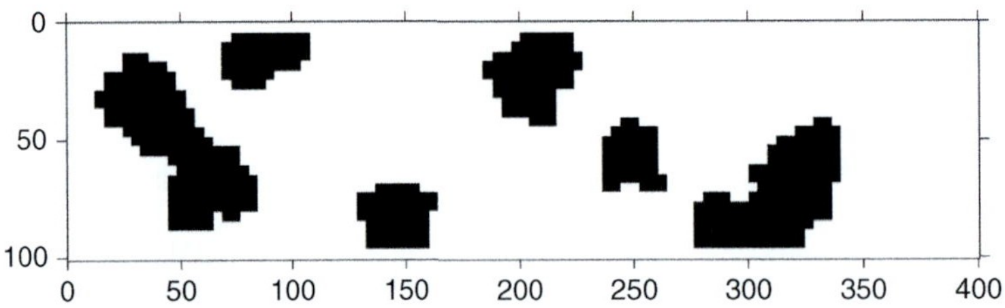

Figure 2.6 Porous structure generated with QSGS. Lattices in black are the solid phase, and lattices in white are void space.

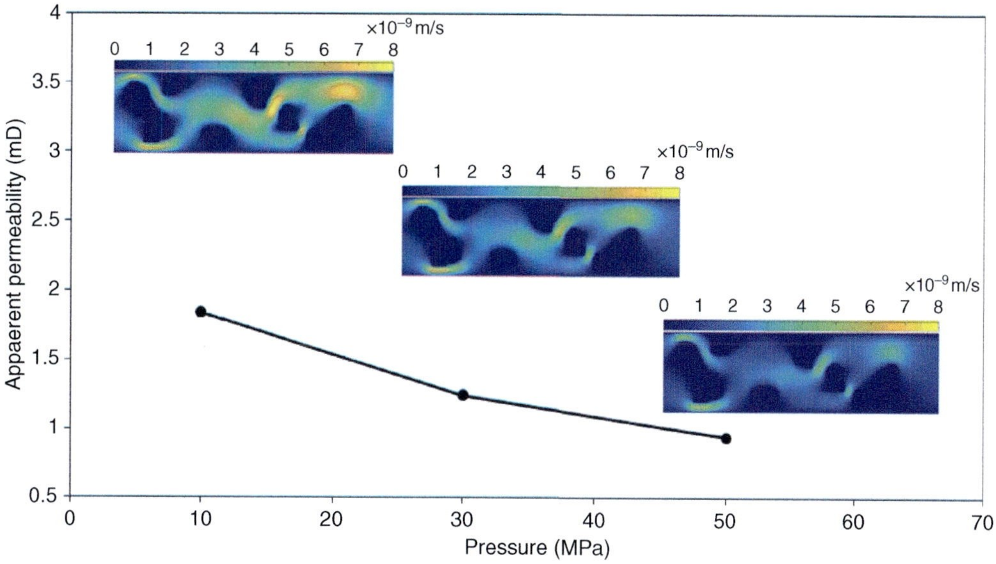

Figure 2.7 Simulated apparent permeability and corresponding velocity fields of structure shown in Figure 2.6. Results indicate that both the gas apparent permeability and gas flow velocities decrease with formation pressure.

To consider the KL effect, the wall function approach is used, and the local relaxation time τ_s satisfies:

$$\tau_s = \frac{1}{2} + \sqrt{\frac{\pi}{6}} \frac{\lambda J(d, \lambda)}{\delta x}. \tag{2.49}$$

To clarify the influence of rarefaction on gas flow, the methane flow in pore structure shown in Figure 2.6 at fixed outlet pressures of 10, 30, and 50 MPa and a constant pressure gradient $\nabla P = 0.1$ MPa/m are discussed. The velocity vector distribution of methane and the corresponding gas apparent permeability at different pressures are presented in Figure 2.7.

As shown in the figure, rarefaction is evident for gas flow in micro-porous media as the gas molecules collide with the solid surface more frequently. The results show that the apparent permeability of methane is not constant with pressure or Kn and increases by a factor around 2 with decreased pressure for current study. Specifically, it takes values ranging from 0.95 mD for outlet pressure of 50 MPa to 1.89 mD for outlet pressure of 10 MPa. This phenomenon is well explained by the changed velocity field. Given that rarefaction dominates gas flow in small pore and/or at low pressures, decreasing pressure gives rise to a higher bulk velocity, especially for flow through narrow pore throats or locations near the outlet with a low pressure, and finally, it leads to an increased apparent permeability.

2.4.1.2 Gas Adsorption

The effect of gas adsorption on gas flow in unconventional reservoir rocks such as organic shale was analyzed by Zhao et al. (2016). The interaction forces between solids and adsorbed gas particles are introduced, and GCMC is applied to obtain the input parameters related to gas adsorption for LB model. Following the procedure illustrated in Section 2.3.5.1. GCMC is firstly applied to simulate the adsorption isotherm of methane (Figure 2.8a), followed by an LB simulation under the same

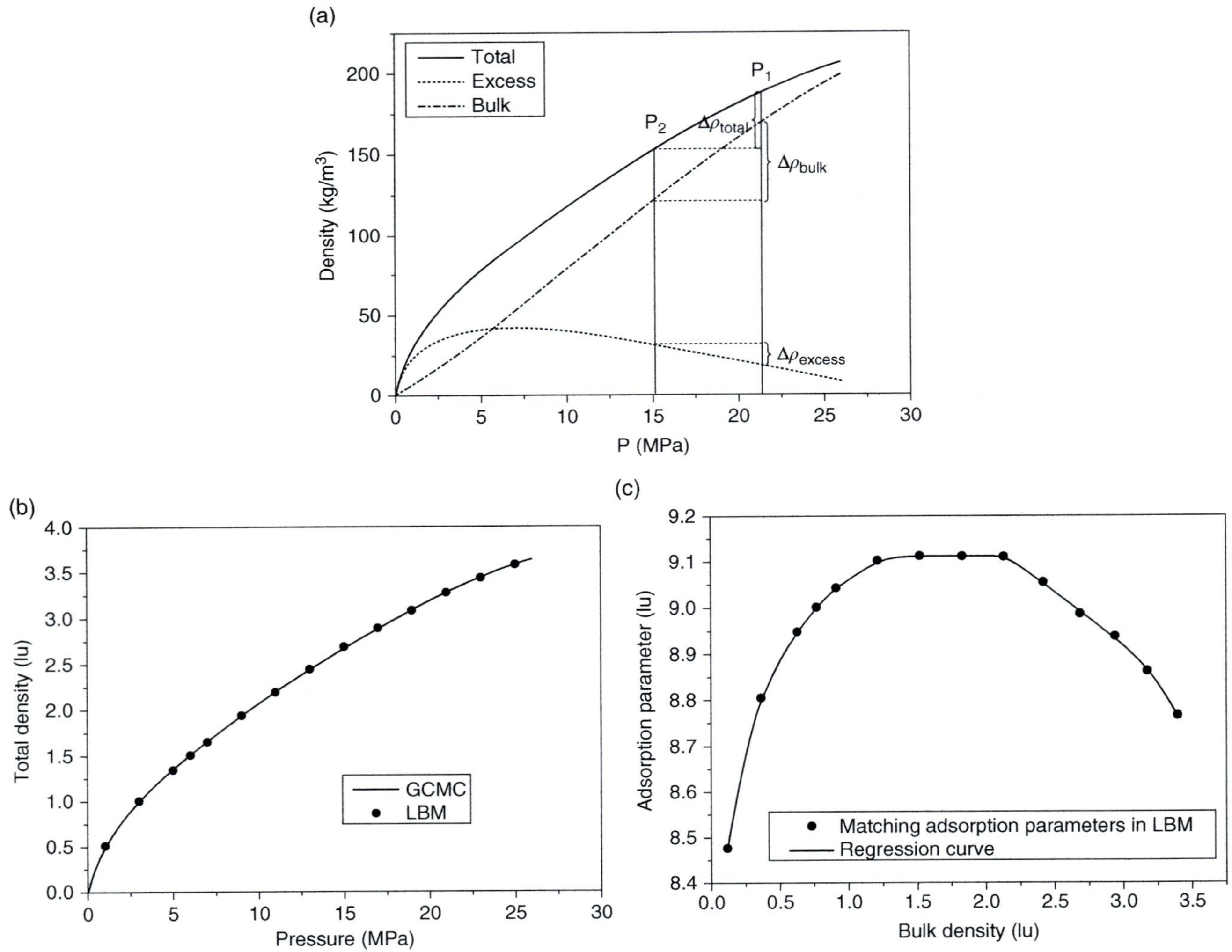

Figure 2.8 Methane adsorption isotherms modeled with GCMC and matching results of LBM with GCMC. (a) Adsorption isotherms of methane in the slit pore obtained by GCMC, (b) relationship between total density and pressure obtained by GCMC and the matching results of LBM, (c) relationship between the matching adsorption parameter and bulk density in LBM.

condition to obtain the adsorption parameter such as density and regression curve (Figure 2.8b and c).

The structure model in this study is shown in Figure 2.9, and the gas is produced by the only outlet (inlet) at the bottom of the porous medium.

The gas density distributions at different pressures are shown in Figure 2.10. It can be seen that a layer of adsorbed gas with high density would form near the organic matter under the adsorption force. By calculating the densities of different types of gas at different pressures, it can be concluded that the total gas density in the whole inorganic matter is always lower than that in organic matter, which means that the organic matter stores more gas than the inorganic matter (Figure 2.11a). Additionally, the bulk gas densities are almost the same in the two types of porous media at different pressures, while the adsorbed gas in organic matter is much higher than that in inorganic matter, especially when the pressure is low. Therefore, the total gas density in the whole pores increases in the organic matter (Figure 2.11b and c).

Figure 2.9 Schematic of the porous medium in lattice unit. Black means solid and white means pore space. The resolution is 0.4 nm per lattice. There is only one inlet/outlet of the porous medium.

2.4.1.3 Surface Diffusion of Adsorbed Gas

In order to qualify the influence of adsorbed gas and its surface diffusion on micro-gaseous flow, the normalized streamwise velocity distribution of the free gas at different adsorbed gas fractional coverage θ_{ads} and surface diffusion velocity $u_{surface}$ are analyzed by (Wang et al. 2017a). The boundary condition explained in Section 2.3.5.2 is implemented with an MRT-LB model. As shown in Figure 2.12, when $\theta_{ads} = 0$, the variation of $u_{surface}$ does not influence the gas flow because there is no adsorbed gas in the system (Figure 2.12a). The gas slip at the solid surface is controlled by the interaction of free gas molecules with solid molecules and the rarefaction dominates the flow, which leads to a large slip velocity at the boundary. When $\theta_{ads} = 1$, the slip velocity of the free gas at the solid surface is completely determined by the desorbed gas, and it vanishes to zero at $u_{surface} = 0$ (Figure 2.12d). This result indicates that the gas flow in confined micro-channel manifests a parabolic velocity profile with zero slip velocity at the solid surface when an immobile adsorption layer is formed, and the slip model based on kinetic theory (Karniadakis et al. 2006) fails to capture the flow behavior. When the solid surface is partially covered by an amount of adsorbed gas, the slippage of the free gas is determined by the interactions with both solid wall and the adsorbed gas, and the "dynamic similarity" of rarefied gas flow and micro-scale confined flow breaks down. At a certain coverage, the gas slip at the wall decreases with the decreased $u_{surface}$ as a result of the premature short-time desorption of adsorbed gas (Figure 2.12b and c). Comparing different sub-figures in Figure 2.12, one can see that when $u_{surface} = u_s$, the increase with coverage does not influence the slip velocity. Otherwise, the slip of the free gas is decreased with the increase of adsorbed gas coverage θ_{ads} at a certain $u_{surface}$.

2.4.2 Simulation of Water Flow in Unconventional Reservoir Rocks

The water flow in unconventional reservoir rock was investigated by Zhao et al. (2018b). In their study, BR slip boundary is applied with MRT-LB model, and the first-order slip boundary condition is used to calculate the slip coefficient. Therefore,

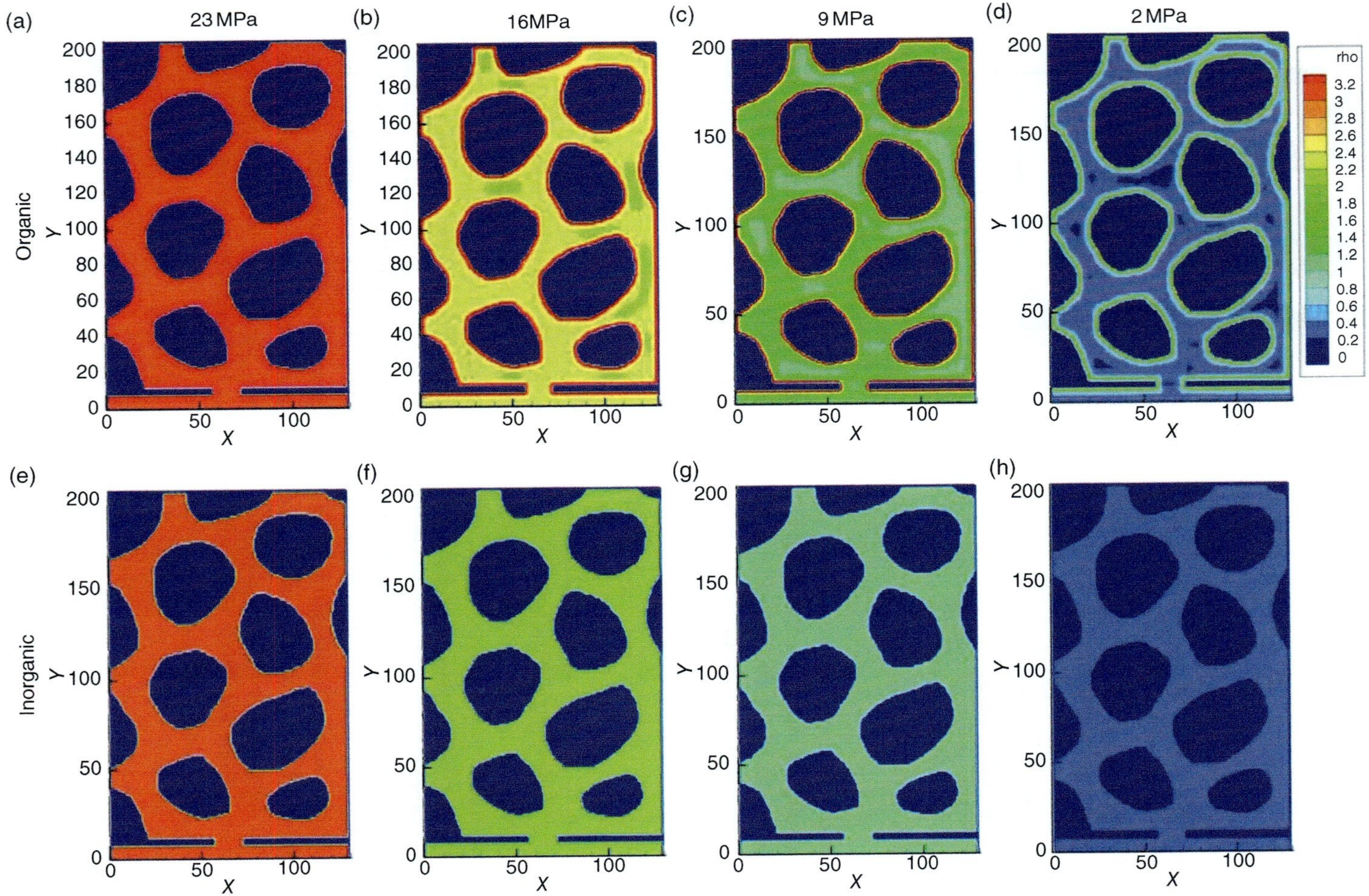

Figure 2.10 Gas density distributions at different pressures. (a–d) Gas density distribution variations of organic matter from 23 to 2 MPa, (e–h) gas density distribution variations of inorganic matter from 23 to 2 MPa.

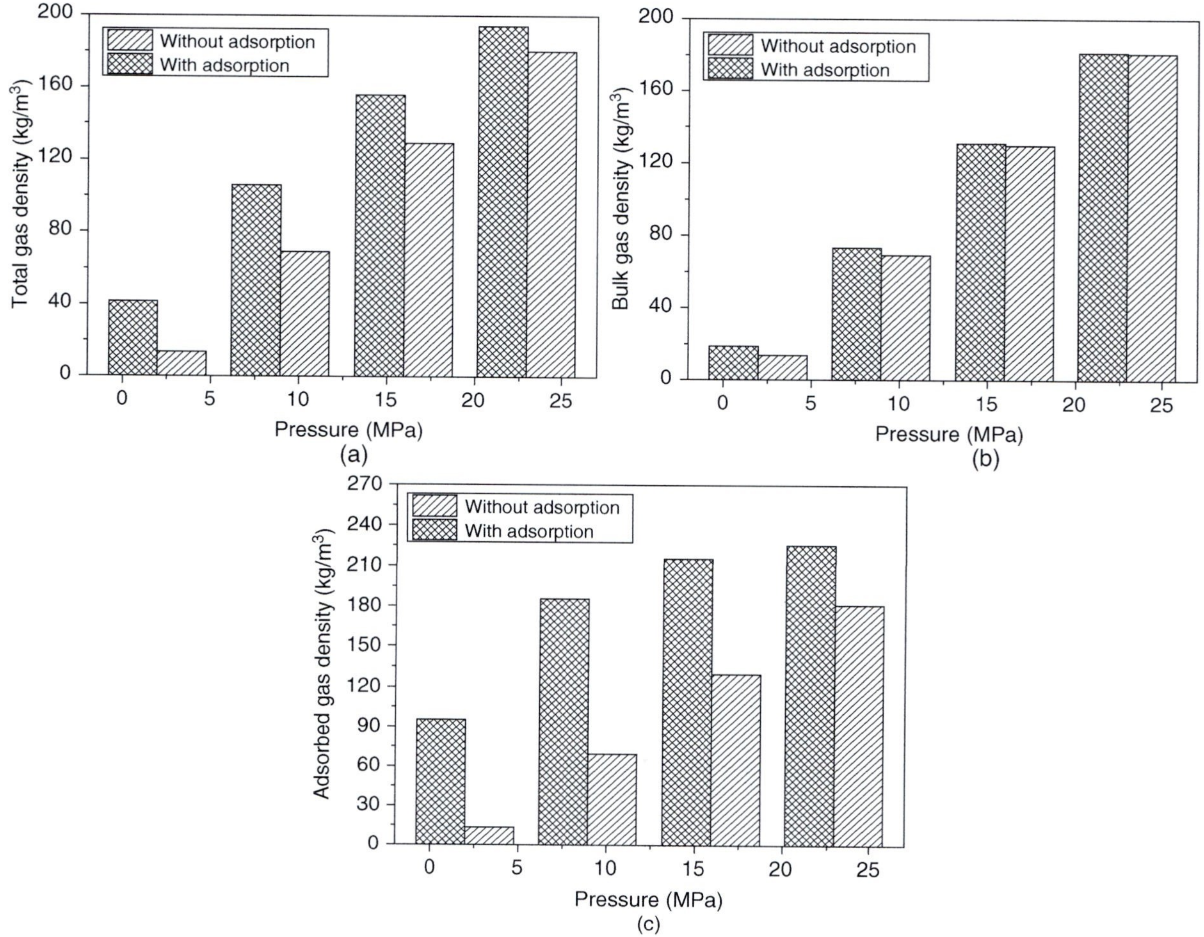

Figure 2.11 The densities of different types of gas at different pressures. (a) Total gas density, (b) bulk gas density, (c) adsorbed gas density.

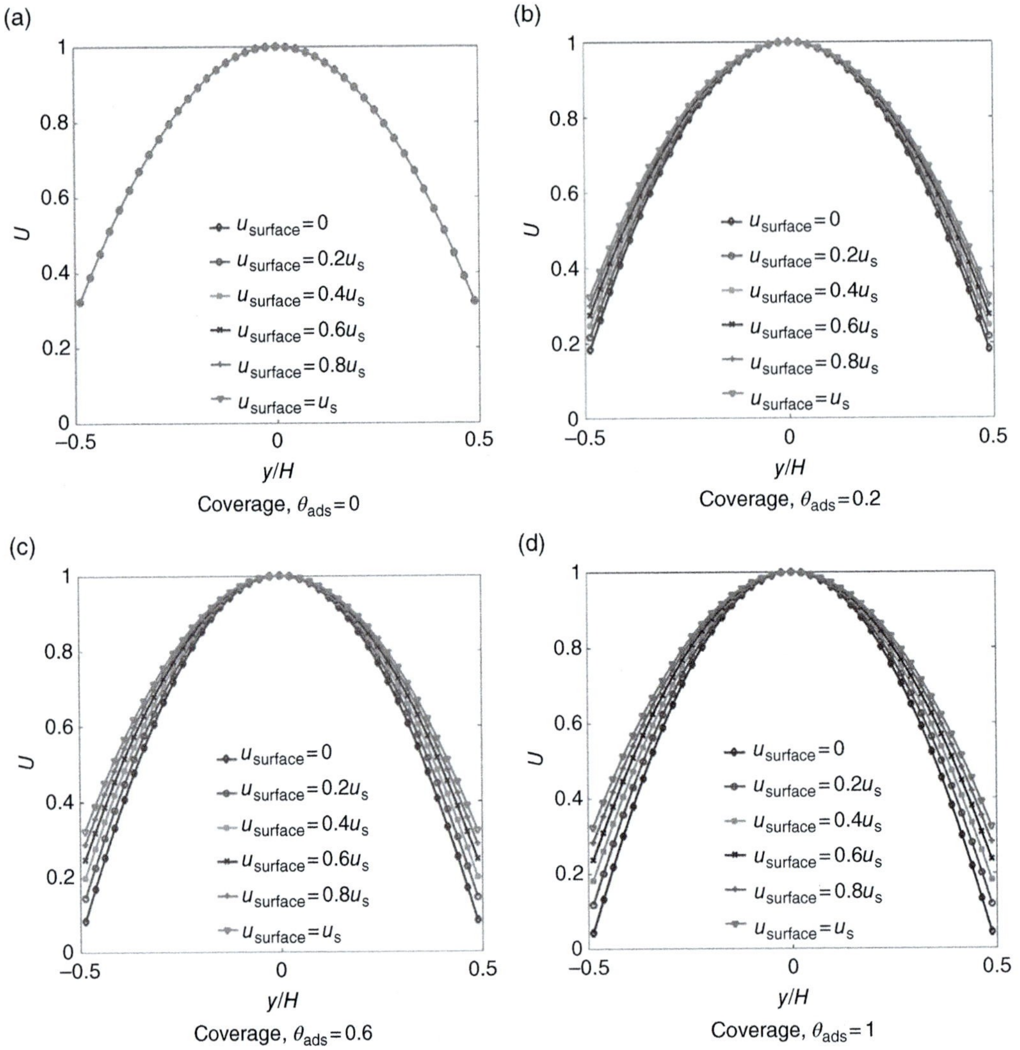

Figure 2.12 Normalized velocity at different surface coverage and surface diffusion velocity (a–d). *Source:* Wang et al. (2017a)/With permission of Elsevier.

$$C_{\mathrm{br}} = \frac{2\tau_{\mathrm{s}} - 1}{2l_{\mathrm{s}}^{\mathrm{w}} + 2\tau_{\mathrm{s}} - 1}, \tag{2.50}$$

$$\tau_{\mathrm{q}} = \frac{8\tau_{\mathrm{s}} - 1}{8(2\tau_{\mathrm{s}} - 1)}. \tag{2.51}$$

A block-filled porous medium as shown in Figure 2.13 is adopted to simulate water flow in strong hydrophilic and strong hydrophobic conditions.

The distribution of dimensionless velocity in y direction across slice $Y = 400$ is shown in Figure 2.14, and the velocity distribution is illustrated in Figure 2.15. As can be seen from Figure 2.14, the slip effect has a significant influence on liquid flow on micro-scale. For the hydrophilic porous medium, because of the strong fluid–solid interaction, the fluid prefers to stick on the solid surface. This increases the fluid viscosity near the solid surface and decreases the flow velocity. For the hydrophobic porous medium, because of the repulsive solid surface, the slip velocity occurs on the solid surface, and the fluid viscosity near the solid surface decreases, which significantly increases the flow velocity. Figure 2.15 qualitatively shows the dimensionless velocity distributions in different porous media. Although the water flow velocity in hydrophilic porous medium is much smaller than that without considering the nano-scale effect (as shown in Figure 2.14), the relative velocity distributions are similar. However, for the strong hydrophobic porous medium, the velocity distribution is quite different. From Figure 2.15, it can be seen that the dimensionless velocities in small pores are higher, and the velocity distribution is more uniform compared with the other two cases.

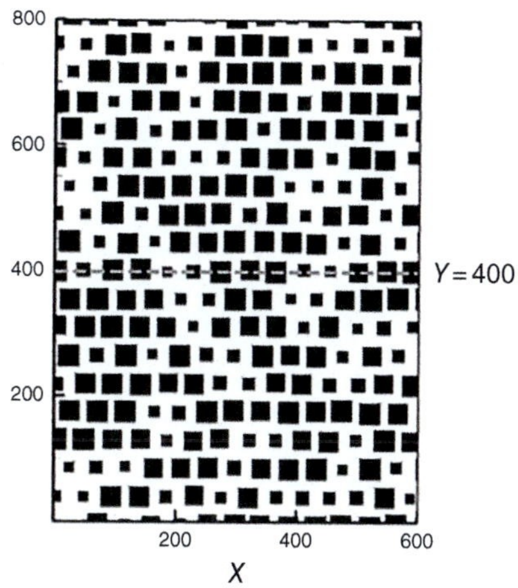

Figure 2.13 Schematic of porous structure.

2.4.3 Simulation of Immiscible Two-Phase Flow

The color-gradient LB model introduced in Section 2.3.2.1 is applied to simulate spontaneous imbibition of water in a straight nano-channel and a nano-porous medium shown in Figure 2.16a. In these simulations, the slip boundary condition and effective viscosity caused by nano-pore confinement effect are considered in both phases (Zhao et al. 2018b). For the water phase, the interface region around the solid surface affected by the fluid–solid interaction is explicitly considered. As shown in Figure 2.16b, the red color represents the interface region with viscosity μ_i and blue color represents the bulk region with viscosity μ_∞. For the octane phase, the local effective viscosity (μ_e^o) dependent on local pore size (H) is considered, as shown in Figure 2.16c, which can be calculated by (Wang et al. 2016b):

$$\mu_e^o = -0.127 \exp\left(\frac{H}{-33.513}\right) + 0.315, \tag{2.52}$$

and the slip length (l_s^o) for octane phase is calculated by:

$$l_s^o = 0.387 \exp\left(\frac{H}{-32.470}\right) + 0.826. \tag{2.53}$$

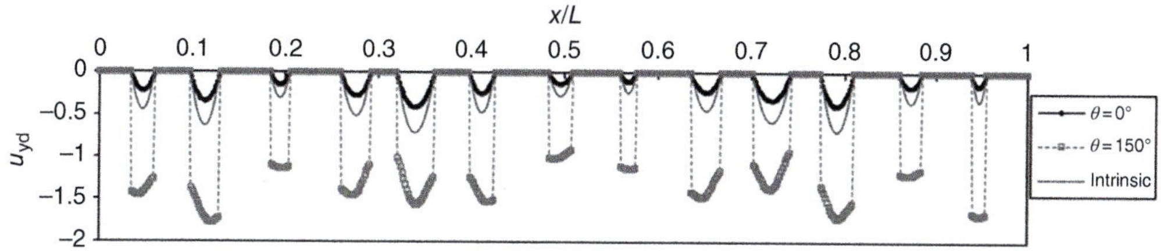

Figure 2.14 Distribution of dimensionless velocity in y direction across slice $Y = 400$.

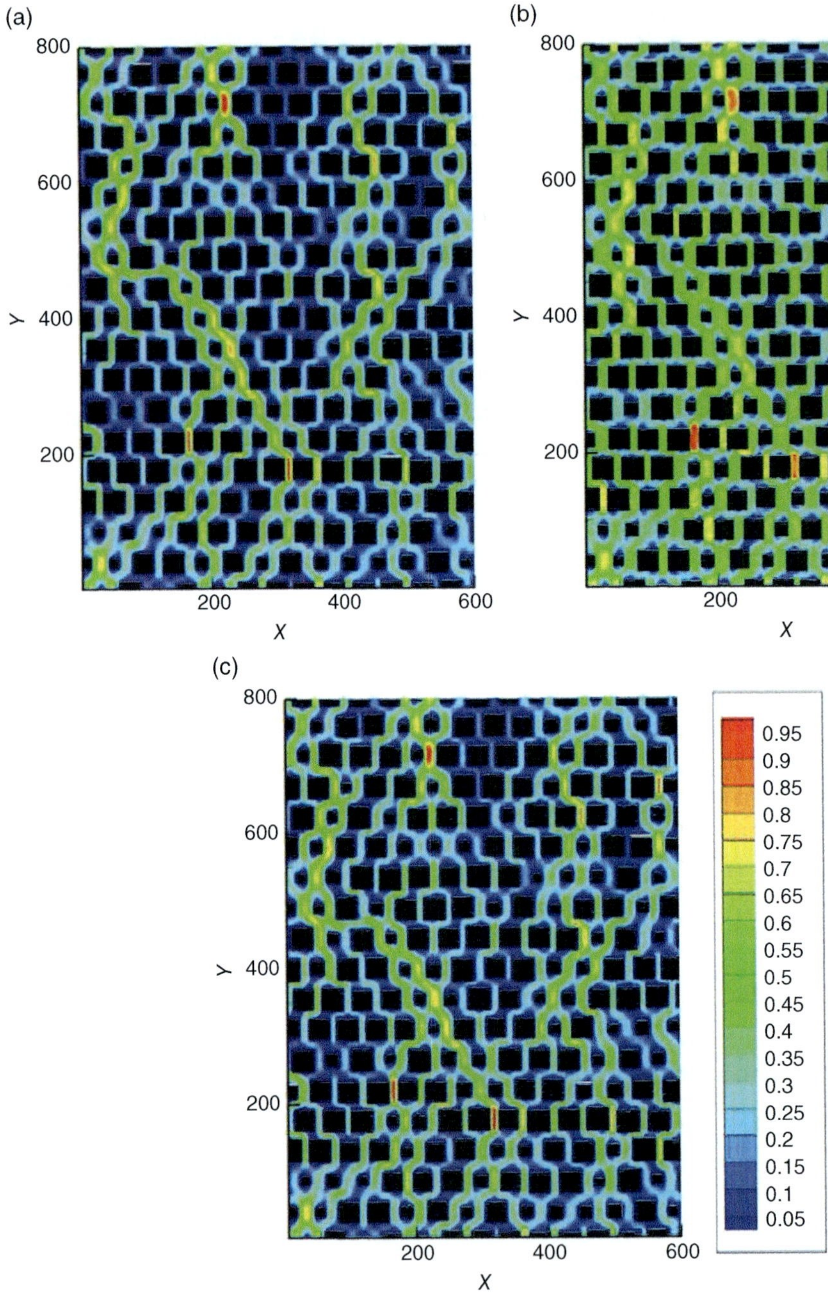

Figure 2.15 Dimensionless normalized velocity distributions in microporous media at different contact angles (from left to right). (a) Hydrophilic, (b) hydrophobic, and (c) intrinsic.

With above effective viscosity and slip length, the octane flow ability would be increased in quartz-based nano-porous medium (Zhao et al. 2018b).

In all the simulations, the water-octane-quartz three-phase contact angle θ_{wo} was 29° and water was the wetting phase. The water contact angle θ_w used to characterize the water-solid interaction force was related to temperature and pressure (Wu et al. 2017), and different values of θ_w were

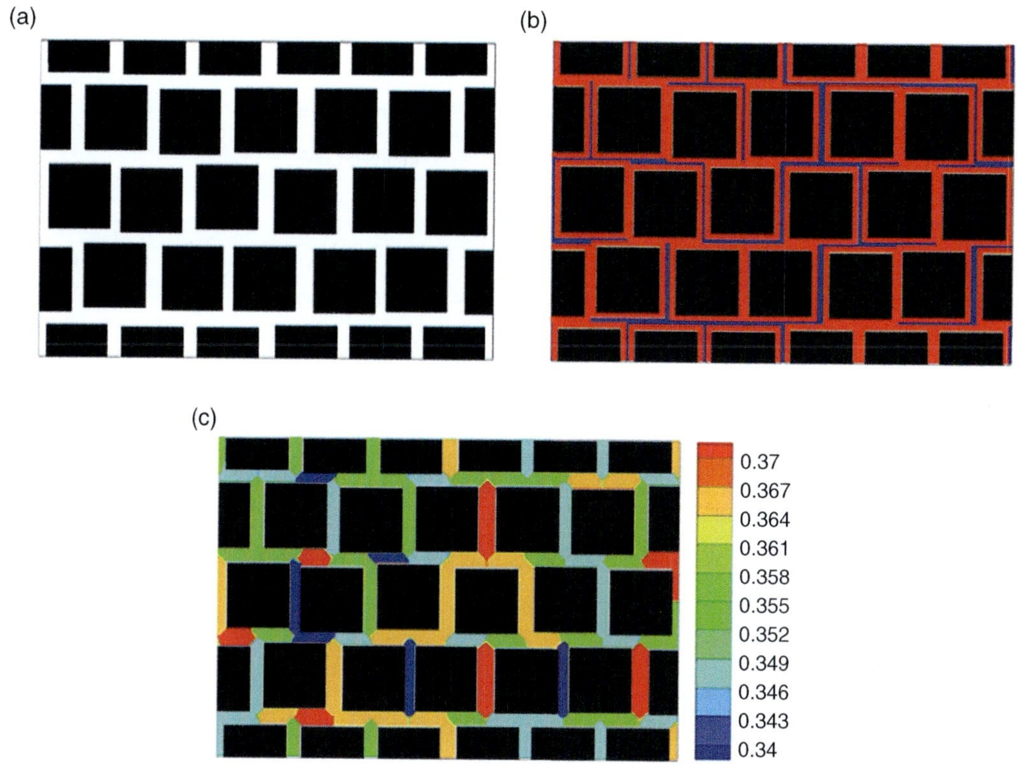

Figure 2.16 (a) Schematic of the nano-porous medium. Black color represents solid and white represents pore space. (b) The spatial distribution of water viscosity. Red color represents interface region and blue represents bulk region. (c) The spatial distribution of octane viscosity in lattice unit.

studied. To verify the accuracy of the proposed LB model, it was first used to simulate water imbibition in a straight nano-channel with 20 lattices width and 1000 lattices length. The resolution was 0.1 nm/lattice; therefore, the pore size was 2 nm in physical unit. The theoretical solution for the spontaneous imbibition in a 2D straight nano-channel is shown below:

$$\frac{2\sigma\cos\theta_{\mathrm{wo}}}{H^2}t = \frac{l_{\mathrm{w}}^2}{2g_{\mathrm{w,nano}}} - \frac{l_{\mathrm{w}}^2}{2g_{\mathrm{n,nano}}} + \frac{Ll_{\mathrm{w}}}{g_{\mathrm{n,nano}}} + c, \tag{2.54}$$

$$g_{\mathrm{w,nano}} = \frac{2a^3}{3\mu_\infty} - \frac{2a^3}{3\mu_i} + \frac{H^3}{12\mu_i} + \frac{l_{\mathrm{s}}{}^{\mathrm{w}}H^2}{2\mu_i}, \tag{2.55}$$

$$g_{\mathrm{n,nano}} = \frac{H^3 + 6l_{\mathrm{s}}^{\mathrm{o}}H^2}{2\mu_{\mathrm{e}}^{\mathrm{o}}}, \tag{2.56}$$

where σ is the interfacial tension coefficient, H is the channel width, L is the channel length, l_{w} is the water imbibition length, c is a constant related to the initial water length, a is half the width of bulk water region. The comparison of the LBM simulation results with theoretical solutions is shown in Figure 2.17a under different conditions in lattice units. The "intrinsic" means the results without considering nano-scale effect. At the initial stage of imbibition process, the nano-scale effect enhances the imbibition velocity of water under different θ_{w}. This is because imbibition

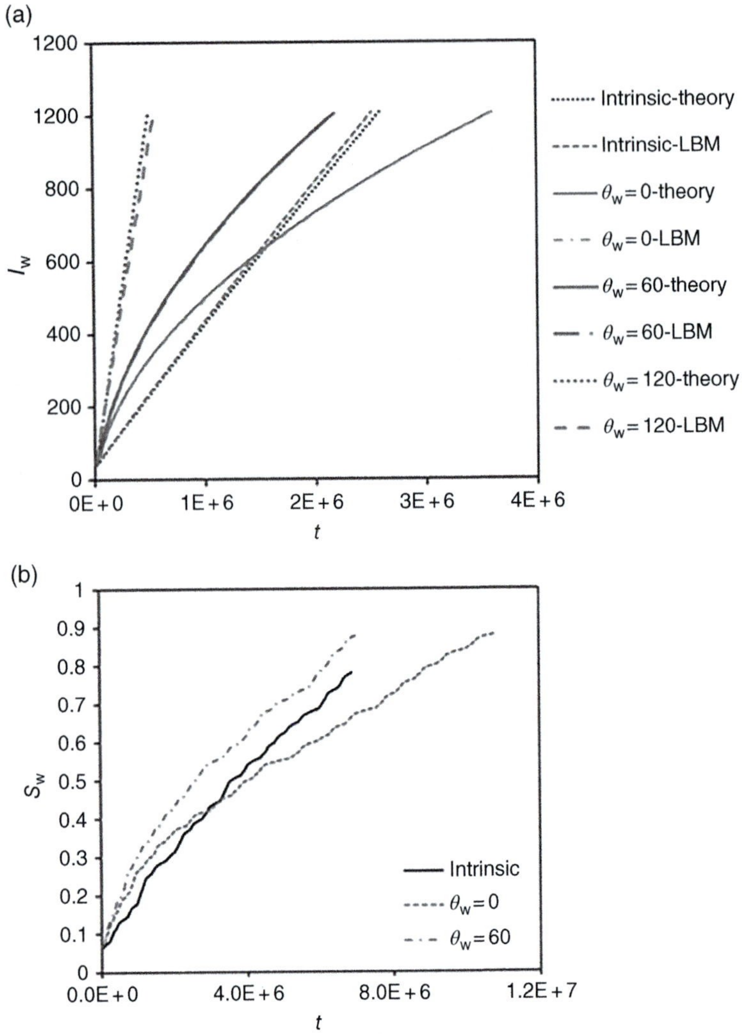

Figure 2.17 Dynamics of Spontaneous imbibition of water in a straight nano-channel (a) and a nano-porous medium (b) shown in Figure 2.15 at different θ_w. The unit is in lattice unit.

in the initial stage is dominated by the non-wetting octane phase, which flows faster under the nano-scale effect. With the increase of l_w, the influence of the wetting phase becomes more significant. Under different θ_w, the imbibition dynamics show different trends. For a weak water-solid interaction force ($\theta_w = 120°$), the nano-scale effect enhances the water flow ability. Therefore, the imbibition of water is always faster compared to the intrinsic case, as shown by the green curve. We note that this θ_w may not exist for the water–quartz system, and this is a theoretical analysis. For the neural ($\theta_w = 60°$) and strong ($\theta_w = 0°$) water-solid interaction forces, the nano-scale effect reduces the water flow ability in nano-channel. As a result, the imbibition process becomes slower with l_w. Under a certain imbibition length, the overall imbibition time considering nano-scale effect may exceed the intrinsic case, as shown by the crossover of red curve and black curve. In all these cases, the LBM simulation results match well with the theoretical solutions, demonstrating the accuracy of the proposed LB model.

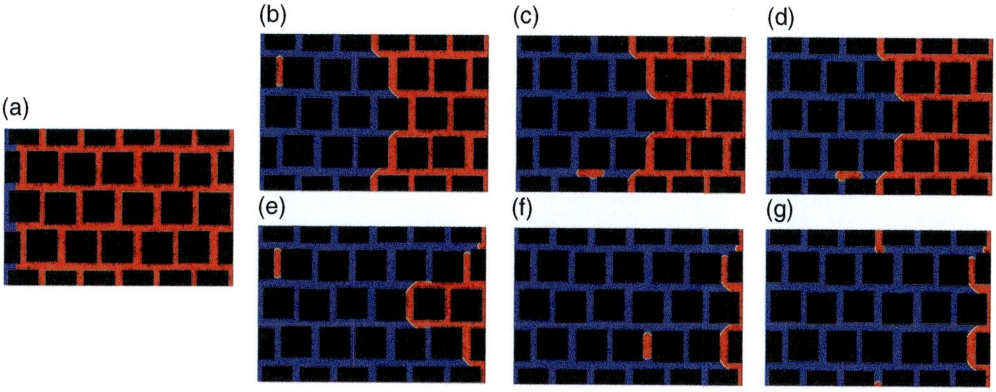

Figure 2.18 Spontaneous imbibition of water in a nano-porous medium. (a) The initial phase distribution. Red color represents the octane phase and blue represents water. (b–d) show the phase distributions when the water saturation is around 0.5; (e–g) show the final phase distributions when the water phase reaches the outlet. From left to right, the figures correspond to the intrinsic situation, $\theta_w = 0°$ and $\theta_w = 60°$, respectively.

Then the proposed LB model is used to simulate spontaneous imbibition of water in an octane-saturated nano-porous medium, as shown in Figure 2.16a. The simulation domain is 720 × 480 lattice^2 with a resolution of 0.1 nm/lattice and the mean pore size is 20 lattices. At the initial condition, a small amount of water phase is placed at the left of the domain, as shown in Figure 2.18a. The simulation stops when the water phase reaches the right outlet boundary. Three conditions are considered: intrinsic, $\theta_w = 60°$ and $\theta_w = 0°$. The obtained water saturation as a function of simulation time is shown in Figure 2.17b. As can be seen, the imbibition in a nano-porous medium shows similar trends with that in a straight nano-channel, which implies similar underlying flow mechanisms. In Figure 2.18b–g, we further compare the phase distributions under different conditions. Figure 2.18b–d show the phase distributions when the water saturation is around 0.5 and Figure 2.18e–g show the final phase distributions when the water phase breakthroughs the outlet. From left to right, the figures correspond to the intrinsic situation, $\theta_w = 0°$ and $\theta_w = 60°$, respectively. As can be seen, the displacement efficiency of water phase is very high due to the capillary valve effect. The nano-scale effect has some influence on the phase distribution, even for this simple geometry. We observe trapped non-wetting bubbles in different positions of the porous medium under different conditions. For the cases considering nano-scale effect ($\theta_w = 60°$ and $\theta_w = 0°$), as the water-solid interaction force increases the interface viscosity of water, the displacement front of water phase becomes more compact. As a result, the final displacement efficiency considering nano-scale effect is higher than the intrinsic case, as shown in Figure 2.18e–g.

2.4.4 Simulation of Vapor Condensation

The work presented in this subsection was performed by the authors of this chapter and has not been published elsewhere. During methane gas production, vapor condensation may occur when pressure changes, which in turn influences gas production. In addition, the nano-scale confinement may affect gas condensation. To simulate the condensation phenomenon at the pore scale, the Shan-Chen model introduced in Section 2.3.2.2 was applied. The original Shan-Chen model developed to simulate multi-phase flow is more suitable for conventional reservoir rocks. To correctly simulate the phase change problems in unconventional reservoir, the effects of gas slippage, effective viscosity, and gas adsorption have to be considered. To realize the gas slippage effect, the combined bounce-back/specular-reflection boundary condition (BR) introduced in Section 2.3.3 was used at the solid surface. The details for choosing free parameter C_{br} can be found in Li et al. (2011). To capture the Knudsen layer and gas denseness effects, the effective relaxation time is expressed as (Zhao et al. 2018b):

$$\tau_e = \frac{1}{2} + \sqrt{\frac{6}{\pi}} Kn \cdot J \cdot N \cdot \frac{(1 + 0.5b\rho\chi)^2}{\chi}, \tag{2.57}$$

with a Bosanquet-type $J = 1/(1 + 2Kn)$ applied. $N = H/\Delta x$ is the number of lattices occupied by the characteristic length H where Δx is the lattice length. $b = 2\pi d^3/3m$ where d, m are the diameter and molecular mass, respectively. χ is the collision correction factor caused by the gas denseness effect expressed as

$$\chi = 1 + \frac{5}{8} b\rho + 0.2869(b\rho)^2 + 0.1103(b\rho)^3 + 0.0386(b\rho)^4. \tag{2.58}$$

And the method introduced in Section 2.3.5.1 is applied for modeling gas adsorption.

2.4.4.1 Model Validations

Since the phase change problem in unconventional reservoir is not well studied, a single validation case considering all the physics is difficult to conduct. Therefore, the model proposed here was validated in two sub-steps, i.e. first by a single-phase slip flow to validate the gas slippage and effective viscosity effects and followed by droplet evaporation to validate phase change.

The slip flow considering gas slippage and effective viscosity was validated by simulating gas flow in a periodic microchannel. A constant pressure gradient was implemented between left and right sides of a channel with fixed height. The size of computational domain is $L \times H = 240 \times 60$ lattices^2 and the pressure gradient is implemented by external force $\boldsymbol{G} = \left(-\frac{\partial p}{\partial x}, 0\right) = (10^{-6}, 0)$ in lattice units. Two cases with Knudsen numbers $Kn = 0.1128, 2.2568$ are studied. The obtained velocity profile against channel height is shown in Figure 2.19, where $U = u/\overline{u}$ and $\overline{u} = 1/H\int_0^H u dy$. It is observed that higher Kn leads to higher slip velocity at the wall surface. Present simulation results agree well with that from Ohwada et al. (1998) and Li et al. (2011), showing the accuracy of current implementation. It is noted that the results from Ohwada et al. (1998) under-estimate the slip velocity at the wall surface to a small extent.

The phase change is validated by evaporation of a suspended droplet. In current simulation, the domain size is $L \times H = 180 \times 180$ lattices^2. A droplet with initial diameter of $D_0 = 98$ lattices is placed in the center of the domain and the four boundaries are set at constant pressure lower than saturation pressure to induce evaporation. A liquid–vapor density ratio of $\rho_l/\rho_v \approx 30$ is used at the temperature ratio of $T/T_c = 0.75$. Under quasi-isothermal condition, the droplet diameter square should decrease linearly with time. As shown in Figure 2.20, the decrease of droplet diameter square generally follows the linear line. At the end of evaporation, the evaporation rate is a little under-estimated. The reason might be the heat transfer and its influence on evaporation rate is not considered (Qin et al. 2019). Overall, the accuracy of model is acceptable for studying phase change problem qualitatively.

2.4.4.2 Vapor Condensation in Two Adjacent Nano-Pores

Similar to the validation case, a liquid–vapor density ratio of $\rho_l/\rho_v \approx 30$ was used at the temperature ratio of $T/T_c = 0.75$. The simulation domain is shown in Figure 2.21, where the left-hand side is the

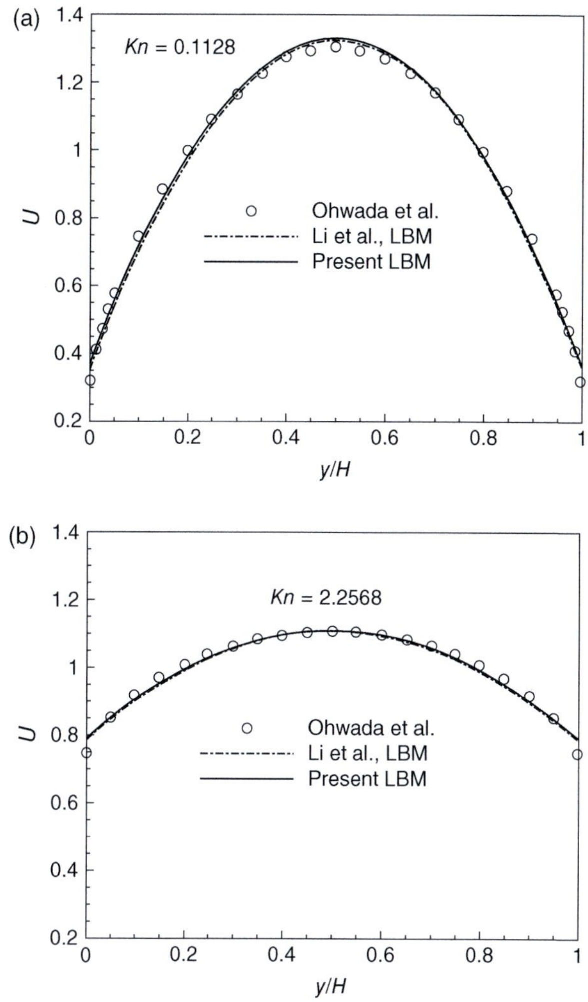

Figure 2.19 Comparison of velocity profiles of gas flow in a periodic microchannel between present results and that from Ohwada et al. (1998) and Li et al. (2011) at Knudsen numbers of (a) 0.1128 and (b) 2.2568. *Source:* Adapted from Ohwada et al. (1998), Li et al. (2011).

vapor inlet with a constant pressure slightly higher than the saturation vapor pressure to trigger condensation. The two pores are of the same length, but with a width ratio of 2, and $Kn = 1.0$, 2.0 for the large and small pores, respectively. In current simulations, the virtual solid density $\rho_{vs} = 0.54\rho_l$ is adopted, to realize a liquid–vapor-solid contact angle of around $\theta_{lw} = 60°$. As explained in Section 2.3, ρ_{vs} also plays the role of the gas adsorption parameter ρ_{ad} to achieve vapor adsorption at the wall. To demonstrate the influence of gas slippage in condensation in nano-pores, simulation results obtained with and without gas slippage effect are compared, while the Knudsen layer and adsorption effects are kept the same. According to current simulation setup, the liquid formed after vapor condensation is at quasi-equilibrium state with its velocity around zero. Thus, the flow in vapor phase is mainly analyzed.

The vapor condensation process of two cases is compared in Figure 2.22, with (a) showing the result considering gas slippage while (b) showing that using no-slip boundary condition. For the

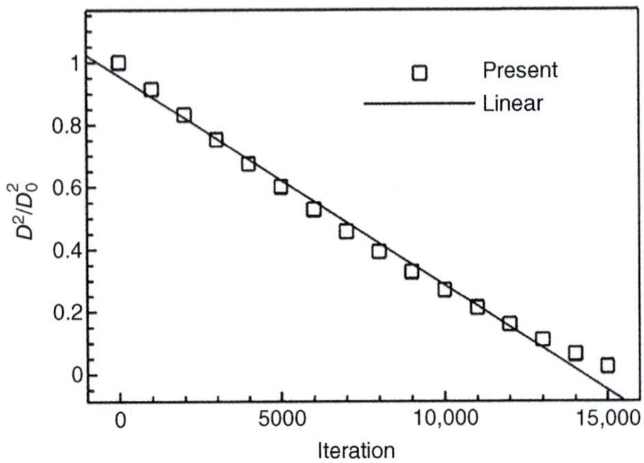

Figure 2.20 The evolution of droplet diameter square against iteration during evaporation.

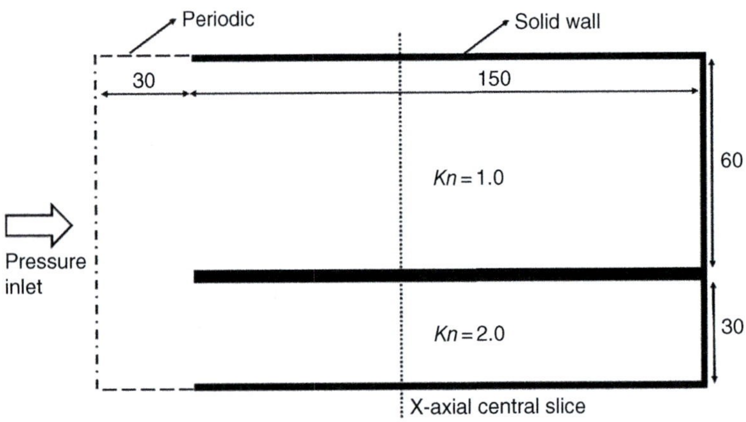

Figure 2.21 Simulation setup of methane vapor condensation in two adjacent nano-pores.

left and right columns, the liquid saturation in the large pore (S_l) is kept almost the same. For both cases, a layer of dense vapor is observed at the solid surface at $t = 10{,}000$ iterations before condensation starts. As time goes by, the vapor starts to condensate at the corners of the large pores. After some time, the liquid also forms at the corner of small pores at $t = 110{,}000$ and 140,000 iterations, respectively. At this stage, the condensed liquid is in partial-droplet shape with convex interfaces. Later at $t = 220{,}000$ and 360,000 iterations, the small convex interfaces merge into concave ones, after enough amount of liquid is formed by condensation. At $t = 920{,}000$ and 1,330,000 iterations, the large pores in both cases are almost fully filled with liquid, while the small pores are only partially filled, indicating the condensation is faster in large pores. According to Figure 2.22, the condensation processes of the two cases are qualitatively similar. However, the amount of liquid against time during condensation shows a significant difference. To reach the same liquid saturation in large pore, the condensation time considering gas slippage (Figure 2.22a) is shorter than that using no-slip boundary condition (Figure 2.22b). Moreover, at the same saturation of large pore (S_l), the saturation of small pore (S_s) is higher when considering gas slippage effect. Both observations indicate the vapor condensation rate is generally higher when considering gas slippage effect.

Figure 2.22 The comparison of vapor condensation process in two adjacent nano-pores (a) considering gas slippage effect and (b) using no-slip boundary.

To better illustrate the difference in condensation rate, the vapor velocity at the x-axial central slice of the computational domain between the two cases at $S_l \approx 19.7\%$ ($t = 220{,}000$ and 360,000 iterations in Figure 2.22a and b) is compared, since the vapor velocity determines the condensation rate. As shown in Figure 2.23a, the vapor velocity considering gas slippage is always larger than that using no-slip boundary. The ratios of slippage velocity to maximum velocity for large and small pores are 15.6% and 22.4%, respectively. The ratio in the small pore is larger because of the higher Kn. The velocity profile is not in a typical parabolic shape near the wall, due to the influence of dense vapor formed (gas adsorption). As shown in Figure 2.23b, the vapor density near the wall is about eight times that in the center of the pore. The adoption considering gas slippage and using no-slip boundaries are very close to each other. The maximum vapor density in large pores is

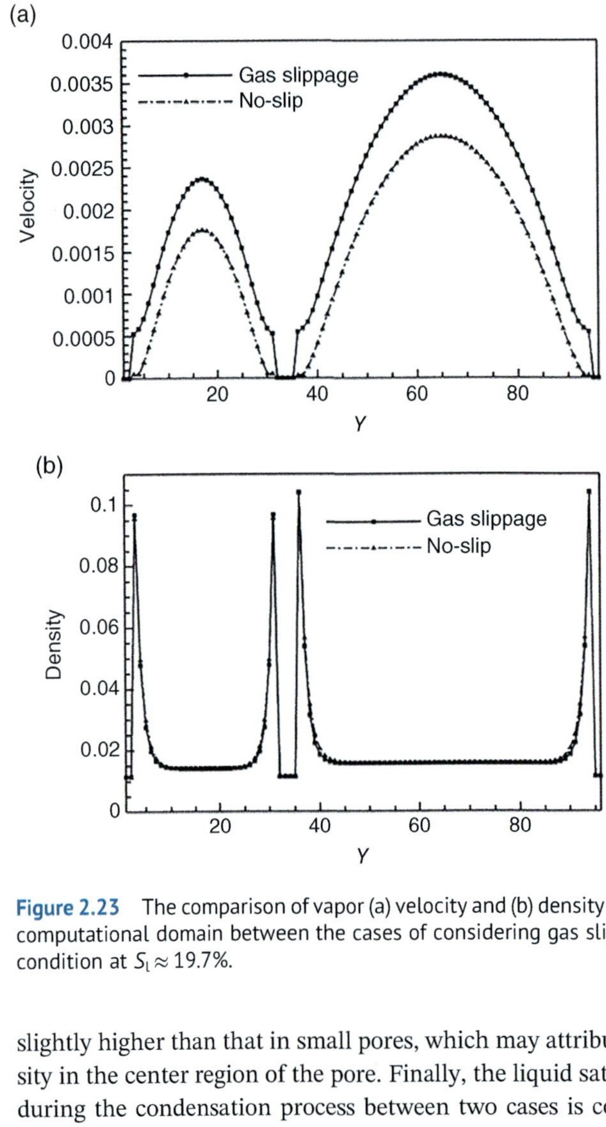

Figure 2.23 The comparison of vapor (a) velocity and (b) density at the x-axial central slice (Figure 2.21) of the computational domain between the cases of considering gas slippage effect and using no-slip boundary condition at $S_l \approx 19.7\%$.

slightly higher than that in small pores, which may attribute to the higher vapor velocity and density in the center region of the pore. Finally, the liquid saturation of large (S_l) and small (S_s) pores during the condensation process between two cases is compared, to quantitatively compare the average condensation rate. Figure 2.24 shows that with the given Kn, when considering gas slippage, the condensation rate in large and small pores are about 45.1 and 100.1% higher than those using no-slip boundary. Compared with that in small pores, the condensation rate in large pores in the two cases are about 40.9 and 98.5% higher. The condensation rate in small pore considering gas slippage is even 15.4% higher than that in large pore using no-slip boundary, indicating the significant influence of gas slippage.

2.5 Conclusion

In this chapter, a comprehensive overview of the recent progress on pore-scale simulations and digital rock physics of fluid flow in unconventional reservoirs are provided. Required extensions of the LB models to simulate nano-scale physics including slippage, effective viscosity, adsorption/desorption, and surface diffusion are presented in detail. Below are our main conclusions.

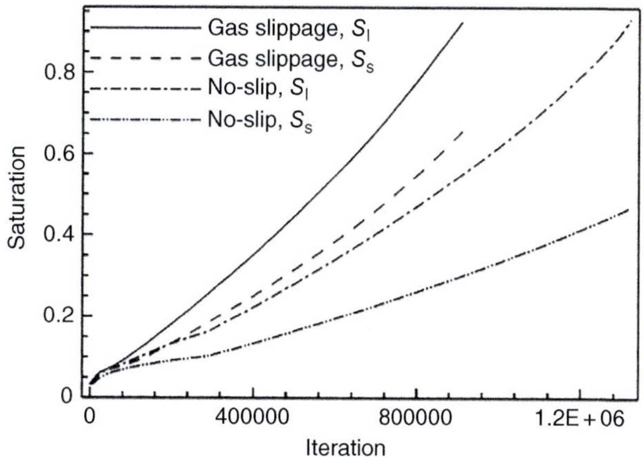

Figure 2.24 The comparison of liquid saturation of large (S_l) and small (S_s) pores during the condensation process between the cases considering gas slippage and using no-slip boundary.

1) Fluid flow in unconventional reservoirs presents very different behavior from those in conventional gas reservoirs. High *Kn*, high total organic content, complex structure, mix wettability, etc., all could introduce more complicated flow patterns.
2) With the scarcity of experimental data, pore-scale simulations can be essential tools to reveal the unique flow characteristics in unconventional gas reservoirs. Pore-scale LB models have been successfully extended to modeling single/multi-phase flow and phase change in unconventional gas reservoirs through modification of boundary conditions, collision operators, or force terms. Current simulations indicate that when the pore size is small enough, slippage appears at the solid interface for both liquid and gas flows and thus results in a faster flow rate and a higher condensation rate, although these effects could be diminished by wettability change or adsorption. Also, surface diffusion of adsorbed gas could provide additional mass flow for gas flow in unconventional gas reservoirs.
3) Although pore-scale simulations have been widely applied to study fluid flow in unconventional reservoirs, it should be noted that phenomenological models or physical approaches are usually introduced in these models, and therefore significant challenges remain to provide accurate predictions of fluid flow in complex porous media at nano- to micro-scales. Further development of specified pore-scale experiments on unconventional gas reservoirs will be useful to validate current numerical models.

References

Ba, Y., Liu, H., Li, Q. et al. (2016). Multiple-relaxation-time color-gradient lattice Boltzmann model for simulating two-phase flows with high density ratio. *Physical Review E* 94: https://doi.org/10.1103/PhysRevE.94.023310.

Barrat, J.L. and Bocquet, L. (1999). Large slip effect at a nonwetting fluid-solid interface. *Physical Review Letters* 82: 4671–4674. https://doi.org/10.1103/PhysRevLett.82.4671.

Barsotti, E., Tan, S.P., Saraji, S. et al. (2016). A review on capillary condensation in nanoporous media: implications for hydrocarbon recovery from tight reservoirs. *Fuel* https://doi.org/10.1016/j.fuel.2016.06.123.

Beskok, A. and Karniadakis, G.E. (1999). Report: a model for flows in channels, pipes, and ducts at micro and nano scales. *Microscale Thermophysical Engineering* 3: 43–77. https://doi.org/10.1080/108939599199864.

Bhatnagar, P.L., Gross, E.P., and Krook, M. (1954). A model for collision processes in gases. I. Small amplitude processes in charged and neutral one-component systems. *Physics Review* 94: 511–525. https://doi.org/10.1103/PhysRev.94.511.

Brunauer, S., Emmett, P.H., and Teller, E. (1938). Adsorption of gases in multimolecular layers. *Journal of the American Chemical Society* 60: 309–319. https://doi.org/10.1021/ja01269a023.

Chen, L., Kang, Q., Mu, Y. et al. (2014). A critical review of the pseudopotential multiphase lattice Boltzmann model: methods and applications keywords: pseudopotential model lattice Boltzmann method multiphase flow porous media slip phenomenon multiple physicochemical processes. *International Journal of Heat and Mass Transfer* 76: 210–236. https://doi.org/10.1016/j.ijheatmasstransfer.2014.04.032.

Chen, L., He, A., Zhao, J. et al. (2022). Pore-scale modeling of complex transport phenomena in porous media. *Progress in Energy and Combustion Science* 88: 100968. https://doi.org/10.1016/J.PECS.2021.100968.

Dongari, N., Zhang, Y., and Reese, J.M. (2011). Modeling of Knudsen layer effects in micro/nanoscale gas flows. *Journal of Fluids Engineering, Transactions of the ASME* 133: https://doi.org/10.1115/1.4004364.

Guo, Z. and Zheng, C. (2008). Analysis of lattice Boltzmann equation for microscale gas flows: relaxation times, boundary conditions and the Knudsen layer. *International Journal of Computational Fluid Dynamics* 22: 465–473. https://doi.org/10.1080/10618560802253100.

Guo, Z., Zheng, C., and Shi, B. (2002). Discrete lattice effects on the forcing term in the lattice Boltzmann method. *Physical Review. E, Statistical Physics, Plasmas, Fluids, and Related Interdisciplinary Topics* 65: 6. https://doi.org/10.1103/PhysRevE.65.046308.

Guo, Z., Zhao, T.S., and Shi, Y. (2006). Physical symmetry, spatial accuracy, and relaxation time of the lattice Boltzmann equation for microgas flows. *Journal of Applied Physics* 99: 074903. https://doi.org/10.1063/1.2185839.

He, X. and Luo, L.S. (1997). A priori derivation of the lattice boltzmann equation. *Physical Review. E, Statistical Physics, Plasmas, Fluids, and Related Interdisciplinary Topics* 55: R6333. https://doi.org/10.1103/PhysRevE.55.R6333.

Higashi, K., Ito, H., and Oishi, J. (1963). Surface diffusion phenomena in gaseous diffusion, (I). *Journal of the Atomic Energy Society of Japan / Atomic Energy Society of Japan* 5: 846–853. https://doi.org/10.3327/jaesj.5.846.

Huang, H., Sukop, M.C., and Lu, X.Y. (2015). *Multiphase Lattice Boltzmann Methods: Theory and Application, Multiphase Lattice Boltzmann Methods: Theory and Application.* Wiley https://doi.org/10.1002/9781118971451.

Inamuro, T., Yoshino, M. and Ogino, F., 1995. A non-slip boundary condition for lattice Boltzmann simulations. Physics of Fluids, 7(12), pp.2928–2930. https://doi.org/10.1063/1.868766.

Javadpour, F., Fisher, D., and Unsworth, M. (2007). Nanoscale gas flow in shale gas sediments. *Journal of Canadian Petroleum Technology* 46: 55–61. https://doi.org/10.2118/07-10-06.

Jing, Y., Rabbani, A., Armstrong, R.T. et al. (2020). A hybrid fracture-micropore network model for multiphysics gas flow in coal. *Fuel* 281: 118687. https://doi.org/10.1016/j.fuel.2020.118687.

Karniadakis, G., Beskok, A., and Aluru, N. (2006). *Microflows and Nanoflows: Fundamentals and Simulation*, vol. 29. Springer Science & Business Media.

Kim, T.H., Cho, J., and Lee, K.S. (2017). Evaluation of CO_2 injection in shale gas reservoirs with multi-component transport and geomechanical effects. *Applied Energy* 190: 1195–1206. https://doi.org/10.1016/j.apenergy.2017.01.047.

Krishna, R. and Wesselingh, J.A. (1997). Review article number 50: the Maxwell-Stefan approach to mass tranfer. *Chemical Engineering Science* https://doi.org/10.1016/S0009-2509(96)00458-7.

Lallemand, P. and Luo, L.S. (2000). Theory of the lattice Boltzmann method: dispersion, dissipation, isotropy, Galilean invariance, and stability. *Physical Review. E, Statistical Physics, Plasmas, Fluids, and Related Interdisciplinary Topics* 61: 6546–6562. https://doi.org/10.1103/PhysRevE.61.6546.

Langmuir, I. (1918). The adsorption of gases on plane surfaces of glass, mica and platinum. *Journal of the American Chemical Society* 40: 1361–1403. https://doi.org/10.1021/ja02242a004.

Levesque, M., Duvail, M., Pagonabarraga, I. et al. (2013). Accounting for adsorption and desorption in lattice Boltzmann simulations. *Physical Review. E, Statistical, Nonlinear, and Soft Matter Physics* 88: 013308. https://doi.org/10.1103/PhysRevE.88.013308.

Li, Q., He, Y.L., Tang, G.H., and Tao, W.Q. (2011). Lattice Boltzmann modeling of microchannel flows in the transition flow regime. *Microfluidics and Nanofluidics* 10: 607–618. https://doi.org/10.1007/s10404-010-0693-1.

Li, Q., Luo, K.H., Kang, Q.J. et al. (2016). Lattice Boltzmann methods for multiphase flow and phase-change heat transfer. *Progress in Energy and Combustion Science* https://doi.org/10.1016/j.pecs.2015.10.001.

Liu, H., Kang, Q., Leonardi, C.R. et al. (2016). Multiphase lattice Boltzmann simulations for porous media applications: a review. *Computational Geosciences* 20: 777–805. https://doi.org/10.1007/s10596-015-9542-3.

Lockerby, D.A., Reese, J.M., and Gallis, M.A. (2005). The usefulness of higher-order constitutive relations for describing the Knudsen layer. *Physics of Fluids* 17: 100609. https://doi.org/10.1063/1.1897005.

Lubbers, N., Agarwal, A., Chen, Y. et al. (2020). Modeling and scale-bridging using machine learning: nanoconfinement effects in porous media. *Scientific Reports* 101 (10): 1–13. https://doi.org/10.1038/s41598-020-69661-0.

Marković, A., Stoltenberg, D., Enke, D. et al. (2009). Gas permeation through porous glass membranes. Part II: transition regime between Knudsen and configurational diffusion. *Journal of Membrane Science* 336: 32–41. https://doi.org/10.1016/j.memsci.2009.02.030.

Martys, N.S. and Chen, H. (1996). Simulation of multicomponent fluids in complex three-dimensional geometries by the lattice Boltzmann method. *Physical Review. E, Statistical Physics, Plasmas, Fluids, and Related Interdisciplinary Topics* 53: 743–750. https://doi.org/10.1103/PhysRevE.53.743.

Maxwell, J. (1879). VII. On stresses in rarified gases arising from inequalities of temperature. *Philosophical Transactions. Royal Society of London* 170: 231–256. https://doi.org/10.1098/rstl.1879.0067.

McNamara, G.R. and Zanetti, G. (1988). Use of the boltzmann equation to simulate lattice-gas automata. *Physical Review Letters* 61: 2332–2335. https://doi.org/10.1103/PhysRevLett.61.2332.

Michalis, V.K., Kalarakis, A.N., Skouras, E.D., and Burganos, V.N. (2010). Rarefaction effects on gas viscosity in the Knudsen transition regime. *Microfluidics and Nanofluidics* 9: 847–853. https://doi.org/10.1007/s10404-010-0606-3.

Middleton, R.S., Carey, J.W., Currier, R.P. et al. (2015). Shale gas and non-aqueous fracturing fluids: opportunities and challenges for supercritical CO_2. *Applied Energy* 147: 500–509. https://doi.org/10.1016/j.apenergy.2015.03.023.

Moon, J.H., Park, Y.J., Kim, M.B. et al. (2005). Permeation and separation of a carbon dioxide/nitrogen mixture in a methyltriethoxysilane templating silica/α-alumina composite membrane. *Journal of Membrane Science* 250: 195–205. https://doi.org/10.1016/j.memsci.2004.10.027.

Ohwada, T., Sone, Y., and Aoki, K. (1998). Numerical analysis of the shear and thermal creep flows of a rarefied gas over a plane wall on the basis of the linearized Boltzmann equation for hard-sphere molecules. *Physics of Fluids A: Fluid Dynamics* 1: 1588. https://doi.org/10.1063/1.857304.

Qin, F., Del Carro, L., Moqaddam, A.M. et al. (2019). Study of non-isothermal liquid evaporation in synthetic micro-pore structures with hybrid lattice Boltzmann model. *Journal of Fluid Mechanics* 866: 33–60. https://doi.org/10.1017/JFM.2019.69.

Raviv, U., Laurat, P., and Klein, J. (2001). Fluidity of water confined to subnanometre films. *Nature* 413: 51–54. https://doi.org/10.1038/35092523.

Shan, X. and Chen, H. (1993). Lattice Boltzmann model for simulating flows with multiple phases and components. *Physical Review E* 47 (3): 1815. https://doi.org/10.1103/PhysRevE.47.1815.

Suga, K., Takenaka, S., Ito, T. et al. (2010). Evaluation of a lattice Boltzmann method in a complex nanoflow. *Physical Review. E, Statistical, Nonlinear, and Soft Matter Physics* 82: 016701. https://doi.org/10.1103/PhysRevE.82.016701.

Theodorou, D. and Wei, J. (1983). Diffusion and reaction in blocked and high occupancy zeolite catalysts. *Journal of Catalysis* 83: 205–224. https://doi.org/10.1016/0021-9517(83)90043-X.

Vanson, J.M., Coudert, F.X., Rotenberg, B. et al. (2015). Unexpected coupling between flow and adsorption in porous media. *Soft Matter* 11: 6125–6133. https://doi.org/10.1039/c5sm01348h.

Verhaeghe, F., Luo, L.S., and Blanpain, B. (2009). Lattice Boltzmann modeling of microchannel flow in slip flow regime. *Journal of Computational Physics* 228: 147–157. https://doi.org/10.1016/j.jcp.2008.09.004.

Wang, J., Chen, L., Kang, Q., and Rahman, S.S. (2016a). The lattice Boltzmann method for isothermal micro-gaseous flow and its application in shale gas flow: a review. *International Journal of Heat and Mass Transfer* https://doi.org/10.1016/j.ijheatmasstransfer.2015.12.009.

Wang, S., Javadpour, F., and Feng, Q. (2016b). Molecular dynamics simulations of oil transport through inorganic nanopores in shale. *Fuel* 171: 74–86. https://doi.org/10.1016/j.fuel.2015.12.071.

Wang, J., Kang, Q., Chen, L., and Rahman, S.S. (2017a). Pore-scale lattice Boltzmann simulation of micro-gaseous flow considering surface diffusion effect. *International Journal of Coal Geology* 169: 62–73. https://doi.org/10.1016/j.coal.2016.11.013.

Wang, J., Kang, Q., Wang, Y. et al. (2017b). Simulation of gas flow in micro-porous media with the regularized lattice Boltzmann method. *Fuel* 205: 232–246. https://doi.org/10.1016/j.fuel.2017.05.080.

Watari, M. (2010). Relationship between accuracy and number of velocity particles of the finite-difference lattice Boltzmann method in velocity slip simulations. *Journal of Fluids Engineering, Transactions of the ASME* 132: https://doi.org/10.1115/1.4002359.

Weijermars, R. (2014). US shale gas production outlook based on well roll-out rate scenarios. *Applied Energy* 124: 283–297. https://doi.org/10.1016/j.apenergy.2014.02.058.

Wu, K., Chen, Z., Li, J. et al. (2017). Wettability effect on nanoconfined water flow. *Proceedings of the National Academy of Sciences of the United States of America* 114: 3358–3363. https://doi.org/10.1073/pnas.1612608114.

Yuan, P. and Schaefer, L. (2006). Equations of state in a lattice Boltzmann model. *Physics of Fluids* 18: 042101. https://doi.org/10.1063/1.2187070.

Zhang, W.M., Meng, G., and Wei, X. (2012). A review on slip models for gas microflows. *Microfluidics and Nanofluidics* https://doi.org/10.1007/s10404-012-1012-9.

Zhao, J., Yao, J., Zhang, L. et al. (2016). Pore-scale simulation of shale gas production considering the adsorption effect. *International Journal of Heat and Mass Transfer* 103: 1098–1107. https://doi.org/10.1016/j.ijheatmasstransfer.2016.08.026.

Zhao, J., Kang, Q., Yao, J. et al. (2018a). The effect of wettability heterogeneity on relative permeability of two-phase flow in porous media: a lattice Boltzmann study. *Water Resources Research* 54: 1295–1311. https://doi.org/10.1002/2017WR021443.

Zhao, J., Kang, Q., Yao, J. et al. (2018b). Lattice Boltzmann simulation of liquid flow in nanoporous media. *International Journal of Heat and Mass Transfer* 125: 1131–1143. https://doi.org/10.1016/j.ijheatmasstransfer.2018.04.123.

3

Digital Rock Modeling

A Review

Yuqi Wu[1,2] *and Pejman Tahmasebi*[3]

[1] *Shandong Provincial Key Laboratory of Deep Oil and Gas, China University of Petroleum (East China), Qingdao, China*
[2] *School of Geosciences, China University of Petroleum (East China), Qingdao, China*
[3] *Colorado School of Mines, Golden 80401, USA*

3.1 Introduction

Digital rock physics (DRP), as an emerging computational simulation technique in recent years, is not only used for the characterization of pore systems and simulation of acoustic and electrical properties (Andrä et al. 2013; Blunt et al. 2013; Karimpouli and Tahmasebi 2016, 2019a; Sahimi 2011; Wu et al. 2019a; Wu et al. 2020c), but are also widely used for simulation of fluid flow, mass transport and understanding of pore-scale mechanisms of An et al. (2020), Bultreys et al. (2016), Karimpouli and Tahmasebi (2016, 2019c), Tahmasebi et al. (2020a), Xia et al. (2019), Zhu et al. (2019), and Wu et al. (2022a). Compared to conventional laboratory experiments, DRP embraces many advantages (Blunt 2017; Bultreys et al. 2016; Fagbemi et al. 2018; Munawar et al. 2018; Tahmasebi et al. 2020a; Wu et al. 2019b, d). Firstly, DRP can ensure that valuable digital information of the cores can be permanently stored and reused in the future (Andrä et al. 2013; Lin et al. 2018). Secondly, the traditional physical experiments, such as the mercury injection capillary pressure experiment, may damage the rock, so the used sample cannot be utilized for other experiments (Golparvar et al. 2018; Zhu et al. 2019). DRP is a non-destructive method that uses numerical simulation to compute all physical and chemical processes on digital rock models (Wu et al. 2022b). Thirdly, DRP is capable of changing the proportions of minerals and pores in the digital core and investigating how each component affects the petrophysical properties (Yang et al. 2021; Wu et al. 2020d). Thus, due to the above-mentioned reasons, DRP has become a game-changer in rock physics.

The first step of DRP is to construct an accurate digital model. Thanks to the recent advances in imaging techniques and computational methods, plenty of modeling methods have been proposed to build such models (Gerke et al. 2015; Lemmens et al. 2019; Mosser et al. 2017; Tahmasebi et al. 2020a; Wu et al. 2019c; Yang et al. 2015; Zhu et al. 2019). Among such methods, some are developed to generate simple porous media with single-scale pore systems in conventional reservoirs, while others construct heterogeneous digital rocks with multiscale pore structures in unconventional reservoirs. Compared to the former, the latter involves micro- and nano-meter pores, and it presents more complex structures. Therefore, modeling methods for conventional and unconventional reservoir rocks are different. This chapter aims to review these methods of digital rocks. The benefits and drawbacks of these methods are also discussed. Furthermore, the challenges and future development paths of the modeling methods are presented as well.

Physics of Fluid Flow and Transport in Unconventional Reservoir Rocks, First Edition.
Edited by Behzad Ghanbarian, Feng Liang, and Hui-Hai Liu.

3.2 Single-Scale Modeling of Digital Rocks

Conventional reservoir rocks usually involve micrometer pores and single-scale structures, which require using single-scale modeling methods. Such techniques can be divided into two groups of experimental and computational methods, as discussed in detail in the following (Blunt et al. 2013; Tahmasebi and Sahimi 2012; Zhu et al. 2019).

3.2.1 Experimental Techniques

Techniques in this group usually scan or continuously photograph samples using experimental equipment to obtain multiple two-dimensional (2D) slices of the core and then superimpose the 2D slices to construct a three-dimensional (3D) digital core. It mainly includes the imaging technique of serial sectioning (Chawla et al. 2006), laser scanning confocal microscopy (Shah et al. 2017), and X-Ray computer tomography scanning technique (Munawar et al. 2018; Tahmasebi et al. 2017; Wu et al. 2019b; Zhang et al. 2020).

3.2.1.1 Imaging Technique of Serial Sectioning

The workflow of the imaging technique of serial sectioning is cumbersome. It contains cutting, polishing, and imaging of a sample. These steps should be repeated until enough 2D sections are obtained. Finally, a 3D digital model can be produced by superimposing these slices (Chawla et al. 2006). This technique was invented by Lymberopoulos and Payatakes (1992), who analyzed the characteristics of pore structures of rocks. In order to enhance the quality and contrast of the images, they injected a dye into the sample, then cut the sample with a microtome, and used the vernier on the microtome to ensure the thickness of the sample was the same every time. Then the sample was polished and photographed with a high-resolution microscope. After photographing, the sample was cut and polished again until a certain number of slices were obtained, and finally, a 3D digital core was obtained through image superposition (Lymberopoulos and Payatakes 1992). Vogel and Roth (2001) used this method to study the pore structures of soil. They put clean soil samples in polyester resin to improve the quality of shooting. After processing, samples were cut and polished, and then a digital camera was used to photograph samples, and they finally superimposed 2D pictures to construct the 3D digital model of soils (Vogel and Roth 2001).

At present, the combination of imaging techniques of serial sectioning and focused ion beam (FIB), namely FIB-SEM, is widely used in the construction of digital cores for tight reservoirs. The advantage of the imaging technique of serial sectioning is that it can obtain high-resolution rock pictures, but it usually spends a long time to obtain a lot of 2D images. Moreover, the captured area of such images may be small, and samples will be demolished in the process of cutting.

3.2.1.2 Laser Scanning Confocal Microscopy

Initially, laser scanning confocal microscopy was used for observing cell structures. Minsky (1988) first proposed the technique and clarified the basic work principle of laser scanning confocal microscopy (Minsky 1988). This technology uses a laser light as the illumination source, puts two pinholes in front of the light source and the detector, and places a dichroic mirror on the path of light propagation; this mirror reflects the light from the source. When the reflected light reaches the surface of an object, it propagates to the detector after reflection. The detector then receives the light with an intensity, which is converted into an electrical signal and sent to a computer. After some processing, a 3D image with a certain thickness can be obtained. In order to improve the imaging quality, a core should be immersed in a low-viscosity epoxy resin intercalated with fluorescent substances as Fredrich (1999) proposed to construct 3D digital cores (Fredrich 1999). After the epoxy resin was completely penetrated into a rock sample, the 3D pore structure can be acquired. Some researchers used laser scanning confocal microscopy to construct 3D digital models of carbonate and shales (Kus 2015; Shah et al. 2017). Due to the limitation of the experiment instrument, this technique cannot meet the requirements of high resolution and large field of view at the same time. Therefore, the constructed digital rock by this technique would be single-scale, not multiscale.

3.2.1.3 X-Ray Computed Tomography Scanning

In 1967, Hounsfield invented the first computed tomography (CT) scanning machine (Dmytriw 2012), which was later applied in the medical field. In the 1980s, this technology was used for soil and rock. With the advances in CT instruments, X-ray CT scanners have become the most widely used technique to build 3D digital rocks (Cnudde and Boone, 2013; Wildenschild and Sheppard 2013; Wu et al. 2020a). This technique uses an X-ray as the ray source, which is absorbed by materials and its intensity attenuates when the X-ray passes through a rock sample. The attenuated X-ray intensity would be different depending on the materials in a sample. Then a certain reconstruction algorithm is used to convert the detected ray intensity into the gray value in an image, and next the gray value is related to the density of minerals. After processing, one can acquire the digital rock. To date, the fast development of CT machines has improved the speed of scanning samples. The CT instrument has been used to monitor the dynamic processes inside the samples (Bultreys et al. 2016), such as fluid flow in porous media (Berg et al. 2013; Singh et al. 2017; Wu et al. 2020a, 2020c), dissolution of minerals (Menke et al. 2018), pore-system evolution with the temperature (Saif et al. 2019). Thus, it is an excellent tool to construct static and dynamic digital rocks.

The commonly used CT scanners can get images of 1000^3–2000^3 voxels, and the resolution can be micrometers or nanometers (Bultreys et al. 2016). Compared with the first two experimental methods, this technique can build 3D digital cores with higher resolution and larger size and is the most accurate experimental method for building digital rocks of conventional reservoirs. Also, the X-ray CT scanning method does not damage the utilized samples. Nevertheless, the technique is not suitable for constructing 3D digital cores with micro- and nano-meter pores since it has to make a balance between the high resolution and large field of view.

3.2.2 Computational Methods

Compared to experimental techniques, computational methods reconstruct 3D digital cores using some algorithms based on 2D images, for example, SEM images or thin sections. This group includes the truncated Gaussian random field method (Yeong and Torquato 1998), phase recovery algorithm (Fullwood et al. 2008; Hasanabadi et al. 2016), simulated annealing method (Eschricht et al. 2005; Wu et al. 2018), Markov chain Monte Carlo method (Wu et al. 2006), sequential indicator simulation method (Keehm et al. 2004), multiple-point statistics method (Okabe and Blunt 2005; Tahmasebi and Sahimi 2012, 2013), quartet structure generation set method (Chen et al. 2015; Wang et al. 2007; Wu et al. 2020c), deep learning method (Feng et al. 2018b), and process-based method (Bakke and Øren 1997), etc. The cost of these techniques is quite low, and the modeling speed is fast, so these methods are quite popular. The truncated Gaussian random field method and phase recovery algorithm are currently less used. Thus, in the following, we put the focus on the other methods.

3.2.2.1 Simulated Annealing

The simulated annealing method was initially proposed to solve global optimization problems (Metropolis et al. 1953). In 1995, Hazlett used this method to reconstruct 3D digital cores of Berea sandstones (Hazlett 1995). Afterward, it has been widely applied to the reconstruction of 3D porous media (Gerke et al. 2015; Hamzehpour et al. 2007; Jiao et al. 2007; Liu et al. 2009; Manwart and Hilfer 1999; Talukdar et al. 2002; Yeong and Torquato 1998). The modeling process of simulated annealing is similar to the annealing and cooling process of metals, and the simulation aims to transform a high "energy" system composed of disordered particles and pores into an ordered low "energy" system (Jiao et al. 2009). As to the workflow of the method, first, a simulation field is considered with a grid size, for example, N^3, and start random simulation under the constraints of porosity and autocorrelation function (both obtained from 2D images). For each iteration, a pore voxel and a particle voxel are exchanged, and this exchange slightly changes the simulation function, so the energy of the system also changes. When the simulation result meets the requirements of constraint conditions, the simulation ends.

If only low-order correlation functions (e.g. porosity or autocorrelation function) are taken as constraints, it would be difficult to accurately generate complex structures in porous media. In order to solve this issue, different additional constraints are introduced, such as pore size distribution (Eschricht et al. 2005), linear path function, two-point cluster correlation function (Karsanina et al. 2015; Zhao et al. 2007), fractal system control function (Ju et al. 2014), and co-occurrence correlation function (Feng et al. 2018b). These constraints improve the accuracy of modeling, but the cost of modeling increases. The simulated annealing method has better modeling accuracy than the truncated Gaussian random field method, but it is still not suitable for generating digital rocks of heterogeneous porous media.

3.2.2.2 Markov Chain Monte Carlo

The Markov Chain Monte Carlo method originates from the Markov random fields. The Markov random fields can predict unknown global image features based on small images because it theoretically considers the correlation of a few pixels and several descriptors of global features (Geman and Geman 1984). Thus, unknown features can be determined from the given prior images. The original algorithm only updated one unknown pixel at a time, which limits the simulation speed. Wu et al. (2004) improved the method to predict a line of unknown pixels and then used it to reconstruct 2D images based on 2D thin sections of soil samples (Wu et al. 2004). Wu et al. (2006) extended their algorithm to 3D reconstruction based on 2D images. They added a multiple-voxel interaction scheme into the method to create the 3D reconstruction, which can reduce the simulation time. When predicting an unknown voxel in a one-dimensional (1D) simulation, the adjacent two- and three-neighborhoods are taken as the conditioning data. Then, the first row can be generated after which the reconstruction of unknown voxels in the second row can be evaluated under the constraints of the three- and four-neighborhoods. Next, the values of all the voxels in the first layer can be predicted row-by-row. When it comes to the pixels in the second layer, more known voxels can be taken into consideration. Finally, all the voxels can be assigned to void or solid. Although Wu et al. (2006) generated several 3D heterogeneous porous media, such as sandstones, mudrocks, and soils, the digital rocks are single-scale and cannot involve the micro-and nanometer digital rocks (Wu et al. 2006).

3.2.2.3 Sequential Indicator Simulation

Keehm et al. (2004) utilized the sequential indicator simulation approach to reconstruct simple porous media from 2D thin sections (Keehm et al. 2004). The sequential indicator simulation algorithm is based on geostatistics. In this method, unknown nodes are predicted on a random path. For every node, the local conditional cumulative distribution function is evaluated by indicator kriging. The function is related to the 2D input image. The procedure of the sequential indicator simulation approach is as follows: (i) Take a value from the local conditional cumulative distribution function; (ii) Keep the number as the conditioning data and visit the next unknown node along a predefined path; (iii) Determine the value of the unknown node using the porosity and autocorrelation function; (iv) Visit all the nodes in the simulation region and produce a binary 3D image. In their study, petrophysical properties of generated digital rocks, including porosity and permeability, were compared with those from laboratory measurements, and the results showed agreement between estimated and measured values (Keehm et al. 2004). Nevertheless, samples in their study were from conventional reservoirs, so the produced digital models were single-scale.

3.2.2.4 Multiple-Point Statistics

The theory of most aforementioned computational modeling algorithms, such as the truncated Gaussian random field method, simulated annealing, and sequential indicator simulation method, is based on the two-point correlation functions. Such methods might not be very efficient for generating 3D digital rocks with long-range connectivity since considering two points is not sufficient, particularly for reconstructing complex structures (Tahmasebi and Sahimi 2012). Thus, multiple-point statistics (MPS) was proposed to overcome this issue (Strebelle 2002; Tahmasebi 2018b; Zhang et al. 2006a). To date, MPS methods have been developed in pixel- and pattern-based forms (Tahmasebi 2018b). The former contains a single normal equation simulation (Strebelle 2002), direct sampling (Mariethoz et al. 2010), and cumulants (Dimitrakopoulos et al. 2010). The latter involves the simulation of a pattern (Arpat and Caers 2007), filter-based simulation (Zhang et al. 2006b), and cross-correlation simulation (Tahmasebi et al. 2012). Afterward, the hybrid pixel- and pattern simulation in the MPS method was proposed by Tahmasebi et al. (2017) where the capabilities of both categories are combined. The MPS method extracts the multiple-point correlation from a conceptual input called a training image, using which new values are estimated. The training image can be 2D or 3D.

Okabe and Blunt (2005) used this method to reconstruct 3D digital rocks, but their simulation was very time-demanding (Okabe and Blunt 2005). Later, many researchers adopted the MPS methods to build 3D porous media (Ding et al. 2018; Hajizadeh et al. 2011; Ji et al. 2017; Tahmasebi and Sahimi 2012; Wu et al. 2018; Zhang et al. 2015). Among the studies, the cross-correlation simulation algorithm is widely used and has been applied to various porous materials, including rock, copper foam, hard-sphere packing, electron detector, stroboscopic optical reflectance image, oxygen-free copper, ferritic steel sample, and fiber concrete material (Ji et al. 2017; Song et al. 2018; Tahmasebi 2018a; Tahmasebi and Sahimi 2012, 2013, 2018). During the 3D reconstruction from a 2D training image, one should employ a sampling method to extract necessary pixels in one layer as conditioning data (or hard data) when generating the next layer since the pore structure should slowly change between two adjacent layers. The sampling strategy is quite significant, which determines the connectivity of the reconstructed porous media. The main modeling procedure of this method is to generate the next layer based on the prior layer. Finally, all the produced layers are piled together to generate an entire 3D model. If the initially built 3D model contains artifacts or discontinuities, one can use optimization strategies to remove these structures (Tahmasebi and Sahimi 2016; Tahmasebi et al. 2020a). In addition, Feng et al. (2019) combined this strategy with a conditional generative adversarial network in deep learning to improve the simulation speed of the conventional MPS method (Feng et al. 2018a).

Compared with other stochastic modeling methods, MPS has some distinct advantages. It can effectively reconstruct 3D digital rocks with great connectivity of pores. Besides, MPS based on cross-correlation simulation also considers multiple mineral components, rather than only pore

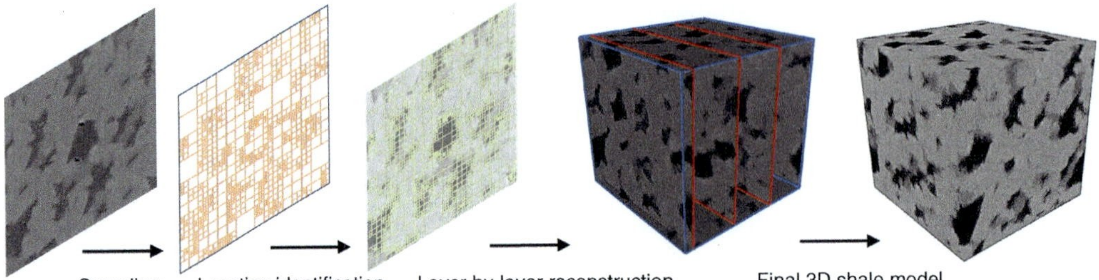

Figure 3.1 The modeling procedure from 2D to 3D image using multiple-point statistics. *Source:* Tahmasebi et al. (2020a)/from American Chemical Society.

space and solid matrix, when it is used to reconstruct 3D multicomponent digital rocks (Tahmasebi 2018a; Tahmasebi et al. 2017); see Figure 3.1. This is of great importance since realistic samples indeed encompass several minerals. However, the drawback of this method is that it relies on training images. This issue was later addressed by including other sources of information, such as histogram matching (Tahmasebi et al. 2016b) and soft data conditioning (Tahmasebi and Sahimi 2015). Clearly, the generated 3D digital rock is single-scale if the training image only involves single-scale structures. If one wants to use it to reconstruct the multiscale digital rocks, multiscale images should be available for the training images.

3.2.2.5 Machine Learning

Machine learning has been used to reconstruct 2D and 3D porous media and calculate their transport and physical properties (Kamrava et al. 2019b, 2021; Karimpouli and Tahmasebi 2019b; Mosser et al. 2017, 2018). For example, one can use generative adversarial networks and produce more realizations. Furthermore, conditional generative networks can also be used to reconstruct the entire images based on an extremely small subarea (Feng et al. 2019). 3D porous media based on 2D images has also been reconstructed using this approach (Feng et al. 2020). Similarly, the physical properties can also be estimated with high accuracy. These previous studies indicate that the deep learning method presents an excellent performance in the reconstruction of porous media. A comprehensive review of all such methods is available elsewhere (Tahmasebi et al. 2020b). However, in such studies, the utilized samples are still single-scale and simple rocks, such as Berea sandstones, Bead packs, and Ketton limestones. If the training images are multiscale, the method may construct multiscale digital rocks.

3.2.2.6 Process-Based Modeling

Sedimentary rocks, such as sandstones, mudrocks, and carbonates, are formed after sedimentary materials experience various diagenesis events including compaction, cementation, and dissolution. By going through such diagenesis events, loose and high-permeability sediments gradually tighten, and rocks become low-permeable. Several attempts have been made to mimic the above natural events. For example, Bryant et al. (1993) used spheres to construct a 3D digital core by simulating the compaction and overgrowth of minerals in sedimentary rocks. However, they only considered two diagenesis events, and their morphological features were not realistic. After that, Bakke and Øren (1997) reconstructed 3D digital cores by mimicking the deposition process, compaction process, and cementation process of sedimentary rocks. In their study, they first obtained the particle size cumulative probability curve from 2D images, randomly selected a value on the cumulative probability curve, and used the value as the particle radius. Then, the particle was deposited in a preset sedimentation box under the gravitational field. Their algorithm searched for the lowest local- or global-energy point, and this point was considered to be the deposition location of the particle. Considering the realistic situation that the location of real particle deposition is affected by horizontal hydrodynamic effects, when the deposition process occurs in a hydrodynamic environment, the particle is deposited at the point of the lowest local potential. In a strongly hydrodynamic environment, particles settle at the point of the lowest global potential. According to the above-described rules, all particles are deposited in the simulation box. When it comes to the simulation of compaction, they hold the viewpoint that compaction is a process of a linear transformation of the ordinates of the deposited particles, so the simulation of compaction is realized by changing the coordinates of the particles. With regard to the simulation of cementation, they only simulated the overgrowth of quartz and clay coating.

There are three main types of clay coating: pore lining, pore filling, and pore bridging (Øren and Bakke 2002). The clustering algorithm is used to simulate pore lining and pore-filling. After simulating the diagenesis events, a 3D digital rock is reconstructed, and the accuracy of the model is verified by comparing the simulated rock properties with the experimental values. Numerical researchers built digital porous media using a similar process-based method (Al Ibrahim et al. 2019; Biswal et al. 1999; Jin et al. 2003; Keehm et al. 2001; Latief et al. 2010; Mousavi and Bryant 2012; Politis et al. 2008; Ratnanabha et al. 2014; Torskaya et al. 2014; Wu et al. 2020b; Zhao et al. 2013; Zhu et al. 2012). Over time, the simulation accuracy of the process-based techniques was improved by adding more physics and complexities.

Compared to other computational modeling methods, process-based approaches consider the geological rules when generating digital rocks (Wu et al. 2020b, 2020e). In addition, they could be used to reconstruct multiscale digital rocks if complex diagenesis events and multiscale pore structures are fully taken into consideration. However, there are some disadvantages. Firstly, the used particles are spherical and not near irregular geometries. The realistic particles are very complex in shape. More textural characteristics including roundness, sphericity, and elongation should be accounted for while such particles are generated (Tahmasebi 2019; Tahmasebi et al. 2020a). Secondly, one should analyze diagenesis events and the history of diagenesis according to thin sections before randomly simulating diagenesis processes. If not, implemented diagenesis events would be totally random without a theoretical basis. Last, since some minerals, particularly clay minerals, have special morphological features, not all the used objects exactly represent the realistic structure of minerals.

3.3 Multiscale Modeling of Digital Rocks

Section 3.2 presented the widely used modeling techniques of single-scale digital rocks appropriate for conventional reservoirs. However, unconventional reservoirs exhibit extremely heterogeneous characteristics and their pore systems involve both the micro- and nano- or centimeter pores (Cai et al. 2018; Jiang et al. 2013; Lai et al. 2018; Loucks et al. 2012; Ma et al. 2014; Tahmasebi 2018c; Tahmasebi et al. 2020a; Wang et al. 2017; Zhang et al. 2017). One single image from X-ray CT or SEM experiment may not necessarily contain multiscale structure features of such reservoir rocks (Prodanović et al. 2015; Tahmasebi and Kamrava 2018; Tahmasebi et al. 2016a; Zeng et al. 2021). On the other hand, conventional computational methods are not able to generate multiscale digital porous media based on one single simulation. To address this issue, several approaches, for example, multiscale imaging techniques (Ma et al. 2017, 2019; Saif et al. 2017), image superposition method, pore-network integration method, object-based reconstruction method, and image resolution enhancement method were proposed to include information from distinct scales.

3.3.1 Multiscale Imaging Techniques

Unconventional reservoirs, tight sandstones, and carbonates manifest a strong level of heterogeneity. Taking shale formations as an example, their length-scale heterogeneity spans involve at least five orders of magnitude (10^{-9} to 10^{-4} m). As to modeling the microstructures at different scales, one should implement different imaging experiments. At the millimeter or centimeter scale, the medical CT or low-resolution X-ray CT can be used to build the digital model of rocks. From micrometer to millimeter scale, micrometer CT is a popular method to construct the 3D digital core at a higher resolution. When it comes to nanometer-scale 3D imaging, the nanometer CT and focused ion beam-scanning electron microscopy (FIB-SEM) are suitable choices. Macro-CT images can capture large-scale structures, such as fractures and silt-rich laminas (Ma et al. 2019). Micro-CT images can obtain a lot of micrometer pores, mainly intergranular pores. In higher-resolution imaging experiments including nano-CT and FIB-SEM, the acquired images mainly include nanopores, especially organic-matter pores. Also, a transmission electron microscope can be used to capture sub-nanometer pores.

The above-discussed multiscale imaging techniques can also be applied to other unconventional reservoirs. These techniques are of significance to characterize and model the structures of porous media at different scales, which can improve our understanding of the complex unconventional reservoirs. Nevertheless, these multiscale techniques have to make a balance between the resolution and the field of view. In other words, current 3D imaging techniques either cover large-scale structures at a lower resolution or cover a small region at a higher resolution (Tahmasebi 2018c; Wu et al. 2019b). Moreover, it is not easy to integrate 3D images at small scales into large-scale ones. Overall, it is of great difficulty to construct large-scale and high-resolution porous media only by multiscale imaging techniques.

3.3.2 Computational Methods

For simple and homogeneous porous media, one may use one of the stochastic methods discussed in Section 3.2.2 to properly reconstruct single-scale 3D digital models. However, for complex and heterogeneous samples, it is difficult to produce multiscale 3D digital rocks. Thus, some other methods are required to generate large-scale and high-resolution images, such as image superposition, pore-network integration, image resolution, and object-based reconstruction methods.

3.3.2.1 Image Superposition

The goal of the image superposition method is to build a multiscale image by superposing images from different scales and resolutions. Such images can cover structures of different scales. The high-resolution images cover a small area but capture ample fine-scale structures. The low-resolution images, however, include large-scale information and may not disclose fine-scale patterns. The image superposition method aims to fuse multiresolution spatial information into a large-scale and high-resolution model.

Okabe and Blunt (2007) used this approach to generate a large-scale and high-resolution digital image of carbonate rocks. They carried out X-ray CT imaging to capture large pores and vugs in carbonates at a low resolution. On the other hand, they randomly generated high-resolution

3D micropores using multiple-point statistics based on a 2D image. Then, the voxel number of the 3D image from the CT technique was scaled to match the number of the stochastically produced image. Finally, the superposition of two images created a multiscale digital rock containing macropores and micropores (Okabe and Blunt 2007). Later, Yao et al. (2013) used the image superposition method to reconstruct multiscale digital images of carbonate rocks. In their study, they utilized simulated annealing to build a digital porous medium containing macropores based on a coarse-scale 2D image, while they randomly produced digital core-covered micropores by the Markov Chain Monte Carlo technique (Yao et al. 2013). In addition to carbonates, multiscale digital images of shale samples were constructed using the superposition method. Tahmasebi and his coauthors (2015, 2016) employed multiple-point statistics, which involves a cross-correlation function, to generate low-resolution and high-resolution 3D digital shale models according to micrometer-scale and nanometer-scale 2D images, respectively. Then, they overlapped the micro- and nano-meter pores to generate multiscale 3D digital shale models (Tahmasebi et al. 2015, 2016b). Similarly, other researchers used the superposition method to reconstruct the multiscale digital rocks of complex porous media (Cao et al. 2018; Gerke et al. 2015; Ji et al. 2019; Lemmens et al. 2019; Li et al. 2020; Rao et al. 2021; Wang et al. 2018a; Wu et al. 2020c; Yang et al. 2015). In the above-mentioned studies, high-resolution 3D digital rocks were generated using one stochastic method.

The superposition method seems to be appropriate for constructing multiscale digital rocks of unconventional reservoirs. It can be applied to the superposition of more than two-scale images. Besides, some researchers also used this method to build multicomponent (multiphase) digital porous media (Ji et al. 2019; Lemmens et al. 2019). One of the drawbacks of the method is computation-demanding to fuse digital images with more than one billion voxels.

3.3.2.2 Pore-Network Integration

Pore space can be represented by the pore networks for large porous systems. From the pore space to the pore network (see Figure 3.2), the pores and throats are simplified as balls and tubes, respectively (Fatt 1956; Tahmasebi and Kamrava 2018). There are numerous methods to extract the pore networks from the images (Bultreys et al. 2016), for instance, the medial axis method (Idowu et al. 2014; Lindquist et al. 2000), maximal ball method (Dong and Blunt 2009), grain-based method (Idowu et al. 2014), and Delaunay-tesselation based method (Bhattad et al. 2011). Bultreys et al. (2016) compared the advantages and disadvantages of different extraction methods. If one only

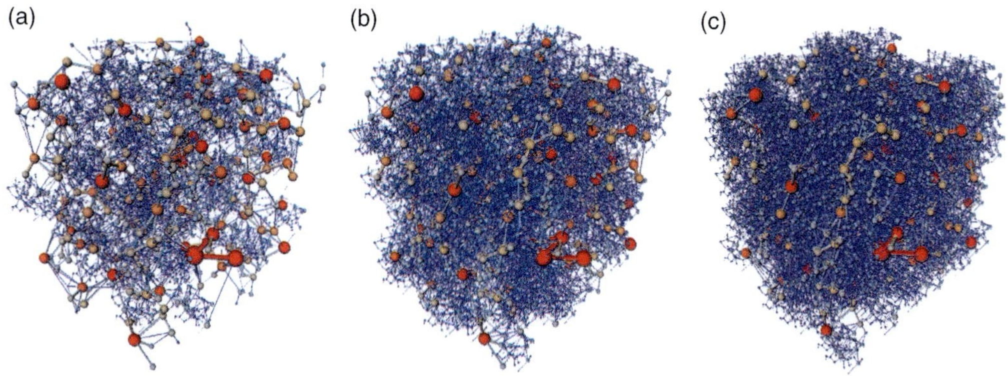

Figure 3.2 The varied fraction of micropores in the multiscale pore networks. (a) 10%, (b) 40%, and (c) 80%. *Source:* The images are taken from Tahmasebi and Kamrava (2018).

focuses on the pore space rather than solid minerals in complex porous media, one can integrate macropores and micropores into a pore network. Then the multiscale pore network can be used to predict the transport properties and study the flow mechanisms.

Bauer et al. (2012) proposed a dual pore network method to predict the petrophysical properties of heterogeneous rocks by combining the macropores and micropores. Jiang et al. (2013) generated multiscale pore networks of carbonate rocks and presented a detailed modeling workflow. Firstly, they scanned a carbonate sample using a multiple-resolution X-ray CT machine and obtained multiscale digital images. Then, they extracted the pore networks and analyzed their properties, such as the pore/throat radius and coordination number distributions. Under the constraint of the pore-system properties, they randomly generated equivalent fine-scale networks, which can be rescaled to be of any size. Finally, they integrated the networks of macropores and micropores into a single pore network. Among the two kinds of networks, the fine-scale throats were added to keep their connection. Eventually, the three-scale pore networks of carbonate rocks were generated (Jiang et al. 2013). Later, Mehmani and Prodanović (2014) used a similar technique to build a two-scale pore network of complex rocks and analyzed the effects of micropores on the transport properties of porous media. In their study, they also utilized a scaling factor to change the size of the micropores. In contrast, the micropores were randomly produced in two forms, namely pore filling and grain filling. For these two cases, the connected forms among macropores are different (Mehmani and Prodanović 2014). Afterward, Bultreys et al. (2015) regarded the micropores as micro-links that can connect the macropores in parallel or series. Then, they generated the dual pore networks of carbonates and validated the accuracy of the proposed method by comparing the produced network and the one from the CT experiment (Bultreys et al. 2015). Yao et al. (2015) also integrated three-scale pore networks from stochastically generated 3D images and constructed multiscale pore networks containing vugs, macropores, and micropores (Yao et al. 2015). de Vries et al. (2017) treated the micropores in heterogeneous porous media as the microscopic aggregate domain. In the simulation domain, the microscopic aggregates are stochastically produced and randomly inserted in the macropore domain in order to construct the multiscale pore network. In each microscopic aggregate, there were numerous micropores. The micropores are connected to adjacent macropores by the small throats. Tahmasebi and Kamrava (2018) presented a mini-review of constructing multiscale pore networks. Moreover, they also put forward a new approach for producing multiscale pore networks. The macropore networks were first extracted from the 3D coarse-scale CT images, while the micropore networks were obtained from the stochastically generated 3D porous media based on the high-resolution 2D SEM images. The technique of reconstructing 3D images from 2D ones was multiple-point statistics based on the cross-correlation function, which was initially developed by Tahmasebi and his coauthors (Tahmasebi 2017; Tahmasebi and Sahimi 2012; Tahmasebi et al. 2012). The generated micropore networks were fused into the specific locations and connected with the previous existing macropores in rocks. In addition, they also took the grain-filling and pore-filling micropores into consideration. In their study, the fraction of micropores was also varied, displayed in Figure 3.2, to investigate its impact on rock properties (Tahmasebi and Kamrava 2018). Wang et al. (2018b, 2019a) produced the multiscale pore networks by integrating the multiple pore networks of different scales and used the pore network to study the transport properties of complex porous media (Wang et al. 2018b, 2019a).

Multiscale pore networks are an alternative strategy for multiscale digital rocks. They can represent the complex pore systems by the simplified geometries, such as spheres and tubes. This is a quite efficient approach to characterizing the inner pore structures of heterogeneous porous media. In addition, it is uncomplicated to simulate the fluid flow in such simplified pore networks. However, the pore networks only signify the pore structures rather than the whole elements. They do not contain any information on solid space, so the multiscale networks cannot be utilized to study the rock properties related to the minerals, for instance, the elastic moduli and other properties of solid mechanics. Moreover, the pore networks are simplified pore systems, which may lead to inaccurate results when they are used to predict the transport properties.

3.3.2.3 Image Resolution Enhancement

Generating high-resolution and large-scale images is of great significance for the accurate evaluation of petrophysical properties. Apart from the image superposition and pore-network integration method, another alternative approach is to enhance the resolution of coarse-scale images (Kamrava et al. 2019a; Tahmasebi et al. 2020a; Wu et al. 2019c).

Tahmasebi (2018c) used a hybrid pattern- and pixel-based simulation algorithm and a similarity map to insert the fine-scale structures into a coarse-scale image. The finally generated large model included both the micro- and nano-meter structures (Tahmasebi 2018c). Wu et al. (2019b) used a template-matching algorithm based on the cross-correlation function to fuse the fine-scale structures from multiscale and multiresolution images into a high-resolution and large-scale model. The workflow of enhancing the image resolution based on the template matching algorithm is shown in Figure 3.3. They searched for the best matching patterns from nanometer-scale training images, then inserted them into the rescaled coarse image. Finally, they obtained a large-scale and high-resolution image that is closer to the actual one (Wu et al. 2019c). Later, the conventional neural networks were applied to enhance the resolution enhancement of CT images (Chen et al. 2020; Ebadi et al. 2021; Janssens et al. 2020; Ying Da Wang et al. 2019; Wang et al. 2019b) and SEM images (Kamrava et al. 2019a). Among these studies, Kamrava et al. (2019a) proposed a hybrid stochastic deep-learning (HSDL) algorithm to improve the resolution of coarse-scale images. They combined the cross-correlation-based simulation (CCSIM) method and image processing techniques to stochastically generate hundreds of images based on several available images as the training images for situations where enough data are not available (Kamrava et al. 2019a).

3.3.2.4 Object-Based Reconstruction

Object-based methods construct digital rocks by stochastically generating relevant geometries (objects) in shale samples that are similar to the existing minerals/pores. Such elements may be represented by rectangles, circles, and ellipses in 2D, and cubes, spheres, and ellipsoids in 3D. When generating these objects, the morphological characteristics, size distributions, and volume fractions of real minerals and pores should be fully taken into account (Wu et al. 2020d). Thus, the multiscale pores or minerals can be reproduced if their properties are available. When more properties and complex geometries are considered, the generated models would be closer to the real rocks (Wu et al. 2020d).

Naraghi and Javadpour (2015) used 2D rectangular patches as the organic matter and randomly put them in the simulation domain. The regions which are not occupied by the rectangles were taken as the inorganic matter. The sizes of pores in organic matter and inorganic matrix were assigned to the cells according to the size distributions of organic and inorganic pores (Naraghi and Javadpour 2015). Naraghi et al. (2018) used more complex 2D objects to signify the different minerals. The ellipses of different sizes represent the fecal pellet, feldspar, quartz, and clay mineral. The hexagon denotes the quartz and they finally generated 2D multiscale digital rocks. Later, Tahmasebi et al. (2020a) extended the 2D objects to 3D ones and used the ellipsoids to produce 3D digital cores. Harland et al. (2015) utilized the spheres and cuboids to produce the 2D and 3D multiscale digital models of carbonates and proved that the micropores had a great impact on the permeability of carbonates (Harland et al. 2015).

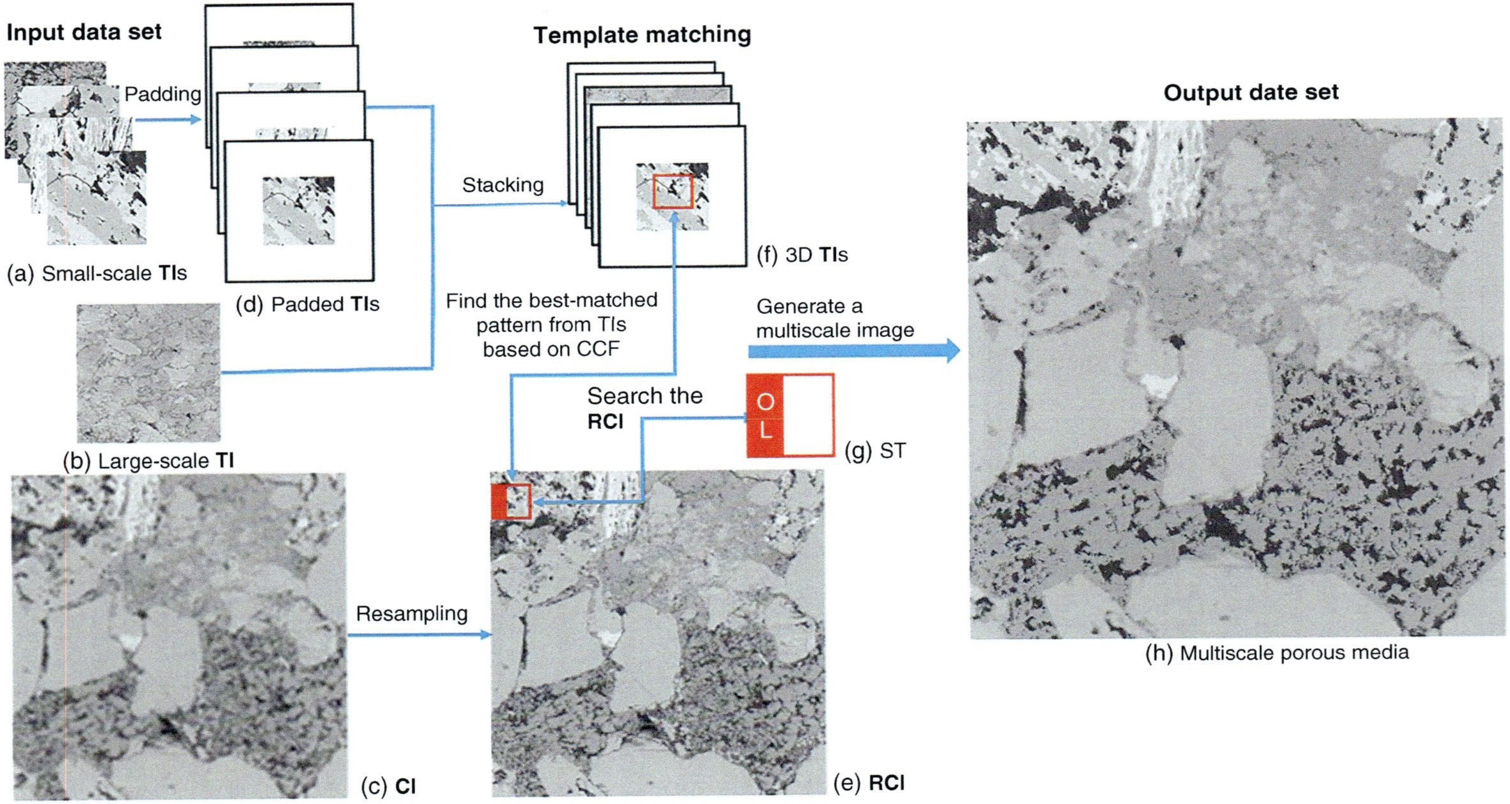

Figure 3.3 The workflow of fusing the fine- and coarse-scale structures into a multiscale and high-resolution image. (a) Small-scale training images, (b) large-scale training image, (c) coarse image, (d) padded training images, (e) resampled coarse image, (f) 3D training images, (g) search template, (h) multiscale porous media. *Source:* Wu et al. (2019b), from Elsevier.

Figure 3.4 Produced two shale models by the quartet structure generation set algorithm and object-based method. *Source:* Adapted from Wu et al. (2020c). Each color represents one kind of mineral or pore.

The above-mentioned shale models are static, which cannot vary with the increase of temperature and pressure, namely the thermal maturation. In fact, the formation of shale samples must experience complex diagenesis events and the internal components greatly change (Curtis et al. 2012; Ko et al. 2018; Loucks and Reed 2014; Pommer and Milliken 2015; Wu et al. 2015). Wu et al. (2020b, 2020e) combined the quartet structure generation set algorithm (Wang et al. 2007) and object-based method to generate dynamic shale models. During the construction of the shale models, the variations of minerals, organic matter, and pores were fully taken into account. For example, once the shale models experience cementation, the calcite and quartz often overgrow, and calcite is then also added. While the organic matter transforms into oil, bitumen, and gas, the organic-matter pores appear (Loucks and Reed 2014). The construction process is indeed process-mimic modeling. The morphological features, size distributions, and volume fractions of all the phases, including minerals, organic matter, interparticle pores, intraparticle pores, and organic-matter pores, should be all accounted for (Wu et al. 2020b, 2020e). Some finally generated shale models are demonstrated in Figure 3.4.

3.4 Conclusions and Future Perspectives

In terms of the simple porous media with homogeneous pore structures, single-scale stochastic modeling or experiment techniques should be enough to generate a representative digital core. For the experimental instruments, the X-ray computer tomography scanner is the most effective tool to build the digital rocks of conventional reservoirs. With regard to the stochastic modeling methods, simulated annealing, multiple-point statistics, and process-based modeling are widely used methods.

As to complex porous media with heterogeneous pore systems, multiscale imaging techniques should be used to capture the characteristics of pore structures at different scales. However, it is quite difficult to fuse the multiresolution and multiscale structures into a large-scale digital image.

Thus, several methods, e.g. image superposition and pore-network integration algorithms, were proposed to achieve such a goal. The former method can generate a multiscale digital rock containing the pore space and solid matrix, but the latter only can produce a multiscale pore network. Thus, the former i.e. the image superposition method may be a more popular technique than the latter in the future. To build the multiscale digital rocks, some algorithms, such as deep learning and template matching, were proposed to enhance the resolution of digital images. In addition, the object-based modeling method is an alternative strategy to generate multiscale and multiphase digital rocks. The morphological features, size distributions, and volume fractions are all fully taken into consideration when using the method to generate the digital cores.

Although numerous algorithms for constructing single- and multiple-scale digital rocks were developed, most of them can only generate static models which do not vary with the increase of temperature and pressure. For sedimentary rocks, especially shale samples, the inner pore structures would greatly change when the temperature and pressure vary. The process-based and object-based modeling techniques show great potential to implement dynamic modeling.

Building the multiphase or multicomponent digital rocks is of great significance for the study of the solid mechanisms of porous media. Most stochastic modeling and experimental techniques cannot easily accomplish such a goal. Object-based modeling methods could generate such models. Nevertheless, the properties of minerals, e.g. morphological features, are partially considered in the method. Thus, developing more powerful techniques to produce accurate multicomponent digital cores can be researched further.

The development of numerous modeling methods of single- and multiple-scale digital cores has greatly improved the accuracy of modeling. The single-scale, static, and two-phase (i.e. pore and solid) models have been perfectly constructed. At present, constructing multiscale, dynamic, and, multiphase (multicomponent) digital rocks would be the development trend in the field.

Acknowledgments

The first author wants to thank the National Natural Science Foundation of China (Nos. 42202146 and U19B2006), Support Program of China Postdoctoral InnovativeTalents (Grant No. BX2021373), Postdoctoral Science Foundation of China (GrantNo. 2022M723495), Natural Science Foundation of Shandong Province (Grant No. ZR2022QD029), Fundamental Research Funds for the Central Universities (Grant No. 22CX06006A), and Postdoctoral Application Research Project of Qingdao City (Grant No. qdyy20210088).

References

Al Ibrahim, M.A., Kerimov, A., Mukerji, T., and Mavko, G. (2019). Particula: a simulator tool for computational rock physics of granular media. *Geophysics* 84 (3): F85–F95. https://doi.org/10.1190/geo2018-0481.1.

An, S., Hasan, S., Erfani, H. et al. (2020). Unravelling effects of the pore-size correlation length on the two-phase flow and solute transport properties: GPU-based pore-network modeling. *Water Resources Research* 56 (8): 1–32. https://doi.org/10.1029/2020WR027403.

Andrä, H., Combaret, N., Dvorkin, J. et al. (2013). Digital rock physics benchmarks – part I: imaging and segmentation. *Computers & Geosciences* 50: 25–32. https://doi.org/10.1016/J.CAGEO.2012.09.005.

Arpat, G.B. and Caers, J. (2007). Conditional simulation with patterns. *Mathematical Geology* https://doi.org/10.1007/s11004-006-9075-3.

Bakke, S. and Øren, P.-E. (1997). 3-D pore-scale modelling of sandstones and flow simulations in the pore networks. *SPE Journal* 2 (2): 136–149. https://doi.org/10.2118/35479-PA.

Bauer, D., Youssef, S., Fleury, M. et al. (2012). Improving the estimations of petrophysical transport behavior of carbonate rocks using a dual pore network approach combined with computed microtomography. *Transport in Porous Media* 94 (2): 505–524. https://doi.org/10.1007/s11242-012-9941-z.

Berg, S., Ott, H., Klapp, S.A. et al. (2013). Real-time 3D imaging of Haines jumps in porous media flow. *Proceedings of the National Academy of Sciences* 110 (10): 3755–3759. https://doi.org/10.1073/pnas.1221373110.

Bhattad, P., Willson, C.S., and Thompson, K.E. (2011). Effect of network structure on characterization and flow modeling using X-ray micro-tomography images of granular and fibrous porous media. *Transport in Porous Media* 90 (2): 363–391. https://doi.org/10.1007/s11242-011-9789-7.

Biswal, B., Manwart, C., Hilfer, R. et al. (1999). Quantitative analysis of experimental and synthetic microstructures for sedimentary rock. *Physica A: Statistical Mechanics and its Applications* 273 (3–4): 452–475. https://doi.org/10.1016/S0378-4371(99)00248-4.

Blunt, M.J. (2017). *Multiphase Flow in Permeable Media: A Pore-Scale Perspective*. Cambridge.

Blunt, M.J., Bijeljic, B., Dong, H. et al. (2013). Pore-scale imaging and modelling. *Advances in Water Resources* 51: 197–216. https://doi.org/10.1016/J.ADVWATRES.2012.03.003.

Bryant, S.L., Mellor, D.W., and Cade, C.A. (1993). Physically representative network models of transport in porous media. *AIChE Journal* 39 (3): 387–396. https://doi.org/10.1002/aic.690390303.

Bultreys, T., Van Hoorebeke, L., and Cnudde, V. (2015). Multi-scale, micro-computed tomography-based pore network models to simulate drainage in heterogeneous rocks. *Advances in Water Resources* 78: 36–49. https://doi.org/10.1016/j.advwatres.2015.02.003.

Bultreys, T., De Boever, W., and Cnudde, V. (2016). Imaging and image-based fluid transport modeling at the pore scale in geological materials: a practical introduction to the current state-of-the-art. *Earth-Science Reviews* 155: 93–128. https://doi.org/10.1016/j.earscirev.2016.02.001.

Cai, J., Lin, D., Singh, H. et al. (2018). Shale gas transport model in 3D fractal porous media with variable pore sizes. *Marine and Petroleum Geology* 98: 437–447. https://doi.org/10.1016/J.MARPETGEO.2018.08.040.

Cao, G., Lin, M., Jiang, W. et al. (2018). A statistical-coupled model for organic-rich shale gas transport. *Journal of Petroleum Science and Engineering* 169: 167–183. https://doi.org/10.1016/J.PETROL.2018.05.033.

Chawla, N., Sidhu, R.S., and Ganesh, V.V. (2006). Three-dimensional visualization and microstructure-based modeling of deformation in particle-reinforced composites. *Acta Materialia* 54: 1541–1548. https://doi.org/10.1016/j.actamat.2005.11.027.

Chen, L., Kang, Q., Dai, Z. et al. (2015). Permeability prediction of shale matrix reconstructed using the elementary building block model. *Fuel* 160: 346–356. https://doi.org/10.1016/J.FUEL.2015.07.070.

Chen, H., He, X., Teng, Q. et al. (2020). Super-resolution of real-world rock microcomputed tomography images using cycle-consistent generative adversarial networks. *Physical Review E* 101 (2): 1–15. https://doi.org/10.1103/PhysRevE.101.023305.

Cnudde, V. and Boone, M.N.N. (2013). High-resolution X-ray computed tomography in geosciences: a review of the current technology and applications. *Earth-Science Reviews* https://doi.org/10.1016/j.earscirev.2013.04.003.

Curtis, M.E., Cardott, B.J., Sondergeld, C.H., and Rai, C.S. (2012). Development of organic porosity in the Woodford Shale with increasing thermal maturity. *International Journal of Coal Geology* 103: 26–31. https://doi.org/10.1016/J.COAL.2012.08.004.

Da Wang, Y., Armstrong, R.T., and Mostaghimi, P. (2019). Enhancing resolution of digital rock images with super resolution convolutional neural networks. *Journal of Petroleum Science and Engineering* 182 (March): 106261. https://doi.org/10.1016/j.petrol.2019.106261.

Dimitrakopoulos, R., Mustapha, H., and Gloaguen, E. (2010). High-order statistics of spatial random fields: exploring spatial cumulants for modeling complex non-Gaussian and non-linear phenomena. *Mathematical Geosciences* 42 (1): 65–99. https://doi.org/10.1007/s11004-009-9258-9.

Ding, K., Teng, Q., Wang, Z. et al. (2018). Improved multipoint statistics method for reconstructing three-dimensional porous media from a two-dimensional image via porosity matching. *Physical Review E* 97 (6): https://doi.org/10.1103/PhysRevE.97.063304.

Dmytriw, A.A. (2012). Godfrey Hounsfield: intuitive genius of CT. *The British Journal of Radiology*. https://doi.org/10.1259/bjr/31610998.

Dong, H. and Blunt, M.J. (2009). Pore-network extraction from micro-computerized-tomography images. *Physical Review E - Statistical, Nonlinear, and Soft Matter Physics* 80 (3): 036307. https://doi.org/10.1103/PhysRevE.80.036307.

Ebadi, M., Orlov, D., Makhotin, I. et al. (2021). Strengthening the digital rock physics, using downsampling for sub-resolved pores in tight sandstones. *Journal of Natural Gas Science and Engineering* 89: 103869. https://doi.org/10.1016/j.jngse.2021.103869.

Eschricht, N., Hoinkis, E., Mädler, F. et al. (2005). Knowledge-based reconstruction of random porous media. 291: 201–213. https://doi.org/10.1016/j.jcis.2005.05.004.

Fagbemi, S., Tahmasebi, P., and Piri, M. (2018). Pore-scale modeling of multiphase flow through porous media under triaxial stress. *Advances in Water Resources* 122 (October): 206–216. https://doi.org/10.1016/j.advwatres.2018.10.018.

Fatt, I. (1956). The network model of porous media. *Petroleum Transactions, AIME* 207: 144–181.

Feng, J., Teng, Q., He, X., and Wu, X. (2018a). Accelerating multi-point statistics reconstruction method for porous media via deep learning. *Acta Materialia* 159: 296–308. https://doi.org/10.1016/j.actamat.2018.08.026.

Feng, J., Teng, Q., He, X. et al. (2018b). Reconstruction of three-dimensional heterogeneous media from a single two-dimensional section via co-occurrence correlation function. *Computational Materials Science* 144: 181–192. https://doi.org/10.1016/j.commatsci.2017.11.030.

Feng, J., He, X., Teng, Q. et al. (2019). Reconstruction of porous media from extremely limited information using conditional generative adversarial networks. *Physical Review E* 100 (3): 33308. https://doi.org/10.1103/PhysRevE.100.033308.

Feng, J., Teng, Q., Li, B. et al. (2020). An end-to-end three-dimensional reconstruction framework of porous media from a single two-dimensional image based on deep learning. *Computer Methods in Applied Mechanics and Engineering* 368: 113043. https://doi.org/10.1016/j.cma.2020.113043.

Fredrich, J.T. (1999). 3D imaging of porous media using laser scanning confocal microscopy with application to microscale transport processes. *Physics and Chemistry of the Earth, Part A: Solid Earth and Geodesy* 24 (7): 551–561. https://doi.org/10.1016/S1464-1895(99)00079-4.

Fullwood, D.T., Niezgoda, S.R., and Kalidindi, S.R. (2008). Microstructure reconstructions from 2-point statistics using phase-recovery algorithms. 56: 942–948. https://doi.org/10.1016/j.actamat.2007.10.044.

Geman, S. and Geman, D. (1984). Stochastic relaxation, Gibbs distributions, and the Bayesian restoration of images. *IEEE Transactions on Pattern Analysis and Machine Intelligence* PAMI-6(6): 721–741. https://doi.org/10.1109/TPAMI.1984.4767596.

Gerke, K.M., Karsanina, M.V., and Mallants, D. (2015). Universal stochastic multiscale image fusion: an example application for shale rock. *Scientific Reports* 5: 1–12. https://doi.org/10.1038/srep15880.

Golparvar, A., Zhou, Y., Wu, K. et al. (2018). A comprehensive review of pore scale modeling methodologies for multiphase flow in porous media. *Advances in Geo-Energy Research* 2 (4): 418–440. https://doi.org/10.26804/ager.2018.04.07.

Hajizadeh, A., Safekordi, A., and Farhadpour, F.A. (2011). A multiple-point statistics algorithm for 3D pore space reconstruction from 2D images. *Advances in Water Resources* 34 (10): 1256–1267. https://doi.org/10.1016/j.advwatres.2011.06.003.

Hamzehpour, H., Rasaei, M.R., and Sahimi, M. (2007). Development of optimal models of porous media by combining static and dynamic data: the permeability and porosity distributions. *Physical Review E* 75 (5): 056311. https://doi.org/10.1103/PhysRevE.75.056311.

Harland, S.R., Wood, R.A., Curtis, A. et al. (2015). Quantifying flow in variably wet microporous carbonates using object-based geological modeling and both lattice-Boltzmann and pore-network fluid flow simulations. *AAPG Bulletin* 99 (10): 1827–1860. https://doi.org/10.1306/04231514122.

Hasanabadi, A., Baniassadi, M., Abrinia, K. et al. (2016). 3D microstructural reconstruction of heterogeneous materials from 2D cross sections: a modified phase-recovery algorithm. *Computational Materials Science* 111: 107–115. https://doi.org/10.1016/j.commatsci.2015.09.015.

Hazlett, R.D. (1995). Simulation of capillary-dominated displacements in microtomographic images of reservoir rocks. *Transport in Porous Media* https://doi.org/10.1007/BF00616924.

Idowu, N.A., Nardi, C., Long, H. et al. (2014). Effects of segmentation and skeletonization algorithms on pore networks and predicted multiphase-transport properties of reservoir-rock samples. *SPE Reservoir Evaluation and Engineering* 17 (4): 473–483. https://doi.org/10.2118/166030-PA.

Janssens, N., Huysmans, M., and Swennen, R. (2020). Computed tomography 3D super-resolution with generative adversarial neural networks: implications on unsaturated and two-phase fluid flow. *Materials* 13 (6): https://doi.org/10.3390/ma13061397.

Ji, L., Lin, M., Jiang, W., and Wu, C. (2017). An improved method for reconstructing the digital core model of heterogeneous porous media. *Transport in Porous Media* 121 (2): 389–406. https://doi.org/10.1007/s11242-017-0970-5.

Ji, L., Lin, M., Cao, G., and Jiang, W. (2019). A core-scale reconstructing method for shale. *Scientific Reports* 9 (1): 4364. https://doi.org/10.1038/s41598-019-39442-5.

Jiang, Z., Van Dijke, M.I.J., Sorbie, K.S., and Couples, G.D. (2013). Representation of multiscale heterogeneity via multiscale pore networks. *Water Resources Research* 49 (9): 5437–5449. https://doi.org/10.1002/wrcr.20304.

Jiao, Y., Stillinger, F.H., and Torquato, S. (2007). Modeling heterogeneous materials via two-point correlation functions: basic principles. *Physical Review E - Statistical, Nonlinear, and Soft Matter Physics* 76 (3): 1–15. https://doi.org/10.1103/PhysRevE.76.031110.

Jiao, Y., Stillinger, F.H., and Torquato, S. (2009). A superior descriptor of random textures and its predictive capacity. *Proceedings of the National Academy of Sciences of the United States of America* 106 (42): 17634–17639. https://doi.org/10.1073/pnas.0905919106.

Jin, G., Patzek, T.W., and Silin, D.B. (2003). Physics-based reconstruction of sedimentary rocks. *SPE Western Regional/AAPG Pacific Section Joint Meeting* (May), 307–320. https://doi.org/10.2118/83587-ms.

Ju, Y., Zheng, J., Epstein, M. et al. (2014). 3D numerical reconstruction of well-connected porous structure of rock using fractal algorithms. *Computer Methods in Applied Mechanics and Engineering* 279 (November): 212–226. https://doi.org/10.1016/j.cma.2014.06.035.

Kamrava, S., Tahmasebi, P., and Sahimi, M. (2019a). Enhancing images of shale formations by a hybrid stochastic and deep learning algorithm. *Neural Networks* 118: 310–320. https://doi.org/10.1016/J.NEUNET.2019.07.009.

Kamrava, S., Tahmasebi, P., and Sahimi, M. (2019b). Linking morphology of porous media to their macroscopic permeability by deep learning. *Transport in Porous Media* https://doi.org/10.1007/s11242-019-01352-5.

Kamrava, S., Tahmasebi, P., and Sahimi, M. (2021). Physics- and image-based prediction of fluid flow and transport in complex porous membranes and materials by deep learning. *Journal of Membrane Science* 622 (December 2020): 119050. https://doi.org/10.1016/j.memsci.2021.119050.

Karimpouli, S. and Tahmasebi, P. (2016). Conditional reconstruction: an alternative strategy in digital rock physics. *Geophysics* 81 (4): D465–D477. https://doi.org/10.1190/geo2015-0260.1.

Karimpouli, S. and Tahmasebi, P. (2019a). 3D multifractal analysis of porous media using 3D digital images: considerations for heterogeneity evaluation. *Geophysical Prospecting* 67 (4): 1082–1093. https://doi.org/10.1111/1365-2478.12681.

Karimpouli, S. and Tahmasebi, P. (2019b). Image-based velocity estimation of rock using convolutional neural networks. *Neural Networks* 111: 89–97. https://doi.org/10.1016/j.neunet.2018.12.006.

Karimpouli, S. and Tahmasebi, P. (2019c). Segmentation of digital rock images using deep convolutional autoencoder networks. *Computers & Geosciences* 126: 142–150. https://doi.org/10.1016/J.CAGEO.2019.02.003.

Karsanina, M., Gerke, K.M., and Karsanina, M.V. (2015). Improving stochastic reconstructions by weighing correlation functions in an objective function. *Europhysics Letters* 111: 56002. https://doi.org/10.1209/0295-5075/111/56002.

Keehm, Y., Mukerji, T., and Nur, A. (2001). Computational rock physics at the pore scale: transport properties and diagenesis in realistic pore geometries. *The Leading Edge* 20 (2): 180–183. https://doi.org/10.1190/1.1438904.

Keehm, Y., Mukerji, T., and Nur, A. (2004). Permeability prediction from thin sections: 3D reconstruction and lattice-Boltzmann flow simulation. *Geophysical Research Letters* 31 (4): 3–6. https://doi.org/10.1029/2003GL018761.

Ko, L.T., Ruppel, S.C., Loucks, R.G. et al. (2018). Pore-types and pore-network evolution in upper Devonian-lower Mississippian Woodford and Mississippian Barnett mudstones: insights from laboratory thermal maturation and organic petrology. *International Journal of Coal Geology* 190 (May 2017): 3–28. https://doi.org/10.1016/j.coal.2017.10.001.

Kus, J. (2015). Application of confocal laser-scanning microscopy (CLSM) to autofluorescent organic and mineral matter in peat, coals and siliciclastic sedimentary rocks – a qualitative approach. *International Journal of Coal Geology* 137: 1–18. https://doi.org/10.1016/j.coal.2014.10.014.

Lai, J., Wang, G., Wang, Z. et al. (2018). A review on pore structure characterization in tight sandstones. *Earth-Science Reviews* 177 (November 2017): 436–457. https://doi.org/10.1016/j.earscirev.2017.12.003.

Latief, F.D.E., Biswal, B., Fauzi, U., and Hilfer, R. (2010). Continuum reconstruction of the pore scale microstructure for Fontainebleau sandstone. *Physica A* 389 (8): 1607–1618. https://doi.org/10.1016/j.physa.2009.12.006.

Lemmens, L., Rogiers, B., Jacques, D. et al. (2019). Nested multiresolution hierarchical simulated annealing algorithm for porous media reconstruction. *Physical Review E* 100 (5): 53316. https://doi.org/10.1103/PhysRevE.100.053316.

Li, X., Teng, Q., Zhang, Y. et al. (2020). Three-dimensional multiscale fusion for porous media on microtomography images of different resolutions. *Physical Review E* 101 (5): 1–12. https://doi.org/10.1103/PhysRevE.101.053308.

Lin, C., Wu, Y., Ren, L. et al. (2018). Review of digital core modeling methods. *Progress in Geophysics* 33 (2): 0679–0689.

Lindquist, W.B., Venkatarangan, A., Dunsmuir, J., and Wong, T.F. (2000). Pore and throat size distributions measured from synchrotron X-ray tomographic images of Fontainebleau sandstones.

Journal of Geophysical Research, Solid Earth 105 (B9): 21509–21527. https://doi.org/10.1029/2000jb900208.

Liu, X., Sun, J., and Wang, H. (2009). Reconstruction of 3-D digital cores using a hybrid method. *Applied Geophysics* 6 (2): 105–112. https://doi.org/10.1007/s11770-009-0017-y.

Loucks, R.G. and Reed, R.M. (2014). Scanning-electron-microscope petrographic evidence for distinguishing organic-matter pores associated with depositional organic matter versus migrated organic matter in mudrock. *Gulf Coast Association of Geological Societies* 3: 51–60.

Loucks, R.G., Reed, R.M., Ruppel, S.C., and Hammes, U. (2012). Spectrum of pore types and networks in mudrocks and a descriptive classification for matrix-related mudrock pores. *AAPG Bulletin* 96 (6): 1071–1098. https://doi.org/10.1306/08171111061.

Lymberopoulos, D. and Payatakes, A. (1992). Derivation of topological, geometrical, and correlational properties of porous media from pore-chart analysis of serial section data. *Journal of Colloid and Interface Science* 150 (1): 61–80. https://doi.org/10.1016/0021-9797(92)90268-Q.

Ma, J., Couples, G., Jiang, Z., and van Dijke, R. (2014). A multi-scale framework for digital core analysis of gas shale at millimeter scales. *Proceedings of the 2nd Unconventional Resources Technology Conference*, 1063–1070. https://doi.org/10.15530/urtec-2014-1934450.

Ma, L., Fauchille, A.L., Dowey, P.J. et al. (2017). Correlative multi-scale imaging of shales: a review and future perspectives. *Geological Society Special Publication* 454 (1): 175–199. https://doi.org/10.1144/SP454.11.

Ma, L., Dowey, P.J., Rutter, E. et al. (2019). A novel upscaling procedure for characterising heterogeneous shale porosity from nanometer-to millimetre-scale in 3D. *Energy* 181: 1285–1297. https://doi.org/10.1016/j.energy.2019.06.011.

Manwart, C. and Hilfer, R. (1999). Reconstruction of random media using Monte Carlo methods. *Physical Review E* 59 (5): 5596.

Mariethoz, G., Renard, P., and Straubhaar, J. (2010). The direct sampling method to perform multiple-point geostatistical simulations. *Water Resources Research* 46 (11): 1–14. https://doi.org/10.1029/2008WR007621.

Mehmani, A. and Prodanović, M. (2014). The effect of microporosity on transport properties in porous media. *Advances in Water Resources* 63: 104–119. https://doi.org/10.1016/j.advwatres.2013.10.009.

Menke, H.P., Reynolds, C.A., Andrew, M.G. et al. (2018). 4D multi-scale imaging of reactive flow in carbonates: assessing the impact of heterogeneity on dissolution regimes using streamlines at multiple length scales. *Chemical Geology* 481: 27–37. https://doi.org/10.1016/J.CHEMGEO.2018.01.016.

Metropolis, N., Rosenbluth, A.W., Rosenbluth, M.N. et al. (1953). Equation of state calculations by fast computing machines. *The Journal of Chemical Physics* 21 (6): 1087–1092. https://doi.org/10.1063/1.1699114.

Minsky, M. (1988). Memoir on inventing the confocal scanning microscope. *Scanning* 10 (4): 128–138. https://doi.org/10.1002/sca.4950100403.

Mosser, L., Dubrule, O., and Blunt, M.J. (2017). Reconstruction of three-dimensional porous media using generative adversarial neural networks. *Physical Review E* 96 (4): 043309. https://doi.org/10.1103/PhysRevE.96.043309.

Mosser, L., Dubrule, O., and Blunt, M.J. (2018). Stochastic reconstruction of an oolitic limestone by generative adversarial networks. *Transport in Porous Media* 125 (1): 81–103. https://doi.org/10.1007/s11242-018-1039-9.

Mousavi, M.A. and Bryant, S.L. (2012). Connectivity of pore space as a control on two-phase flow properties of tight-gas sandstones. *Transport in Porous Media* 94 (2): 537–554. https://doi.org/10.1007/s11242-012-0017-x.

Munawar, M.J., Lin, C., Cnudde, V. et al. (2018). Petrographic characterization to build an accurate rock model using micro-CT: case study on low-permeable to tight turbidite sandstone from Eocene Shahejie formation. *Micron* 109: 22–33. https://doi.org/10.1016/j.micron.2018.02.010.

Naraghi, M.E. and Javadpour, F. (2015). A stochastic permeability model for the shale-gas systems. *International Journal of Coal Geology* 140: 111–124. https://doi.org/10.1016/j.coal.2015.02.004.

Naraghi, M.E., Javadpour, F., and Ko, L.T. (2018). An object-based shale permeability model: non-Darcy gas flow, sorption, and surface diffusion effects. *Transport in Porous Media* 125 (1): 23–39. https://doi.org/10.1007/s11242-017-0992-z.

Okabe, H. and Blunt, M.J. (2005). Pore space reconstruction using multiple-point statistics. *Journal of Petroleum Science and Engineering* 46 (1–2): 121–137. https://doi.org/10.1016/j.petrol.2004.08.002.

Okabe, H. and Blunt, M.J. (2007). Pore space reconstruction of vuggy carbonates using microtomography and multiple-point statistics. *Water Resources Research* 43 (12): https://doi.org/10.1029/2006WR005680.

Øren, P.-E. and Bakke, S. (2002). Process based reconstruction of sandstones and prediction of transport properties. *Transport in Porous Media* 46 (2/3): https://doi.org/10.1023/A.

Politis, M.G., Kikkinides, E.S., Kainourgiakis, M.E., and Stubos, A.K. (2008). A hybrid process-based and stochastic reconstruction method of porous media. *Microporous and Mesoporous Materials* 110 (1): 92–99. https://doi.org/10.1016/j.micromeso.2007.09.024.

Pommer, M. and Milliken, K. (2015). Pore types and pore-size distributions across thermal maturity, Eagle Ford Formation, southern Texas. *AAPG Bulletin* 99 (9): 1713–1744. https://doi.org/10.1306/03051514151.

Prodanović, M., Mehmani, A., and Sheppard, A.P. (2015). Imaged-based multiscale network modelling of microporosity in carbonates. *Geological Society, London, Special Publications* 406 (1): 95–113. https://doi.org/10.1144/SP406.9.

Rao, Y., Fu, L.-Y., Wang, Z.-W., and Fu, B.-Y. (2021). Multiscale reconstructions, effective elastic properties, and ultrasonic responses of kerogen matter based on digital organic shales. *IEEE Access* 1-1: https://doi.org/10.1109/access.2021.3058944.

Ratnanabha, S., Mukerji, T., and Mavko, G. (2014). How computational rock-physics tools can be used to simulate geologic processes, understand pore-scale heterogeneity, and refine theoretical models. *The Leading Edge* 33 (3): 324–334.

Sahimi, M. (2011). *Flow and Transport in Porous Media and Fractured Rock.* Weinheim: Wiley-VCH https://doi.org/10.1002/9783527636693.

Saif, T., Lin, Q., Butcher, A.R. et al. (2017). Multi-scale multi-dimensional microstructure imaging of oil shale pyrolysis using X-ray micro-tomography, automated ultra-high resolution SEM, MAPS mineralogy and FIB-SEM. *Applied Energy* 202: 628–647. https://doi.org/10.1016/j.apenergy.2017.05.039.

Saif, T., Lin, Q., Gao, Y. et al. (2019). 4D in situ synchrotron X-ray tomographic microscopy and laser-based heating study of oil shale pyrolysis. *Applied Energy* 235: 1468–1475. https://doi.org/10.1016/J.APENERGY.2018.11.044.

Shah, S.M., Crawshaw, J.P., and Boek, E.S. (2017). Three-dimensional imaging of porous media using confocal laser scanning microscopy. *Journal of Microscopy* 265 (2): 261–271. https://doi.org/10.1111/jmi.12496.

Singh, K., Menke, H., Andrew, M. et al. (2017). Dynamics of snap-off and pore-filling events during two-phase fluid flow in permeable media. *Scientific Reports* 7 (1): 1–13. https://doi.org/10.1038/s41598-017-05204-4.

Song, W., Yao, J., Ma, J. et al. (2018). Pore-scale numerical investigation into the impacts of the spatial and pore-size distributions of organic matter on shale gas flow and their implications on multiscale characterisation. *Fuel* 216: 707–721. https://doi.org/10.1016/j.fuel.2017.11.114.

Strebelle, S. (2002). Conditional simulation of complex geological structures using multiple-point statistics. *Mathematical Geology* 34 (1): 1–21.

Tahmasebi, P. (2017). HYPPS: a hybrid geostatistical modeling algorithm for subsurface modeling. *Water Resources Research* https://doi.org/10.1002/2017WR021078.

Tahmasebi, P. (2018a). Accurate modeling and evaluation of microstructures in complex materials. *Physical Review E* 97 (2): 1–13. https://doi.org/10.1103/PhysRevE.97.023307.

Tahmasebi, P. (2018b). Multiple point statistics: a review. In: *Handbook of Mathematical Geosciences*, 613–643. Cham: Springer International Publishing https://doi.org/10.1007/978-3-319-78999-6_30.

Tahmasebi, P. (2018c). Nanoscale and multiresolution models for shale samples. *Fuel* 217: 218–225. https://doi.org/10.1016/j.fuel.2017.12.107.

Tahmasebi, P. (2019). An optimization-based approach for modeling of complex particles. *Powder Technology* 356: 342–352. https://doi.org/10.1016/J.POWTEC.2019.08.027.

Tahmasebi, P. and Kamrava, S. (2018). Rapid multiscale modeling of flow in porous media. *Physical Review E* 98 (5): 052901. https://doi.org/10.1103/PhysRevE.98.052901.

Tahmasebi, P. and Sahimi, M. (2012). Reconstruction of three-dimensional porous media using a single thin section. *Physical Review E - Statistical, Nonlinear, and Soft Matter Physics* 85 (6): 1–13. https://doi.org/10.1103/PhysRevE.85.066709.

Tahmasebi, P. and Sahimi, M. (2013). Cross-correlation function for accurate reconstruction of heterogeneous media. *Physical Review Letters* 110 (7): 1–5. https://doi.org/10.1103/PhysRevLett.110.078002.

Tahmasebi, P. and Sahimi, M. (2015). Geostatistical simulation and reconstruction of porous media by a cross-correlation function and integration of hard and soft data. *Transport in Porous Media* 107 (3): 871–905. https://doi.org/10.1007/s11242-015-0471-3.

Tahmasebi, P. and Sahimi, M. (2016). Enhancing multiple-point geostatistical modeling: 2. Iterative simulation and multiple distance function. *Water Resources Research* 52 (3): 2099–2122. https://doi.org/10.1002/2015WR017807.

Tahmasebi, P. and Sahimi, M. (2018). A stochastic multiscale algorithm for modeling complex granular materials. *Granular Matter* 123: https://doi.org/10.1007/s10035-018-0816-z.

Tahmasebi, P., Hezarkhani, A., and Sahimi, M. (2012). Multiple-point geostatistical modeling based on the cross-correlation functions. *Computational Geosciences* 16 (3): 779–797. https://doi.org/10.1007/s10596-012-9287-1.

Tahmasebi, P., Javadpour, F., and Sahimi, M. (2015). Multiscale and multiresolution modeling of shales and their flow and morphological properties. *Scientific Reports* 5: 1–11. https://doi.org/10.1038/srep16373.

Tahmasebi, P., Javadpour, F., Sahimi, M., and Piri, M. (2016a). Multiscale study for stochastic characterization of shale samples. *Advances in Water Resources* 89: 91–103. https://doi.org/10.1016/j.advwatres.2016.01.008.

Tahmasebi, P., Javadpour, F., and Sahimi, M. (2016b). Stochastic shale permeability matching: three-dimensional characterization and modeling. *International Journal of Coal Geology* 165: 231–242. https://doi.org/10.1016/j.coal.2016.08.024.

Tahmasebi, P., Sahimi, M., Kohanpur, A.H., and Valocchi, A. (2017). Pore-scale simulation of flow of CO2 and brine in reconstructed and actual 3D rock cores. *Journal of Petroleum Science and Engineering* 155: 21–33. https://doi.org/10.1016/j.petrol.2016.12.031.

Tahmasebi, P., Javadpour, F., and Enayati, S.F. (2020a). Digital rock techniques to study shale permeability: a mini-review. *Energy and Fuels* 34 (12): 15672–15685. https://doi.org/10.1021/acs.energyfuels.0c03397.

Tahmasebi, P., Kamrava, S., Bai, T., and Sahimi, M. (2020b). Machine learning in geo- and environmental sciences: from small to large scale. *Advances in Water Resources* 142 (May): 103619. https://doi.org/10.1016/j.advwatres.2020.103619.

Talukdar, M.S., Torsaeter, O., Ioannidis, M.A., and Howard, J.J. (2002). Stochastic reconstruction, 3D characterization and network modeling of chalk. 35: 1–21.

Torskaya, T., Shabro, V., Torres-Verdín, C. et al. (2014). Grain shape effects on permeability, formation factor, and capillary pressure from pore-scale modeling. *Transport in Porous Media* 102 (1): 71–90. https://doi.org/10.1007/s11242-013-0262-7.

Vogel, H.J. and Roth, K. (2001). Quantitative morphology and network representation of soil pore structure. *Advances in Water Resources* 24 (3–4): 233–242. https://doi.org/10.1016/S0309-1708(00)00055-5.

de Vries, E.T., Raoof, A., and van Genuchten, M.T. (2017). Multiscale modelling of dual-porosity porous media; a computational pore-scale study for flow and solute transport. *Advances in Water Resources* 105: 82–95. https://doi.org/10.1016/j.advwatres.2017.04.013.

Wang, M., Wang, J., Pan, N., and Chen, S. (2007). Mesoscopic predictions of the effective thermal conductivity for microscale random porous media. *Physical Review E* 75 (3): 1–10. https://doi.org/10.1103/PhysRevE.75.036702.

Wang, L., Tian, Y., Yu, X. et al. (2017). Advances in improved/enhanced oil recovery technologies for tight and shale reservoirs. *Fuel* 210: 425–445. https://doi.org/10.1016/J.FUEL.2017.08.095.

Wang, Y., Arns, C.H., Rahman, S.S., and Arns, J.Y. (2018a). Porous structure reconstruction using convolutional neural networks. *Mathematical Geosciences* 50 (7): 781–799. https://doi.org/10.1007/s11004-018-9743-0.

Wang, Y., Yuan, Y., Rahman, S.S., and Arns, C. (2018b). Semi-quantitative multiscale modelling and flow simulation in a nanoscale porous system of shale. *Fuel* 234: 1181–1192. https://doi.org/10.1016/j.fuel.2018.08.007.

Wang, D., Yao, J., Chen, Z. et al. (2019a). Image-based core-scale real gas apparent permeability from pore-scale experimental data in shale reservoirs. *Fuel* 254: 115596. https://doi.org/10.1016/j.fuel.2019.06.004.

Wang, Y., Teng, Q., He, X. et al. (2019b). CT-image of rock samples super resolution using 3D convolutional neural network. *Computers and Geosciences* 133: https://doi.org/10.1016/j.cageo.2019.104314.

Wildenschild, D. and Sheppard, A.P. (2013). X-ray imaging and analysis techniques for quantifying pore-scale structure and processes in subsurface porous medium systems. *Advances in Water Resources* 51: 217–246. https://doi.org/10.1016/j.advwatres.2012.07.018.

Wu, K., Nunan, N., Crawford, J.W. et al. (2004). An efficient Markov chain model for the simulation of heterogeneous soil structure. *Soil Science Society of America Journal* 68 (2): 346–351. https://doi.org/10.2136/sssaj2004.0346.

Wu, K., Van Dijke, M.I.J., Couples, G.D. et al. (2006). 3D stochastic modelling of heterogeneous porous media – applications to reservoir rocks. *Transport in Porous Media* 65 (3): 443–467. https://doi.org/10.1007/s11242-006-0006-z.

Wu, S., Zhu, R., Cui, J. et al. (2015). Characteristics of lacustrine shale porosity evolution, Triassic Chang 7 Member, Ordos Basin, NW China. *Petroleum Exploration and Development* 42 (2): 167–176. https://doi.org/10.11698/PED.2015.02.05.

Wu, Y., Lin, C., Ren, L. et al. (2018). Reconstruction of 3D porous media using multiple-point statistics based on a 3D training image. *Journal of Natural Gas Science and Engineering* 51: 129–140. https://doi.org/10.1016/j.jngse.2017.12.032.

Wu, Y., Tahmasebi, P., Lin, C. et al. (2019a). A comprehensive study on geometric, topological and fractal characterizations of pore systems in low-permeability reservoirs based on SEM, MICP, NMR, and X-ray CT experiments. *Marine and Petroleum Geology* 103: 12–28. https://doi.org/10.1016/j.marpetgeo.2019.02.003.

Wu, Y., Tahmasebi, P., Lin, C. et al. (2019b). Effects of micropores on geometric, topological and transport properties of pore systems for low-permeability porous media. *Journal of Hydrology* 575: 327–342. https://doi.org/10.1016/j.jhydrol.2019.05.014.

Wu, Y., Tahmasebi, P., Lin, C. et al. (2019c). Multiscale modeling of shale samples based on low- and high-resolution images. *Marine and Petroleum Geology* 109: 9–21. https://doi.org/10.1016/J.MARPETGEO.2019.06.006.

Wu, Y., Tahmasebi, P., Lin, C. et al. (2020a). Quantitative characterization of non-wetting phase in water-wet porous media based on multiphase flow experiment and numerical simulation. *Journal of Petroleum Science and Engineering* 188: https://doi.org/10.1016/j.petrol.2020.106914.

Wu, Y., Tahmasebi, P., Yu, H. et al. (2020b). Pore-scale 3D dynamic modeling and characterization of shale samples: considering the effects of thermal maturation. *Journal of Geophysical Research, Solid Earth* 125 (1): e2019JB018309. https://doi.org/10.1029/2019JB018309.

Wu, Y., Lin, C., Yan, W. et al. (2020c). Pore-scale simulations of electrical and elastic properties of shale samples based on multicomponent and multiscale digital rocks. *Marine and Petroleum Geology* 117: 104369. https://doi.org/10.1016/j.marpetgeo.2020.104369.

Wu, Y., Tahmasebi, P., Lin, C., and Dong, C. (2020d). A comprehensive investigation of the effects of organic-matter pores on shale properties: a multicomponent and multiscale modeling. *Journal of Natural Gas Science and Engineering* 81: 103425. https://doi.org/10.1016/j.jngse.2020.103425.

Wu, Y., Tahmasebi, P., Lin, C., and Dong, C. (2020e). Process-based and dynamic 2D modeling of shale samples: considering the geology and pore-system evolution. *International Journal of Coal Geology* 218: 103368. https://doi.org/10.1016/J.COAL.2019.103368.

Wu, Y., Tahmasebi, P., Lin, C., and Dong, C. (2022). Using digital rock physics to investigate the impacts of diagenesis events and pathways on rock properties. *Journal of Petroleum Science and Engineering* 208: 108025. https://doi.org/10.1016/j.petrol.2020.108025.

Wu, Y., Tahmasebi, P., Liu, K., et al. Two-phase flow inheterogeneous porous media: A multiscale digital model approach [J]. International Journal of Heat and Mass Transfer, 2022a, 194: 123080. https://doi.org/10.1016/j.ijheatmasstransfer.2022.123080.

Xia, Y., Cai, J., Perfect, E. et al. (2019). Fractal dimension, lacunarity and succolarity analyses on CT images of reservoir rocks for permeability prediction. *Journal of Hydrology* 579 (September): 124198. https://doi.org/10.1016/j.jhydrol.2019.124198.

Yang, Y., Yao, J., Wang, C. et al. (2015). New pore space characterization method of shale matrix formation by considering organic and inorganic pores. *Journal of Natural Gas Science and Engineering* 27: 496–503. https://doi.org/10.1016/J.JNGSE.2015.08.017.

Yang, Y., Zhou, Y., Blunt, M.J. et al. (2021). Advances in multiscale numerical and experimental approaches for multiphysics problems in porous media. *Advances in Geo-Energy Research* 5 (3): 233–238. https://doi.org/10.46690/AGER.2021.03.01.

Yao, J., Wang, C., Yang, Y. et al. (2013). The construction of carbonate digital rock with hybrid superposition method. *Journal of Petroleum Science and Engineering* 110: 263–267. https://doi.org/10.1016/J.PETROL.2013.10.005.

Yao, J., Hu, R., Wang, C., and Yang, Y. (2015). Multiscale pore structure analysis in carbonate rocks. *International Journal for Multiscale Computational Engineering* 13 (1): 1–9. https://doi.org/10.1615/IntJMultCompEng.2014010500.

Yeong, C.L.Y. and Torquato, S. (1998). Reconstructing random media. II. Three-dimensional media from two-dimensional cuts. *Physical Review E - Statistical Physics, Plasmas, Fluids, and Related Interdisciplinary Topics* 58 (1): 224–233. https://doi.org/10.1103/PhysRevE.58.224.

Zeng, F., Dong, C., Lin, C. et al. (2021). Analyzing the effects of multi-scale pore systems on reservoir properties – a case study on Xihu depression, East China Sea Shelf Basin, China. *Journal of Petroleum Science and Engineering* 203: 108609. https://doi.org/10.1016/j.petrol.2021.108609.

Zhang, T., Bombarde, S., Strebelle, S.B., and Oatney, E. (2006a). 3D porosity modeling of a carbonate reservoir using continuous multiple-point statistics simulation. *SPE Journal* 11 (03): 375–379. https://doi.org/10.2118/96308-PA.

Zhang, T., Switzer, P., and Journel, A. (2006b). Filter-based classification of training image patterns for spatial simulation. *Mathematical Geology* 38 (1): 63–80. https://doi.org/10.1007/s11004-005-9004-x.

Zhang, T., Du, Y., Huang, T., and Li, X. (2015). GPU-accelerated 3D reconstruction of porous media using multiple-point statistics. *Computational Geosciences* 19 (1): 79–98. https://doi.org/10.1007/s10596-014-9452-9.

Zhang, L., Lu, S., Xiao, D., and Li, B. (2017). Pore structure characteristics of tight sandstones in the northern Songliao Basin, China. *Marine and Petroleum Geology* 88: 170–180. https://doi.org/10.1016/j.marpetgeo.2017.08.005.

Zhang, X., Lin, C., Wu, Y. et al. (2020). Experimental investigation of the effects of water and polymer flooding on geometric and multifractal characteristics of pore structures. *Energies* 13 (20): 5288. https://doi.org/10.3390/en13205288.

Zhao, X., Yao, J., and Yi, Y. (2007). A new stochastic method of reconstructing porous media. *Transport in Porous Media* 69 (1): 1–11. https://doi.org/10.1007/s11242-006-9052-9.

Zhao, J., Sun, J., Liu, X. et al. (2013). Numerical simulation of the electrical properties of fractured rock based on digital rock technology. *Journal of Geophysics and Engineering* 10 (5): 1–6. https://doi.org/10.1088/1742-2132/10/5/055009.

Zhu, W., Yu, W., and Chen, Y. (2012). Digital core modeling from irregular grains. *Journal of Applied Geophysics* 85: 37–42. https://doi.org/10.1016/j.jappgeo.2012.06.013.

Zhu, L., Zhang, C., Zhang, C. et al. (2019). Challenges and prospects of digital core-reconstruction research. *Geofluids* 2019 (7814180): 1–29. https://doi.org/10.1155/2019/7814180.

4

Scale Dependence of Permeability and Formation Factor

A Simple Scaling Law

Behzad Ghanbarian and Misagh Esmaeilpour

Porous Media Research Lab, Department of Geology, Kansas State University, Manhattan, KS, USA

4.1 Introduction

In geological and petroleum engineering, flow and transport in porous rocks are typically investigated at three scales: (i) pore, (ii) core, and (iii) reservoir. Accordingly, scaling has been an active subject of research. In fact, the effect of scale (or sample volume) on petrophysical properties of rocks has been known for decades (Brace 1984; Ewing et al. 2010; Garbesi et al. 1996; Hyun et al. 2002; Rovey and Cherkauer 1995; Yoon and Dewers 2013), and various scaling approaches have been proposed to study the scale dependence of flow and transport in porous media. For example, based on experimental measurements from different formations, Schulze-Makuch et al. (1999) proposed an empirical power–law scaling relationship to relate permeability k to sample volume V_{s} as follows:

$$k = CV_{\mathrm{s}}^{\mathrm{m}} \tag{4.1}$$

where the constant C and scaling exponent m characterizing the formation's heterogeneity are empirical. For different values of the exponent m in various types of porous media, see Esmaeilpour et al. (2021). Although Eq. (4.1) has been frequently applied in the literature, it is purely empirical, and because of its empiricism the interpretation of its parameters and their variations from one sample to another is not clear.

Ghanbarian et al. (2015) applied a machine learning method called contrast pattern-aided regression and proposed scale-dependent functions to estimate permeability from other porous media properties. They showed that by including sample internal diameter and height (or length), k estimations were substantially improved (Ghanbarian et al. 2015, 2017).

In the context of percolation theory (Hunt et al. 2014), Hunt (2006) assumed that pore-throat size distribution followed the power–law probability density function and proposed a theoretical relationship that estimates the scale dependence of permeability from pore space characteristics as well as the permeability value at the representative elementary volume (REV). Using experimental data of Schulze-Makuch (1996), collected from various sites within a carbonate-rock aquifer in southeastern Wisconsin, Hunt (2006) found generally well agreement between his model estimations and experiments. More recently, Esmaeilpour et al. (2021) generalized Hunt's approach (Hunt 2006) to be independent of the shape of pore-throat size distribution, compared their proposed generalized model with pore-network simulations, and extended the percolation-based model to formation factor and its scale dependence in porous media. By comparing with pore-scale

Physics of Fluid Flow and Transport in Unconventional Reservoir Rocks, First Edition.
Edited by Behzad Ghanbarian, Feng Liang, and Hui-Hai Liu.

simulations, Esmaeilpour et al. (2021) found relative errors between −3.7 and 3.8% for permeability and between 0.21 and 4.04% for formation factor.

Despite recent progress (Ghanbarian 2022a, 2022b; Ghanbarian et al. 2021), we are still far from fully understanding the effect of length scale on petrophysical properties of rocks. The main objectives of this study are to (i) apply a simple theoretical scaling law developed by Machta and Guyer (1987) for random resistor networks, (ii) adopt it to networks of pores, and (iii) evaluate it using pore-network simulations.

4.2 Theory

In this section, we first present the concepts of the theoretic model proposed by Machta and Guyer (1987). Then, its applications to porous media and scale-dependent permeability and formation factor are discussed.

4.2.1 Funnel Defect Approach

Machta and Guyer (1987) studied the expected value of the largest current flowing in a random resistor network. They assumed that the network was composed of low- and high-conductance (g_l and g_h) resistors whose distribution conformed to the following probability density function:

$$f(g) = p\delta(g - g_h) + (1 - p)\delta(g - g_l) \tag{4.2}$$

where $f(g)$ is the conductance distribution, δ represents the delta function, p is the occupation probability of high-conductance resistors, and $0 < g_l/g_h < 1$. Equation (4.2) is schematically shown in Figure 4.1.

Machta and Guyer (1987) considered the most effective defect for producing large currents and argued that the average largest current over configurations of the conductance $\langle i_{max} \rangle$ should conform to the following scaling law

$$\langle i_{max} \rangle \propto [\ln(L)]^{\alpha} \tag{4.3}$$

where L is the linear size of the network, and α is a scaling exponent. The value of α depends on g_l/g_h and network dimensionality d. In two dimensions ($d = 2$), Machta and Guyer (1987) found the following relationship:

$$\alpha = \frac{1 - \frac{4}{\pi}\arctan\left[(g_l/g_h)^{0.5}\right]}{2} \tag{4.4}$$

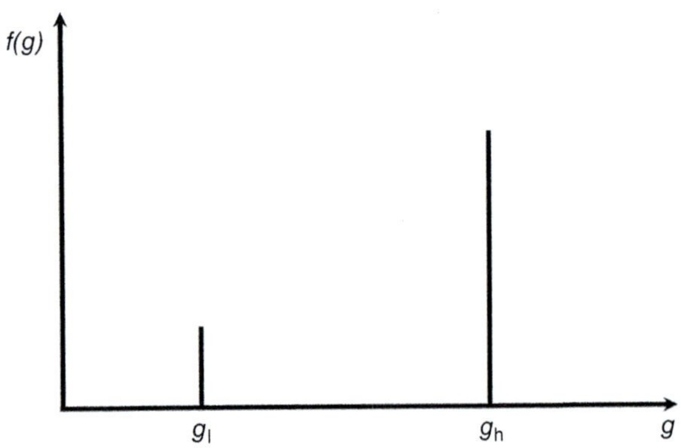

Figure 4.1 The delta-based conductance distribution assumed by Machta and Guyer (1987), Eq. (4.2), for a random network composed of low- and high-conductance (g_l and g_h) resistors.

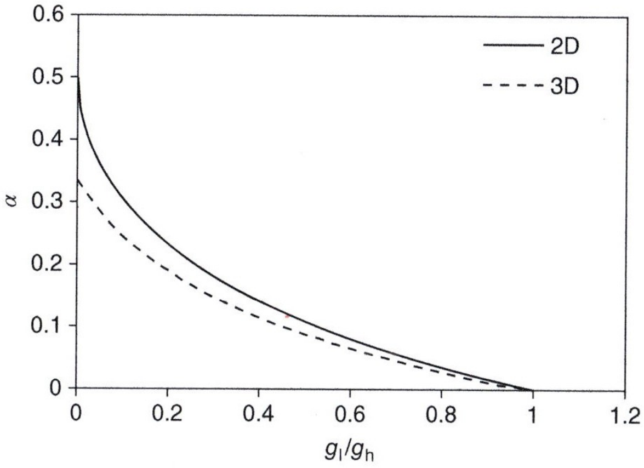

Figure 4.2 Scaling exponent α versus g_l/g_h for both two and three dimensions.

In three dimensions, however, the relationship between α and g_l/g_h is more complicated (see equation 3.2 in Machta and Guyer 1987). In Figure 4.2, we show α versus g_l/g_h for two and three dimensions. The 2D line represents Eq. (4.4), while the 3D one was digitized from figure 2 of Machta and Guyer (1987). As can be seen, α ranges between 0 and 0.5 in two dimensions. However, its value varies from 0 to nearly 0.33 in three dimensions. Theoretically, $0 < \alpha < 1/d$, so that the average maximum expected current diverges with L, if g_l/g_h is near 1 (Machta and Guyer 1987).

Chan et al. (1989) simulated the electric current in two-dimensional resistor networks and evaluated the predictability of Eq. (4.3) for $g_l/g_h = 0.25$, 0.5, and 0.75. By comparing with numerical simulations, they found reasonably well agreement between theory and simulation. Chan et al. (1989) also showed that the configuration of resistors in the neighborhood of large currents had a funnel shape.

4.2.2 Application to Porous Media

Unlike random resistor networks constructed of low- and high-conductance values (g_l and g_h) presumed by Machta and Guyer (1987), natural porous media are typically composed of pore conductances following a broad distribution shown in Figure 4.3. The term "broad," however, has not been satisfactorily defined in the literature. This means that pore conductance distributions in porous rocks probably do not conform to the delta function given in Eq. (4.1). Nonetheless, when $g_l = 0$ and if $f(g)$ is narrow enough, we demonstrate that the proposed scaling law, Eq. (4.3), as a phenomenological approach, results in accurate estimations of permeability and formation factor below the REV. Recall that when $g_l = 0$, $g_l/g_h = 0$ and thus $\alpha = 0.33$ in three dimensions (Fig. 4.2).

To apply Eq. (4.3) to estimate the scale dependence of permeability and formation factor, we assume that the average largest current is approximately the main contributor to hydraulic or electrical flow, and, thus, $k \approx \langle i_{\max} \rangle$ and $F \approx \langle i_{\max} \rangle^{-1}$. Accordingly, by normalizing Eq. (4.3), one has:

$$k(L) = k(L_{\min}) \left(\frac{\ln(L)}{\ln(L_{\min})} \right)^{\alpha} \tag{4.5}$$

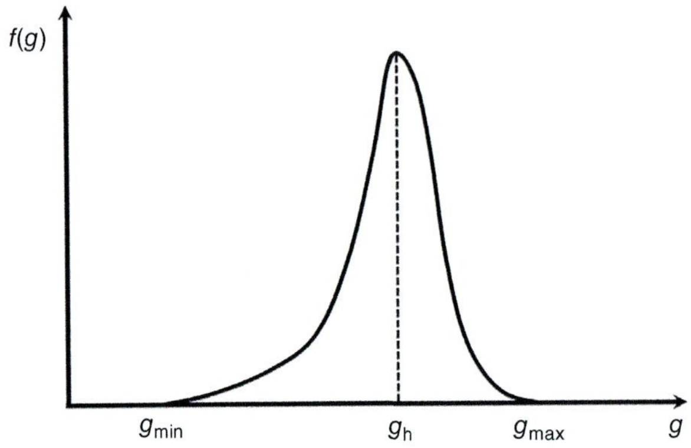

Figure 4.3 The schematic conductance distribution in a porous medium. The conductance g varies around g_h between g_{min} and g_{max}.

$$F(L) = F(L_{\min})\left(\frac{\ln(L)}{\ln(L_{\min})}\right)^{-\alpha} \tag{4.6}$$

where $k(L_{min})$ and $F(L_{min})$ are, respectively, the permeability and formation factor at the smallest scale.

4.3 Pore-network Simulations

To construct pore networks composed of spherical pore bodies and cylindrical pore throats, we used the open-access code developed by Valvatne (2004). The code generates pore networks based on a truncated Weibull distribution with two shape parameters δ and γ as well as lower and upper cutoffs. In this study, three different pore-throat radius ranges, i.e. 0.1–10, 1–50, and 10–75 μm, were used to have different levels of pore-scale heterogeneity (Table 4.1). Within each range, we varied γ from 12 to 30 ($\gamma = 12$, 18, 24, and 30), while set $\delta = 0.2$. Overall, 12 synthetic pore networks with constant pore-throat length of 100 μm and pore coordination number of 6 were generated. The network size varied from 1130 to 6770 μm ($L = 1130$, 2250, 3380, 4510, and 6770 μm). This means that $L_{min} = 1130$ μm.

Figure 4.3 shows the pore-throat radius distributions for the12 pore networks. As can be seen, the pore-throat radius distribution becomes broader from Network 1 to Network 3. Permeability and formation factors were simulated using the "poreflow" simulator, also developed by Valvatne (2004). The value of permeability was determined using Darcy's law, and the total flow rate was calculated by solving for pressure throughout the network under the steady-state flow conditions. Formation factor F, defined as the ratio of saturated medium resistivity to brine resistivity, was determined based on Ohm's law. Simulations of permeability and formation factor were iterated a hundred times and averaged over all realizations.

Table 4.1 Salient characteristics of the 12 pore networks constructed in this study.

Network	r_b (μm)	r_t (μm)	γ	δ	Z	l_t (μm)	ϕ (%)
1.1	0.1–10	0.1–10	12	0.2	6	100	4.3
1.2	0.1–10	0.1–10	18	0.2	6	100	4.7
1.3	0.1–10	0.1–10	24	0.2	6	100	4.9
1.4	0.1–10	0.1–10	30	0.2	6	100	5.1
2.1	1–50	1–50	12	0.2	6	100	32.1
2.2	1–50	1–50	18	0.2	6	100	34.2
2.3	1–50	1–50	24	0.2	6	100	35.4
2.4	1–50	1–50	30	0.2	6	100	36.1
3.1	10–75	10–75	12	0.2	6	100	43.9
3.2	10–75	10–75	18	0.2	6	100	46
3.3	10–75	10–75	24	0.2	6	100	47.1
3.4	10–75	10–75	30	0.2	6	100	47.8

r_b is pore-body radius, r_t is pore-throat radius, γ and δ are Weibull distribution parameters, l_t is pore-throat length, Z is pore coordination number, and ϕ is porosity.

4.4 Results and Discussion

Figure 4.5 presents the results of pore-scale numerical simulations of k and F as well as the theoretical estimations by Eqs. (4.5) and (4.6) for Network 1. Recall that this is the pore network with the narrowest pore-throat radius distribution studied here (Figure 4.4). As reported in Figure 4.5, we found an average relative error between −0.67 and −0.44% for permeability, which demonstrates accurate theoretical estimations. The negative average relative errors indicate that Eq. (4.5) generally underestimated the scale dependence of permeability. Figure 4.5 also shows that the permeability increases with increase in the network size, although the pore-throat radius distribution does not statistically vary with the scale.

The estimated scale-dependent formation factors using Eq. (4.6) are also shown in Figure 4.5 for Network 1. We found the average relative error between 0.04 and 0.15% showing precise theoretical estimations. In contrast to the results of $k(L)$, the positive average relative errors reported in Figure 4.5 indicate that, generally speaking, Eq. (4.6) slightly overestimated the scale dependence of the formation factor.

Results for Network 2 are shown in Figure 4.6. Similar to Network 1, the value of permeability increases, while the formation factor value decreases as the network size increases. Compared to Network 1, k is nearly two orders of magnitude greater, and F is one order of magnitude smaller in Network 2. Based on the permeability plots shown in Figure 4.6, it seems that Network 2 has a greater REV than Network 1. Although the simulated k in Network 1 does not vary with the scale at a network size of 6770 μm (Figure 4.5), its value tends to keep increasing with the scale in Network 2. This shows that Network 2 should be more heterogeneous than Network 1, in accordance with the broader pore-throat radius distribution in the former (Figure 4.4).

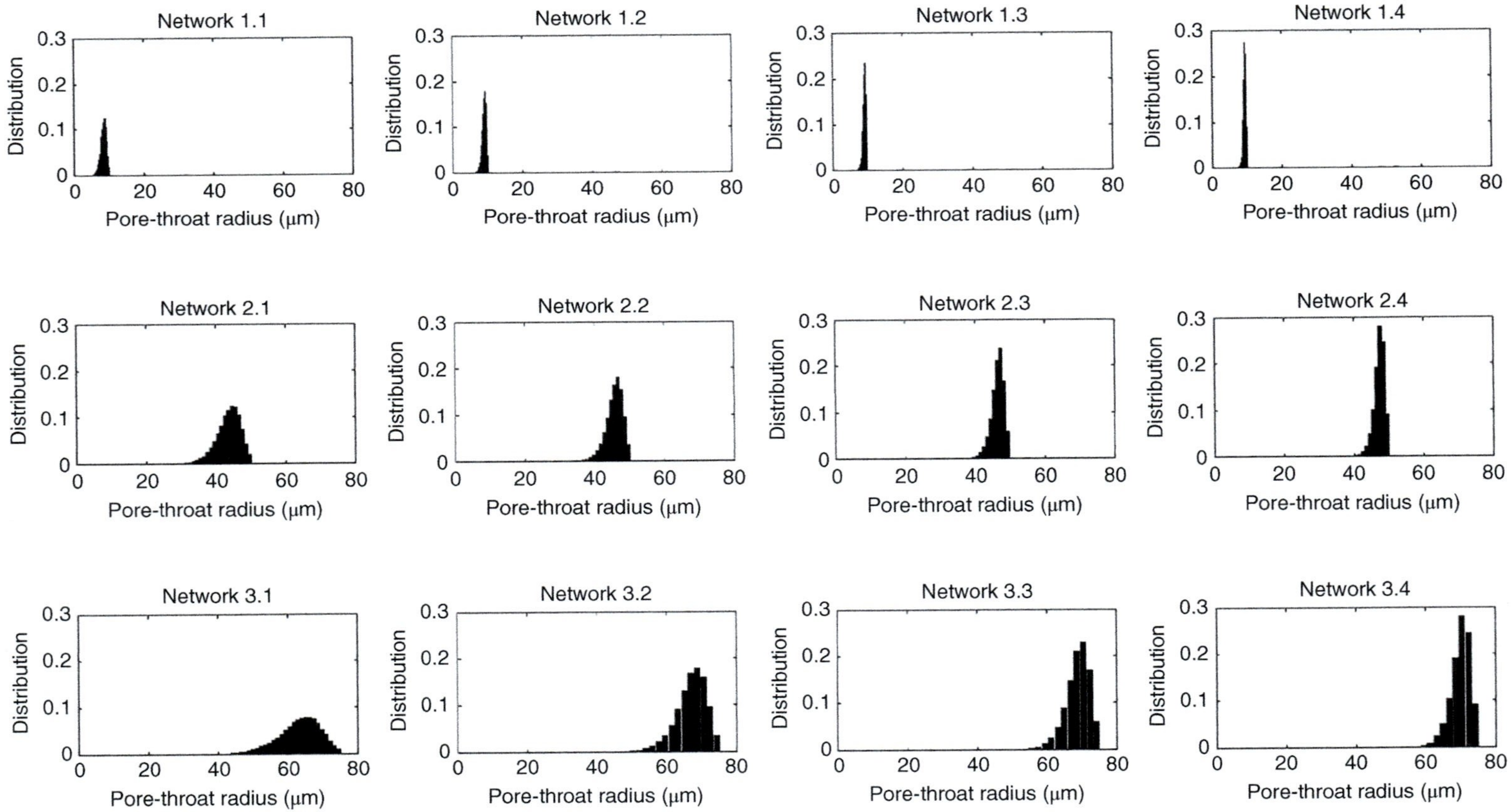

Figure 4.4 Pore-throat radius distributions for 12 pore networks constructed in this study.

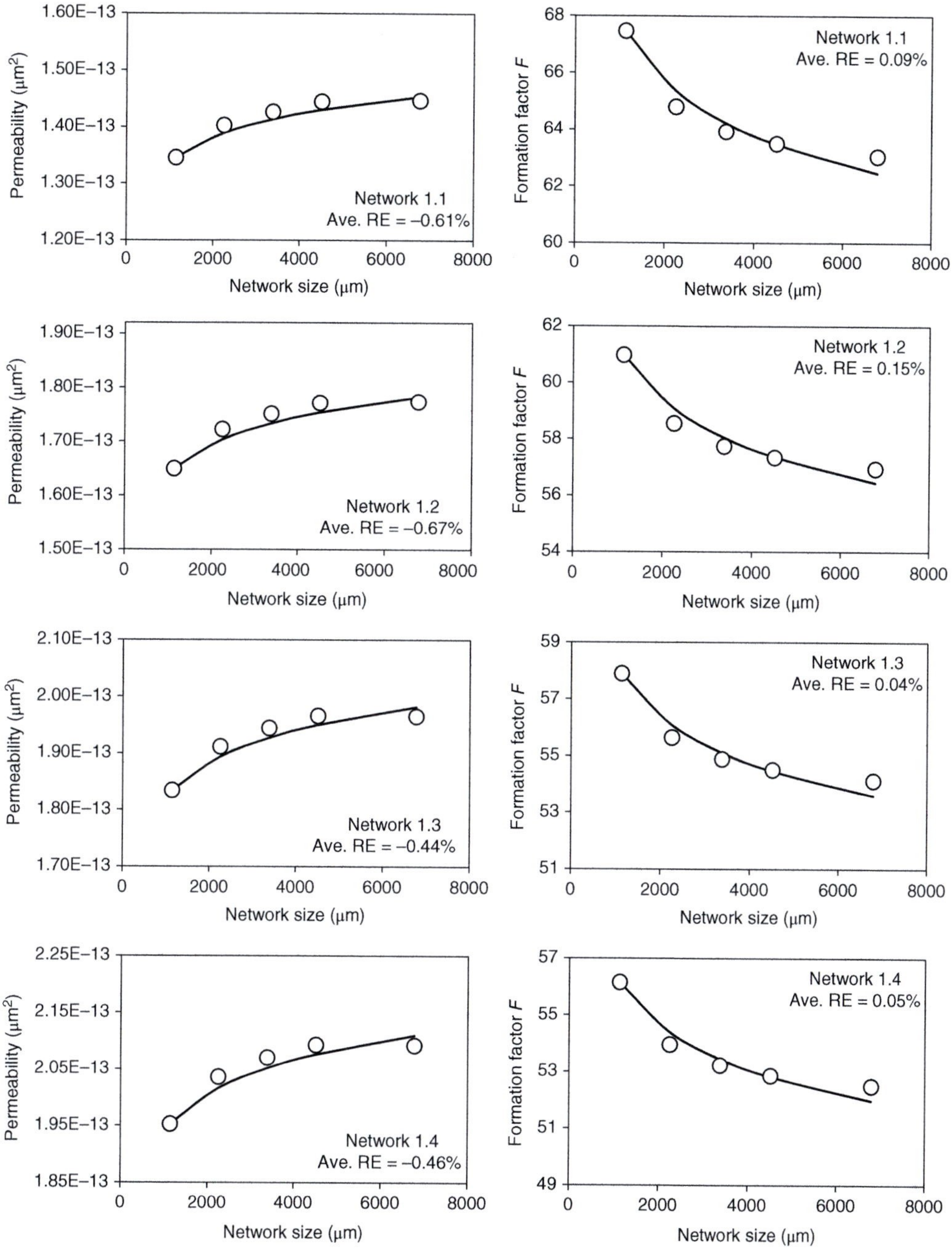

Figure 4.5 Simulated permeability, k, and formation factor, F, against the network size, L, for the pore Network 1. Solid lines represent theoretical estimations by Eqs. (4.5) and (4.6) with $\alpha = 0.33$.

As can be seen in Figure 4.6, Eq. (4.5) estimated the scale-dependent permeability in Network 2 reasonably well with average relative errors between 1.67 and 1.86%. We found that Eq. (4.6) estimated $F(L)$ with average relative errors ranged from −1.93 to −1.75% (Figure 4.6). In contrast to the results of Network 1, Eq. (4.5) overestimated $k(L)$, while Eq. (4.6) underestimated $F(L)$ in

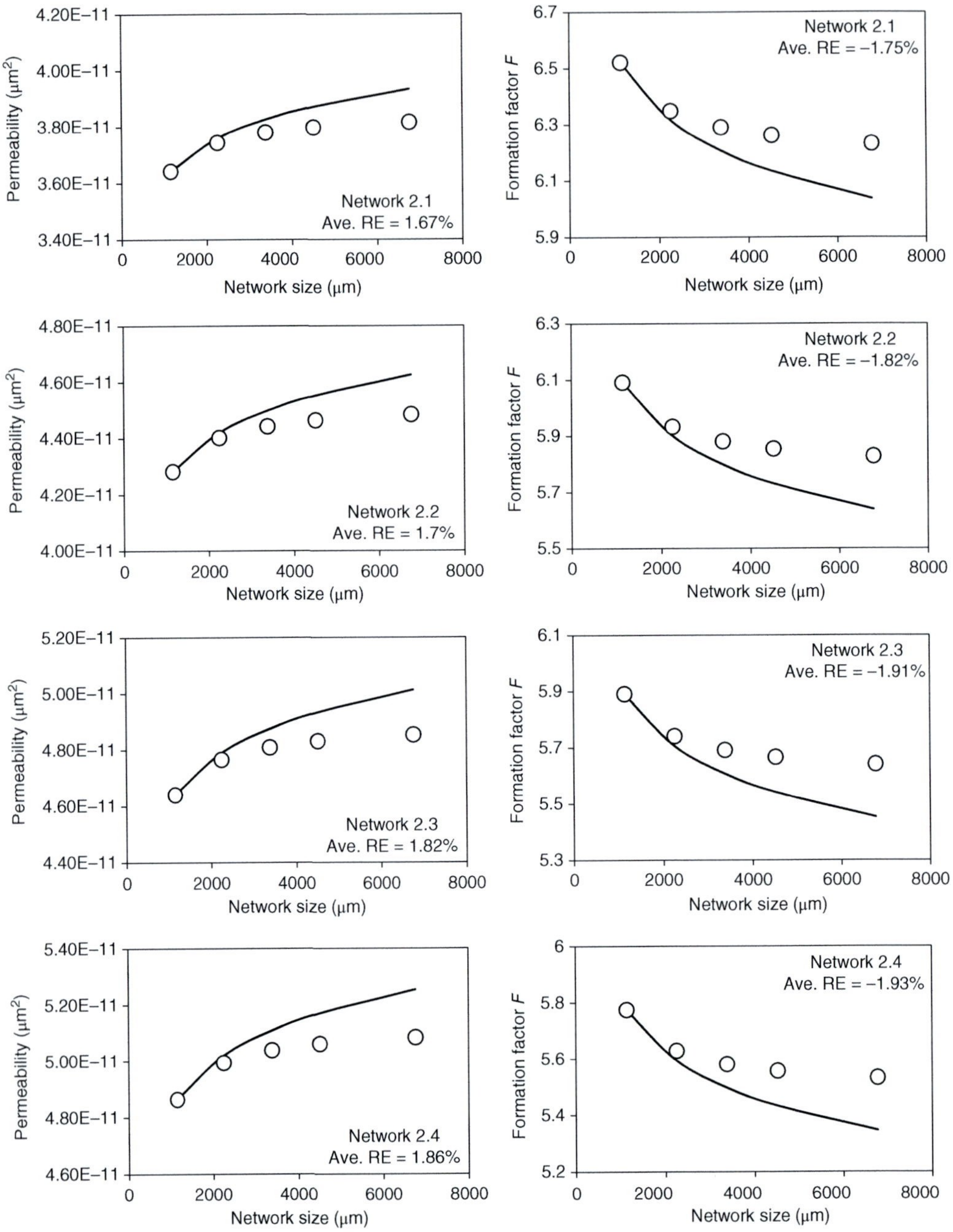

Figure 4.6 Simulated permeability, k, and formation factor, F, against the network size, L, for the pore Network 2. Solid lines represent theoretical estimations by Eqs. (4.5) and (4.6) with $\alpha = 0.33$.

Network 2. We found that the value $\alpha = 0.23$ resulted in more accurate estimations of the permeability and formation factor in Network 2 compared to $\alpha = 0.33$ (results not shown).

Results of the permeability, formation factor, and their scale-dependent estimations for Network 3 are presented in Figure 4.7. In Network 3, the pore-throat radius distributions are broader than those in Networks 1 and 2, and thus, Network 3 is the most heterogeneous network among the three cases studied here. The increasing trend in the permeability at larger network sizes (e.g. 6770 μm) indicates that the REV has not reached (Figure 4.7), which is similar to the results of Network 2 (Figure 4.6). As reported in Figure 4.7, we found the average relative errors ranged from 2.45 to

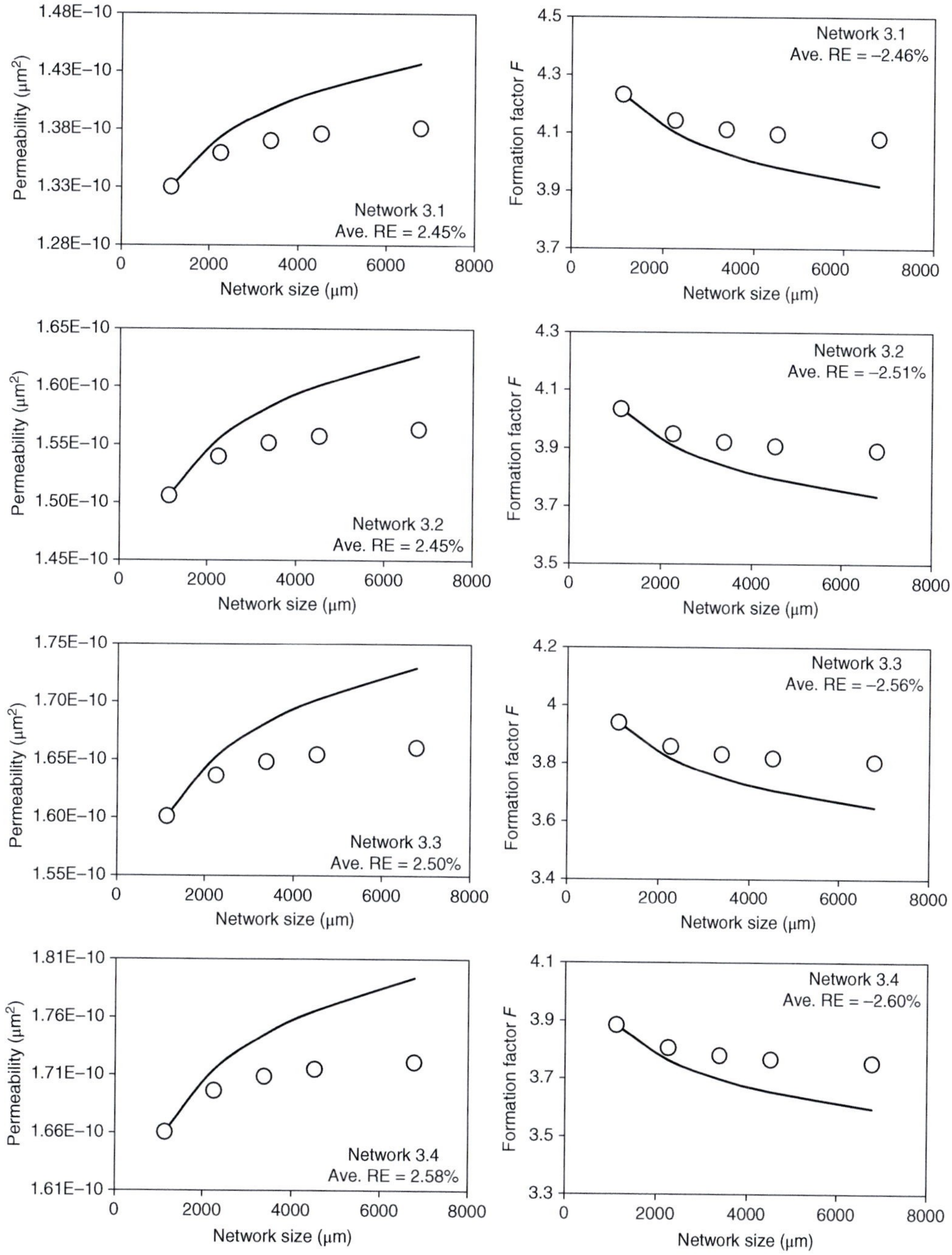

Figure 4.7 Simulated permeability, k, and formation factor, F, against the network size, L, for the pore Network 3. Solid lines represent theoretical estimations by Eqs. (4.5) and (4.6) with $\alpha = 0.33$.

2.58% for Eq. (4.5) and −2.60 to −2.46% for Eq. (4.6) in the estimation of the scale dependence of permeability and formation factor in Network 3. We also found that $\alpha = 0.18$ led to more accurate estimations in Network 3 compared to $\alpha = 0.33$ (results not shown).

The average relative error values for Networks 2 and 3 are greater than those reported for Network 1. This is most probably because the funnel defect approach and its subsequent derivations, Eqs. (4.5) and (4.6), work best in networks whose conductances conform to a delta function. Accordingly, as the conductance (or pore-throat radius) distribution becomes broader, the accuracy of the funnel defect approach is expected to decrease. Comparing the average relative errors reported in Figures 4.5–4.7 clearly demonstrates that with the increase in the pore-scale heterogeneity, the model accuracy decreases.

4.5 Limitations

The effect of pore coordination number on flow and transport in porous media has been well documented in the literature (Hunt et al. 2014; Jerauld et al. 1984; Sahimi 2011). Machta and Guyer (1987) proposed their scaling law, Eq. (4.3), for simple square and cubic lattices in two and three dimensions with coordination numbers 4 and 6, respectively. The average pore coordination number in rocks is typically around 3 or 4 (Blunt 2017; Liang et al. 2000; Vogel and Roth 2001). Therefore, generalizing their approach for any average pore coordination number would make it applicable to a wide range of porous media.

The funnel defect approach and its subsequent derivations, Eqs. (4.5) and (4.6), are not a function of REV and do not incorporate its effect. Recall that REV is the smallest size above which permeability and/or formation factor does not change with scale. Accordingly, applying Eqs. (4.5) and (4.6) near and above the REV may result in significant errors and uncertainties.

4.6 Conclusion

Modeling the scale dependence of permeability and formation factor in porous rocks has been a challenge in petrophysics because understanding the effect of small- and large-scale heterogeneities and capturing them are essential. In this study, we applied a simple scaling law developed based on the funnel defect approach in physics. By adopting the theory to networks of pores represented porous rocks, we investigated the effect of scale on permeability and formation factor. To evaluate the proposed theoretic approach, we simulated permeability and formation factor in 12 pore networks with different levels of heterogeneities reflected in pore-throat radius distributions. Comparing pore-network simulations with theoretical estimations showed that the proposed theoretic approach estimated k and F accurately with relative errors less than 2.60%. However, further investigations are required to generalize the proposed approach and evaluate it both numerically and experimentally using a broader range of porous rocks.

Acknowledgment

The authors are grateful to Jon Machta, Department of Physics and Astronomy, the University of Massachusetts at Amherst, for fruitful discussions. BG acknowledges Kansas State University for supports through faculty startup funds.

References

Blunt, M.J. (2017). *Multiphase Flow in Permeable Media*. Cambridge University Press https://doi.org/10.1017/9781316145098.

Brace, W.F. (1984). Permeability of crystalline rocks: new in situ measurements. *Journal of Geophysical Research, Solid Earth* 89: 4327–4330.

Chan, S.K., Machta, J., and Guyer, R.A. (1989). Large currents in random resistor networks. *Physical Review B* 39: 9236–9239.

Esmaeilpour, M., Ghanbarian, B., Liang, F., and Liu, H.H. (2021). Scale-dependent permeability and formation factor in porous media: applications of percolation theory. *Fuel* 301: 121090.

Ewing, R.P., Hu, Q., and Liu, C. (2010). Scale dependence of intragranular porosity, tortuosity, and diffusivity. *Water Resources Research* 46 (6): 1–12. https://doi.org/10.1029/2009WR008183.

Garbesi, K., Sextro, R.G., Robinson, A.L. et al. (1996). Scale dependence of soil permeability to air: measurement method and field investigation. *Water Resources Research* 32: 547–560.

Ghanbarian, B. (2022a). Estimating the scale dependence of permeability at pore and core scales: incorporating effects of porosity and finite size. *Advances in Water Resources* 161: 104123.

Ghanbarian, B. (2022b). Scale dependence of tortuosity and diffusion: finite-size scaling analysis. *Journal of Contaminant Hydrology* 245: 103953.

Ghanbarian, B., Taslimitehrani, V., Dong, G., and Pachepsky, Y.A. (2015). Sample dimensions effect on prediction of soil water retention curve and saturated hydraulic conductivity. *Journal of Hydrology* 528: 127–137. https://doi.org/10.1016/j.jhydrol.2015.06.024.

Ghanbarian, B., Taslimitehrani, V., and Pachepsky, Y.A. (2017). Accuracy of sample dimension-dependent pedotransfer functions in estimation of soil saturated hydraulic conductivity. *Catena* 149: 374–380. https://doi.org/10.1016/j.catena.2016.10.015.

Ghanbarian, B., Esmaeilpour, M., Ziff, R.M., and Sahimi, M. (2021). Effect of pore-scale heterogeneity on scale-dependent permeability: pore-network simulation and finite-size scaling analysis. *Water Resources Research* 57: e2021WR030664.

Hunt, A.G. (2006). Scale-dependent hydraulic conductivity in anisotropic media from dimensional cross-over. *Hydrogeology Journal* 14 (4): 499–507. https://doi.org/10.1007/s10040-005-0453-6.

Hunt, A., Ewing, R., and Ghanbarian, B. (2014). *Percolation Theory for Flow in Porous Media*. Springer.

Hyun, Y., Neuman, S.P., Vesselinov, V.V. et al. (2002). Theoretical interpretation of a pronounced permeability scale effect in unsaturated fractured tuff. *Water Resources Research* 38: 1092.

Jerauld, G.R., Hatfield, J.C., Scriven, L.E., and Davis, H.T. (1984). Percolation and conduction on Voronoi and triangular networks: a case study in topological disorder. *Journal of Physics C: Solid State Physics* 17 (9): 1519–1529. https://doi.org/10.1088/0022-3719/17/9/010.

Liang, Z., Ioannidis, M.A., and Chatzis, I. (2000). Geometric and topological analysis of three-dimensional porous media: pore space partitioning based on morphological skeletonization. *Journal of Colloid and Interface Science* 221 (1): 13–24. https://doi.org/10.1006/jcis.1999.6559.

Machta, J. and Guyer, R.A. (1987). Largest current in a random resistor network. *Physical Review B* 36 (4): 2142–2146. https://doi.org/10.1103/PhysRevB.36.2142.

Rovey, C.W. and Cherkauer, D.S. (1995). Scale dependency of hydraulic conductivity measurements. *Groundwater* 33 (5): 769–780. https://doi.org/10.1111/j.1745-6584.1995.tb00023.x.

Sahimi, M. (2011). *Flow and Transport in Porous Media and Fractured Rock: From Classical Methods to Modern Approaches*, 2nde. Weinheim, Germany: Wiley-VCH.

Schulze-Makuch, D. (1996). Facies dependent scale behavior of hydraulic conductivity and longitudinal dispersivity in the carbonate aquifer of Southeastern Wisconsin. PhD dissertation. University of Wisconsin, Milwaukee.

Schulze-Makuch, D., Carlson, D.A., Cherkauer, D.S., and Malik, P. (1999). Scale dependency of hydraulic conductivity in heterogeneous media. *Ground Water* 37: 904–919.

Valvatne, P.H. (2004). *Predictive Pore-Scale Modelling of Multiphase Flow*. Imperial College London.

Vogel, H.J. and Roth, K. (2001). Quantitative morphology and network representation of soil pore structure. *Advances in Water Resources* 24 (3–4): 233–242. https://doi.org/10.1016/S0309-1708(00)00055-5.

Yoon, H. and Dewers, T.A. (2013). Nanopore structures, statistically representative elementary volumes, and transport properties of chalk. *Geophysical Research Letters* 40: 4294–4298.

Part II

Core-Scale Heterogeneity

5

Modeling Gas Permeability in Unconventional Reservoir Rocks

Behzad Ghanbarian[1], Feng Liang[2], and Hui-Hai Liu[2]

[1] *Porous Media Research Lab, Department of Geology, Kansas State University, Manhattan, KS, USA*
[2] *Aramco Americas: Aramco Research Center–Houston, Houston, TX, USA*

5.1 Introduction

Precise estimations of petrophysical properties of rocks, such as gas permeability, have numerous practical applications, particularly in determining gas production from unconventional reservoirs. Shale and tight formations have been successfully explored and produced not only in the United States and North America but also in China (Jia et al. 2019; Wang et al. 2017). Accordingly, they became one of the major contributors to energy supplies. Although investigating flow and transport in unconventional reservoir rocks has made significant progress in the past decade, there is still a long way to fully understand all mechanisms and key factors that control gas transport in such formations.

Literature on single-phase gas permeability and its modeling in mudrocks is extensive and vast (Allan and Mavko 2013; Civan 2010; Darabi et al. 2012; Ghanbarian and Javadpour 2017; Javadpour 2009; Landry et al. 2016; Nazari Moghaddam and Jamiolahmady 2016; Sakhaee-Pour and Bryant 2012), while investigations of gas relative permeability are still very limited. In what follows, we summarize some applications from theoretic approaches and numerical simulations to the modeling of gas transport in shales and tight porous rocks.

5.1.1 Theoretical Models

Early theoretical models of gas permeability were developed based on first- and then higher-order slip equations (Zhang et al. 2012). For example, Beskok and Karniadakis (1999) accounted for slip flow and modified the Hagen-Poiseuille equation for gas flow in a single cylindrical pore tube. Javadpour (2009) incorporated first-order slip flow and Knudsen diffusion mechanisms and proposed another model for gas permeability in a cylindrical nanopore. To scale up from pore scale to core scale, theoretic models based on a bundle of capillary tubes have been developed to study single-phase gas permeability and gas-relative permeability in shales and tight porous rocks. For instance, Zhang et al. (2018) presumed that nanoscale inorganic pores in shales followed fractal scaling, and their size distribution conformed to a power–law probability density function. They proposed a gas-relative permeability model by considering effects of gas slippage in the entire Knudsen regime, stress dependence, and water film thickness. Zhang et al. (2018) found that

Physics of Fluid Flow and Transport in Unconventional Reservoir Rocks, First Edition.
Edited by Behzad Ghanbarian, Feng Liang, and Hui-Hai Liu.

Knudsen diffusion and slippage effects boosted the gas-relative permeability considerably at low to intermediate pressures (e.g., 0.1–1 MPa). They also stated that as pore space or tortuosity fractal dimension increased, gas relative permeability slightly increased, while water relative permeability decreased (see their figure 10).

Liu et al. (2018) proposed power–law functions to model gas relative permeability in unconventional reservoir rocks based on the bundle of capillary tubes approach. Although theoretically developed, their models (see their eqs. (31) and (32)) assume that gas relative permeability vanishes at zero gas saturation meaning that critical gas saturation is zero. By fitting their one-parameter model i.e. $k_{rg} = (1 - S_w)^{\gamma}$ to four experimental data, Liu et al. (2018) found γ ranged between 3.9 and 40.9.

5.1.2 Pore-Network Models

Pore-network models (PNMs) have been used to study gas transport in nanoporous membranes by incorporating effects of different mechanisms, such as viscous flow, Knudsen diffusion, and hindered diffusion. Mehmani et al. (2013) were among the first to propose three-dimensional PNMs to study single-phase gas permeability in shales. Their models were composed of two domains: (i) pores in nanometer size range (~100–575 nm), and (ii) pores in micrometer size range (~10–57.5 μm). In both domains, the pore coordination number was four ($Z = 4$), and slip flow and Knudsen diffusion were important contributors to gas flow. Results of Mehmani et al. (2013) demonstrated the non-trivial contribution of Knudsen diffusion and slip flow, particularly at low pressures and in pore networks with dominant fraction of nanopores. Mehmani et al. (2013) also showed that gas permeability significantly depended on the fraction of nanoscale pores. However, they did not address how the connectivity between organic patches and inorganic matrix would affect gas permeability of the entire pore network.

Ma et al. (2014) derived the pore network of a shale sample, contained near 20,000 pore bodies and 30,000 pore throats, from a three-dimensional model reconstructed from serial SEM images. The porosity of the network was 2.9%, the average pore coordination number was near 3, and 80% of the pores were less than 10 nm in radius. Considering slip flow and Knudsen diffusion mechanisms, Ma et al. (2014) simulated gas permeability in PNMs using methane and nitrogen under operational and laboratory conditions, respectively. They stated that "... the nitrogen permeability can only approximate the methane permeability within a certain range of field operational conditions when the slip flow is not dominating."

In another study, Zhang et al. (2015a) developed a three-dimensional multi-scale PNM in which organic pores, inorganic pores, and pore coordination number conformed to normal distributions with means and standard deviations obtained from actual shale samples. The average organic and inorganic pore sizes were 6 and 300 nm, respectively. Results of Zhang et al. (2015a) showed that gas permeability was greatly dependent on pore pressure and organic pore sizes. More specifically, they found that as pore pressure decreases, the effect of viscous flow vanishes, while slip flow and Knudsen diffusion become gradually dominant flow regimes.

Song et al. (2017) simulated gas permeability via PNMs constructed based on a section of a high-resolution SEM image of an organic-rich shale sample. Their PNM simulations, accordingly, represent gas flow contributions in organic pores. Song et al. (2017) developed three PNMs with different average pore radii – 1.56, 3.2, and 15.6 nm – and determined gas permeability for pore pressures from 5 to 70 MPa. They found that for pores smaller than 5 nm, gas permeability was remarkably sensitive to gas properties and surface diffusion. They stated that "... when the pore radius is less than 2 nm, the gas permeability will be significantly underestimated if the surface diffusion is neglected."

5.1.3 Gas Transport Mechanisms

In addition to slip flow and Knudsen diffusion, more complex mechanisms, such as sorption and surface diffusion, were incorporated to model gas flow in shales (Allan and Mavko 2013; Naraghi et al. 2018; Song et al. 2016; Zhang et al. 2015b). For instance, Wu et al. (2016) proposed a unified theoretical model by coupling various mechanisms including slip flow, Knudsen diffusion, sorption, and surface diffusion. The authors stated that "... surface diffusion is an important transport mechanism, and its contribution cannot be negligible and even dominates in nanopores with less than 2 nm in diameter." In contrast, Naraghi et al. (2018) claimed that surface diffusion has little impact on total gas flow. Their results, however, are based on two-dimensional stochastic models of shale samples at 13.8 MPa (2000 psi) pore pressure. Remarkable discrepancies between two- and three-dimensional models, representing shales, and their gas permeability estimations have been recently highlighted in Cao et al. (2017). Those authors considered slip flow and Knudsen diffusion mechanisms and demonstrated that deviations between two- and three-dimensional models are negligible only if pore pressure is greater than 10 MPa. Interestingly, Finkbeiner et al. (2001) reported that pore pressure in shale reservoirs is typically greater than 20 MPa (see their tables 5 and 6). In another study, Gherabati et al. (2016) indicated that for Eagle Ford the pore pressure varied from near 14 to about 97 MPa. A recent three-dimensional stochastic model based on an embedded discrete fracture model by Xu et al. (2017) included organic and inorganic pores as well as fractures to investigate permeability anisotropy of shale formations. As expected, their results showed that the contribution of fractures to gas permeability was more prominent than that of matrix. They also found that the contribution of inorganic pores to permeability was greater than that of organic pores.

Table 5.1 summarizes several studies as well as mechanisms used to investigate gas permeability of unconventional reservoir rocks by means of bundle-of-tubes, pore-network, and stochastic models in the literature.

5.1.4 Objectives

In this book chapter, we aim to present applications from effective-medium theory (EMT) to modeling single-phase gas permeability and relative gas permeability in unconventional reservoir rocks. The main objectives of this study are to (i) develop upscaling theoretical frameworks based upon EMT to model gas transport under variably saturated conditions and (ii) to compare the proposed models with experimental data, PNMs, and lattice-Boltzmann simulations from the literature.

5.2 Effective-Medium Theory

EMT (Bruggeman 1935; Landauer 1952) is an upscaling technique from statistical physics in which a heterogeneous medium is replaced by a homogeneous one with an effective permeability that is supposed to be an estimate of the original medium (Sahimi 2011). Based on EMT, the effective conductance g_e of the homogeneous medium is calculated by requiring that the spatially averaged perturbations, taken with respect to the pore conductance distribution $f(g)$ in the heterogeneous medium, to be zero as follows

$$\int_{g_{min}}^{g_{max}} \frac{g_e - g}{g + (Z/2 - 1)g_e} f(g)\mathrm{d}g = 0 \tag{5.1}$$

Table 5.1 Various studies as well as mechanisms proposed in the literature to investigate gas transport in shales.

Reference	System dimension	Slip flow	Knudsen diffusion	Surface diffusion	Sorption	Bimodal PSD[a]	Monomodal PSD	Pore connectivity	Single pore	PNM	Bundle of tubes	Stochastic model
Civan (2010)	3	✓	✗	✗	✗	✗	✓	✗	✓	✗	✓	✗
Javadpour (2009)	3	✓	✓	✗	✗	✗	✗	✗	✓	✗	✗	✗
Nazari Moghaddam and Jamiolahmady (2016)	3	✓	✗	✗	✗	✗	✓	✗	✓	✗	✓	✗
Beskok and Karniadakis (1999)	3	✓	✗	✗	✗	✗	✗	✗	✓	✗	✗	✗
Mehmani et al. (2013)	3	✓	✓	✗	✗	✓	✗	✓	✗	✓	✗	✗
Ma et al. (2014)	3	✓	✓	✗	✗	✗	✓	✓	✗	✓	✗	✗
Zhang et al. (2015a)	3	✓	✓	✗	✗	✗	✓	✓	✗	✓	✗	✗
Song et al. (2017)	3	✓	✓	✓	✓	✗	✓	✓	✓	✓	✗	✗
Naraghi et al. (2018)	2	✓	✓	✓	✓	✓	✗	✓	✗	✗	✗	✓
Wu et al. (2016)	3	✓	✓	✓	✓	✗	✗	✗	✓	✗	✗	✗
Cao et al. (2017)	3	✓	✓	✗	✗	✓	✗	✓	✗	✓	✗	✓
Zheng and Yu (2012)	3	✗	✓	✗	✗	✓	✗	✓	✓	✗	✓	✗
Behrang et al. (2016)	3	✓	✓	✗	✗	✗	✓	✗	✓	✗	✓	✗
Behrang and Kantzas (2017)	3	✓	✓	✓	✓	✗	✓	✗	✓	✗	✓	✗

[a] PSD is pore-throat size distribution and PNM is pore-network model.

where g is the pore conductance, g_{min} and g_{max} are the smallest and largest pore conductances in the heterogeneous medium, respectively, and Z is the average pore coordination number. Recall that pore conductance, g, is related to pore-throat radius, r, via the Hagen-Poiseuille's law. So, one can determine the pore conductance distribution $f(g)$ from $f(r)/(dg/dr)$. For further detail, see Hunt et al. (2014).

Following Ghanbarian et al. (2016), one may replace the percolation threshold $p_c = 2/Z$ in Eq. (5.1) by the critical saturation S_c, the minimum amount of saturation required to form sample-spanning cluster for percolation, and accordingly $(Z/2 - 1)$ by $(1 - S_c)/S_c$. Therefore, Eq. (5.1) changes to

$$\int_{g_{min}}^{g_{max}} \frac{(g_e - g)}{g + [S_c/(1 - S_c)]g_e} f(g)\mathrm{d}g = 0 \tag{5.2}$$

In Eq. (5.2), S_c can be estimated from mercury saturation S_{Hg} at the inflection point on the mercury intrusion porosimetry curve (Ghanbarian and Javadpour 2017). The greater the critical saturation S_c, the smaller the effective pore conductance g_e, and accordingly, the smaller the permeability.

It is well documented in the literature that estimations of EMT are accurate as long as the pore conductance distribution is not too broad (Adler and Berkowitz 2000; Doyen 1988; Ghanbarian and Daigle 2016; Ghanbarian and Javadpour 2017; Ghanbarian et al. 2016, 2019, 2020; Koplik 1981). For example, numerical simulations of Adler and Berkowitz (2000) in three-dimensional lattices, randomly composed of bonds chosen from a lognormal distribution, demonstrated that the EMA results under fully saturated conditions would be accurate as long as the lognormal distribution standard deviation $\sigma < 1$.

5.3 Single-Phase Gas Permeability

5.3.1 Gas Permeability in a Cylindrical Tube

Let us assume that gas transport at nanoscales is mainly controlled by slip flow and Knudsen diffusion, and these two mechanisms contribute in parallel. Accordingly, gas permeability in a single, straight, and perfectly cylindrical nanotube with radius R is given by (Javadpour 2009)

$$k_p = \frac{2R\mu M}{3R_g\rho}\left(\frac{8R_gT}{\pi M}\right)^{0.5} + F\frac{R^2}{8} \tag{5.3}$$

in which k_p is the permeability of a single pore, and the theoretical dimensionless coefficient for slippage effect F is (Brown et al. 1946)

$$F = 1 + \left(\frac{8\pi R_gT}{M}\right)^{0.5}\frac{\mu}{pR}\left(\frac{2}{\mathrm{TMAC}} - 1\right) \tag{5.4}$$

where M is molar mass, T is temperature, R_g is gas constant, μ is gas viscosity, ρ is gas density, p is pore pressure, and TMAC is tangential momentum accommodation coefficient whose value theoretically ranges between 0 and 1, depending on gas type, temperature, pressure and surface roughness (Arkilic et al. 1997).

5.3.2 Pore Pressure-Dependent Gas Permeability in Tight Rocks

To estimate gas permeability from pore-throat radius distribution, porosity, tortuosity, and gas characteristics, one may apply concepts from EMT. Combining the EMT permeability with the Javadpour (2009) model gives

$$k = \frac{2r_e\mu M}{3R_g\rho}\frac{\phi}{\tau}\left(\frac{8R_gT}{\pi M}\right)^{0.5} + F\frac{\phi}{C_s\tau}\frac{r_e^4}{\langle r_b^2\rangle} \tag{5.5}$$

where C_s is the pore shape factor equal to 8 in cylindrical pores, r_b is the pore-body radius, r_e is the effective pore-throat radius, τ is the tortuosity (> 1), and φ is total porosity.

The value of TMAC required to determine F may be approximated using the following empirical relationship (Agrawal and Prabhu 2008)

$$\text{TMAC} = 1 - \log\left(1 + K_n^{0.7}\right) \tag{5.6}$$

in which K_n, Knudsen number, is determined from (Heidemann et al. 1984)

$$K_n = \frac{k_B T}{2\sqrt{2}\pi d_m^2 p r_e} \tag{5.7}$$

where k_B is the Boltzmann constant and d_m is the collision diameter of gas molecule ($d_m = 0.3$ and 0.38 nm for nitrogen and methane, respectively).

To evaluate the accuracy of our pore pressure-dependent permeability model, we compare it with experimental data, PNMs, and lattice-Boltzmann simulations presented in the literature listed in Table 5.2.

5.3.3 Comparison with Experiments

Experimental data used in this study are from Nazari Moghaddam and Jamiolahmady (2016b), who reported measurements on three shale rock samples from the Eagle Ford, Pierre, and Barnett formations. Mercury intrusion porosimetry data measured in the range of 0–380 MPa (0–55,000 psi) were used to determine the pore throat-size distribution for each sample. The critical saturation S_c for percolation determined from mercury saturation at the inflection point of the mercury intrusion porosimetry curve was 0.13, 0.35, and 0.27 for the Eagle Ford, Pierre, and Barnett samples, respectively.

The lognormal probability density function fitted to the pore throat-size distribution and results of the pore pressure-dependent permeability estimations are presented in Figure 5.1 for the Eagle Ford, Pierre, and Barnett shales.

As can be observed, EMT, Eq. (5.5), underestimated the pore pressure dependence of the permeability for Eagle Ford with a factor of 6, while it estimated k more accurately, within a factor of three of the measured value, for the Pierre and Barnett samples. One possible source of error in the permeability estimation may be rough approximation of tortuosity whose value was set equal to 2.

We should note that the permeability of the Eagle Ford sample is in the order of milliDarcies (mD), while it is in the order of microDarcies (μD) for the Pierre and Barnett samples. The critical saturation in the Eagle Ford sample is remarkably less than that in the Pierre and Barnett samples, which might be an indication of microfractures in the Eagle Ford sample, associated with microfossils, fragmentary fossil material, or pyrite framboids that are not reflected in the pore throat-size distribution (Ghanbarian and Javadpour 2017).

Table 5.2 Salient properties of experiments, pore-network model, and lattice-Boltzmann simulations used to evaluate the proposed model based on effective-medium theory.

Reference	Method	Remark	Gas	T (K)	ϕ (cm³/cm³)	τ	Lognormal distribution parameters		S_c	r_e (μm)
							μ	σ		
Nazari Moghaddam and Jamiolahmady (2016b)	Exp[a]	Eagle Ford	N_2	333	0.134	2	−2.81	0.44	0.13	0.069
	Exp	Pierre	N_2	333	0.316	2	−3.47	0.35	0.35	0.030
	Exp	Barnett	N_2	333	0.086	2	−4.75	0.85	0.27	0.023
Mehmani et al. (2013)	PNM	PNM	CH_4	400	0.100	2	—	—	0.50	12.7
Landry et al. (2016)	LB	Reconstruction 1	CH_4	400	0.583	2	−5.45	0.15	0.25	0.004
Chen et al. (2015)	LB	Sichuan Basin	CH_4	383	0.191	12.2	−4.25	0.49	0.02	0.018

[a] Exp: experiment, PNM: pore-network model, LB: lattice-Boltzmann, N_2: nitrogen, CH_4: methane, T: temperature, ϕ: porosity, τ: tortuosity, μ: mean, σ: standard deviation, S_c: critical saturation for percolation, r_e: effective pore radius.

Source: Ghanbarian and Javadpour (2017)/John Wiley & Sons.

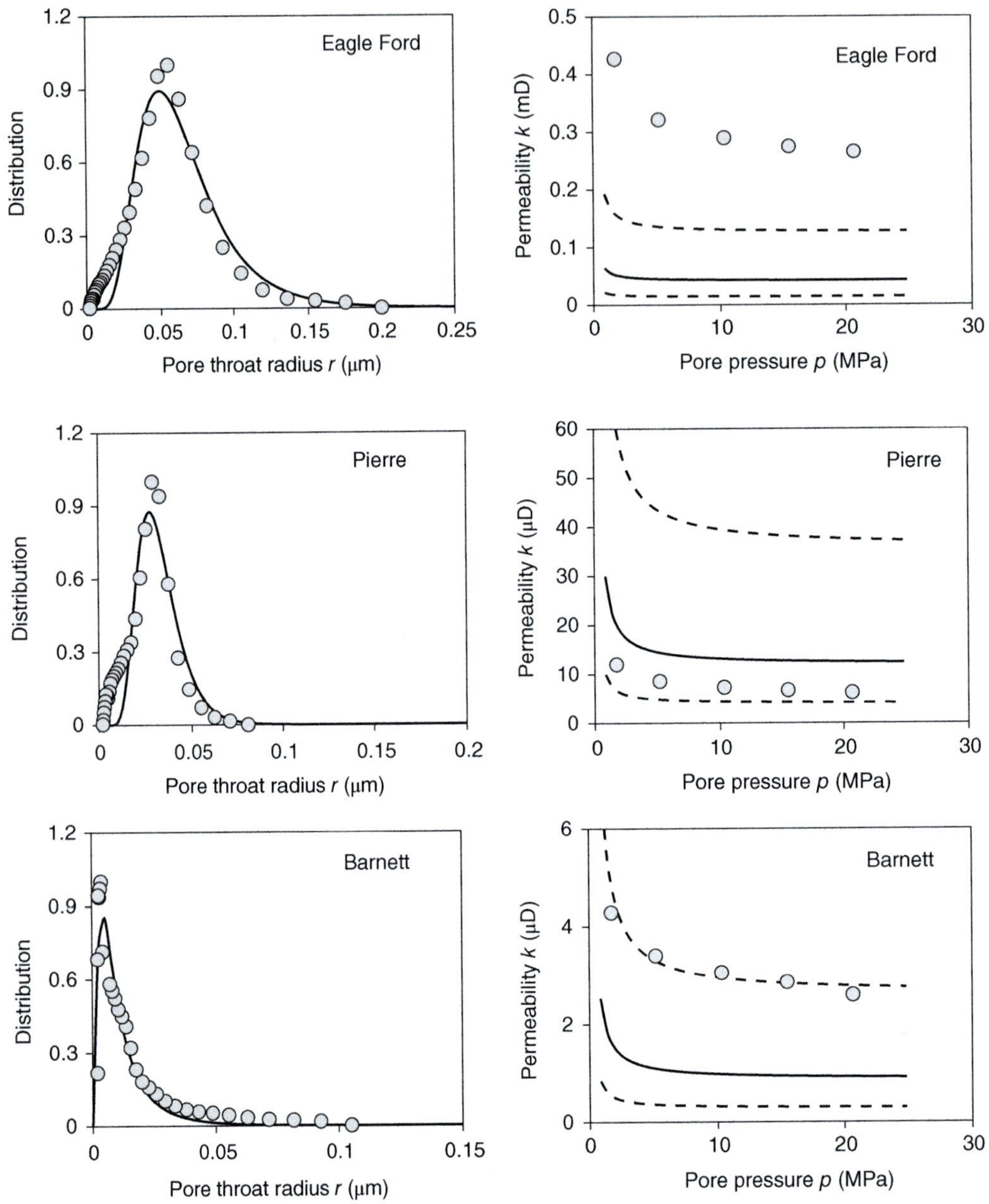

Figure 5.1 The pore throat-size distribution and the fitted lognormal distribution (on the left) and the measured and estimated permeability versus pore pressure (on the right) for three experiments from Nazari Moghaddam and Jamiolahmady (2016b). In the right plots, the solid black line denotes the estimated permeability, and the dashed lines represent the factor of three boundaries. *Source:* Data from Nazari Moghaddam and Jamiolahmady (2016b).

5.3.4 Comparison with Pore-Network Simulations

The pore-network simulations are from Mehmani et al. (2013), who generated a complex pore network with an average coordination number of 4 ($Z = 4$) and different fractions of nanoscale pore throats. The pore throat-size distribution and the estimated permeability as a function of pore pressure are shown in Figure 5.2. As can be seen, the size of pores is in the order of 5–25 μm. Well

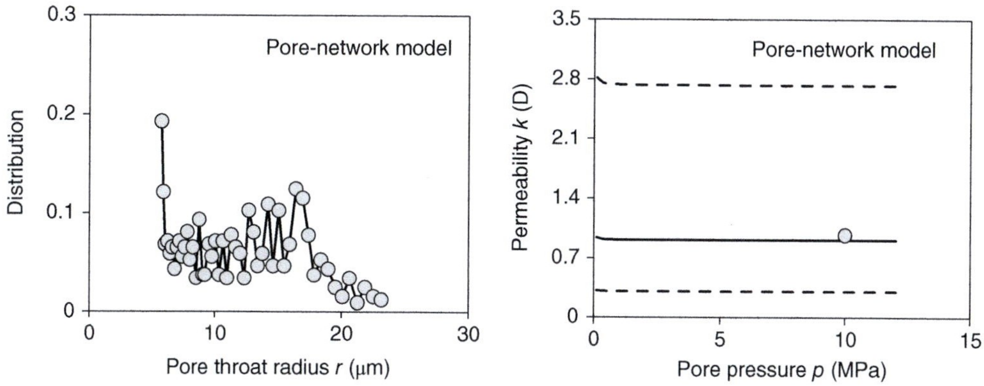

Figure 5.2 The pore throat-size distribution (on the left) and the simulated permeability versus pore pressure (on the right) for the pore-network model of Mehmani et al. (2013). In the right plot, the solid black line denotes the estimated permeability, and the dashed lines represent the factor of three boundaries. *Source:* Data from Mehmani et al. (2013).

agreement between $k = 0.91$ D, the estimated permeability, and $k = 0.97$ D, the simulated one, at $p = 10$ MPa, indicates that Eq. (5.5) accurately upscaled permeability in the three-dimensional network constructed of cylindrical pores whose size varies in a limited range ($5 \leq r < 25$ μm).

5.3.5 Comparaison with Lattice-Boltzmann Simulations

We used Reconstruction 1 from Landry et al. (2016), composed of spheres of radii ranged from 4.5 to 6 nm with a porosity of 0.583. Landry et al. (2016) applied the local-effective-viscosity multi-relaxation-time lattice-Boltzmann method to simulate methane permeability at 400 K temperature and various pore pressures between 5 and 35 MPa (725–5076 psi). The critical saturation $S_c = 0.25$ for percolation was determined from the mercury intrusion porosimetry data simulated by the full morphology model flooding algorithm.

The pore throat-size distribution as well as the simulated permeabilities versus pore pressure are presented in Figure 5.3. The estimated permeabilities via Eq. (5.5) and its factor of three boundaries (denoted by dashed lines) are also shown in Figure 5.3. Results indicate that the estimations are in good agreement with the lattice-Boltzmann simulations, although Eq. (5.5) slightly overestimated k.

Another set of lattice-Boltzmann simulation was selected from Chen et al. (2015), who reconstructed a shale medium using the Markov Chain Monte Carlo method based on SEM images of organic content of a shale sample from Sichuan Basin, China. The pore size in the reconstructed sample was in the range of 5–80 nm with porosity $\phi = 0.191$. Methane permeability at $T = 383$ K was simulated by the lattice-Boltzmann method adapted to model fluid flow and Knudsen diffusion. Chen et al. (2015) applied a connected phase labeling algorithm, checked the local connectivity, and reported a connectivity of 98% for the Sichuan Basin sample indicated $S_c \approx 0.02$.

The pore throat-size distribution of the Sichuan Basin sample as well as the lattice-Boltzmann permeability simulations are shown in Figure 5.4. As can be seen, the estimated permeabilities via Eq. (5.5) agree well with the simulated values. The reported tortuosity value by Chen et al. (2015) is greater than that for other samples listed in Table 5.2. However, $\tau = 12.2$ resulted in accurate estimation of gas permeability (Figure 5.4). This clearly confirms that precise quantification of the pore pressure dependence of gas permeability in shales requires accurate characterization of the pore structure and microscopic topological properties of the pore space.

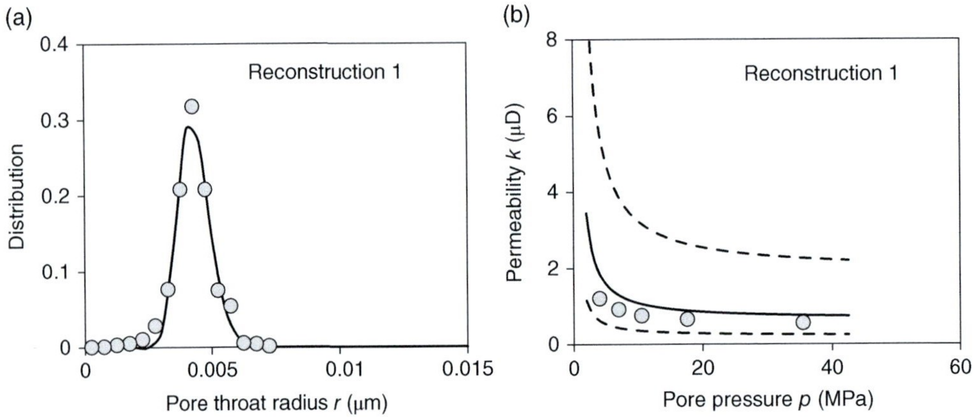

Figure 5.3 (a) The pore throat-size distribution and (b) the LB simulated permeability versus pore pressure for Reconstruction 1 from Landry et al. (2016). In (b), the solid black line denotes the estimated permeability, and the dashed lines represent the factor of three boundaries. *Source:* Data from Landry et al. (2016).

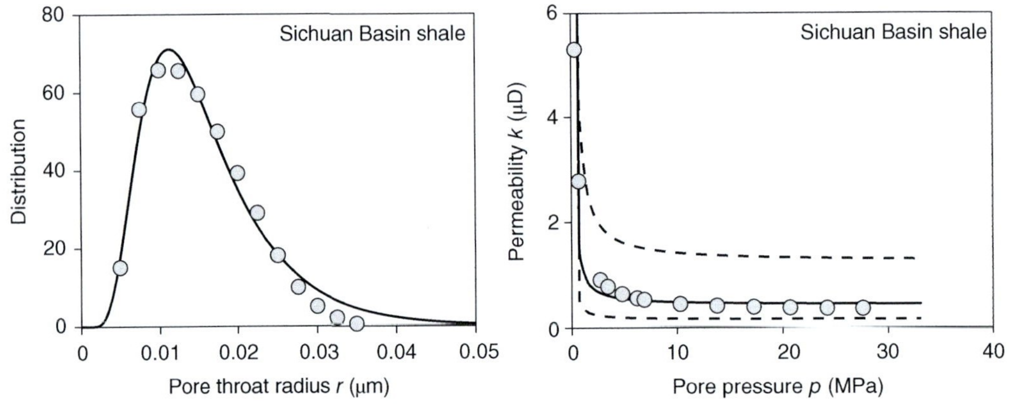

Figure 5.4 The pore throat-size distribution (on the left) and the LB simulated permeability versus pore pressure (on the right) for the Sichuan Basin shale from Chen et al. (2015). In the right plot, the solid black line denotes the estimated permeability, and the dashed lines represent the factor of three boundaries. *Source:* Data from Chen et al. (2015).

5.4 Gas Relative Permeability

5.4.1 Hydraulic Flow in a Cylindrical Pore

Non-slip boundary conditions may not accurately capture the physics of flow, particularly at the nano- and micro-scale levels (Karniadakis et al. 2006). In unconventional reservoir rocks, slippage of gas molecules on surface walls facilitates gas flow, and its effect should be incorporated into the modeling of gas transport. The hydraulic conductance (g_h) of a cylindrical pore of radius R and length l filled with a gas of viscosity μ is described in Eq. (5.8) (Brown et al. 1946; Javadpour 2009):

$$g_h = F\frac{\pi R^4}{8\mu l} \tag{5.8}$$

in which F is the dimensionless coefficient for slippage and given by Eq. (5.4). Agrawal and Prabhu (2008) proposed an empirical relationship to estimate TMAC from Knudsen number K_n. However, it returns negative TMAC values for $K_n > 23.1$. For the sake of simplicity, one may approximately set TMAC = 0.8 (Javadpour 2009; Karniadakis et al. 2006).

Ghanbarian and Javadpour (2017) employed Eq. (5.8) in combination with Knudsen diffusion and the EMA to model single-phase gas permeability in shales and found good match between theory and measurement/simulation.

5.4.2 Molecular Flow in a Cylindrical Pore

When the mean free path of gas molecules is of the same order or greater than the pore diameter, the majority of molecules move along straight trajectories until hitting a wall. Under the range of relatively large Knudsen number conditions, gas molecules move independently of each other, and collisions between the molecules happen rarely. This type of flow, called molecular, is only caused by the kinetic energy of the molecules (Halwidl 2016; Livesey 1998).

The molecular conductance (g_m) in a cylindrical pore of radius R and length l is (Knudsen 1909; Livesey 1998)

$$g_m = \alpha \pi R^2 \sqrt{\frac{R_g T}{2\pi M_m}} \tag{5.9}$$

in which α is

$$\alpha = 1 + y^2 - y\sqrt{1+y^2} - \frac{\left[(2-y^2)\sqrt{1+y^2} + y^3 - 2\right]^2}{4.5y\sqrt{1+y^2} - 4.5\ln\left(y + \sqrt{1+y^2}\right)} \tag{5.10}$$

where $y = l/2R$.

5.4.3 Total Gas Flow in a Cylindrical Pore

Assuming that gas flow is mainly controlled by hydraulic flow and molecular flow (two mechanisms contributing in parallel) in a single cylindrical tube with radius R and length l, the total conductance (g_t) at the pore scale is given by

$$g_t = g_h + g_m = F\frac{\pi R^4}{8\mu l} + \alpha \pi R^2 \sqrt{\frac{R_g T}{2\pi M_m}} \tag{5.11}$$

Although the linear superposition of transport mechanisms i.e. molecular flow and hydraulic flow employed here to determine the total gas conductance is a first-order approximation, it has been previously validated in the literature to be accurate enough (see Xu et al. (2018) and references therein).

5.4.4 Gas Relative Permeability in Tight Rocks

We presume that each pore is occupied by either water (or oil) or gas, and all pores with size r and greater are accessible to gas. If capillary pressure curve is not available, then the gas saturation, S_g, corresponding to pore radius r can be determined by integrating $r^3 f(r)$ between r and r_{max} and normalizing that as follows:

$$\frac{S_g - S_{gr}}{1 - S_{gr}} = \frac{\int_{r}^{r_{max}} r^3 f(r) dr}{\int_{r_{min}}^{r_{max}} r^3 f(r) dr} \tag{5.12}$$

where S_{gr} is the residual gas saturation formed from gas trapped in some subset of the pores in the range $[r_{min}, r_{max}]$.

Under fully saturated conditions ($S_g = 1$), the effective conductance g_e is given by

$$\int_{g_{tmin}}^{g_{tmax}} \frac{g_e(S_g = 1) - g_t}{g_t + \left[\frac{1 - S_{gc}}{S_{gc}}\right] g_e(S_g = 1)} f(g_t) dg_t = 0 \tag{5.13}$$

where g_{tmin} and g_{tmax} are, respectively, the minimum and maximum total conductances in the pore network. Recall that the actual pore size distribution and the total pore conductance distribution are interrelated via $f(g_t)dg_t = f(r)dr$.

Under partially saturated conditions, one has

$$\int_{g_{tmin}}^{g_t} \frac{g_e(S_g) - 0}{0 + \left[\frac{1 - S_{gc}}{S_{gc}}\right] g_e(S_g)} f(g_t) dg_t + \int_{g_t}^{g_{tmax}} \frac{g_e(S_g) - g_t}{g_t + \left[\frac{1 - S_{gc}}{S_{gc}}\right] g_e(S_g)} f(g_t) dg_t = 0 \tag{5.14}$$

$g_e(S_g = 1)$ and $g_e(S_g)$ are computed by numerically solving Eqs. (5.13) and (5.14), respectively.

Since gas permeability is proportional to the effective conductance, one may define k_{rg} as

$$k_{rg} = \frac{k_g(S_g)}{k_g(S_g = 1)} = \frac{g_e(S_g)}{g_e(S_g = 1)} \tag{5.15}$$

Theoretically, one expects k_{rg} to vanish as S_g tends to its critical value, S_{gc}. It is well-known that EMT is not accurate near the critical gas saturation. Hence, one may apply EMT at high to intermediate gas saturations and use the universal power–law scaling from percolation theory at low gas saturations close to S_{gc}. The crossover between the two occurs at some gas saturation S_{gx}.

The universal power–law scaling from percolation theory is

$$k_{rg} = k_0 (S_g - S_{gc})^t, S_g > S_{gc} \tag{5.16}$$

where the scaling exponent $t = 2$ in three dimensions, and k_0 is a numerical prefactor.

The value of k_0 is determined by setting Eqs. (5.15) and (5.16) equal at S_{gx} for which one needs to set the first derivative of the two equations equal. Since Eq. (5.15) is an implicit function of S_g, one should numerically compute the values of k_0 and S_{gx} (Ghanbarian et al. 2020).

It is noted that Eq. (5.16) has been successfully applied to model gas relative permeability over the entire range of gas saturation in conventional porous media (Ghanbarian-Alavijeh and Hunt 2012; Ghanbarian et al. 2015a; Hunt 2005; Hunt et al. 2014). It was also combined with the linear scaling law from EMT by Ghanbarian et al. (2015b, 2018) to model diffusion in porous materials.

To evaluate the accuracy of the proposed theoretic gas relative permeability model, we compare it with experimental data and pore-network simulations summarized in Table 5.3.

5.4.5 Comparison with Experiments

We first present the obtained results for the Bennion and Bachu (2007) dataset. Figure 5.5 displays the pore size distributions and S_g-k_{rg} curves for both Calmar and Colorado shale samples. As shown, pore sizes span near one order of magnitude for Calmar shale, while more than three orders

Table 5.3 Salient properties of the eight cases used in this study.

Case	Reference	Method	Remark	ϕ	S_{gr}	S_{gc}
I	Bennion and Bachu (2007)	Exp	Calmar shale	0.039	0	0^a
II	Bennion and Bachu (2007)	Exp	Colorado shale	0.044	0	0.30
III	Dacy (2010)	Exp	Tight-gas sand	0.175	0	0.08
IV	Yassin et al. (2016)	Exp	Tight-gas siltstone	0.062	0	0.26
V	Nazari Moghaddam and Jamiolahmady (2019b)	Exp	Eagle Ford Shale	0.135	0	0.13
VI	Nazari Moghaddam and Jamiolahmady (2019b)	Exp	Pierre Shale	0.318	0	0.08^a
VII	Song et al. (2018)	PNM	Shale	0.150	0	0.05
VIII	Wang and Sheng (2019)	PNM	Shale	0.075	0	0.43^a

Exp: experiment, PNM: pore-network model, ϕ: porosity (fractional), S_{gr}: residual gas saturation, S_{gc}: critical gas saturation.

[a] S_{gc} was determined from the measured/simulated k_{rg} curve.

Source: Ghanbarian et al. 2020/With permission of ELSEVIER.

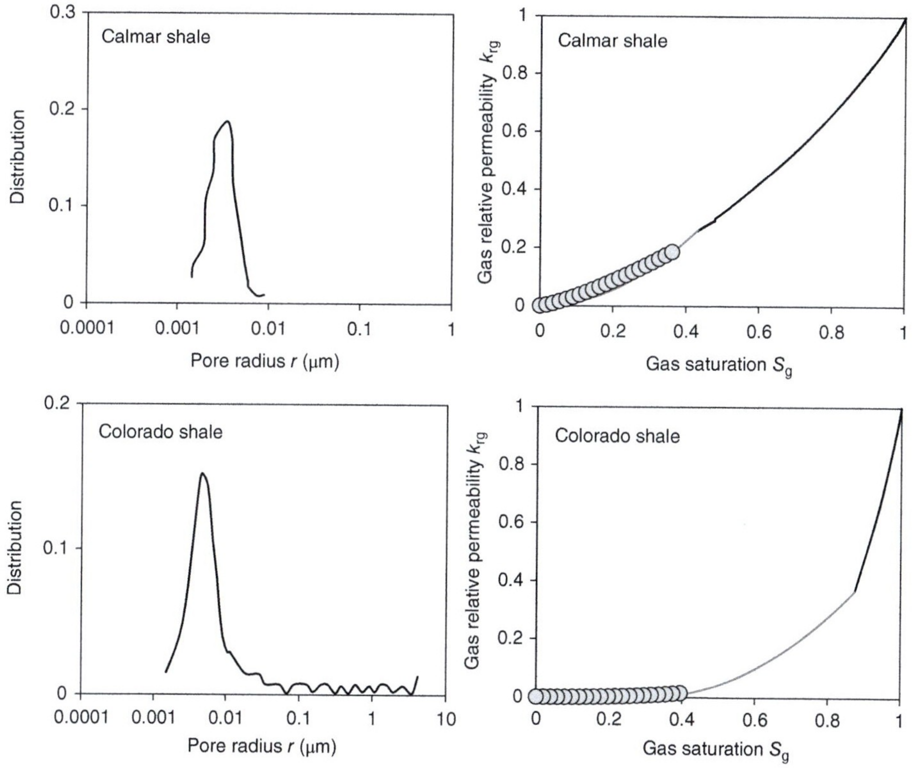

Figure 5.5 The pore size distribution, derived from the measured mercury intrusion porosimetry curves (on the left) and the estimated gas relative permeability curve (on the right) for Calmar and Colorado shales from Bennion and Bachu (2007). The black and gray lines represent the results of effective-medium approximation (Eq. 5.15) and universal power–law scaling from percolation theory (Eq. 5.16), respectively. Unfilled circles represent k_{rg} measured at p = 12.25 MPa (Calmar) and 20 MPa (Colorado). Note that S_{gc} = 0 for Calmar was determined from the actual k_{rg} measurements, while for Colorado, S_{gc} = 0.30 was estimated from the inflection point of the measured MICP curve. *Source:* Data from Bennion and Bachu (2007).

of magnitude for Colorado shale. This means that the latter is more complex in terms of pore-scale heterogeneity than the former. Accordingly, one should expect k_{rg} for Colorado shale to be less than that for Calmar shale, demonstrated in Figure 5.5.

The k_{rg} estimations via the EMA (Eq. 5.15) and power–law scaling from percolation theory (Eq. 5.16) are also given in Fig. 5.5. Results of the EMA and universal power–law scaling from percolation theory are denoted respectively by black and gray lines. From the measured capillary pressure curve reported by Bennion and Bachu (2007) (see their figure 2), we found that the inflection point corresponds to some mercury saturation greater than zero. However, it is clear from the k_{rg} measurements (Figure 5.5) that $S_{gc} = 0$ for Calmar shale. Results (not shown) indicated that $S_{gc} > 0$ would cause k_{rg} underestimation via our model. Although in the following we show that estimating S_{gc} from the inflection point of mercury intrusion porosimetry curve results into accurate prediction of k_{rg}, further investigation is required to address alternative methods for estimating S_{gc} in unconventional reservoir rocks.

Figure 5.5 also shows results for the Colorado shale sample for which we found $S_{gc} = 0.3$, determined from the inflection point of MICP curve. As can be seen, the estimated k_{rg} curve agrees well with the measured data. For Colorado shale, we also found $S_{gx} = 0.87$ – the crossover gas saturation at which the EMA scaling switches to the universal power–law scaling from percolation theory – nearly twice greater than that for Calmar shale ($S_{gx} = 0.43$). The crossover point is non-universal and its value is a function of pore space characteristics such as pore size distribution broadness as well as percolation threshold (or equivalently here, critical gas saturation) (Hunt et al. 2014). This means that one should expect S_{gx} to vary from one rock sample to another.

For the Colorado shale sample with $S_{gc} = 0.3$ (Figure 5.5, bottom right), we found a sharp switch from the percolation theory scaling to the EMA one. As stated earlier, in this case, S_{gx} corresponds to the gas saturation at which the difference in slopes of Eqs. (5.15) and (5.16) is minimal. Ghanbarian et al. (2018) also reported such an abrupt transition for gas relative permeability measurements in sand packs.

Figure 5.6 shows the estimated k_{rg} curves for the tight-gas sand sample from Dacy (2010). As can be observed, pore sizes span two orders of magnitude. Recall that in this dataset, gas permeability

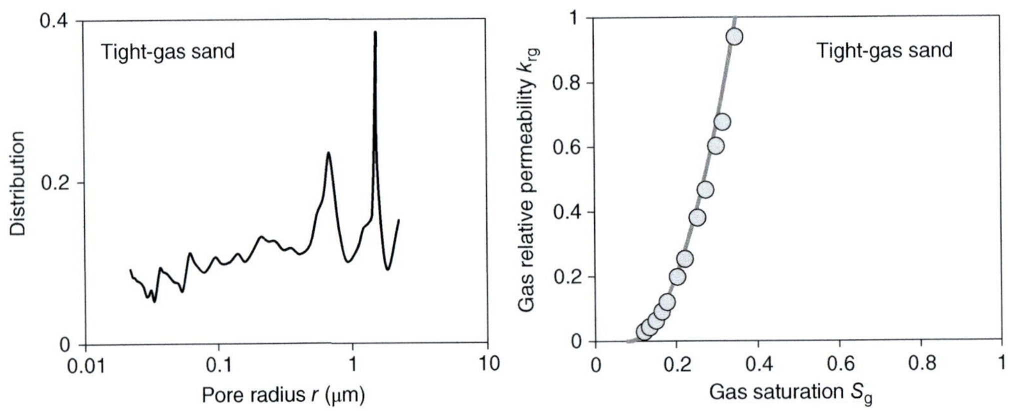

Figure 5.6 The pore size distribution, derived from the air-brine capillary pressure curve measured via centrifuge (on the left) and the estimated gas relative permeability curve (on the right) for tight-gas sand from Dacy (2010). For this dataset, the proposed model reduced to the power–law scaling from percolation theory (Eq. 5.16) represented by the gray line. Unfilled circles represent k_{rg} measured at p = 6.9 MPa. Note that S_{gc} = 0.08 was estimated from the inflection point of the measured MICP curve. *Source:* Data from Dacy (2010).

values were normalized to that measured at $S_g = 0.35$. Our results indicate that the proposed model, Eq. (5.15) in combination with Eq. (5.16), reduced to the universal power–law scaling from percolation theory shown in red in Figure 5.6. We also found that estimating S_{gc} from the mercury saturation corresponding to the inflection point on the MICP curve resulted into accurate estimation of k_{rg} for this dataset.

Results for the tight-gas siltstone from Yassin et al. (2016) are presented in Figure 5.7. Pore radius ranges between 0.0023 and 0.24 μm, spanning near two orders of magnitude, and $r = 0.1$ μm corresponds to the mode of the pore size distribution. Comparison of the estimated k_{rg} curve with the measured one obtained from three Montney samples is also shown in Figure 5.7. We found that our model with $S_{gc} = 0.26$ estimated k_{rg} accurately from the pore size distribution. Yassin et al. (2016) developed a dual-wettability k_{rg} model based on the bundle of capillary tubes approach and, more specifically, the Purcell (1949) model. However, they overestimated k_{rg} and had to replace the tortuosity factor $(1 - S_w)^2$ with $(1 - S_w)^3$ to obtain better match between the measured and estimated k_{rg} (see their figure 9b–d). Interestingly, our model estimated k_{rg} precisely without adjusting/tuning any parameters. We also emphasize that all parameters in our theoretic approach are physically meaningful.

In Figure 5.8, we present the obtained results for the Eagle Ford and Pierre samples from Nazari Moghaddam and Jamiolahmady (2019a, 2019b). Similar to the tight-gas siltstone sample shown in Figure 5.4, the Eagle Ford and Pierre pore size distributions are negatively skewed, indicating the non-trivial contribution of smaller pores in such unconventional reservoir rocks. Yassin et al. (2016) also noticed a long tail at the lower end of the pore size distribution. For Eagle Ford shale, the pore radius ranges between 0.002 and 0.2 μm, spanning two orders of magnitude. However, for Pierre shale, r ranges between 0.002 and 0.08 μm, spanning less than two orders of magnitude.

Figure 5.8 shows the estimated and measured gas relative permeabilities against gas saturation for the Eagle Ford and Pierre shale samples. Similar to the Colorado shale and tight-gas siltstone samples, the crossover gas saturation S_{gx} is around 0.8 ($S_{gx} = 0.79$ and 0.84, respectively, for Eagle Ford and Pierre).

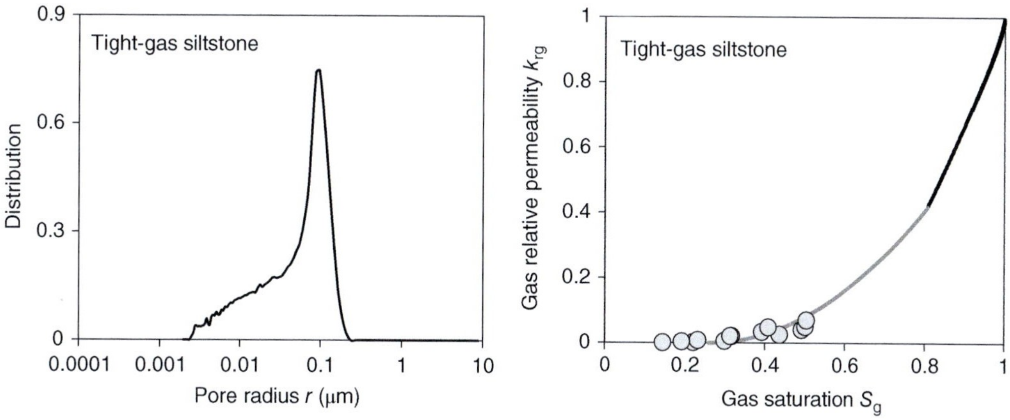

Figure 5.7 The pore size distribution, derived from the measured mercury intrusion porosimetry curve (on the left) and the estimated gas relative permeability curve (on the right) for tight-gas siltstone from Yassin et al. (2016). The black and gray lines in the gas relative permeability plot represent the results of effective-medium approximation (Eq. 5.15) and universal power–law scaling from percolation theory (Eq. 5.16), respectively. Unfilled circles denote k_{rg} measured at $p = 6.9$ MPa. Note that $S_{gc} = 0.26$ was estimated from the inflection point of the measured MICP curve. *Source:* Data from Yassin et al. (2016).

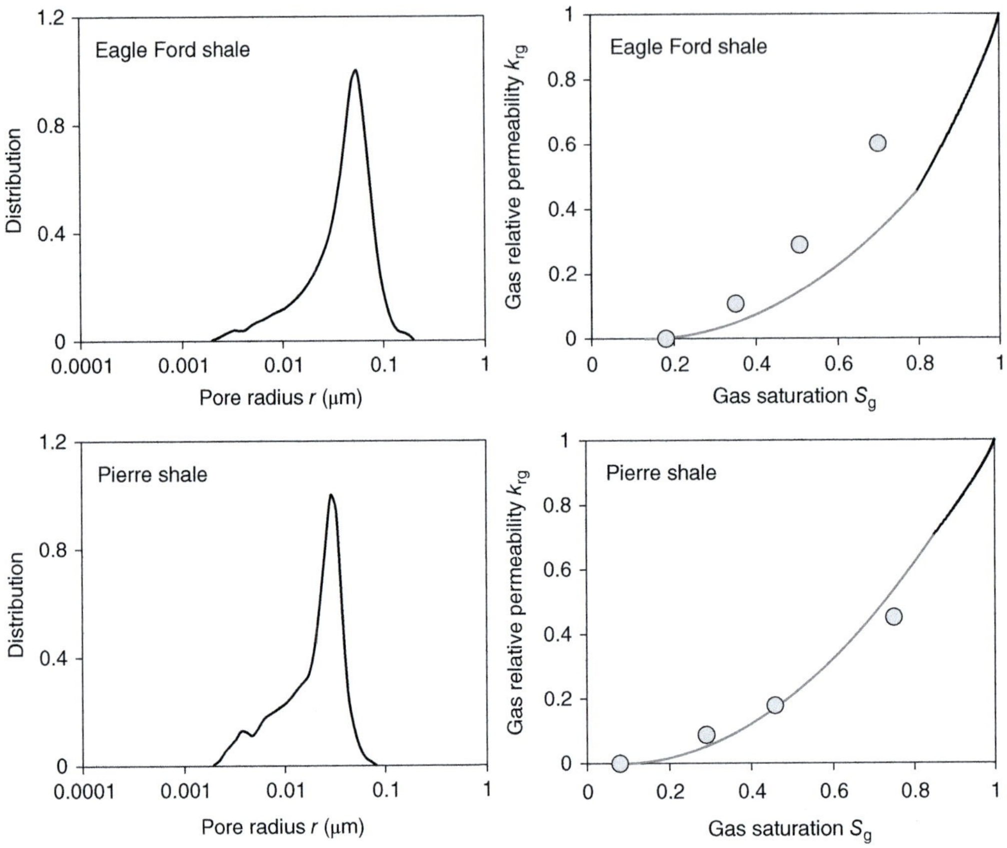

Figure 5.8 The pore size distribution, derived from the measured mercury intrusion porosimetry curve (on the left) and the estimated gas relative permeability curve (on the right) for Eagle Ford and Pierre shale samples from Nazari Moghaddam and Jamiolahmady (2019b). The black and gray lines in the gas relative permeability plots represent the results of effective-medium approximation (Eq. 5.15) and universal power–law scaling from percolation theory (Eq. 5.16), respectively. Unfilled circles denote k_{rg} measured at $p = 10.3$ MPa. Note that $S_{gc} = 0.13$ for Eagle Ford was estimated from the inflection point of the measured MICP curve, while for Pierre $S_{gc} = 0.08$ was determined from the actual k_{rg} measurements. *Source:* Data from Nazari Moghaddam and Jamiolahmady (2019b).

As can be seen in Figure 5.8, for Eagle Ford shale, k_{rg} estimates are relatively accurate at low gas saturations. However, deviations between theory and experiment are observed at intermediate S_g values. Overall, the proposed model, Eq. (5.15) in combination with Eq. (5.16), underestimated k_{rg} for this case. This underestimation is in accord with the findings from Ghanbarian and Javadpour (2017), who estimated single-phase permeability using the EMT on the same sample. Those authors found that the EMT underestimated gas permeability and attributed that to the presence of microfractures in the sample that was not captured via MICP. They stated that "... the critical saturation in the Eagle Ford sample is remarkably less than that in the Pierre and Barnett samples, which might be an indication of microfractures in the Eagle Ford sample, associated with coring processes that are not reflected in the pore throat-size distribution." However, our theoretic model estimated k_{rg} accurately over the entire range of gas saturation for Pierre shale. This means that if S_{gc} can be estimated with high accuracy and the pore size distribution can be determined precisely, and our model would be able to estimate k_{rg} accurately.

In most datasets studied here, k_{rg} was experimentally measured at low to intermediate gas saturations e.g. $S_g < 0.5$. Therefore, model evaluation at high S_g values is still required. We address the predictability of the proposed model, Eq. (5.15) in combination with Eq. (5.16), at high gas saturations using the pore-network simulations in the following section.

5.4.6 Comparison with Pore-Network Simulations

In Figure 5.9, we show the pore size distribution, estimated k_{rg} curve at $p = 25$ MPa, and pore-network simulations from Song et al. (2018). This figure demonstrates that the proposed k_{rg} model estimated gas relative permeability accurately over the entire range of gas saturation. More specifically, we found that $S_{gc} = 0.05$, estimated from the fraction of dead-end pores, is in well agreement with the simulations.

Figure 5.9 also shows that although the pore-network simulations vary by pore pressure, differences in k_{rg} for various pore pressure values are not substantially remarkable, particularly at low and high gas saturations. This is consistent with two-dimensional simulations of Wang and Sheng (2018), who reported similar k_{rg} values for $p = 10$ and 20 MPa (see their figures 7 and 8). Similar results were also theoretically obtained by Zhang et al. (2018). Huang et al. (2016) also found that the relative permeability curves for the boundary pressure difference of 0.05 and 5 MPa were similar. However, they reported that the curves for 5 and 500 MPa cases were different, particularly at low and intermediate gas saturations (see their figure 12). As Huang et al. (2016) stated, this might be because at lower pore pressures gas transport is dominated by capillary forces, while at high p values viscous forces dominate.

Figure 5.10 shows the pore size distribution as well as the estimated and simulated gas relative permeability curves for the pore-network model of Wang and Sheng (2019). Pore size spans near one order of magnitude with a weighted arithmetic mean of 98 nm. This value is very close to that Wang and Sheng (2019) reported (i.e., 110 nm). As can be seen, the EMT, shown in black, estimated gas relative permeability at high gas saturations accurately. The universal power–law scaling from

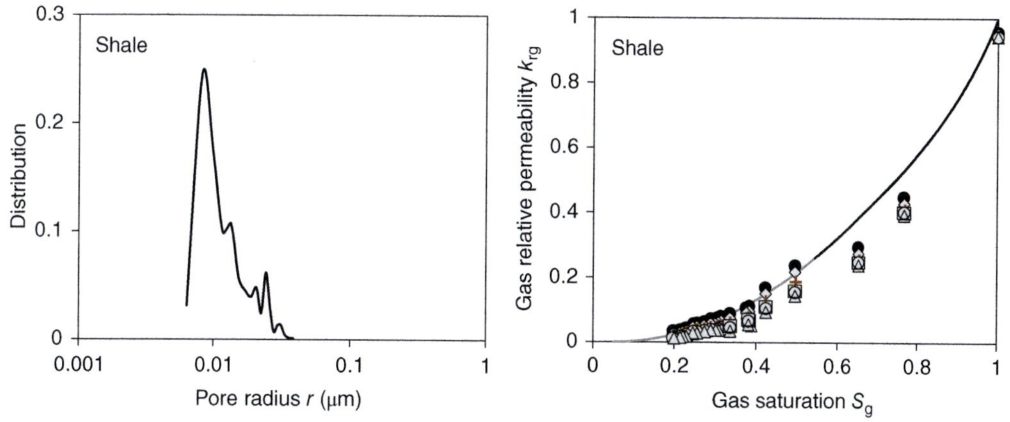

Figure 5.9 The pore size distribution, constructed from two-dimensional SEM images of the organic-rich shale sample (on the left) and the estimated gas relative permeability curve at $p = 25$ MPa (on the right) from Song et al. (2018). The black and gray lines represent the results of effective-medium approximation (Eq. 5.15) and universal power–law scaling from percolation theory (Eq. 5.16), respectively. Symbols denote k_{rg} simulated via the pore-network model at various p values. Note that $S_{gc} = 0.05$ was determined from the fraction of dead-end pores in the network. *Source:* Data from Song et al. (2018).

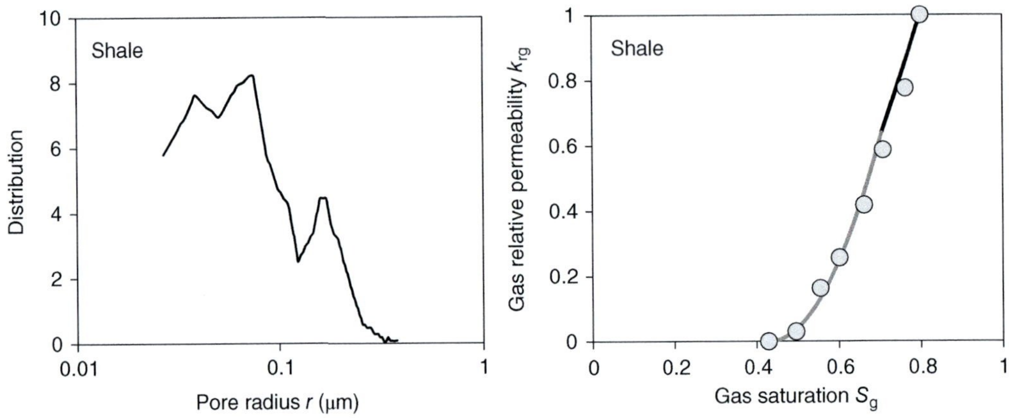

Figure 5.10 The pore size distribution, synthetically generated from a Berea sandstone to mimic the actual pore size distribution in a shale sample (on the left) and the estimated gas relative permeability curve at $p = 6.9$ MPa (on the right) for the pore-network simulations from Wang and Sheng (2019). The black and gray lines in gas-relative permeability plot represent the results of effective-medium approximation (Eq. 5.15) and universal power–law scaling from percolation theory (Eq. 5.16), respectively. Note that $S_{gc} = 0.43$ was determined from the k_{rg} simulations. *Source:* Data from Wang and Sheng (2019).

percolation theory, shown in gray, also estimated k_{rg} at low to intermediate S_g values precisely. We should point out that the value of critical gas saturation for this dataset was not available. We, accordingly, had to determine its value ($S_{gc} = 0.43$) from the simulated k_{rg} curve.

5.5 Conclusions

In this study, we used concepts from the effective-medium approximation to upscale gas permeability k in shales. Given that permeability is mainly controlled by slip flow and Knudsen diffusion, we modified the Doyen (1988) model and proposed Eq. (5.5) to calculate k from pore-size distribution, pore connectivity, tortuosity, porosity, and gas characteristics such as gas type, temperature, molar mass, and viscosity. Comparison with six datasets including three experiments, one pore-network model, and two lattice-Boltzmann simulations showed that the proposed upscaling model estimated gas permeability in shales accurately, within a factor of three of the measurements. We should, however, caution that our model provides accurate results as long as pore throat-size distribution is not too broad. Results also showed that our model is most sensitive to the effective pore radius r_e that was determined using the effective-medium approximation, Eq. (5.3), and tortuosity that was approximated by 2 ($\tau = 2$) in all datasets except in the Chen et al. (2015) dataset in which $\tau = 12.2$ was reported by those authors.

In this study, we applied concepts from the effective-medium approximation and universal power–law scaling from percolation theory to develop a theoretical gas relative permeability k_{rg} model in shales and tight porous rocks. For this purpose, we assumed that gas transport is mainly controlled by two mechanisms contributing in parallel: (i) hydraulic flow and (ii) molecular flow. Our proposed model estimates k_{rg} from pore space properties such as pore size distribution, pore connectivity (reflected in critical gas saturation), and porosity as well as gas characteristics e.g. gas type, viscosity, temperature, and molar mass. To evaluate the proposed theoretic approach, we compared model estimations with six experimental measurements and two pore-network simulations from the literature. Results showed that k_{rg} was satisfactorily estimated for the entire range of gas saturation in shales and tight porous rocks except Case VI, which might have some microfractures.

Acknowledgment

BG is grateful to Kansas State University for its support through the faculty startup fund.

References

Adler, P.M. and Berkowitz, B. (2000). Effective medium analysis of random lattices. *Transport in Porous Media* 40 (2): 145–151. https://doi.org/10.1023/A:1006611011600.

Agrawal, A. and Prabhu, S.V. (2008). Survey on measurement of tangential momentum accommodation coefficient. *Journal of Vacuum Science & Technology A: Vacuum, Surfaces, and Films* 26 (4): 634–645. https://doi.org/10.1116/1.2943641.

Allan, A.M. and Mavko, G. (2013). The effect of adsorption and Knudsen diffusion on the steady-state permeability of microporous rocks Adam M. Allan and Gary Mavko. *Geophysics* 78 (2): D75–D83.

Arkilic, E.B., Schmidt, M.A., and Breuer, K.S. (1997). Gaseous slip flow in long microchannels. *Journal of Microelectromechanical Systems* 6 (2): 167–178. https://doi.org/10.1109/84.585795.

Behrang, A. and Kantzas, A. (2017). A hybrid methodology to predict gas permeability in nanoscale organic materials; a combination of fractal theory, kinetic theory of gases and Boltzmann transport equation. *Fuel* 188: 239–245. https://doi.org/10.1016/j.fuel.2016.10.014.

Behrang, A., Mohammadmoradi, P., Taheri, S., and Kantzas, A. (2016). A theoretical study on the permeability of tight media; effects of slippage and condensation. *Fuel* 181: 610–617. https://doi.org/10.1016/j.fuel.2016.05.048.

Bennion, D.B. and Bachu, S. (2007). Permeability and relative permeability measurements at reservoir conditions for CO2-water systems in ultra low permeability confining caprocks. Paper presented at the EUROPEC/EAGE Conference and Exhibition, London, UK (11 June), SPE 106995-MS. Society of Petroleum Engineers.

Beskok, A. and Karniadakis, G.E. (1999). A model for flows in channels, pipes, and ducts at micro and nano scales. *Microscale Thermophysical Engineering* 3 (1): 43–77. https://doi.org/10.1080/108939599199864.

Brown, G.P., Dinardo, A., Cheng, G.K., and Sherwood, T.K. (1946). The flow of gases in pipes at low pressures. *Journal of Applied Physics* 17 (10): 802–813. https://doi.org/10.1063/1.1707647.

Bruggeman, D.A.G. (1935). Berechnung verschiedener physikalischer Konstanten von heterogenen Substanzen. I. Dielektrizitätskonstanten und Leitfähigkeiten der Mischkörper aus isotropen Substanzen. *Annalen der Physik* 416: 636–664.

Cao, G., Lin, M., Jiang, W. et al. (2017). A 3D coupled model of organic matter and inorganic matrix for calculating the permeability of shale. *Fuel* 204: 129–143. https://doi.org/10.1016/j.fuel.2017.05.052.

Chen, L., Zhang, L., Kang, Q. et al. (2015). Nanoscale simulation of shale transport properties using the lattice Boltzmann method: permeability and diffusivity. *Scientific Reports* 5: 1–8. https://doi.org/10.1038/srep08089.

Civan, F. (2010). Effective correlation of apparent gas permeability in tight porous media. *Transport in Porous Media* 82 (2): 375–384. https://doi.org/10.1007/s11242-009-9432-z.

Dacy, J.M. (2010). Core tests for relative permeability of unconventional gas reservoirs. Paper presented at the SPE Annual Technical Conference and Exhibition, Florence, Italy (19 September), SPE 135427-MS. Society of Petroleum Engineers.

Darabi, H., Ettehad, A., Javadpour, F., and Sepehrnoori, K. (2012). Gas flow in ultra-tight shale strata. *Journal of Fluid Mechanics* 710: 641–658. https://doi.org/10.1017/jfm.2012.424.

Doyen, P.M. (1988). Permeability, conductivity, and pore geometry of sandstone. *Journal of Geophysical Research* 93 (B7): 7729. https://doi.org/10.1029/JB093iB07p07729.

Finkbeiner, T., Zoback, M., Flemings, P., and Stump, B. (2001). Stress, pore pressure, and dynamically constrained hydrocarbon columns in the South Eugene Island 330 field, northern Gulf of Mexico. *AAPG Bulletin* 85 (6): 1007–1031.

Ghanbarian, B. and Daigle, H. (2016). Permeability in two-component porous media: effective-medium approximation compared with lattice-Boltzmann simulations. *Vadose Zone Journal* 15 (2): https://doi.org/10.2136/vzj2015.05.0071.

Ghanbarian, B. and Javadpour, F. (2017). Upscaling pore pressure-dependent gas permeability in shales. *Journal of Geophysical Research, Solid Earth* 122 (4): 2541–2552. https://doi.org/10.1002/2016JB013846.

Ghanbarian, B., Hunt, A.G., Skinner, T.E., and Ewing, R.P. (2015a). Saturation dependence of transport in porous media predicted by percolation and effective medium theories. *Fractals* 23 (1): 1540004. https://doi.org/10.1142/S0218348X15400046.

Ghanbarian, B., Daigle, H., Hunt, A.G. et al. (2015b). Gas and solute diffusion in partially saturated porous media: percolation theory and effective medium approximation compared with lattice Boltzmann simulations. *Journal of Geophysical Research, Solid Earth* 120 (1): 182–190. https://doi.org/10.1002/2014JB011645.

Ghanbarian, B., Sahimi, M., and Daigle, H. (2016). Modeling relative permeability of water in soil: application of effective-medium approximation and percolation theory. *Water Resources Research* 52 (7): 5025–5040. https://doi.org/10.1002/2015WR017903.

Ghanbarian, B., Hamamoto, S., Kawamoto, K. et al. (2018). Saturation-dependent gas transport in sand packs: experiments and theoretical applications. *Advances in Water Resources* 122 (August): 139–147. https://doi.org/10.1016/j.advwatres.2018.10.011.

Ghanbarian, B., Torres-Verdín, C., Lake, L.W., and Marder, M. (2019). Gas permeability in unconventional tight sandstones: scaling up from pore to core. *Journal of Petroleum Science and Engineering* 173 (October 2018): 1163–1172. https://doi.org/10.1016/j.petrol.2018.10.057.

Ghanbarian, B., Liang, F., and Liu, H.H. (2020). Modeling gas relative permeability in shales and tight porous rocks. *Fuel* 272: 117686.

Ghanbarian-Alavijeh, B. and Hunt, A.G. (2012). Comparison of the predictions of universal scaling of the saturation dependence of the air permeability with experiment. *Water Resources Research* 48: W08513.

Gherabati, S.A., Browning, J., Male, F. et al. (2016). The impact of pressure and fluid property variation on well performance of liquid-rich Eagle Ford shale. *Journal of Natural Gas Science and Engineering* 33: 1056–1068. https://doi.org/10.1016/j.jngse.2016.06.019.

Halwidl, D. (2016). *Development of an Effusive Molecular Beam Apparatus*. Springer Spektrum.

Heidemann, R.A., Jeje, A.A., and Mohtadi, F. (1984). *An Introduction to the Properties of Fluids and Solids*. University of Calgary Press.

Huang, X., Bandilla, K.W., and Celia, M.A. (2016). Multi-physics pore-network modeling of two-phase shale matrix flows. *Transport in Porous Media* 111 (1): 123–141. https://doi.org/10.1007/s11242-015-0584-8.

Hunt, A.G. (2005). Continuum percolation theory for saturation dependence of air permeability. *Vadose Zone Journal* 4 (1): 134–138. https://doi.org/10.2136/vzj2005.0134.

Hunt, A., Ewing, R., and Ghanbarian, B. (2014). *Percolation Theory for Flow in Porous Media*. Springer.

Javadpour, F. (2009). Nanopores and apparent permeability of gas flow in mudrocks (shales and siltstone). *Journal of Canadian Petroleum Technology* 48 (8): 16–21.

Jia, B., Tsau, J.-S., and Barati, R. (2019). Review article a review of the current progress of CO_2 injection EOR and carbon storage in shale oil reservoirs. *Fuel* 236 (July 2018): 404–427. https://doi.org/10.1016/j.fuel.2018.08.103.

Karniadakis, G., Beskok, A., and Aluru, N. (2006). *Microflows and Nanoflows: Fundamentals and Simulation*. Springer Science & Business Media.

Knudsen, M. (1909). Die Gesetze der Molekularströmung und der inneren Reibungsströmung der Gase durch Röhren. *Annalen der Physik* 333 (1): 75–130.

Koplik, J. (1981). On the effective medium theory of random linear networks. *Journal of Physics C: Solid State Physics* 14 (32): 4821–4837.

Landauer, R. (1952). The electrical resistance of binary metallic mixtures. *Journal of Applied Physics* 23: 779–784.

Landry, C.J., Prodanović, M., and Eichhubl, P. (2016). Direct simulation of supercritical gas flow in complex nanoporous media and prediction of apparent permeability. *International Journal of Coal Geology* 159: 120–134. https://doi.org/10.1016/j.coal.2016.03.015.

Liu, A., Wang, K., Zang, J. et al. (2018). Relative permeability of gas for unconventional reservoirs. *Transport in Porous Media* 1–19: https://doi.org/10.1007/s11242-018-1064-8.

Livesey, R.G. (1998). Flow of gases through tubes and orifices. In: *Vacuum Science and Technology* (ed. J.M. Lafferty), 728. Wiley.

Ma, J., Sanchez, J.P., Wu, K. et al. (2014). A pore network model for simulating non-ideal gas flow in micro- and nano-porous materials. *Fuel* 116: 498–508. https://doi.org/10.1016/j.fuel.2013.08.041.

Mehmani, A., Prodanović, M., and Javadpour, F. (2013). Multiscale, multiphysics network modeling of shale matrix gas flows. *Transport in Porous Media* 99 (2): 377–390. https://doi.org/10.1007/s11242-013-0191-5.

Naraghi, M.E., Javadpour, F., and Ko, L.T. (2018). An object-based shale permeability model: non-Darcy gas flow, sorption, and surface diffusion effects. *Transport in Porous Media* 1–17. https://doi.org/10.1007/s11242-017-0992-z.

Nazari Moghaddam, R. and Jamiolahmady, M. (2016). Slip flow in porous media. *Fuel* 173: 298–310. https://doi.org/10.1016/j.fuel.2016.01.057.

Nazari Moghaddam, R. and Jamiolahmady, M. (2019a). Impacts of stress and capillary end effects on relative permeability of shale rocks. Paper presented at the SPE Europec Featured at 81st EAGE Conference and Exhibition, London, England, UK (3 June 2019), SPE-195541-MS. Society of Petroleum Engineers.

Nazari Moghaddam, R. and Jamiolahmady, M. (2019b). Steady-state relative permeability measurements of tight and shale rocks considering capillary end effect. *Transport in Porous Media* 128: 75–96.

Purcell, W.R. (1949). Capillary pressures - their measurement using mercury and the calculation of permeability therefrom. *Journal of Petroleum Technology* 1 (2): 39–48. https://doi.org/10.2118/949039-G.

Sahimi, M. (2011). *Flow and Transport in Porous Media and Fractured Rock: From Classical Methods to Modern Approaches*, 2nde. Wiley-VCH.

Sakhaee-Pour, A. and Bryant, S. (2012). Gas permeability of shale. *SPE Reservoir Evaluation & Engineering* 15 (4): 401–409. https://doi.org/10.2118/146944-PA.

Song, W., Yao, J., Li, Y. et al. (2016). Apparent gas permeability in an organic-rich shale reservoir. *Fuel* 181: 973–984. https://doi.org/10.1016/j.fuel.2016.05.011.

Song, W., Yao, J., Ma, J. et al. (2017). Assessing relative contributions of transport mechanisms and real gas properties to gas flow in nanoscale organic pores in shales by pore network modelling. *International Journal of Heat and Mass Transfer* 113: 524–537. https://doi.org/10.1016/j.ijheatmasstransfer.2017.05.109.

Song, W., Yao, J., Ma, J. et al. (2018). Numerical simulation of multiphase flow in Nanoporous organic matter with application to coal and gas shale systems. *Water Resources Research* 54: 1077–1092. https://doi.org/10.1002/2017WR021500.

Wang, X. and Sheng, J.J. (2018). Pore network modeling of the non-Darcy flows in shale and tight formations. *Journal of Petroleum Science and Engineering* 163 (March): 511–518. https://doi.org/10.1016/j.petrol.2018.01.021.

Wang, X. and Sheng, J.J. (2019). Multi-scaled pore network modeling of gas-water flow in shale formations. *Journal of Petroleum Science and Engineering* 177: 899–908. https://doi.org/10.1016/j.petrol.2019.03.005.

Wang, L., Tian, Y., Yu, X. et al. (2017). Advances in improved/enhanced oil recovery technologies for tight and shale reservoirs Energy Information Administration of the United States. *Fuel* 210 (May): 425–445. https://doi.org/10.1016/j.fuel.2017.08.095.

Wu, K., Chen, Z., Li, X. et al. (2016). A model for multiple transport mechanisms through nanopores of shale gas reservoirs with real gas effect-adsorption-mechanic coupling. *International Journal of Heat and Mass Transfer* 93: 408–426. https://doi.org/10.1016/j.ijheatmasstransfer.2015.10.003.

Xu, S., Feng, Q., Wang, S., and Zha, M. (2017). A stochastic permeability model for shale formations based on embedded discrete fracture model. Paper presented at the SPE/AAPG/SEG Unconventional Resources Technology Conference, Austin, Texas, USA (24–26 July 2017).

Xu, J., Wu, K., Li, R. et al. (2018). Real gas transport in shale matrix with fractal structures. *Fuel* 219: 353–363.

Yassin, M.R., Dehghanpour, H., Wood, J., and Lan, Q. (2016). A theory for relative permeability of unconventional rocks with dual-wettability pore network. *SPE Journal* 21 (6): 1970–1980. https://doi.org/10.2118/178549-PA.

Zhang, W.M., Meng, G., and Wei, X. (2012). A review on slip models for gas microflows. *Microfluidics and Nanofluidics* 13 (6): 845–882. https://doi.org/10.1007/s10404-012-1012-9.

Zhang, P., Hu, L., Meegoda, J.N., and Gao, S. (2015a). Micro/Nano-pore network analysis of gas flow in shale matrix. *Scientific Reports* 5 (1): 1–11. https://doi.org/10.1038/srep13501.

Zhang, L., Li, D., Lu, D., and Zhang, T. (2015b). A new formulation of apparent permeability for gas transport in shale. *Journal of Natural Gas Science and Engineering* 23: 221–226. https://doi.org/10.1016/j.jngse.2015.01.042.

Zhang, T., Li, X., Li, J. et al. (2018). A fractal model for gas – water relative permeability in inorganic shale with nanoscale pores. *Transport in Porous Media*.

Zheng, Q. and Yu, B. (2012). A fractal permeability model for gas flow through dual-porosity media. *Journal of Applied Physics* 111 (2): https://doi.org/10.1063/1.3679070.

6

NMR and Its Applications in Tight Unconventional Reservoir Rocks

Jin-Hong Chen[1], Mohammed Boudjatit[2], and Stacey M. Althaus[1]

[1] *Aramco Americas: Aramco Research Center–Houston, Houston, TX, USA*
[2] *Saudi Aramco, Dhahran, Saudi Arabia*

6.1 Introduction

Nuclear magnetic resonance (NMR) is a resonance phenomenon between electromagnetic waves and atomic nuclei. After the discovery of NMR in condensed matter in 1945 (Bloch et al. 1946; Purcell et al. 1946), NMR has been widely used to investigate matter in a multitude of areas, encompassing physics, chemistry, biology, and medicine. NMR was also quickly adapted to petrophysical laboratory research and subsequently developed downhole logging tools for in situ reservoir evaluation (Brown and Gamson 1960; Dunn et al. 2002).

A rigorous discussion of NMR physics is deeply rooted in quantum mechanics, especially when the fundamental of NMR relaxation mechanism is considered (Abragam 1961); however, for most applications of NMR in petroleum industry, classical physics using magnetism vectors suffices. In this chapter, we first give a brief introduction about basic NMR physics, including both quantum and classical mechanical descriptions. We then show a few examples of NMR applications in unconventional source rock reservoirs. NMR applications in formation rocks comprise two major branches: downhole logging and laboratory measurement of retrieved rock samples. Here we do not intend for a comprehensive review of NMR and its applications in measurements of reservoir properties, but to present some recent NMR technology development that extend the application of NMR for the investigation of unconventional source rocks.

6.2 Basic NMR Physics

This section describes the fundamentals most relevant to the application of NMR in formation rocks. The focus is to provide a correct understanding of the basic physics for interested readers. Sections 6.2.1–6.2.3 are the quantum mechanics description of basic NMR physics and can be skipped for the uninterested reader.

Physics of Fluid Flow and Transport in Unconventional Reservoir Rocks, First Edition.
Edited by Behzad Ghanbarian, Feng Liang, and Hui-Hai Liu.

6.2.1 Nuclear Spin

The atomic nucleus consists of protons and neutrons. Each proton and neutron possess a spin 1/2. Each nucleus has a specific *spin quantum number S*, which is determined by the enclosed specific combination of protons and neutrons. Spin is an intrinsic and a purely quantum mechanics property. It has no classical analog and does not originate from particle rotation. Spin is a type of angular momentum, expressed as a vector $\vec{S}$ for a nucleus with *spin quantum number S*. The magnitude of the nuclear spin angular momentum $\vec{S}$ is given by

$$\left|\vec{S}\right| = \hbar\sqrt{S(S+1)}, \tag{6.1}$$

in which $\hbar = 1.054\times10^{-34}$J/S is the reduced Planck's constant. Quantum mechanics restricts such that only one of the three Cartesian components of $\vec{S}$ can be specified simultaneously with $\left|\vec{S}\right|$. Normally, the z-component of $\vec{S}$ is specified

$$S_Z = m\hbar, \tag{6.2}$$

with m taking quantum number, $m = (-S, -S+1, \ldots, S-1, S)$, for a total $2S+1$ states. Thus, $\vec{S}$ is quantized: its magnitude is a constant and z-component takes several discrete values.

For applications in formation rocks, the most important nucleus is hydrogen (^{1}H), whose nucleus contains a single proton and has a total *spin quantum number* $S = 1/2$. Hydrogen is the most sensitive isotope and a major element of both water and hydrocarbon. Other nuclei, such as ^{13}C, ^{23}Na, and ^{31}P can also be used to investigate rock formations but are not generally used in unconventional source rocks. Hereafter, only hydrogen is discussed. For hydrogen, the z-component of angular momentum has two states, $-1/2\hbar$ and $1/2\hbar$.

6.2.2 Nuclear Zeeman Splitting and NMR

The nuclear spin angular momentum $\vec{S}$ possesses a magnetic moment $\vec{\mu}$ which is proportional to the angular momentum

$$\vec{\mu} = \gamma\vec{S} \tag{6.3}$$

where γ is the gyromagnetic ratio, a constant for a specific nucleus. For ^{1}H, $\gamma_H = 4.258\times10^7$ Hz/T. Because the spin angular momentum is quantized, so is the nuclear magnetic moment. For hydrogen, the z-component of spin magnetic moment

$$\mu_z = \begin{cases} -1/2\hbar\gamma \\ 1/2\hbar\gamma \end{cases} \tag{6.4}$$

The two quantum states have the same energy and thus are twofold degenerated. The degeneracy breaks when the nucleus is placed in and interacts with a magnetic field B_0. Usually, the magnetic field B_0 is chosen to along z-axis, and the energy from the interaction is

$$E_m = -\vec{\mu}\cdot\overrightarrow{B_0} = -m\hbar\gamma B_0 \tag{6.5}$$

The splitting between the nuclear spin levels is called nuclear Zeeman splitting, which is the source and subject of all the NMR spectroscopy (Abragam 1961). Considering Eqs. (6.4) and (6.5), the energy difference between the two energy levels for spin 1/2 is

$$\Delta E = \hbar\gamma B_0 \tag{6.6}$$

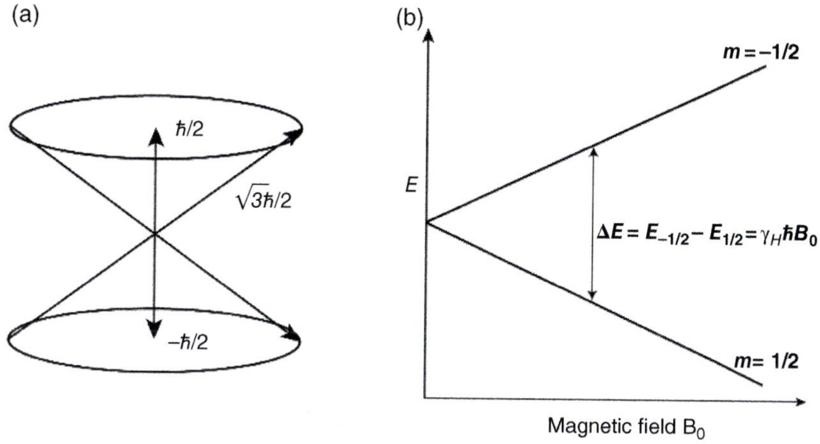

Figure 6.1 (a) The angular momentum vector $\vec{S}$ and the quantized z-components for hydrogen nuclear spin. (b) Illustration of the two corresponding Zeeman energy levels in a magnetic field.

The quantization of hydrogen angular momentum is shown in Figure 6.1a, and the hydrogen nuclear Zeeman levels as a function of strength of the applied magnetic field B_0 is illustrated in Figure 6.1b. Note that Figure 6.1 also shows that the nuclear spin angular momentum is neither parallel nor anti-parallel to the magnetic field.

Transitions between the two energy levels can be induced by applying an electromagnetic wave. The photon energy $\hbar\omega_0$ required to excite a transition must be equal to the difference of the two energy levels, aka, nuclear magnetic resonance (NMR). The frequency of the electromagnetic wave is then required as

$$\omega_0 = \gamma B_0 \tag{6.7}$$

in unit rad/s. This frequency, also called the Larmor frequency, is the frequency at which the nuclear spin angular momentum processes on the cone shown in Figure 6.1a. The frequency of all current NMRs falls into the range of radio frequency (*r.f.*) and the electromagnetic wave applied to excite NMR is often called *r.f.* field.

6.2.3 Nuclear Magnetization

For an ensemble of nuclear spins in macroscopic condensed matter, for example ^{1}H in a water sample, in a static magnetic field, the net magnetic moment $\vec{M}$ is the vector summation of the magnetic moment of individual nuclei $\vec{\mu}$. At equilibrium, the transverse components of different nuclei are uncorrelated and sum to zero. For the z-component in the direction of the B_0, more protons are in the lower energy state than in the higher energy state, thus resulting in a net equilibrium magnetization. According to Boltzmann's fundamental distribution law, the population of the energy levels are proportional to $\exp(-E_\text{m}/k_\text{B}T) = \exp(\gamma\hbar m\ B_0/k_\text{B}T)$ with $k_\text{B} = 1.38 \times 10^{-23}/\text{m}^2/\text{kg}/\text{s}^{-2}/\text{k}^{-1}$ the Boltzmann constant. The net magnetization of a sample containing N_s spins in unit volume (number density) at equilibrium is then (Abragam 1961)

$$M_0 = \frac{N_\text{s}\gamma\hbar \sum\limits_{\text{m}=-\text{S}}^{\text{S}} m\exp\left(-\dfrac{E_\text{m}}{k_\text{b}T}\right)}{\sum\limits_{\text{m}=-\text{S}}^{\text{S}} \exp\left(-\dfrac{E_\text{m}}{k_\text{b}T}\right)} = \frac{N_\text{s}\gamma^2\hbar^2}{4k_\text{B}T}B_0 \tag{6.8}$$

The magnetization in Eq. (6.8) shows that the NMR signal is proportional to the strength of static magnetic field B_0, the number density N_s, and inversely to the temperature T.

6.2.4 Bloch Equations and NMR Relaxation

For noninteracting spin 1/2 nuclei, a vector model based on the Bloch Equations is sufficient to describe the behavior of the macroscopic magnetization for all the NMR experiments (Bloch 1956). A detailed introduction of complete Bloch Equations is beyond the scope of this chapter (interested readers are referred to Abragam (1961)).

Magnetization as a vector in a coordinate frame has three components, M_x, M_y, and M_z. At equilibrium, the magnetization in Eq. (6.8) lies along the applied static magnetic field B_0 or the z-axis. In this notation, the z-component of magnetization $M_z = M_0$ and $M_x = M_y = 0$, as illustrated in Figure 6.2a. M_z is also referred as longitudinal magnetization, M_x and M_y as transverse magnetization. In all current NMR experimental setups, only the transverse components of the magnetization are measured. Therefore, no signal can be detected at equilibrium state.

Applying an *r.f.* pulse to the sample with a frequency equal or close to the Larmor frequency excites the nuclear spins and rotate the macroscopic magnetization away from the z-axis. In general, an *r.f.* pulse with intensity B_1 and length τ is applied along x-axis that rotates the magnetization away from z-axis with an angle $\theta = \gamma B_1 \tau$, as shown in a rotating reference frame of Larmor frequency in Figure 6.2b. In the following discussions, all the transverse magnetizations, or signals are in the rotating reference frame. Since only the transverse component of the magnetization can be detected, the maximum NMR signal is then $M_0 \sin\theta$.

After the excitation by an *r.f.* pulse, the nuclear spins need to re-establish thermal equilibrium, as shown in Figure 6.2a, which is the most stable state. In the vector model, two relaxation time constants were introduced to describe how the excited state returns to equilibrium: longitudinal relaxation time T_1 and transverse relaxation time T_2 (Bloch 1956). T_1 and T_2 are also called spin-lattice relaxation time and spin-spin relaxation time, respectively. Nuclear spin relaxation

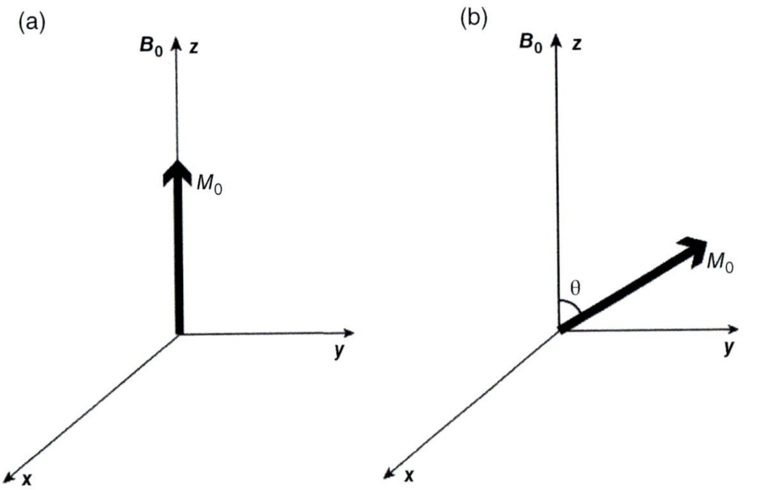

Figure 6.2 Magnetization vector model in one-pulse experiment in a rotating reference frame of Larmor frequency. (a) magnetization at equilibrium along z-axis. (b) an *r.f.* pulse along x-axis rotates the magnetization toward y-axis. In current NMR instrument, only the transverse component of the magnetization can be detected. The maximum signal detected is $M_0 \sin\theta$.

is the result of a fluctuating local magnetic field on the nuclear spin by the neighboring nuclear spins (Bloembergen et al. 1948). The mobility of the molecules has a determinant effect on the relaxation time T_1 and T_2 (Abragam 1961). Nuclear spins in solids have very short T_2 relaxation times and cannot be studied at low magnetic fields. NMR for petrophysical measurement is mainly used to investigate the mobile molecules in liquid and gas phases in formation rocks.

The equations governing the magnetizations as a function of time *t* to establish equilibrium after excitation are

$$M_z = M_0 \left[1 - [1 - \cos\theta] \exp\left[-\frac{t}{T_1}\right]\right] \tag{6.9}$$

$$M_y = M_0 \sin\theta \exp\left(-\frac{t}{T_2}\right) \tag{6.10}$$

The relaxation process ensures that at sufficiently long times, the net magnetization in the *z*-axis grows to M_0 and the transverse components go to zero, i.e. the equilibrium state. In practical applications, *t* is considered sufficiently long for the spin system to re-established equilibrium when $t \geq 5T_1$. In the majority of NMR experiments, the *r.f.* pulse is chosen to such that $\theta = 90°$, for maximized transverse signal and the largest detection sensitivity. This *r.f.* pulse is called a 90°-pulse or $\pi/2$-pulse. Similarly, an *r.f.* pulse that rotates the magnetization by $\theta = 180°$ is called a 180°-pulse or π-pulse.

6.2.5 Simple NMR Experiments: Free Induction Decay and CPMG Echoes

The inhomogeneity of static magnetic field B_0 also induces decay of the transverse magnetization of nuclear spins. As a result, the apparent transverse relaxation time T_2^* is

$$\frac{1}{T_2^*} = \frac{1}{T_2} + \frac{1}{T_2^{B_0\text{inhom}}} \tag{6.11}$$

This inhomogeneity has no effects on the longitudinal relaxation T_1.

After a 90° pulse, the NMR signal is

$$M_y = M_0 \exp\left(-\frac{t}{T_2^*}\right) \tag{6.12}$$

This detected NMR signal is called Free Induction Decay (FID). In pure liquid samples at high magnetic field strength the inhomogeneity of the B_0 is small and therefore $1/T_2^{B_0\text{inhom}}$ is small or negligible in Eq. (6.11). However, for application of low-field NMR to fluids in formation rocks, two sources can generate significant static field homogeneity. First, the applied static field can be inhomogeneous, and this is especially true for NMR downhole logging tools. Second, magnetic susceptibility effects from the rock minerals can generate additional spatial inhomogeneity. This is the main contributor of inhomogeneity in laboratory NMR experiments. In both cases, $1/T_2^{B_0\text{inhom}}$ can become the dominant term in Eq. (6.11).

The most vital NMR parameter is the intrinsic T_2, which contains petrophysical information. A modified spin-echo (Hahn 1950) pulse sequence, called CPMG (Carr and Purcell 1954; Meiboom and Gill 1958), can eliminate the magnetization decay caused by static field inhomogeneity. The CPMG pulse sequence applies a carefully timed series of 180°-pulse after the 90°-pulse excitation of the equilibrium magnetization, as

$$t_w - 90^o_x - \left(\frac{\tau}{2}\right) - 180^o_y - \left(\frac{\tau}{2}\right)_N \tag{6.13}$$

In this sequence, t_w is a waiting time to allow the nuclear spin to establish equilibrium with a net z-magnetization of M_0, and it should satisfy $t_w \geq 5T_1$. The 90°-pulse applied on x-axis rotates the magnetization to y-axis. A duration $\tau/2$ allows the magnetization to decay with the apparent transverse relaxation time T_2^*. However, the $180^o_y - \tau/2$ refocus the decay from the field inhomogeneity and generate a spin-echo (Hahn 1950). Right at the end of the second $\tau/2$, one data point is acquired. At time τ, the magnetization decays with intrinsic T_2. The $\tau/2 - 180^o_y - \tau/2$ is repeated N times and N data points are acquired for the NMR signal. The acquired NMR CPMG echoes are

$$M_y(n\tau) = M_0 \exp\left(-\frac{n\tau}{T_2}\right) \tag{6.14}$$

with $n = 1, 2, \ldots, N$. Almost all the current NMR applications in formation rocks are acquired using CPMG pulse sequence. As a result, the inhomogeneity of the static field can be ignored.

6.2.6 NMR Relaxation of a Pure Fluid in a Rock Pore

NMR transverse relaxation time, T_2, of a fluid in a pore can be expressed as (Brown and Gamson 1960; Kleinberg et al. 1994):

$$\frac{1}{T_2} = \rho_2 \frac{A}{V} + \frac{1}{T_{2b}} + \frac{1}{T_{2D}} \tag{6.15}$$

The three terms on the right in Eq. (6.15) represent three different relaxation mechanisms, which are summed to give the overall transverse relaxation time T_2 at the fast diffusion regime (Brownstein and Tarr 1979). The first term is from surface relaxation in which ρ_2, A, and V are surface relaxivity, surface area, and pore volume, respectively. The second term is the bulk relaxation time of fluid, which comes from the interaction of intramolecular spins. It is important to note that nanoconfined fluids also have additional relaxation mechanisms from intermolecular spin interaction on the bulk relaxation (Chen et al. 2020). However, the second term is generally on the order of magnitude of seconds for non-viscous liquids and high-pressure gas. It is typically two orders of magnitude smaller than the first term in unconventional source rocks and can generally be ignored. The third term is from signal dephasing in a gradient magnetic field. It may be worth noting that the third term is not a traditional relaxation mechanism, which is always the result of a fluctuating interaction to a nuclear spin (Abragam 1961). While this term's contribution can be ignored in a laboratory setup, it is important in downhole logging tools because of the large gradient in the static magnetic field. Consequently, for laboratory studies Eq. (6.15) is simplified to

$$\frac{1}{T_2} = \rho_2 \frac{A}{V} \tag{6.16}$$

In this case, the relaxation is only sensitive to the pore surface-to-volume ratio A/V and pore surface relaxivity ρ_2. For some simple, well-defined pore shapes, Eq. (6.16) can be further expressed as

$$\frac{1}{T_2} = \rho_2 \frac{\alpha}{R} \tag{6.17}$$

where R is the pore size and α is 1, 2, or 3 for slit, cylinder, or spherical pore, respectively. For rocks formed with packed grains of different sorting and shapes that have undergone diagenesis, α is clearly not a constant; but it should not change by an order of magnitude either. Therefore, the measured relaxation time of fluid in a pore is largely determined by the surface relaxivity ρ_2 and pore-body size R. Surface relaxivity ρ_2 depends on the surface mineralogy, especially paramagnetic minerals. If ρ_2 is approximately constant for a homogeneous rock, T_2 responds to the pore-body size, and the NMR measurement reflects the pore-size distribution in a rock.

6.2.7 Measured NMR CPMG Echoes in a Formation Rock

CPMG is the mostly widely used method of NMR application in formation rocks. In a real rock, the pores are nonuniform and have a distribution of different sizes. As a result, the measured NMR signal is a summation of fluids in different pores with different relaxation times. The CPMG echoes in a rock sample are a superposition of all pores and can be expressed as

$$M_y(n\tau) = \sum_i M_0^i \exp\left(-\frac{n\tau}{T_2^i}\right) \tag{6.18}$$

in which M_0^i and T_2^i are the magnetization and transverse relaxation time of the ith pore, respectively. M_0^i is determined by the total amount of fluid in the pore and T_2^i depends on the pore size. The summation in Eq. (6.18) extends to all the pores in a sample. The distribution of the magnetization according to the transverse relaxation time then gives the distribution of fluids in different-sized pores, which is obtained by inversion of the measured NMR signal in Eq. (6.18), discussed in the following section 6.2.8.

6.2.8 Inversion

The task is to obtain the distribution of magnetization according to different transverse relaxation time $M_0^j\left(T_2^j\right)$ or the so-called transverse relaxation time spectrum from the measured NMR CPMG echoes. The acquired NMR data $s(t)$ always contains thermal noise from the electronics of an NMR instrument and can be expressed as

$$s(n\tau) = \sum_i M_0^i \exp\left(-\frac{n\tau}{T_2^i}\right) + \delta(n\tau) \tag{6.19}$$

where $\delta(n\tau)$ represents the thermal noise. Obtaining $M_0^j\left(T_2^j\right)$ from measured data expressed in Eq. (6.19) is a classic inversion problem. Over the years, various inversion approaches have been developed in NMR data processing. A commonly used method is to find a least square solution or a set (K number) of $M_0^j\left(T_2^j\right) = M_0^j \exp\left(-n\tau/T_2^j\right)$ to satisfy the following equation (Provencher 1982)

$$\text{var} \equiv \sum_{n=1}^{N} w_n \left[s[n\tau] - \sum_{j=1}^{K} M_0^j \exp\left[-\frac{n\tau}{T_2^j}\right]\right]^2 = \min \tag{6.20}$$

where N is the total number of acquired data point, w_n is an optional weighting factor that can be assigned, and K is the number of variables. Note that in the calculation, a predefined set of T_2^j values are used. The exponential term in Eq. (6.20) can be formally replaced with x variables defined as

$$x_j(n\tau) \equiv \exp\left(-\frac{n\tau}{T_2^j}\right) \tag{6.21}$$

Then Eq. (6.20) becomes a set of linearized equations as

$$\text{var} \equiv \sum_{n=1}^{N} w_n \left[s(n\tau) - \sum_{j=1}^{K} M_0^j x_j(n\tau) \right]^2 = \min \tag{6.22}$$

Therefore, NMR inversion becomes a classic linear least squares problem and can be routinely done.

6.2.8.1 Regularized Linear Least Squares

To increase the stability of the solution to Eq. (6.22), regularized linear least squares method is used (Vogel 2002)

$$\text{var} \equiv \sum_{n=1}^{N} w_n \left[s(n\tau) - \sum_{j=1}^{K} M_0^j x_j(n\tau) \right]^2 + \alpha^2 \sum_{n=1}^{N} \left[\sum_{j=1}^{K} x_j(n\tau) \right]^2 = \min \tag{6.23}$$

The second term in Eq. (6.23) is called the regularization term. Its strength is determined by regularization factor α.

6.2.8.2 Constrains of the Resulted NMR Spectrum in Inversion

The least squares fitting method can result in many different solutions for the same tolerance in the minimum in Eq. (6.23). Complex constrains based on practical or prior knowledge of the solution can be incorporated into choosing the best possible solution. The normally used constrains are

$$M_0^j \left(T_2^j \right) \geq 0 \tag{6.24}$$

$$M_0^1 \left(T_2^1 \right) = M_0^K \left(T_2^K \right) = 0 \tag{6.25}$$

Figure 6.3 shows an example of an unconventional source rock plug (1 in. diameter and 1 in. length): (a) acquired NMR CPMG echoes using an Oxford 2 MHz spectrometer and (b) the inverted NMR T_2 spectrum. The large fluctuation in the CPMG echoes in Figure 6.3a comes from the thermal noise. Figure 6.3b is a typical NMR T_2 spectrum that plots the fluid content distribution according to the transverse relaxation time T_2.

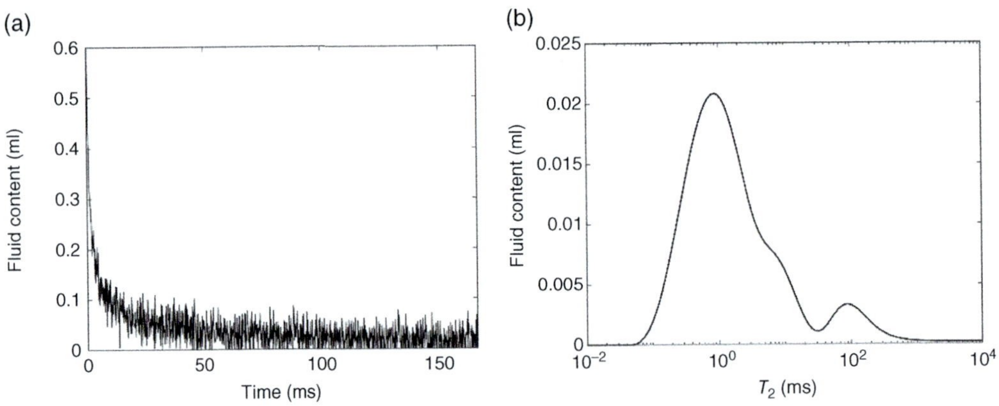

Figure 6.3 (a) acquired CPMG echoes and (b) inverted spectrum of an unconventional source rock sample.

6.2.9 Data from NMR Measurement

The measured NMR signals must be calibrated using a standard sample to determine the absolute amount of fluid in a sample. For most current laboratory or downhole logging NMR instruments, the calibration is done semiautomatically, i.e. data reported from the NMR instruments are calibrated to absolute fluid volume using a standard sample and procedure. For simplicity, here, we do not distinguish the nuclear magnetization and the specific volume of fluid it represents and directly use M_0 as the measured volume in NMR experiments hereafter. In this case, T_2 spectrum $M_0^j\left(T_2^j\right)$ is the NMR-measured fluid distribution according to T_2 and the total amount of fluid in the sample is $V = \sum_{j=1}^{K} M_o^j$. The NMR-measured porosity is then

$$\phi = \sum_{j=1}^{K}\left(\frac{M_o^j}{V_r}\right) \tag{6.26}$$

where V_r is the rock volume under study, which is normally known or can be easily measured from its shape and size.

In addition to porosity, NMR also measures the fluid properties and sizes of the pores containing these fluids through the measured T_2 distribution. From this information, NMR has been used to estimate or infer (Coates et al. 1999; Freedman 2006)

- Volume and distribution of the rock pore space
- Permeability index
- Type and quantity of fluid hydrocarbons
- Rock composition
- Hydrocarbon producibility

6.3 NMR Logging for Unconventional Source Rock Reservoirs

6.3.1 Brief Introduction of Unconventional Source Rocks

Unconventional source rocks are formations where light hydrocarbons were generated and then partially trapped in rock pore system. Hydrocarbon generation depends on the quality and content of organic matter and burial history (Vandenbroucke and Largeau 2007). The organic-rich source rock reservoirs were once considered impossible to produce commercially but now contribute more than 50% of oil and gas production in the United States (U.S.EIA 2016a, 2016b). The rising global importance of hydrocarbon production from source rocks is due to the combination of horizontal drilling techniques and hydraulic fracturing.

The characterization of the pore system in unconventional source rock formations is essential for estimating the original hydrocarbon-in-place and production. Pores have been observed in both organic matter and the inorganic matrix from focused ion beam (FIB) scanning electron microscopy (SEM) (Loucks et al. 2009). Figure 6.4 shows a gray-scale three-dimensional (3D) FIB-SEM of a source rock sample. In the image, the pores, organic matter, and inorganic matrix are identifiable by their distinct colors (Durand et al. 1980). The 3D FIB-SEM illustrates an important characteristic of source rock: the porosity predominantly occurs within the organic matter with very limited matrix porosity in the sample. Organic pores are likely hydrocarbon wet and inorganic pores water wet. The organic porosity is the result of the maturation process as hydrocarbons are being generated and stored (Loucks et al. 2009). The inorganic porosity related to the matrix is mainly controlled by the burial and compaction processes.

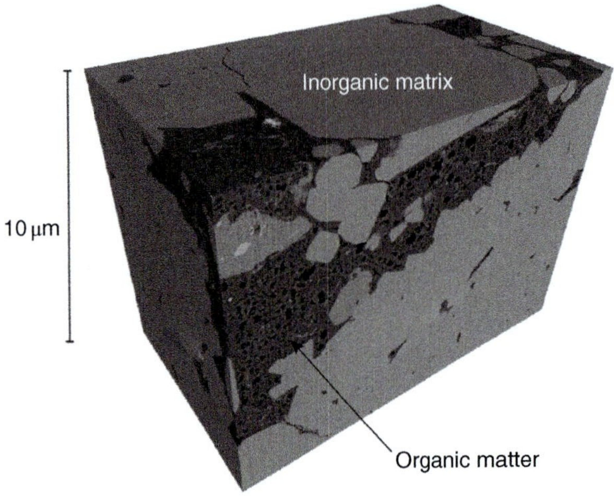

Figure 6.4 FIB-SEM of a representative source rock sample, illustrating the majority of pores (black spots) are in the organic matter (dark gray, labeled in the image as organic matter). The FIB-SEM is a compilation of 1000 two-dimensional images at 5 nm resolution. The resolution per slice is 8 nm. (Image dimensions: 14 × 10 × 8 μm).

Two additional characteristics of source rock pore system can also be observed or deduced from the 3D FIB-SEM. First, the majority of the pores are on the nanometer scale. As a result, adsorption and capillary condensation are important physical mechanisms in estimating the storage and hydrocarbon-in-place (Chen et al. 2012, 2013; Thommes et al. 2015). Second, the pore throats connecting pores are even smaller, which lead to matrix permeability of source rocks on the scale of nano-Darcy. These are also the reason that drilling long horizontal wells and hydraulic fracturing are necessary to produce from unconventional source rock reservoirs.

6.3.2 NMR Measurement of Source Rocks

Many traditional measurement methods developed for conventional reservoirs need to be revisited for application to unconventional source rocks because of the existence of organic matter as part of the rock matrix, multiple pore types (organic pores, inorganic pores, and fractures (Loucks et al. 2012)) with varying pore wettability, and complex fluids. For example, traditionally the hydrocarbon content in a conventional gas reservoir can be estimated from the cross-plot of density and neutron logs (Schön 2015). For unconventional shale gas reservoirs, the presence of organic matter, in addition to the complex mineralogical composition, complicates both logs: the uncertainty in quantity and density of organic matter and other heavy minerals, such as pyrite, makes the density and porosity inaccurate. The large quantity of hydrogen in organic matter and clay-bound water lead to the porosity estimation from neutron log much higher than the real value.

Low-field NMR is lithology independent as it only measures the mobile hydrogen, i.e. oil, gas, and water, but not the solid organic matter. As a result, NMR has become the key technology for evaluating unconventional source rocks including measurement of porosity; identification of fluid types, such as water and/or hydrocarbon; and determination of the confinement environment for fluids such as organic pores or inorganic mineral pores and bound versus mobile fluid volume.

6.3.2.1 NMR Log of a Source Rock Reservoir

NMR logs are obtained by moving an NMR logging tool along the borehole and acquiring NMR data of the surrounding reservoir (Coates et al. 1991, 1999; Chen et al. 2003). A simple type of

NMR log is the transverse relaxation T_2 spectra measured at different depths along a well (Coates et al. 1999; Dunn et al. 2002).

Figure 6.5 shows an example of an NMR log of a source rock section along a well. A T_2 spectrum at some selected depth is shown on the right. The T_2 spectrum is characterized by a bimodal pore system. The NMR T_2 spectrum can be separated into different sections using cutoffs, as illustrated with the dot-dashed line in Figure 6.5. This cutoff was determined based on data from core measurement. The left and right of this cutoff are attributed to the formation of water and light hydrocarbons in the formation, respectively, for this formation.

The integration of the normalized NMR spectrum at each depth of the NMR log gives the formation porosity at depth. Although the studied reservoir produce gas, the phase of the gas in the nanopore system in the formations is in liquid state due to the effect of capillary condensation (Chen et al. 2012, 2013; Barsotti et al. 2016). This makes the normalization of the NMR measurement easier for this reservoir because the hydrogen density of liquid hydrocarbon is similar to that of water and the hydrogen index of fluid in this reservoir is thus close to 1 (Hirasaki et al. 2003). The obtained log results can be compared and validated with laboratory-measured data on retrieved cores from the same well. Unlike logs, which provide continuous data along a well or a section of a well, the core measurements are only done on selected sample at certain depths from the well. Figure 6.6 shows the porosity from NMR log versus the porosity measured from the retrieved cores using Gas Research Institute (GRI) method in the laboratory (Luffel and Guidry 1992). Company confidential requirement does not allow disclosure of the specific porosity data, but in Figure 6.6, the NMR log porosity and the laboratory-measured porosity are distributed quite evenly along the 1 : 1 dash-dot line. This indicates that NMR indeed provides reliable subsurface porosity measurement.

The cutoff method of the NMR T_2 spectra allows further quantification of partial porosity occupied hydrocarbon and water from NMR log. In source rocks, hydrocarbon is largely stored in the pores of organic matter, which can be estimated from 3D FIB-SEM images as shown in

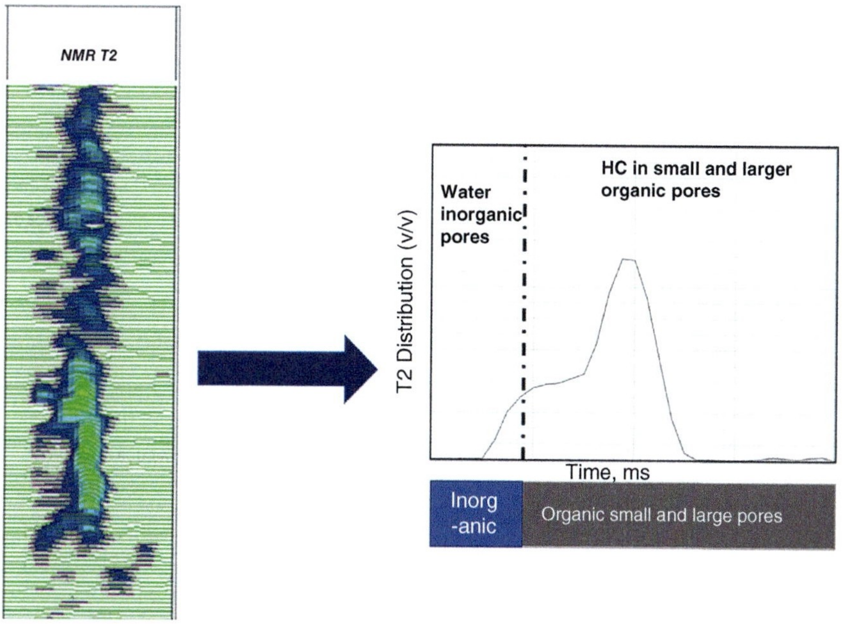

Figure 6.5 NMR log of a source rock section in a well. A T_2 spectrum from a selected depth was shown as an example for detailed view on the information provided from the NMR measurement.

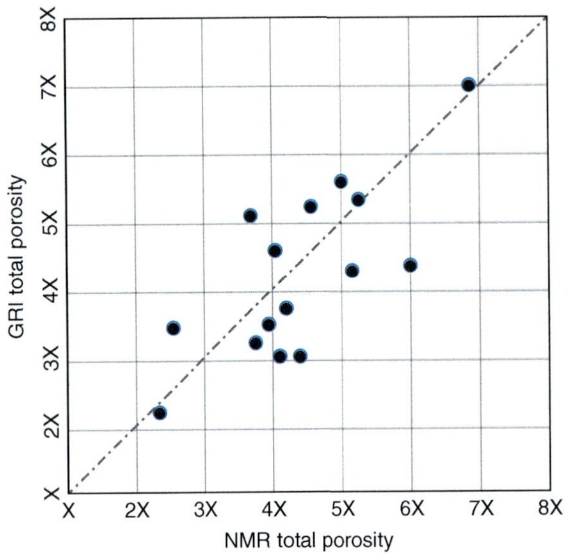

Figure 6.6 GRI total porosity against the NMR log total porosity from the same depth of a well of a source rock reservoir. The dahs-dot line indicates a 1 : 1 line. (Company confidential policy does not allow disclosure of porosity data).

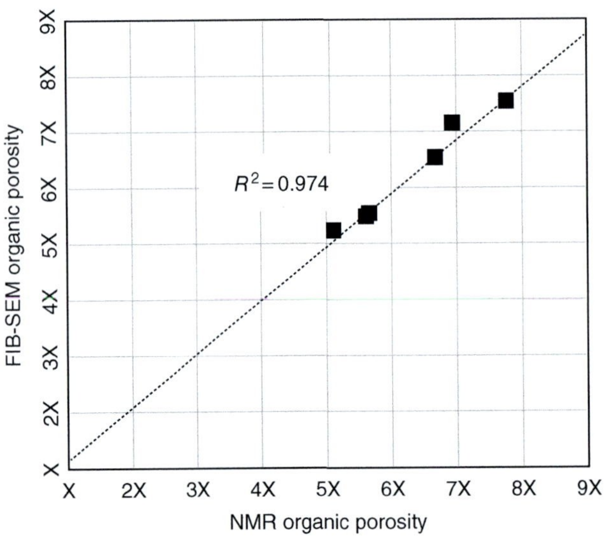

Figure 6.7 Field NMR organic porosity and FIB-SEM organic porosity for a source rock. (Company confidential policy does not allow disclosure of porosity data).

Figure 6.4. In Figure 6.7, the hydrocarbon occupied porosity obtained from the NMR log was plot against the porosity in the organic matter estimated from FIB-SEM at the same depth. The dotted line is a linear fit to the two measurements with a correlation coefficient $R^2 = 0.974$. This indicates that indeed: (i) pores in the organic matter are occupied by hydrocarbon; (ii) cutoff method is a reasonable approach to obtain hydrocarbon saturation from NMR log.

6.3.3 Pore Size Distribution in a Shale Gas Reservoir

Determining the pore size distribution in shale gas source rocks is very useful to understand fluids distribution and flow. As shown in Eq. (6.17), the NMR transverse relaxation T_2 is directly related to the pore size. A geometric mean of the T_2 spectrum expressed by $M_0^j\left(T_2^j\right)$ can be calculated

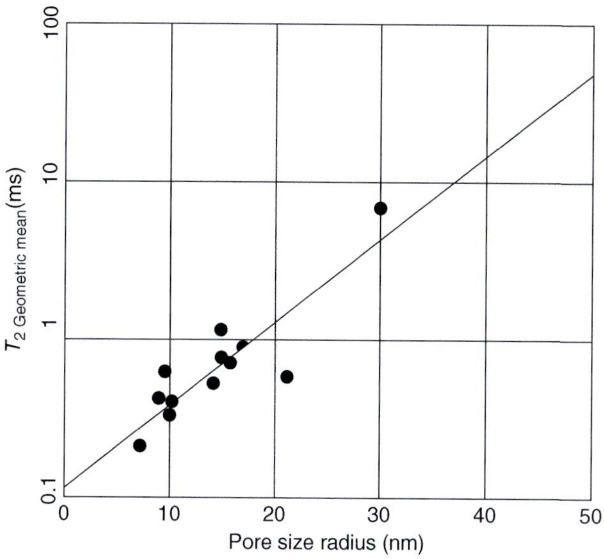

Figure 6.8 The plot of average transverse relaxation T_2 verse the average pore size from FIB-SEM on samples from shale gas reservoir.

$$T_{2\mathrm{gm}} = \left(\prod_{j=1}^{n} T_{2,j}{}^{M_0^j} \right)^{\frac{1}{\sum_{j=1}^{n} M_0^j}} = \exp \left(\frac{\sum_{i=1}^{n} M_0^j \ln T_{2,j}}{\sum_{j=1}^{n} M_0^j} \right) \tag{6.27}$$

where $T_{2,j}$ is the jth bin in the inverted T_2 spectrum of a sample and the weighing factor M_0^j is the amplitude of this bin in the T_2 spectrum. The pore size distribution can also be independently obtained from a sample using FIB-SEM. Figure 6.8 shows a plot of T_2 geometric mean against the average pore radius, which is the mode value of the pore size distribution from FIB-SEM. The distribution is a largely linear relationship, illustrating that the NMR transverse relaxation time is indeed a reflection of the pore size distribution for the studied reservoir.

6.4 NMR Measurement of Long Whole Core

6.4.1 Issues of NMR Instrument for Long Sample

Core retrieval, preservation, and analysis are essential to reservoir development. It provides direct evidence the presence, quantity, distribution, and deliverability of hydrocarbons. The knowledge gained from cores enhances our ability to predict reservoir performance and to select optimal procedures to maximize profitable hydrocarbon recovery. Continuous coring is usually most desirable to obtain complete information of a reservoir section. Cores are collected at the wellsite and returned to the laboratory for measurement. Typically, at wellsite, drilled cores are sectioned into 1 ft length and wrapped in multiple protective layers to preserve the sample fluid in the samples.

The fluids in these whole cores are a prime target for NMR measurement. NMR has been widely used in the laboratory to investigate core samples. However, it has only been used to study small plugs because it has been challenging for NMR to directly acquire data on long whole cores. Here a newly developed technology is described using NMR to study whole core samples (Chen et al.

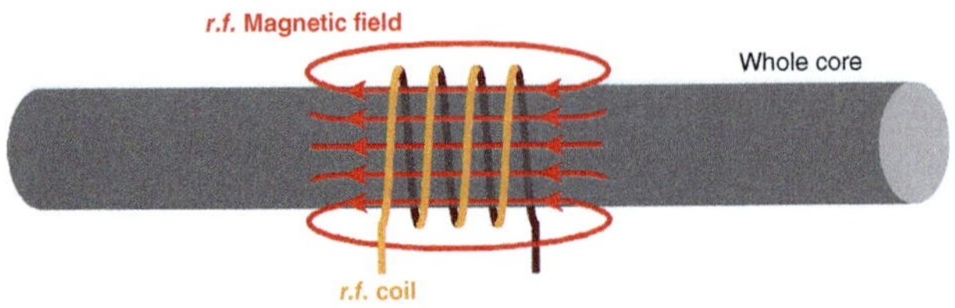

Figure 6.9 Schematic illustration of a long rock sample (gray cylinder) in a sinusoidal *r.f.* coil (yellow wires). The coil generates a magnetic field, relatively homogeneous inside the coil and significantly heterogeneous on the two ends, shown as red vector lines.

2021). The method is especially useful for preserved whole cores from tight unconventional reservoirs.

The length of an *r.f.* coil of a standard low-field NMR instrument usually is about 3 in. In typical NMR measurements of small plugs, the sample is shorter than the coil. The coil is assumed to be infinite and therefore the *r.f.* field strength and measurement sensitivity across the sample is homogenous. The measured NMR data is for the entire sample.

Figure 6.9 illustrates the problems that arise in NMR measurements of long rock samples (typically 1 ft or longer, such as the whole core) in a short helical *r.f.* coil used in a typical NMR instrument. First, the spatial resolution of acquired data is limited by the finite *r.f.* coil length. High-spatial-resolution NMR (HSR-NMR) results are often desired because the reservoir, and therefore the long cores, can be significantly laminated with layers much thinner than 3 in. Second, the end effect of the finite length *r.f.* coil causes inaccuracy of the measurement. For a finite coil, the *r.f.* field is not homogeneous at the two ends of the coil, as illustrated by the red vectors in Figure 6.9. The measured signal now includes signals from sample at the two ends of the *r.f.* coil, in addition to the homogeneous part of the sample within the coil. In this case, the routine NMR experiments cannot provide accurate evaluation of the fluid in the samples.

6.4.2 HSR-NMR of Long Core

A method has been developed to overcome these limits and to achieve HSR-NMR measurement of long cores with a standard *r.f.* coil (Chen et al. 2021). For example, for a core of length L and a desired spatial resolution l, the HSR-NMR of the whole core can be expressed as

$$\boldsymbol{A} = \begin{bmatrix} a_{1,1} & \cdots & a_{1,m} \\ \vdots & \ddots & \vdots \\ a_{k,1} & \cdots & a_{k,m} \end{bmatrix} \tag{6.28}$$

Each row in Eq. (6.28) represents the complete NMR signal, for example the CPMG echoes, of a section with length l of the core. In this method, we also need to determine the response map of the *r.f.* coil $\{r_1, r_2, ..., r_n\}$, which is the relative sensitivity of each section of the coil with the same spatial resolution l. It can be easily measured experimentally (Chen et al. 2021).

In the HSR-NMR method, the data are acquired step wisely by moving the whole core sample through the *r.f.* coil with the step size equal to the desired resolution, as illustrated in Figure 6.10. Specifically, a series of NMR data acquisitions are made, beginning with the top of the core against the start of the sensitive window of the *r.f.* probe. This way, the initial scan should always be a largely noise signal. NMR acquisitions are continued along the length of the core by moving it through the *r.f.* coil at the desired resolution. The acquisition continues until the core completely passes through the sensitive window, once again resulting in a signal which is largely

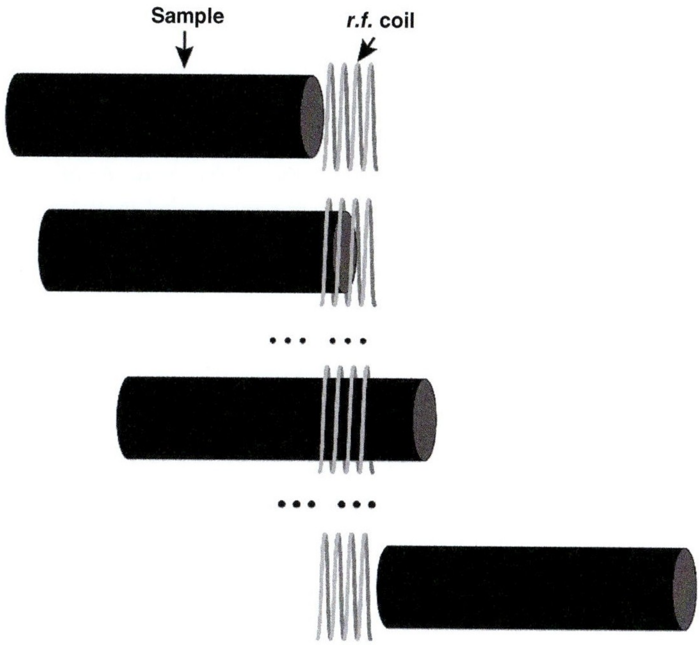

Figure 6.10 Illustration of step-by-step moving the sample through the *r.f.* coil.

noise. At each step, the measured NMR signal is a convolution of the response map of the coil with true NMR signal of the sample.

The measured NMR signal can be written as

$$\boldsymbol{S} = \boldsymbol{R}\boldsymbol{A}, \tag{6.29}$$

where

$$\boldsymbol{S} = \begin{bmatrix} s_{1,1} & \cdots & s_{1,m} \\ \vdots & \ddots & \vdots \\ s_{k+n-1,1} & \cdots & s_{k+n-1,m} \end{bmatrix}, \tag{6.30}$$

and

$$\boldsymbol{R} = \begin{bmatrix} r_1 & & & & & & & & & \\ r_2 & r_1 & & & & & 0 & & & \\ r_3 & r_2 & r_1 & & & & & & & \\ & & & \ddots & & & & & & \\ & & & & r_n & \cdots & r_3 & r_2 & r_1 & \\ & & & & & & & & \ddots & \\ & & & & & & & r_n & r_{n-1} & r_{n-2} \\ & & & 0 & & & & & r_n & r_{n-1} \\ & & & & & & & & & r_n \end{bmatrix}, \tag{6.31}$$

Note that here the spatial resolution is the same for the fluid distribution and for the response map. For optimized measurement, we should make sure: $m \geq k+n-1$. When $m > k+n-1$, the acquired NMR data include situations where the sample is completely outside of the *r.f.* coil and therefore the acquired data only contains noise.

Estimation of NMR fluid distribution $\boldsymbol{A}$ from the measured data $\boldsymbol{S}$ from Eq. (6.2) is an inversion problem. In order to obtain reliable information of the fluid content from the noisy data and mitigate the impacts from the end effects of the measured data, we adopted a robust estimation technique developed in the geophysical exploration industry to suppress the impacts of the measurements with large errors (outliers). This is achieved by iterative reweighted least square (IRLS) approach (Darche 1989; Menke 2018). IRLS estimates the fluid content $\boldsymbol{A}$ from the measured data $\boldsymbol{S}$ by minimizing the following misfit function

$$\min_{A} \frac{1}{p} \|\boldsymbol{RA} - \boldsymbol{S}\|^{\mathrm{p}} \tag{6.32}$$

where p is a predefined constant with values reported in [1, 2]. The impact of $\boldsymbol{S}$ on the solution of Eq. (6.32) varies with the choice of p. When $p = 2$, Eq. (6.32) gives the LS solution, in which case the solution to Eq. (6.32) is simply the inversion of the averaged data, which is taken from all the measurements including those with large errors. When p approaches one, IRLS progressively relies more on the majority of data, which follows the main trend of the fluid content distribution. In the extreme case of $p = 1$, IRLS behaves as a median filter. In this particular case, as long as the number of measurements with very large errors is less than half of the total measurements, the inversion from IRLS is reliable.

One further inversion is then used on the echo trains $a_{i,\ j}(j = 1, \ldots, m)$ to obtain the NMR T_2 spectrum of each section of the sample.

6.4.3 Application Example

An application example is shown here using the method to acquire HSR-NMR data on an unconventional organic-rich whole core, 4 in. in diameter and 12 in. in length, as shown in the photo in Figure 6.11. The whole core was acquired from a carbonate source rock reservoir. The photo clearly shows a significant difference in color. The darker color shows the section having higher TOC while the lighter section with lower TOC.

On a whole core NMR 2 MHz spectrometer from Ecotek Corp., the data were acquired step wisely moving the sample through the *r.f.* coil with 1 in. steps, as illustrated in Figure 6.10. A series of data acquisitions were made, beginning with the top of the standard sample flush against the start of the sensitive window of the *r.f.* coil, continued along the length of the sample by moving it through the *r.f.* coil, 1 in. at a time, until the core completely passes through the sensitive window of the *r.f.* coil. At each step, a train of 1000 echoes was acquired using the CPMG method (Section 6.2.7) with an

Figure 6.11 Photo of a whole core of diameter 4 in and length 12 in.

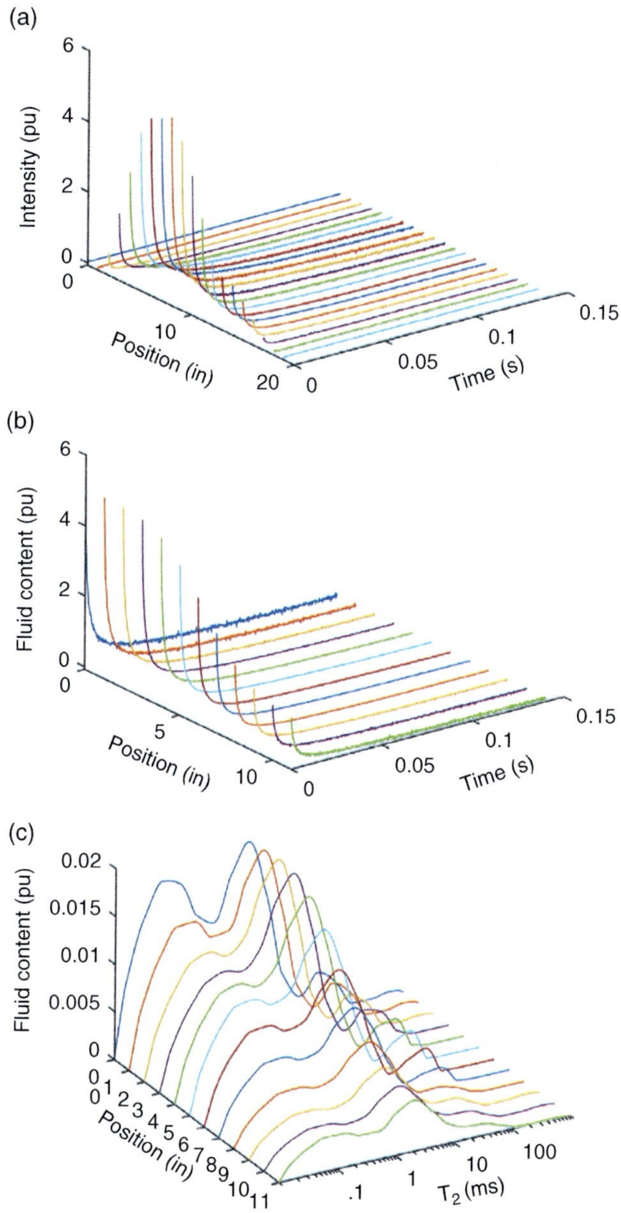

Figure 6.12 (a) Acquired NMR echo trains of a whole cores passing through the *r.f.* coil of whole core NMR spectrometer. The intensity of the signal was normalized according to the signal intensity of 100% water of the same size as the whole core. (b) Inverted CPMG echo trains that are the true NMR measurement of each 1 in. section of the whole core. (c) NMR spectra for the whole core at 1 in. resolution using the HSR-NMR method.

echo time of 140 μs (Carr and Purcell 1954; Meiboom and Gill 1958). The 256 scans were accumulated for each echo train with interscan delay of 0.5 seconds.

A total of 20 echo trains were acquired, as shown in Figure 6.12a. The first two and the last two echo trains were acquired when the cores were outside the *r.f.* coil. The signals were expressed as percent unit (pu) after it was normalized to a same size sample containing 100% water. Each of the

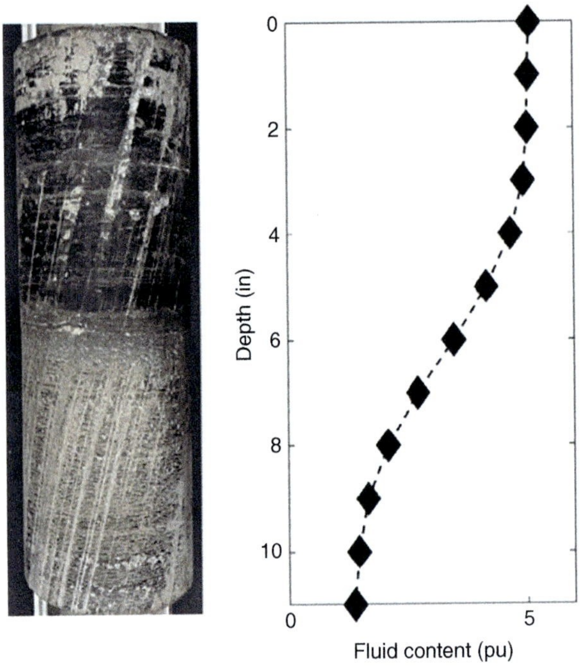

Figure 6.13 Image of the core studied here and the HSR-NMR detected fluid content against the depth.

echo trains in Figure 6.12a is the signal of the whole *r.f.* coil, about 3 in. long with end effects, as discussed in section 6.3.1.

Using the IRLS inversion Eq. (6.32) and the measured response map of the *r.f.* coil of the NMR spectrometer (Figure 3 in ref Chen et al. (2021)), the HSR-NMR echo trains were obtained for each inch of the rock, as shown in Figure 6.12b. Now there are only 12 echo trains. Further inversion of the echo trains using the method shown in Section 6.2.8 generates the NMR T_2 spectra for each inch of the whole core, shown in Figure 6.12c.

The overall total fluid content of each inch of the whole core is the integration of the T_2 spectrum in Figure 6.12c. In Figure 6.13, the measured HSR-NMR fluid content is plotted again the photo of the whole core. It clearly shows that higher TOC has higher total fluid in the sample. Considering that the majority of porosity in source rocks is in organic matter, this result is expected.

6.5 NMR Measurement on Drill Cuttings

Sometimes running an NMR log or collecting cores can be too costly. In this case, drill cuttings can be used to help characterize a reservoir. When drilling a borehole into a reservoir, the drilling bit generates small pieces of rock, which are called drill cuttings. These cuttings flow back to the surface with the drilling mud, where they are separated from the excess mud at the shale shaker and can be collected based on the arrival time and used for petrophysical measurements. The analysis of the cuttings, and the drilling mud, is often referred to as mud logging (Loermans et al. 2011). Mudlogging to determine petrophysical parameters has often been considered a valuable resource, especially in horizontals wells where many traditional logging tools (including NMR) are unavailable. This mudlogging done in conventional reservoir commonly includes measurement techniques such as optical microscopy (Vocke et al. 2018), CT (Siddiqui et al. 2005), gamma-ray, XRD, XRF, porosimetry (Egermann et al. 2002), NMR (Mirotchnik et al. 2004; Kesserwan et al. 2017; Mitchell et al. 2019), or a combination of these (Loermans et al. 2011). A few studies have also been done in unconventional reservoir to study cuttings including infrared Fourier transform (Craddock et al. 2018) and NMR (Fellah et al. 2018; Green et al. 2018; Althaus et al. 2020).

Since drill cuttings are produced regardless of direction, type of mud, or drill bit type, they provide a more inexpensive method to analyze peterophysical properties of shales compared to logs or traditional coring. This makes them valuable, especially in well-drilled after the initial exploration phase. Many measurements can also be done in trailers at the wellsite, making the data near real-time. Since NMR can often be collected quickly and with little sample shape constraints, it lends itself well to cuttings analysis. Herein we focus on an NMR method used to study drill cutting in source rock reservoirs.

6.5.1 Measurement Method

6.5.1.1 Preparation of Drill Cuttings

When drill cuttings are removed from downhole, they are mixed initially with drilling mud. The excess of drilling mud is removed at the shale shaker before the cuttings are collected and stored in jars and/or bags based on the arrival time. This arrival time is then used to calculate a depth.

The cuttings are then typically sent to the lab for analysis (or measurements are performed at an onsite lab). The cuttings still have additional mud and cavings, pieces of rock chipped away from the drill bit, which should be removed for accurate measurements. The preparation depends on the type of mud used in drilling. The cleaning fluid is based on solubility where diesel or water would be used for oil and water-based muds, respectively. This fluid is then used to soak the cutting to solvate the mud particles, and the mixture is wet-sieved to separate mud, cavings, and clean cuttings. The type of drill bit does determine the size of the cuttings pieces, but for a PDC bit (currently the most commonly used) drill cuttings appear to be between the range of 0.5–3 mm, where anything smaller is most likely mud particles and anything larger may be cavings. Althaus et al. reported that using sonication and shaking was the most effective method at sieve separating different particles (Althaus et al. 2020).

After cleaning samples, the cuttings need to be saturated with the fluid of choice. This is sometimes the cleaning fluid, such as diesel or water, but other commonly used saturation fluids such as dodecane (Fellah et al. 2018) or isopar can be used. It is important to note that after saturation, fluid will also remain on the outside of the cuttings.

6.5.1.2 Measurements

The challenge of NMR studies on cuttings samples is to separate the fluid in the sample from the fluid surrounding it. This can be accomplished in different ways. In previous papers, the cuttings and fluid volumes were determined through an NMR Archimedes technique (Althaus et al. 2020). In this method, the cuttings are measured on the NMR using CPMG (section 6.2.7) to determine the T_2 and volume. A cutoff was chosen to determine which fluids were inside the cuttings (V_ϕ) and which were outside of the sample (V_l). The sample is weighed in air (m_air) and then submerged in saturation fluid and weighed (m_f), see Figure 6.14. These results are combined to determine the volumes, porosity, and densities.

Figure 6.14 Picture of a sample being weighed in the air on the left and the sample being weighed well immersed in fluid on the right -drill cuttings in a Teflon bottle. These weights are used to determine density and porosity.

The bulk density can be calculated by

$$\rho_c = \frac{m_{air} - V_l\rho_l}{m_{air} - m_f - V_l\rho_l}\rho_l \tag{6.33}$$

where ρ_l is the density of the liquid.

The grain density can be calculated as

$$\rho_m = \frac{m_{air} - (V_\varphi + V_l)\rho_l}{(m_{air} - m_f - V_l\rho_l)/\rho_l - V_\phi} \tag{6.34}$$

The porosity can be calculated as

$$\phi = \frac{V_\phi}{(m_{air} - m_f - V_l\rho_l)/\rho_l} \tag{6.35}$$

6.5.2 Results

Example results from the NMR-Archimedes technique is discussed here on cleaned drill cutting and crushed imitation cuttings. Drill cuttings were collected at the wellsite from the shale-shakers of an organic-rich source rock section of a well. The imitation samples were obtained by crushing cores collected from the same source rock reservoir. The crushed imitation sample allows comparison with industry-standard GRI measurement.

Real cuttings samples were split into two batches to determine the precision of the measurement method based on the reproducibility since there is no other data available to verify the results. Representative T_2 NMR spectra from the cleaned cuttings saturated with diesel are shown in Figure 6.15. The spectra were then analyzed and for the particular well a cutoff of 44 ms was

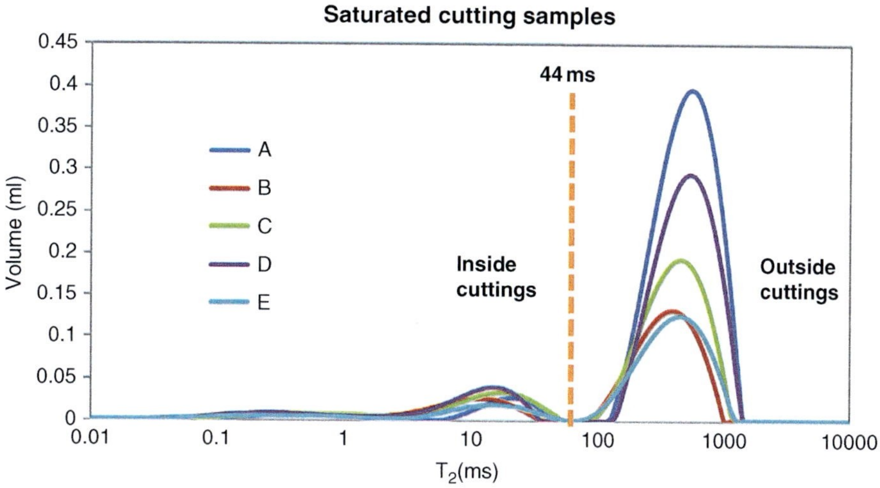

Figure 6.15 Example T_2 spectra of five drill cutting samples after saturation with diesel. The spectrum was collected on a 12 MHz Oxford NMR using CPMG pulse sequence with typical parameters of: echo time $2\tau = 110\ \mu s$, 22,727 echoes, pulse lengths for $\pi/2$ and π of 9.8 μs and 19.8 μs, respectively, recycle delay of 3.75 second, and 32 scans. A cutoff of 44 ms was used to separate the fluids inside the cuttings from outside.

determined. The amount of fluid inside the cuttings V_ϕ and outside of the cuttings was measured V_l. These values were then used to calculate the petrophysical properties as described in 6.5.1.2.

Figure 6.16 shows a comparison of the data from the two sample batches for bulk density, matrix density, and porosity. The density results are in good agreement with less than 5% difference in all samples. The porosity has slightly more variation, however, most of the samples are within 1 pu difference. This variation in porosity may be due to the deformation caused by the drilling bit. Overall, NMR measurements of cutting have been found to provide reliable data on density and porosity in unconventional samples (Althaus et al. 2020).

The artificial cuttings were compared to GRI measurements on crushed cores taken from the same reservoir as the dill cuttings. The bulk and matrix density measurements are compared to GRI 1 : 1 line in Figure 6.17 on the left. The density measured by NMR is in good agreement with those done by GRI. On the right in Figure 6.17, the NMR porosity is plotted versus GRI results. Again, we see fairly good agreement.

6.6 Conclusions

NMR has become a key technology for evaluating unconventional source rock reservoirs because it directly measures the mobile hydrogen in samples and is largely independent of complex rock lithology presented in source rocks. An introduction was given to the NMR basics physics and its measurement concept on samples of formation rocks. Some essential petrophysical parameters of formation rocks from NMR measurements includes porosity, volume and the distribution of rock pore space, permeability index, type and quantity of fluid hydrocarbons, and hydrocarbon producibility.

A downhole NMR logging example was given. The porosity, fractional porosity occupied by hydrocarbon and water, and capability of NMR log data in estimating pore size distribution was shown. The log results were validated with retrieved cores.

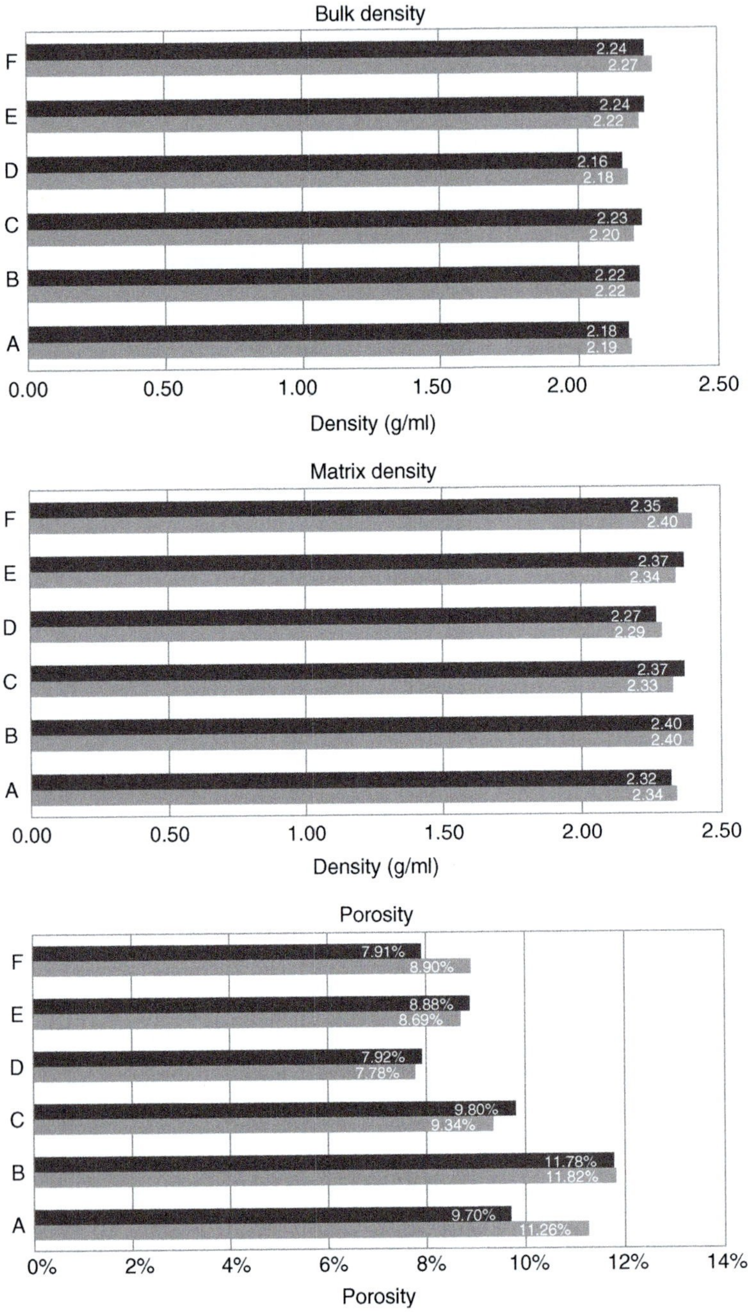

Figure 6.16 Illustrating measurement method reproducibility: bulk density, matrix density, and porosity of two batches of cleaned drill cuttings samples.

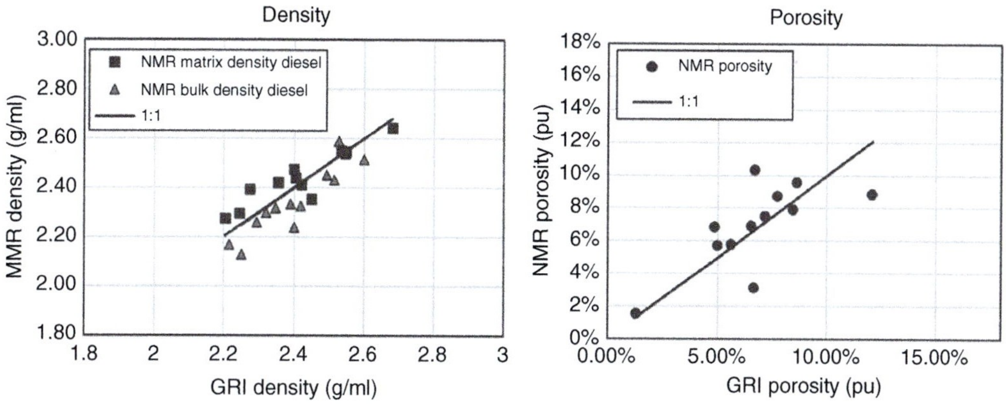

Figure 6.17 Cross plots of artificial cuttings measured using GRI and the NMR Archimedes technique. The left plot shows the data for matrix and bulk density. The right plot compares the porosity measurements.

The HSR-NMR method to analyze the fluids in large source rock samples was presented. The method overcame challenges generated by the finite length of the *r.f.* detection coil and obtained HSR-NMR data with spatial resolution much smaller than the length of the coil. The measured results on preserved whole cores demonstrated that the HSR-NMR measurement can provide critical information about the heterogeneity or lamination of the reservoir.

A method was developed to measure porosity and density from drill cuttings for unconventional source rock reservoirs. Results were shown to be highly reproducible across a set of samples from one well. This measurement can be done quickly, with a run time per sample of approximately two minutes and thus can be easily adopted to wellsite measurement. This allows for real-time management of the well. Once the cuttings have been measured via the NMR Archimedes analysis, they can be cleaned with water and surfactants in a similar manner to the diesel cleaning outlined here. The cleaned cuttings can then be further used for other petrophysical, geochemical, and imaging analyses.

References

Abragam, A. (1961). *The Principles of Nuclear Magnetism*. London: Clarendon Press.

Althaus, S., Chen, J.-H., Zhang, J., and Eppler, G. (2020). Low-field nuclear magnetic resonance methodology for analysis of drill cuttings from unconventional tight reservoirs. *Energy & Fuels* 34: 15990–15994.

Barsotti, E., Tan, S.P., Saraji, S. et al. (2016). A review on capillary condensation in nanoporous media: implications for hydrocarbon recovery from tight reservoirs. *Fuel* 184: 344–361.

Bloch, F. (1956). Dynamical theory of nuclear induction. II. *Physical Review* 102: 104–135.

Bloch, F., Hansen, W.W., and Packard, M. (1946). Nuclear induction. *Physical Review* 69: 127.

Bloembergen, N., Purcell, E.M., and Pound, R.V. (1948). Relaxation effects in nuclear magnetic resonance absorption. *Physical Review* 73: 679–712.

Brown, R.J.S. and Gamson, B.W. (1960). *Nuclear Magnetism Logging*. Society of Petroleum Engineers.

Brownstein, K.R. and Tarr, C.E. (1979). Importance of classical diffusion in NMR studies of water in biological cells. *Physical Review A* 19: 2446–2453.

Carr, H.Y. and Purcell, E.M. (1954). Effects of diffusion on free precession in nuclear magnetic resonance experiments. *Physical Review* 94: 630–638.

Chen, J.-H., Haghmoradi, A., and Althaus, S.M. (2020). NMR intermolecular dipolar cross-relaxation in nanoconfined fluids. *The Journal of Physical Chemistry B* 124: 10237–10244.

Chen, J.-H., Mehmani, A., Li, B. et al. (2013). *Estimation of Total Hydrocarbon in the Presence of Capillary Condensation for Unconventional Shale Reservoirs*. Society of Petroleum Engineers.

Chen, J.-H., Zhang, H., Althaus, S.M., and Boudjatit, M. (2021). High-spatial-resolution nuclear-magnetic-resonance method for investigation of fluid distribution in whole cores. *Fuel* 297: 120777.

Chen, J.-H., Zhang, J., Jin, G. et al. (2012). *Capillary Condensation and NMR Relaxation Time In Unconventional Shale Hydrocarbon Resources*. Society of Petrophysicists and Well-Log Analysts.

Chen, S., Beard, D., Gillen, M. et al. (2003). MR explorer log acquisition methods: petrophysical-objective-oriented approaches. *SPWLA 44th Annual Logging Symposium*. SPWLA-2003-ZZ.

Coates, G.R., Miller, M., Gillen, M., and Henderson, C. (1991). *The MRIL In Conoco 33-1 An Investigation of a New Magnetic Resonance Imaging Log*. Society of Petrophysicists and Well-Log Analysts.

Coates, G.R., Xiao, L., and Prammer, M.G. (1999). *NMR Logging Principles and Applications*. Elsevier Science.

Craddock, P.R., Mossé, L., Prioul, R. et al. (2018). Integrating measured Kerogen properties with log analysis for petrophysics and geomechanics in unconventional resources. *SPWLA 59th Annual Logging Symposium*, London, UK: Society of Petrophysicists and Well-Log Analysts (4–6 June 2018).

Darche, G. (1989). Iterative l1 deconvolution. *SEP-Report*, 281–301.

Dunn, K.J., Bergman, D.J., and Latorraca, G.A. (2002). *Nuclear Magnetic Resonance: Petrophysical and Logging Applications*. Elsevier Science.

Durand, B., Nicaise, G., and Alpern, B. (1980). *Kerogen: Insoluble Organic Matter from Sedimentary Rocks*. Editions Technip.

Egermann, P., Lenormand, R., Longeron, D., and Zarcone, C. (2002). A fast and direct method of permeability measurements on drill cuttings. *SPE Annual Technical Conference and Exhibition*, San Antonio, Texas: Society of Petroleum Engineers (29 September–2 October 2002).

Fellah, K., Utsuzawa, S., Song, Y.-Q., and Kausik, R. (2018). Porosity of drill-cuttings using multinuclear 19F and 1H NMR measurements. *Energy & Fuels* 32: 7467–7470.

Freedman, R. (2006). Advances in NMR logging. *Journal of Petroleum Technology* 58: 60–66.

Green, D.P., Dick, M.J., Kenney, T., and Veselinovic, D. (2018). *Methods of nuclear magnetic resonance measurement of crushed porous media*. U.S. Patent Application 16/608,110.

Hahn, E.L. (1950). Spin echoes. *Physical Review* 80: 580–594.

Hirasaki, G.J., Lo, S.W., and Zhang, Y. (2003). NMR properties of petroleum reservoir fluids. *Magnetic Resonance Imaging* 21 (3–4): 269–277.

Kesserwan, H., Alshanqaiti, E., Jin, G., and Nuic, I. (2017). Porosity measurements on drill cuttings – comprehensive inputs to formation evaluation near real-time while drilling. *Abu Dhabi International Petroleum Exhibition & Conference* (13–16 November 2017), Abu Dhabi, UAE: Society of Petroleum Engineers.

Kleinberg, R.L., Kenyon, W.E., and Mitra, P.P. (1994). Mechanism of NMR relaxation of fluids in rock. *Journal of Magnetic Resonance, Series A* 108: 206–214.

Loermans, T., Bradford, C.M., Kimour, F. et al. 2011. Advanced Mud Logging (AML) aids formation evaluation and drilling, and yields precise hydrocarbon fluid composition. *SPE Middle East Oil and Gas Show and Conference* (25–28 September 2011), Manama, Bahrain: Society of Petroleum Engineers.

Loucks, R.G., Reed, R.M., Ruppel, S.C., and Hammes, U. (2012). Spectrum of pore types and networks in mudrocks and a descriptive classification for matrix-related mudrock pores. *AAPG Bulletin* 96: 1071–1098.

Loucks, R.G., Reed, R.M., Ruppel, S.C., and Jarvie, D.M. (2009). Morphology, genesis, and distribution of nanometer-scale pores in siliceous mudstones of the Mississippian Barnett Shale. *Journal of Sedimentary Research* 79: 848–861.

Luffel, D.L. and Guidry, F.K. (1992). New core analysis methods for measuring reservoir rock properties of devonian shale. Journal of Petroleum Technology 44: 1184–1190. https://doi.org/10.2118/20571-PA

Meiboom, S. and Gill, D. (1958). Modified spin-echo method for measuring nuclear relaxation times. *Review of Scientific Instruments* 29: 688–691.

Menke, W. (2018). *Geophysical Data Analysis: Discrete Inverse Theory*. Academic Press.

Mirotchnik, K., Kryuchkov, S., and Strack, K. (2004). A novel method to determine NMR petrophysical parameters from drill cuttings. *SPWLA 45th Annual Logging Symposium* (6–9 June 2004), Noordwijk, Netherlands: Society of Petrophysicists and Well-Log Analysts.

Mitchell, J., Valori, A., and Fordham, E.J. (2019). A robust nuclear magnetic resonance workflow for quantitative determination of petrophysical properties from drill cuttings. *Journal of Petroleum Science and Engineering* 174: 351–361.

Provencher, S.W. (1982). A constrained regularization method for inverting data represented by linear algebraic or integral equations. *Computer Physics Communications* 27: 213–227.

Purcell, E.M., Torrey, H.C., and Pound, R.V. (1946). Resonance absorption by nuclear magnetic moments in a solid. *Physical Review* 69: 37–38.

Schön, J.H. (2015). *Physical Properties of Rocks: Fundamentals and Principles of Petrophysics*. Elsevier Science.

Siddiqui, S., Grader, A.S., Touati, M. et al. (2005). Techniques for extracting reliable density and porosity data from cuttings. *SPE Annual Technical Conference and Exhibition* (9–12 October 2005), Dallas, Texas: Society of Petroleum Engineers.

Thommes, M., Kaneko, K., Neimark, A. et al. (2015). Physisorption of gases, with special reference to the evaluation of surface area and pore size distribution (IUPAC Technical Report). *Pure and Applied Chemistry* 87: 1051–1069.

U.S.EIA (2016a). Hydraulic fracturing accounts for about half of current U.S. crude oil production. http://www.eia.gov/todayinenergy/detail.cfm?id=25372 (accessed 15 March 2016).

U.S.EIA (2016b). Hydraulically fractured wells provide two-thirds of U.S. natural gas production. http://www.eia.gov/todayinenergy/detail.cfm?id=26112 (accessed 6 May 2016).

Vandenbroucke, M. and Largeau, C. (2007). Kerogen origin, evolution and structure. *Organic Geochemistry* 38: 719–833.

Vocke, C.P., Deglint, H.J., Clarkson, C.R. et al. (2018). Estimation of petrophysical properties of tight rocks from drill cuttings using image analysis: an integrated laboratory-based approach. *SPE Canada Unconventional Resources Conference* (13–14 March 2018), Calgary, Alberta, Canada: Society of Petroleum Engineers.

Vogel, C.R. (2002). *Computational Methods for Inverse Problems*. Society for Industrial and Applied Mathematics.

7

Tight Rock Permeability Measurement in Laboratory

Some Recent Progress

Hui-Hai Liu[1], *Jilin Zhang*[1], *and Mohammed Boudjatit*[2]

[1] *Aramco Americas: Aramco Research Center–Houston, Houston, TX, USA*
[2] *Saudi Aramco, Dhahran, Saudi Arabia*

7.1 Introduction

Permeability is an important parameter for almost all flow and transport processes in the subsurface. The laboratory measurement of tight rock permeability is especially relevant to hydrocarbon recovery from unconventional reservoirs, caprock integrity associated with CO_2 geological sequestration in saline aquifers, and geological disposal of high-level nuclear wastes in tight rock formations.

Unconventional reservoirs, characterized by low to ultra-low permeabilities, have become one of the main sources of oil and gas in the United States and worldwide, resulting from the development of innovative technologies, such as horizontal drilling and hydraulic fracturing. In an unconventional reservoir, hydrocarbon flows from rock matrix to a production well through natural and hydraulic fractures. The low-permeability rock matrix, where the majority of hydrocarbon resides, is really the bottleneck for hydrocarbon recovery (Ozkan et al. 2011; Liu et al. 2018). Thus, the accurate determination of matrix permeability is critical for predicting hydrocarbon production from an unconventional reservoir and optimizing the reservoir management process.

To mitigate the global increase in temperature due to climate change, the world needs to limit the emission of anthropogenic greenhouse gases, particularly carbon dioxide CO_2 (e.g. IPCC 2007, 2014). In order to reduce CO_2 emissions into the atmosphere, several mitigation strategies have been suggested, including CO_2 capture and storage (CCS) that involves capture of CO_2 from industrial units (e.g. power plants and cement factories) and then injection of the captured CO_2 into underground reservoirs, such as saline aquifers, for geological storage. The integrity of caprock overlying the storage reservoir is important for the long-term safe containment of stored CO_2 (IEAGHG 2011). For many existing and potential CO_2 storage sites, clay-rich shales and mudstones form the primary caprock with ultra-low permeability. There are two kinds of potential CO_2 leakage pathways; one through the reactivated faults and newly generated fractures in the caprock owing to pore pressure increase in the reservoir, and the other associated with slow CO_2 flow and diffusion through the non-fractured portion of the caprock. While the latter leakage is slow, it is still significant for the long-term safety of CO_2 storage because it could happen through the large contact area between the caprock and the aquifer. The assessment of leakage risk for both potential pathways requires information on caprock permeability.

Physics of Fluid Flow and Transport in Unconventional Reservoir Rocks, First Edition.
Edited by Behzad Ghanbarian, Feng Liang, and Hui-Hai Liu.

Nuclear energy generates about 20% of the US national electricity that is baseload and carbon-emission free (NEI 2020). However, a major issue related to nuclear energy development is the disposal of high-level nuclear wastes produced from energy generation. It is well-accepted that the deep geological disposal of wastes is the preferred permanent disposal option that has been investigated for many years in several countries worldwide (Besnard et al. 2019). Potential geological disposal sites generally target tight rocks, such as granite, salt, and shale formations (Nutt et al. 2011). The low permeability of these formations minimizes or even eliminates the risk of nuclear waste leakage to locations that might impact the health of future generations. Thus, rock permeability is an important parameter for assessing the feasibility to store nuclear wastes in a geological formation. As will be discussed later, one of the widely used laboratory methods to measure tight rock permeability was initially developed in the community of nuclear waste management (Brace et al. 1968; Hsieh et al. 1981).

While the laboratory permeability measurement is largely a trivial task for high-permeability porous media, the accurate determination of tight rock permeability in the laboratory remains a technical challenge largely because its value is very low and, in many cases, on the order of nano-Darcy (nD). The major objective of this chapter is to review the selected recent advances in developing new methodologies for measuring tight rock permeability in laboratory. The rest of this chapter is organized as follows. Section 7.2 provides a brief introduction to the commonly used methods for tight rock permeability measurements in the laboratory to set the stage for further discussion of the new developments. Sections 7.3 and 7.4 focus on the two key elements of the Aramco Nano Permeameter (ANP) method developed at the Aramco Research Center, including simultaneous measurement of fracture and matrix permeability from fractured core samples and the direct measurement of permeability-pore pressure function from non-fractured core samples.

7.2 Commonly Used Laboratory Methods

This section briefly presents some most widely used methods for laboratory measurement of tight rock permeability, including the steady-state flow, pressure pulse-decay (PPD), and Gas Research Institute (GRI) methods. More detailed reviews of these methods can be found in Cui et al. (2009) and Sander et al. (2017). The non-reactive gases, such as nitrogen, are generally employed as working fluids for permeability measurements because they have lower viscosities compared to liquids and thus can make the permeability measurement faster. As a general note, permeability cannot be directly measured in the laboratory but is inferred through the linkage between permeability and directly measurable variables (e.g. flow rate, pressure, temperature, and others). This linkage is either the known physical laws (e.g. Darcy's law) or analytical solutions to fluid flow through a rock sample under the designed test conditions. It is important to keep this in mind for discussing the currently available methods and the related new developments.

7.2.1 Steady-State Flow Method

The steady-state flow method is the simplest and most straightforward approach to measure rock permeability in the laboratory (API 1998; Sinha et al. 2013). Figure 7.1 is the schematic experiment setup for this method. It consists of a core holder, three pumps (upstream, downstream, and confining pumps), and pressure and temperature transducers. During the steady-state flow test, a rock sample is placed within the core holder, and a confining pump is applied to generate the required confining pressure using the confining fluid (within the core holder) separated from the core

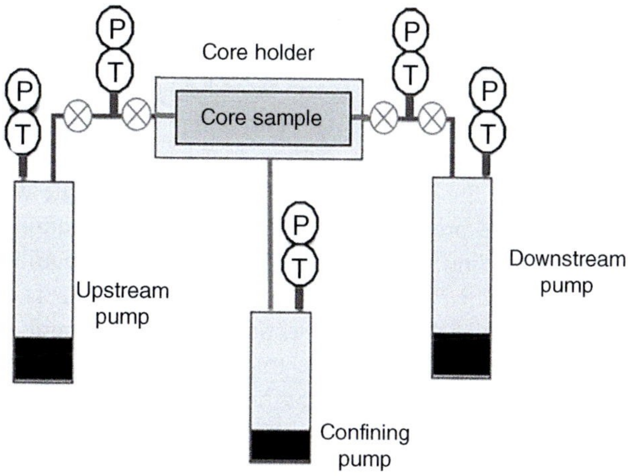

Figure 7.1 The schematic experiment setup for the permeability measurement with steady-state flow method. The symbols "P" and "T" refer to pressure and temperature transducers.

sample with seals. The inlet and outlet of the rock sample are connected to upstream and downstream pumps, respectively. These pumps maintain the required constant upstream and downstream pressures and provide flow-rate measurements to and from the pumps. Pressures and temperatures at the inlet, outlet, and confining pump are monitored with pressure and temperature transducers. The whole system is placed in an oven with constant temperature to ensure isothermal test conditions such that the fluid flow is purely owing to the pressure gradient across the rock sample. When the mass flow rate to the inlet and that out of the outlet are practically identical, the steady-state flow has been reached.

Then permeability can be estimated with Darcy's law expressed as

$$Q = \frac{kA\rho}{\mu}\frac{\Delta p}{L} \tag{7.1}$$

where Q is mass flow rate, k is permeability, ρ is fluid density, μ is fluid viscosity, Δp is pressure difference between the inlet and outlet of the rock sample, and A and L are the sample's cross-sectional area and length, respectively. Volumetric flow rates from the upstream pump and to the downstream pump are measured. The corresponding mass flow rates are determined by multiplying the volumetric flow rates with local fluid densities that depend on pressure and temperature.

Since under steady-state flow condition, values for all the parameters in Eq. (7.1), except permeability, are known or measured, permeability can thus be estimated from the equation (Sinha et al. 2013). For stress-sensitive rocks, such as shale, the pressure difference Δp should be small and generally on the order of few percentages of the average of the upstream and downstream pressures such that permeability can be treated as a constant along the core sample under the test condition.

7.2.2 Pressure Pulse-Decay Method

While the steady-state flow method is conceptually simple and straightforward to use, it generally takes a long time to achieve steady-state flow for tight rock samples with extremely low permeability. To overcome this, the PPD method, based on transient flow test conditions, was introduced to measure permeability of crystalline rocks for nuclear waste disposal (Brace et al. 1968; Hsieh et al. 1981). It was then successfully applied to measure tight rock gas permeability in the oil and gas industry (Dicker and Smits 1988; Jones 1997).

Figure 7.2 shows the experimental setup scheme for the permeability measurement using the PPD method. The system includes a core holder where a rock sample is located during the test and subject to confining pressure provided by the confining fluid, and upstream and downstream reservoirs connected to the inlet and outlet of the rock sample, respectively.

As the first step of the PPD test, the working fluid, used for measuring permeability, is injected from the inlet and outlet of the rock sample with upstream and downstream pumps until the pore pressure within the rock sample equilibrates and reaches the desired initial value. Depending on the permeability value, this equilibrium process can take several hours or days for tight rocks. Then, the upstream reservoir is disconnected from the rock sample by closing the valve for the connection (Figure 7.2), and a gas pressure pulse is created in the upstream gas reservoir by injecting a certain amount of the working fluid. A small pressure pulse is generally used such that in the upstream reservoir pressure increase is a few percent of the initial gas pressure (e.g. Dicker and Smits 1988; Jones 1997). This is because the PPD method relies on the analytical solutions to linear or linearized gas flow equations that require small disturbances to the test system such that fluid density and viscosity can be approximated as constants within the rock sample.

After the pressure increase in the upstream reservoir reaches the desired value, the reservoir is reconnected to the core sample by opening the valve for the connection (Figure 7.2). The gas then flows from the upstream gas reservoir to the downstream one through the rock sample. Evolutions of the gas pressures from the two gas reservoirs are monitored as a function of time. The whole test system is maintained under isothermal conditions during the PPD testing process.

The permeability is finally determined by fitting the pressure data with the relevant analytical solution to fluid flow within the test system, with permeability being a fitting parameter. Since fluid and rock properties could be approximated as constants, the governing equation for working fluid flow is linear and can be solved analytically. The solution allows the permeability to be related to

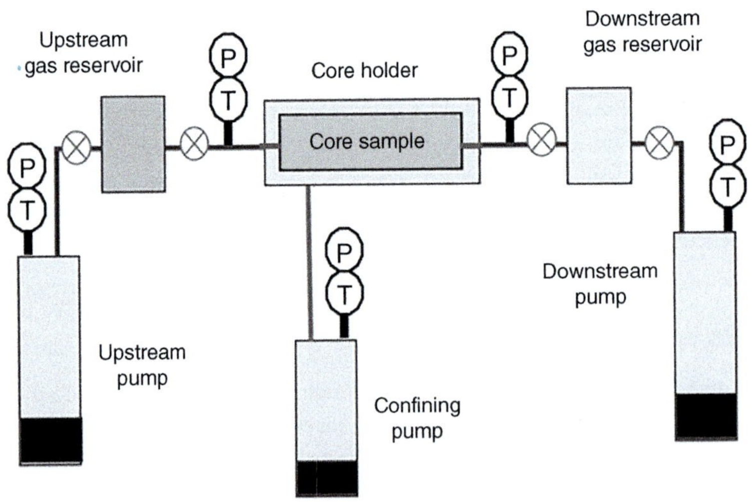

Figure 7.2 The schematic experiment setup for the permeability measurement with PPD method. The symbols “P” and “T” refer to pressure and temperature transducers.

pressures from the upstream and downstream reservoirs by (Brace et al. 1968; Hsieh et al. 1981; Dicker and Smits 1988; Jones 1997; Liu et al. 2016).

$$\ln(p_u - p_d) = -fkt + C \tag{7.2}$$

where p_u and p_d are pressures at the upstream and downstream reservoirs, respectively. In Eq. (7.2), f is a known function of geometry of the rock sample (length and cross-sectional area), rock porosity, and properties of the working fluid (density, viscosity, and compressibility). Fitting the pressure data with Eq. (7.2) gives the permeability value of the rock sample being tested for the given pore pressure and confining stress (or pressure).

7.2.3 Gas Research Institute Method

The process to take a rock sample from the deep reservoir to surface can result in significant stress release for the sample, which, in many cases, generates fractures in it. Fractures can also be induced during sample transportation and preparation (such as drilling a plug to be used for permeability measurements with steady-state and PPD methods). Since the tight rock permeability is very low, these induced fractures can dominate measured laboratory permeability values and make the laboratory permeability measurements irrelevant to the true permeability in the reservoir.

Partially motivated by overcoming the impact of induced features on permeability measurement in core samples, the GRI method was developed and has been widely used in the oil and gas industry (Luffel et al. 1993; GRI 1996). While it is believed that induced fractures do not exist at very small scales, the GRI method uses crushed rock samples with a particle size diameter of about 1 mm and works as follows. The GRI experimental setup includes two cells: a reference cell and a sample cell (Figure 7.3). The sample cell contains crushed rock particles and the reference cell includes work fluid only. One should first let equilibrium occur between the gas pressures in the sample cell and pores in the rock samples and then connect the sample cell to the reference cell with a higher gas pressure by opening the valve for the connecting tubing (Figure 7.3). The gas pressures within two cells become equal almost simultaneously, but the gas pressure continues to decline because of the penetration of the gas into rock particles. Pressure declines more quickly for the rock particles with high permeability values. The rock permeability is finally determined by fitting the observed pressure decline to the associated analytical solutions with permeability as a fitting parameter. The analytical solutions were obtained by assuming the diameters of rock particles to be uniform, the particles to be spherical, and rock properties to be isotropic and homogeneous (Cui et al. 2009).

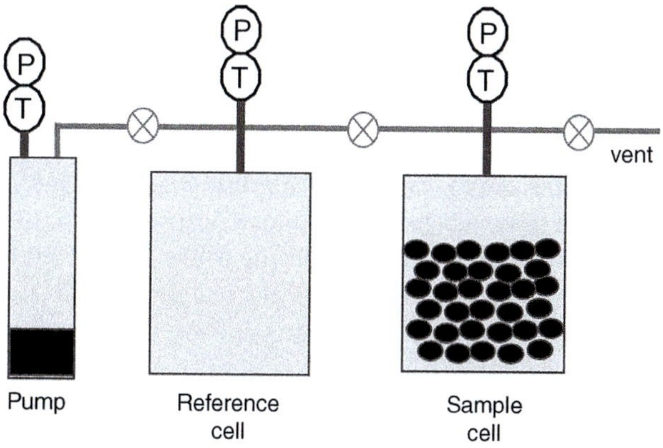

Figure 7.3 The schematic experiment setup for the permeability measurement with the GRI method. The symbols "P" and "T" refer to pressure and temperature transducers.

We refer the readers to Cui et al. (2009) for the details of the analytical solutions used in the GRI method.

While the GRI method has been widely used to measure tight rock permeability in laboratory, it suffers from several well-known limitations. Firstly, the GRI method is only conducted under the zero confining stress condition and thus cannot be used for investigating the stress dependency of shale permeability (Luffel et al. 1993; GRI 1996). Secondly, several researchers (e.g. Cui et al. 2009; Tinni et al. 2012) reported that GRI permeability measurements have been found to be dependent on the particle size of crushed samples; larger particles correspond to higher measured permeability values. Consequently, the representativeness of the permeability measurements at the millimeter scale can be questioned. Thirdly, the GRI test is generally conducted at small pore pressures that results in a significant impact of diffusion on the measured permeability values as well (Liu et al. 2018). The diffusion effect will be discussed later in this chapter.

7.3 Simultaneous Measurement of Fracture and Matrix Permeabilities from Fractured Core Samples

The stress release of a rock sample, when taken from a deep reservoir to the surface, and sample preparation process may induce artificial fractures within the sample. To overcome the impact of the induced fractures on laboratory permeability measurements, the GRI method, as previously indicated, has been widely used in the oil and gas industry for measuring rock permeability with crushed samples. However, it suffers from several major limitations. To deal with this technology gap, we have developed a new method, as part of the ANP method, that simultaneously measures fracture and matrix permeabilities from fractured core samples under different stress conditions (Zhang et al. 2020). This section discusses this method and its validation.

7.3.1 Estimation of Fracture and Matrix Permeability from PPD Data for Two Flow Regimes

Our approach is based on the generalization of the conventional PPD method, which assumes that the core sample being tested is a single continuum, and thus the gas flow is uniform at a given cross-section along the sample. In the PPD test of a single-continuum sample, the upstream reservoir pressure goes down, and the downstream reservoir pressure goes up when gas flows from the upstream to the downstream. As soon as pressures from both reservoirs become identical, they remain constant and the system (rock core sample or reservoirs) reaches equilibrium.

Figure 7.4 shows the concept of fluid flow in a fractured core sample that needs to be characterized as a dual-continuum system, rather than a single continuum. Note that in Figure 7.4, we use one fracture for the demonstration purpose; our discussion here can be applied to fracture networks as well. During the early stage of the PPD test, the flow process is dominated by flow in fractures because fracture permeability is much higher than matrix permeability. Consequently, pressures from both reservoirs become the same relatively quickly and then flow within the fractures stops. In this flow regime, we can apply the conventional PPD data analysis method, as previously discussed, to determine the fracture permeability. At this stage, flow within the rock matrix plays a negligible role in determining the evolution of reservoir pressures.

As demonstrated in Figure 7.5 for a fractured core sample, reservoir pressures, after becoming the same (at Point A), will continue to decline because of the gas flow into rock matrix from both reservoirs shown in Figure 7.4. The pressures eventually become constant and do not change with time

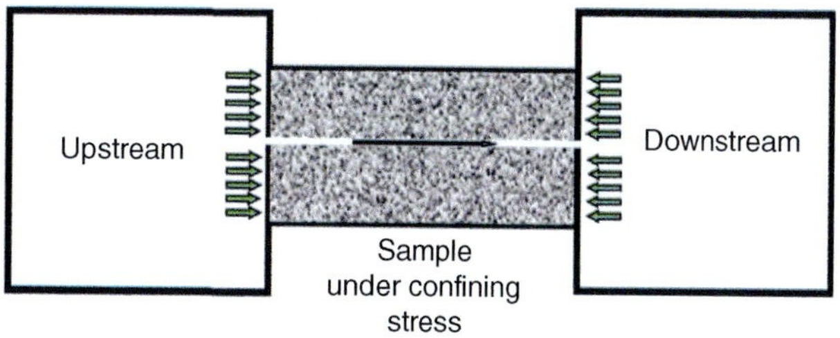

Figure 7.4 A schematic of two flow regimes caused by fractures and the matrix in an unconventional core sample.

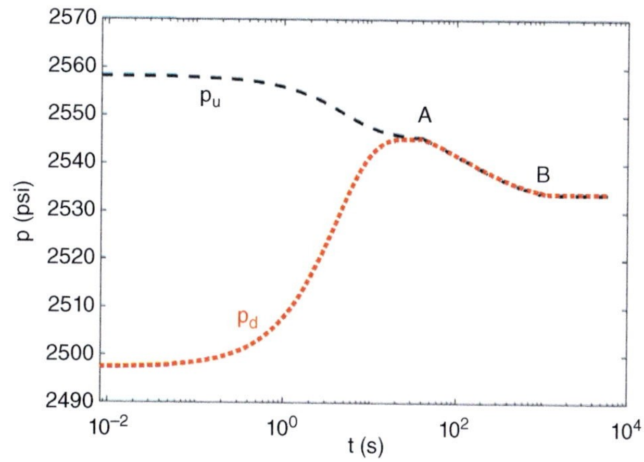

Figure 7.5 Pressure recordings at upstream and downstream gas reservoirs versus time. Note that time is in the logarithmic scale.

any longer, resulting from that the system becomes equilibrium. Recall that for a single continuum system the equilibrium is reached as soon as the pressures from both upstream and downstream gas reservoirs become the same. This second flow regime (beyond Point A in Figure 7.5) is dominantly controlled by flow to and within the matrix. Thus, the pressure data in the flow regime can be used for estimating matrix permeability, as will be discussed later. We herein ignore the flow from fractures to the matrix for a horizontal core sample, given that matrix permeability along the horizontal or bedding direction is significantly higher than that in the vertical direction (e.g. Mokhtari et al. 2013; Zhang et al. 2020). The horizontal permeability is of more interest because it is more relevant to hydrocarbon production from unconventional reservoirs.

To the best of our knowledge, Ning (1992) and Ning et al. (1993) conducted the early work to estimate both fracture and matrix permeability from fractured core samples by history matching of pressure pulse decay tests with numerical methods. However, their method needs detailed geometry of fractures within a test sample and ignores anisotropy of matrix permeability. The use of numerical methods, in many cases, also suffers from the non-uniqueness problem in parameter identifications. Consequently, their methods have not been widely used in practice.

Some researchers assumed fractures to be completely closed at a certain confining stress in order to use conventional PPD and likewise for measuring the matrix permeability from fractured samples (e.g. Gan et al. 2018). As indicated by Liu et al. (2009) and Rutqvist et al. (2002), among others, fractures cannot be completely closed even at very high confining stresses, as a result of the damage of cementation between grain particles and other relevant factors. The remaining fracture opening is called the residual fracture aperture (e.g. Rutqvist et al. 2002; Liu et al. 2009). The residual aperture may not be a significant contributor to fluid flow in conventional rock samples with relatively high permeabilities. It is obviously not the case for tight rock samples that have extremely low permeabilities; even a small residual aperture can dominate permeability measurement results.

7.3.2 Mathematical Model

The measurement of the matrix permeability requires an analytical solution to fluid flow within the rock matrix under the test conditions corresponding to the second flow regime (corresponding to Point A and beyond that is shown in Figure 7.5). Like all the other methods used to estimate permeability in the laboratory, the core sample is considered to be homogeneous. It is a reasonable treatment, given that the heterogeneity at that scale, if it exists, is insignificant compared with large-scale heterogeneity. The flow along the core sample is treated as one-dimensional flow. As previously indicated, flow from the fracture faces to the matrix is negligible because matrix permeability in the horizontal (or bedding) direction is much higher than that in the vertical direction. Consequently, the flow into the sample in the second flow regime is mainly from the end faces of a horizontal core sample.

Based on Darcy's law and the mass conservation, the governing equation of fluid flow in a core sample, with fluid pressure as the dependent variable, is given as:

$$\frac{\partial^2 p(x,t)}{\partial x^2} = \frac{c\mu\varphi_m}{k_m}\frac{\partial p(x,t)}{\partial t} \quad \left(t > t_1;\ 0 < x < \frac{L}{2}\right) \tag{7.3}$$

where p is the fluid pressure, x is the location (Figure 7.6), t is time, c is the total compressibility approximated by fluid compressibility herein that is much larger than solid compressibility for the given test condition, μ is the viscosity, φ_m is the matrix porosity, k_m is the matrix permeability, t_1 is the time when the upstream pressure is equal to the downstream pressure or when the second flow regime begins, and L is the total length of the core sample. Note that the above equation is a linearized one because the rock properties and fluid viscosity and density are constant.

The initial conditions for the second flow regime are

$$p(x,t_1) = p(t_1) \quad \left(0 < x < \frac{L}{2}\right) \tag{7.4}$$

$$p(x,t_1) = p_u(t_1) \quad \left(x = \frac{L}{2}\right) \tag{7.5}$$

where subscript u again refers to the upstream reservoir. Note that pressures in both upstream and downstream reservoirs are identical at a given time in this flow regime. Since the time period of the first flow regime is short (before reaching Point A in Figure 7.5), the pressure within the matrix at

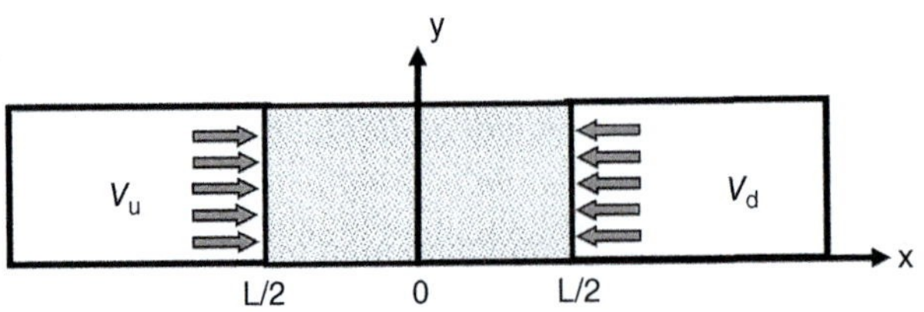

Figure 7.6 The coordination system associated with the rock sample for deriving the analytical solution to fluid flow to the rock matrix from the upstream and downstream reservoirs. The length and cross-section area of the sample are L and A, respectively.

Point A is largely unchanged and should be close to the initial pressure for the first regime. Thus, at $t = t_1$, the pressure in the matrix is approximately considered to be spatially uniform Eq. (7.4).

The corresponding boundary conditions for the problem are given by

$$\frac{\partial p(x,t)}{\partial t} = -\frac{k_\mathrm{m}}{c\mu}\frac{A}{V_\mathrm{e}}\frac{\partial p(x,t)}{\partial x} \quad \left(t > t_1;\ x = \frac{L}{2}\right) \tag{7.6}$$

$$\frac{\partial p(x,t)}{\partial x} = 0 \quad (t > t_1;\ x = 0) \tag{7.7}$$

where A is the cross-sectional area of the core sample and V_e is the total volume including the volume of the upstream and downstream reservoirs as well as the fractures within the sample. In most cases, the fracture volume can be ignored compared with the summation of the volumes of upstream and downstream reservoirs. The symmetry condition also exists in the middle of the sample (Eq. (7.7)).

The above problem was analytically solved by Chen et al. (2018) using Laplace transformation. Their solution is expressed in terms of dimensionless time, τ, and pressure, $\tilde{p}$, that are defined, respectively, as

$$\tau = \frac{4k_\mathrm{m}}{c\mu\varphi_\mathrm{m}L^2}(t - t_1) \tag{7.8}$$

$$\tilde{p} = \frac{p - p_\infty}{p_\mathrm{u}(t_1) - p_\infty} \tag{7.9}$$

where p_∞ is the pore pressure when the whole system reaches equilibrium. The volume ratio, ω, is expressed as

$$\omega = \frac{v_m}{v_e} \tag{7.10}$$

where V_m is the pore volume in the matrix of a rock sample. Then, the solution to the dimensionless pressure in the upstream (and downstream) reservoir is given by (Chen et al. 2018)

$$\tilde{p}_\mathrm{u} = \frac{p_\mathrm{u}(t) - p_\infty}{p_\mathrm{u}(t_1) - p_\infty} = 2\sum_{n=1}^{\infty} \exp(-\delta_n^2\tau)\left(\frac{1}{\omega + \frac{\delta_n^2}{\omega} + 1}\right)\left(1 - \frac{1}{1+\omega}\right)^{-1} \tag{7.11}$$

where $\delta_n > 0$ is the nth solution of the following equation:

$$\tan(\delta) = -\frac{\delta}{\omega} \tag{7.12}$$

Chen et al. (2018) demonstrated that the right-hand side of Eq. (7.11) is dominated by the first term, and the remaining terms can be safely ignored for practical applications. Thus, Eq. (7.11) can be further simplified as (Chen et al. 2018):

$$\log(\tilde{p}_\mathrm{u}) \approx \log(f_1) - \delta_1^2\frac{4k_\mathrm{m}}{c\mu\varphi_\mathrm{m}L^2}(t - t_1) \tag{7.13}$$

As an example, Figure 7.7 shows an observed linear relationship between $\log(\tilde{p}_\mathrm{u})$ and time for a rock sample from the Eagle Ford formation, supporting the validity of Eq. (7.13). Matrix permeability, k_m, is then estimated from the absolute value of the slope, S_m, with Eq. (7.13) by fitting the observed data of $\log(\tilde{p}_\mathrm{u})$ versus time. Specifically, the matrix permeability is calculated with

$$k_\mathrm{m} = \frac{S_\mathrm{m}c\mu\varphi_\mathrm{m}L^2}{4\delta_1^2} \tag{7.14}$$

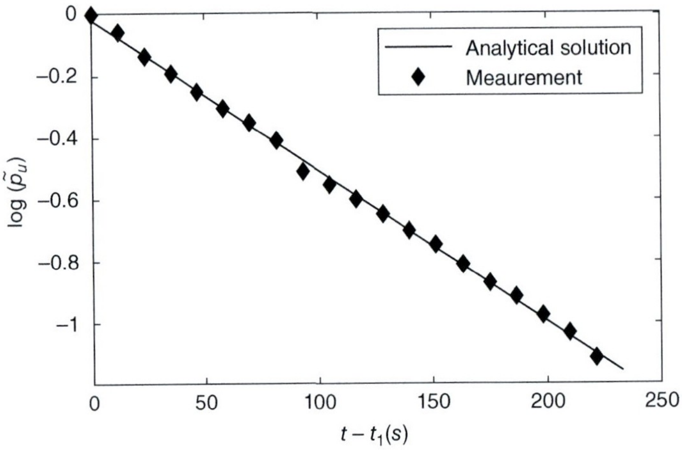

Figure 7.7 A linear relationship between of $\log(\tilde{p}_u)$ and time for a rock sample from the Eagle Ford formation. The matrix permeability is determined with the slope of the fitted line.

Note that in the above and all the other equations discussed in this chapter, the international system of units is required. It is practically of interest to indicate that only linear curve fitting is needed for determining the matrix permeability herein.

7.3.3 Method Validation and Discussion

Observed pressures from the upstream and downstream reservoirs of a PPD test system, as previously discussed, are characterized by the two flow regimes. We herein focus on the evaluation of the method to estimate matrix permeability using data from the second flow regime.

A tight pyrophyllite sample, a metamorphic rock that is quite homogeneous and readily available from commercial sources (Profice et al. 2016), was used for the validation purpose. We first performed the PPD test on the sample to measure its permeability which is considered the true matrix permeability of the sample. We then artificially fractured the sample by cutting the sample into two pieces and measured its matrix permeability again using our method discussed here. During the PPD tests, a relatively high pore pressure of 2500 psi in the sample was used as the initial pressure to minimize the diffusion effect. After the PPD system equilibrated at the initial pressure, a pressure pulse with a magnitude of around 100 psi was applied to the upstream reservoir and then flow occurred across the sample. The pressures as functions of time were monitored from both the upstream and downstream reservoirs with high-precision pressure transducers. Nitrogen gas was used as the working fluid in the tests that stopped when gas pressures from both gas reservoirs did not decrease anymore. The pressure decay data were analyzed with an in-house software written in MATLAB, and the gas adsorption effect was negligible and thus was not considered.

Table 7.1 shows the comparison of matrix permeability measurements for the pyrophyllite sample obtained before and after artificial fracturing. Before fracturing, the measured matrix permeability is 232 nD for the pore pressure of 2500 psi and confining pressure of 3500 psi. The matrix permeability obtained with our method from the artificially fractured core sample is 238 nD for the same pore and confining pressures. The two matrix permeability values are practically identical (only 2.6% different).

Table 7.1 Comparison of the matrix permeability results for a pyrophyllite sample that are obtained before and after artificial fracturing. The PPD and the new method are used for measuring the permeability for non-fractured and fractured sample, respectively.

Before fracturing			After fracturing		
Pore pressure (psi)	Confining pressure (psi)	k_ms (nD)	Pore pressure (psi)	Confining pressure (psi)	k_m (nD)
2500	3500	232	2500	3500	238

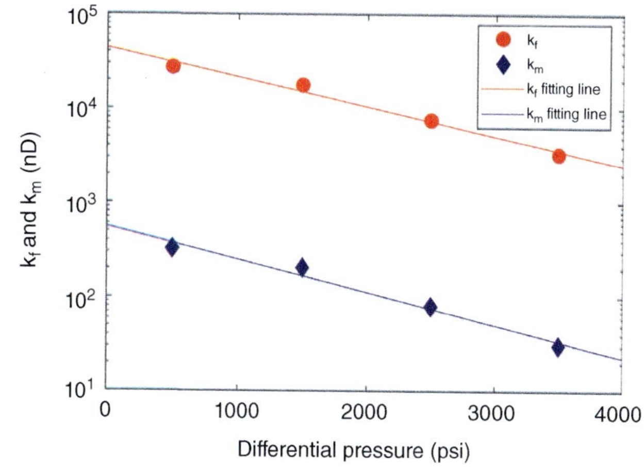

Figure 7.8 Pressure dependence of both fracture and matrix permeabilities (k_f and k_m) for the Eagle Ford shale sample studied here. The differential pressure is defined as the difference between confining and pore pressures. The vertical axis is in log scale and one unit corresponds to one order of magnitude.

The method has also been used for determining matrix permeability for some unconventional reservoirs. As an example, Figure 7.8 shows the measured fracture and matrix permeabilities for a shale sample collected from the Eagle Ford formation at the pore pressure of 2500 psi and different confining pressures. For a given differential pressure, defined as the difference between confining and pore pressures, the fracture permeability is about two orders of magnitude higher than the matrix permeability. Note that the traditional PPD method can only estimate the bulk permeability that is essentially the same as the fracture permeability because the core sample was treated as a single-continuum system, rather than a dual-continuum system. Figure 7.8 also indicates that the stress dependency of permeability can be fitted by exponential functions for both fractures and matrix, which is consistent with previous studies (e.g. Liu 2017).

It may be useful to emphasize that our method to estimate both fracture and matrix permeability is really a generalization of the conventional PPD method. Compared with the latter, our method has at least three advantages. Firstly, it allows estimating both fracture and matrix permeability values from fractured core samples under different stress conditions. This is practically important because we can recover the true matrix permeability even when rock samples contain artificially induced fractures. Secondly, it is based on analytical solutions to gas flow across the core sample and thus easier to use in practice than the analysis of the test results with numerical models. The determination of the matrix permeability with our method involves linear curve fitting only. Thirdly, it uses the same experimental setup as the traditional PPD method and thus is easily adopted.

7.4 Direct Measurement of Permeability-Pore Pressure Function

The conventional PPD method is time-consuming for determining the relation between permeability and pore pressure or stress because that requires multiple permeability data points measured at different pore and confining pressures. For a given confining pressure and pore pressure, one PPD test run is used to measure one permeability point only. In order to measure core sample permeability at a different pore pressure and/or confining pressure, one needs to re-equilibrate the system under the new test conditions and then do another PPD test run. The re-equilibrium process, depending on the permeability, can take a long time (e.g. days) (Jones 1997). Since this procedure needs one test run to measure a permeability point and requires multiple test runs for determining permeability-pore pressure relation, it is called the "point-by-point" approach by Liu and Zhang (2020).

The root cause for needing multiple test runs to determine permeability-pressure relationship is that the "point-by-point" method is based on the linear theory of gas flow within core samples. Because the theory requires the constant fluid and rock properties (gas density, gas viscosity, porosity, and permeability), only small pressure disturbances (or pressure pulses) are allowed during the PPD tests. The linear theory would become invalid, if there were significant variations of fluid and rock flow properties along the core sample during the PPD test run. Consequently, one can only measure one permeability data point at a given pore pressure and confining pressure condition.

To overcome the inefficiency of the "point-by-point" measurement approach, we have developed another method, also as part of the ANP method, that can directly measure the pore pressure dependence of permeability for a non-fractured tight rock sample with a single test run (Liu and Zhang 2020). Unlike currently available methods, it is based on nonlinear theory of fluid flow within core samples and not limited by the small pressure disturbance across the test system. This section is devoted to the discussion of this new development.

7.4.1 Knudsen Diffusion, Slippage Flow, and Effective Gas Permeability

The tight rock gas permeability is a function of pore pressure for several reasons. For a given confining pressure, an increase in pore pressure would enlarge the rock pore size and thus enhance the rock permeability, as indicated in Figure 7.8. The second mechanism is so-called slippage flow. Because of the low density for low gas pressures, the gas does not have zero velocity at the pore walls but has a slippage flow. This makes the measured permeability, or effective permeability, larger than the intrinsic permeability that is defined under the condition of zero fluid velocity at pore walls. The above two mechanisms exist for both conventional rocks with relatively high permeability and tight rocks, simply because they are not directly related to the pore size. The third mechanism is Knudsen diffusion which becomes significant for extremely small pore sizes and thus is unique for tight rocks. The relation between the gas permeability and pore pressure for the tight rocks results from all the above three mechanisms.

The gas diffusion is a result of the thermal motion of gas molecules. The key parameter for describing the diffusion process at the continuum scale, the gas diffusion coefficient D, is given in classic statistical physics by (e.g. Javadpour et al. 2014):

$$D = \frac{1}{3} l u^* \tag{7.15}$$

where l is the mean free path or the average distance for a molecule to travel before it hits other molecules, and u^* is the averaged magnitude of thermal motion velocity. They are calculated by (e.g. Civan 2010; Ziarani and Aguilera 2012; Javadpour et al. 2014)

$$l = \frac{k_B T}{\sqrt{2}\pi d_c^2 p} \tag{7.16}$$

and

$$u^* = \sqrt{\frac{8k_B T}{\pi w}} \tag{7.17}$$

where k_B is the Boltzmann Constant (1.3805×10^{-23} J/K), T is the temperature, p is the pressure, d_c (m) is the collision diameter for gas that is 0.358 nm for N_2 (often used as a working fluid to measure tight rock permeability in the laboratory), and W is the gas molecular weight.

The conventional rocks have much larger pore sizes than the mean free path and thus collisions mainly occur among gas molecules themselves in rock pores. In this case, fluid flow is dominated by the viscous flow process owing to collisions among gas molecules, and the diffusion effect on fluid flow is negligible. In tight rocks, the gas mean free path could be greater than the pore radius on the order of nm under low pore pressures. In this case, the collisions mainly occur between the pore walls and gas molecules in rock pores, which reduces the resistance to the bulk movement of gas molecules compared to the regime of viscous fluid flow that corresponds to the dominant collisions among gas molecules. This diffusion process for the smaller mean free path than the pore size is called Knudsen diffusion and characterized by Knudsen number defined as

$$K_n = \frac{l}{r} \tag{7.18}$$

where r is the pore radius.

When Knudsen diffusion and slippage flow become important, the measured or effective permeability is larger than the intrinsic permeability that is purely determined by pore geometry and has nothing to do with the working fluid. As an example, Figure 7.9 shows the dependency of enhancement factor of nitrogen permeability, f_c, on pore pressure for a microtube with a radius of 5 nm, which is the typical size of pore throat in shale (e.g. Liu et al. 2018). The enhancement factor herein refers to the ratio of the effective permeability to its intrinsic value. Note that the factor results from the effects of both Knudsen diffusion and slippage flow. Beskok and Karniadakis (1999) reported a mathematical formulation of f_c for microtubes:

$$f_c = (1 + \alpha K_n)\left[1 + \frac{4K_n}{1 - bK_n}\right] \tag{7.19}$$

where b is a constant equal to -1 and α is the dimensionless rarefaction coefficient that can be empirically determined from the Knudsen number (Civan 2010). The above formulation has been validated with a large number of numerical experiments and demonstrated to be able to accurately calculate the enhancement factor for micro-tubes (Beskok and Karniadakis 1999). However, the development of the corresponding formulation for porous media is much more difficult because of the complexity of pore geometry in porous media (e.g. Civan 2010; Ziarani and Aguilera 2012; Javadpour et al. 2014). At this point, the experimental method remains the only reliable way to determine the dependency of tight rock permeability on pore pressure.

The measured (effective) permeability can be much higher than the intrinsic one, as indicated in Figure 7.9 by the much higher enhancement factor than one at low pore pressures, while gas flow in an unconventional reservoir is controlled by the effective permeability. This permeability enhancement is desirable for improving gas production rate during the late stage of production when gas pressure in the reservoir is relatively low, which highlights the importance of accurately measuring the effective permeability in the laboratory as a function of pore pressure within the context of predicting the gas production from tight rock reservoirs.

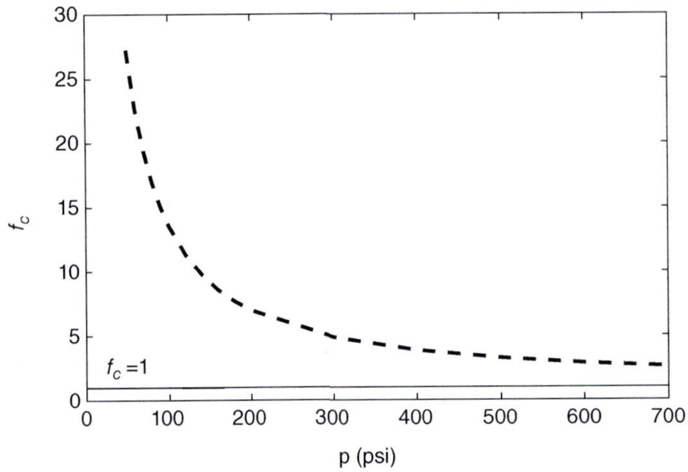

Figure 7.9 Enhancement factor of permeability, f_c, for a micro-tube with a radius of 5 nm as a function of pore pressure, p, for nitrogen that is calculated with Eq. (7.19) using T = 298.15 K, b = −1, and α determined from the Knudsen number. *Source:* Adapted from Civan (2010). The Knudsen number is calculated from Eq. (7.18) together with Eqs. (7.16) and (7.17) for a given pore pressure.

7.4.2 Methodology for Directly Measuring Permeability-Pore Pressure Function

As previously stated, permeability cannot be directly measured in the laboratory. Instead, its value is inferred from the link between rock sample permeability and other directly measurable variables. In many cases, analytical solutions to fluid flow in the core sample play the role of linkage. Thus, new laboratory methods for measuring rock permeability have often been developed side by side with the development of new analytical solutions of fluid flow or the new application of the existing analytical solutions. The limitations of the "point-by-point" methods result from their origin in the application of the linear theory of fluid flow. Here we present a method that is based on a nonlinear theory of fluid flow presented by Liu and Zhang (2020).

Based on Darcy's law and mass conservation, the nonlinear governing equation for one-dimensional non-reactive gas flow along a rock column is

$$\frac{\partial m}{\partial t} = \frac{\partial}{\partial x}\left(\frac{k_{\mathrm{m}}\rho}{\mu}\frac{\partial p}{\partial x}\right) \tag{7.20}$$

where m is the total gas mass per unit volume of the porous medium (or apparent gas density) defined by

$$m = \varphi_{\mathrm{m}}\rho \tag{7.21}$$

Note Eq. (7.3) is only a special case of Eq. (7.20) for constant fluid and rock properties. In Eq. (7.20), all the fluid and rock properties are functions of pore pressure.

The righthand side of Eq. (7.20) can be rearranged as

$$\frac{\partial m}{\partial t} = \frac{\mathrm{d}m}{\mathrm{d}p}\frac{\partial p}{\partial t} \tag{7.22}$$

Under the isothermal conditions, m is considered a function of pressure only.

Consider a problem regarding gas flow through an infinitely long rock column with the inlet at $x = 0$ and under the following boundary and initial conditions:

$$\begin{aligned} p(x,t) &= p_i \quad (x \geq 0, \quad t = 0) \\ p(x,t) &= p_0 \quad (x = 0, \ t > 0) \\ p(x,t) &= p_i \quad (x \to \infty, \quad t > 0) \end{aligned} \tag{7.23}$$

Using the following Boltzmann transformation

$$\lambda = xt^{-\frac{1}{2}} \tag{7.24}$$

we can rewrite Eq. (7.23) as

$$\begin{aligned} p(\lambda) &= p_i \qquad (\lambda \to \infty) \\ p(\lambda) &= p_0 \qquad (\lambda = 0) \end{aligned} \tag{7.25}$$

and then transform Eq. (7.20) to the following ordinary differential equation:

$$-\frac{\lambda}{2}\frac{\mathrm{d}m}{\mathrm{d}p}\frac{\mathrm{d}p}{\mathrm{d}\lambda} = \frac{\mathrm{d}}{\mathrm{d}\lambda}\left[D(p)\frac{\mathrm{d}p}{\mathrm{d}\lambda}\right] \tag{7.26}$$

where

$$D(p) = \frac{k_{\mathrm{m}}\rho}{\mu} \tag{7.27}$$

Integrating both sides of Eq. (7.26) within the interval (λ, ∞) gives

$$D(p) = -\frac{\displaystyle\int_{p_i}^{p} \frac{\lambda \mathrm{d}m}{2\,\mathrm{d}p}\mathrm{d}p}{\dfrac{\mathrm{d}p}{\mathrm{d}\lambda}} \tag{7.28}$$

The derivation of the above equation uses the following condition deduced from Eq. (7.25) for $\lambda \to \infty$:

$$\frac{\mathrm{d}p}{\mathrm{d}\lambda} = 0 \quad (\lambda \to \infty) \tag{7.29}$$

Equation (7.28) indicates that $D(p)$ can be fully determined with $p(\lambda)$ being known. More specifically, if we determine $p(\lambda)$ in one test run, the function $D(p)$ can be calculated from Eq. (7.28). Given that the relations among density, viscosity, and gas pressure are known for a given test fluid (e.g. nitrogen gas), the pore pressure dependence of permeability can then be determined from $D(p)$ with Eq. (7.27).

Based on the mass conservation, the cumulative gas flow into the sample at $x = 0$, $M(t)$, is given by

$$M(t) = \left(A\int_{p_i}^{p_0} \lambda \frac{\mathrm{d}m}{\mathrm{d}p}\mathrm{d}p\right)t^{\frac{1}{2}} \tag{7.30}$$

If B is the slope of the $M(t)$ versus $t^{1/2}$ data, we can then determine the average porosity of the core sample as

$$\varphi_{\mathrm{m}} = \frac{B}{A\displaystyle\int_{p_i}^{p_0} \lambda \frac{\mathrm{d}\rho}{\mathrm{d}p}\mathrm{d}p} \tag{7.31}$$

To use the above analytical results to efficiently and accurately measure the permeability-pore pressure function with a single test run, an experimental setup shown in Figure 7.10 was built. The key hardware elements of the setup are like the PPD or steady-state flow test system, including upstream and downstream gas reservoirs, core holder, pumps to provide gas pressures and confining pressure, respectively, and pressure and temperature transducers. However, the major difference between the current setup and conventional ones is that pressure transducers are used to measure pressure history at one or more locations along the core sample in order to determine $p(\lambda)$ (Figure 7.10). At each location, the x value is known and fixed. Then, the measured $p(t)$ at that location can be easily converted to $p(\lambda)$ using the Boltzmann transformation (Eq. (7.24)). In theory, $p(t)$ from one location is needed only to calculate $p(\lambda)$ because the latter does not depend on the location. The transducers were used at three locations in the setup (Figure 7.10) because the comparison between measurements from different transducers, as will be discussed later, provides a powerful way to validate the proposed methodology and setup. Extra care needs to be taken to minimize the disturbance of the dead volume of these transducers to the gas flow in the core sample. The dead volume refers to the volume of open space inside a pressure transducer connected to the core sample's pore space. In the setup shown in Figure 7.10, analog pressure transducers are used because they have almost zero dead volumes (Liu and Zhang 2020).

The test is conducted in an isothermal condition by placing the setup within an oven with a constant temperature. The test also requires a constant confining pressure making permeability only a function of pore pressure along the core sample. The major steps of the test procedure are as follow. Firstly, following the initial conditions for the theoretical model discussed above, the setup is initially equilibrated with a given constant pore pressure p_i (Eq. (7.23)). Secondly, to be consistent with the required boundary conditions, the pressure in the upstream gas reservoir is then increased to p_0 (Eq. (7.23)) and kept fixed throughout the test. Since the current method is not limited by the small pressure disturbance (or pulse) to the test sample, p_0 can be much greater than p_i. The two pressures are determined in such a way that their difference covers the pressure range of interest for a given practical application. Thirdly, let gas flow from the upstream reservoir through a core sample to the

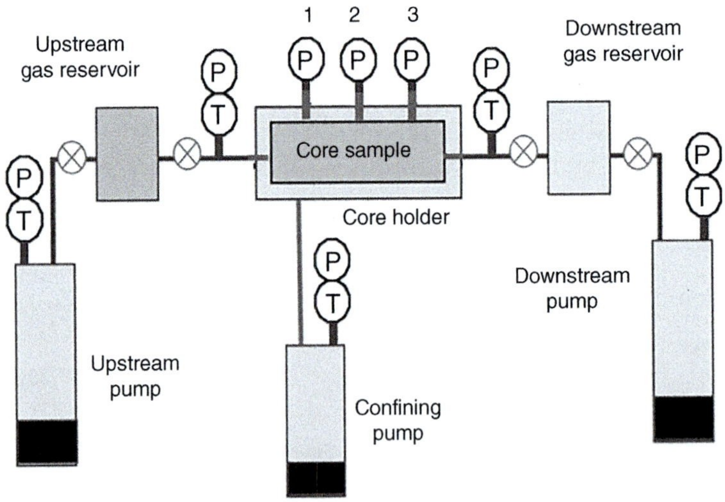

Figure 7.10 The conceptual diagram of the experimental setup to measure the pore pressure dependence of permeability in non-fractured core samples from unconventional reservoirs. The symbols "P" and "T" refer to pressure and temperature transducers.

downstream reservoir and monitor the pressure histories at the desired locations (Figure 7.10). Finally, use the pressure history to calculate $p(\lambda)$. The latter is further used to determine the permeability-pore pressure function, as described above.

The analytical solution (Eq. (7.28)) was developed for an infinitely long rock core sample. This condition can never be met in practice. Liu and Zhang (2020) proposed a concept of critical time for dealing with this; before the critical time, the gas flow at a given location is equivalent to the flow occurring in an infinitely long core sample. During the test, the critical times at the inlet and outlet of a core sample can be determined based on experimental observations. The critical time for the outlet is when the pressure increase is observed in the downstream gas reservoir. Then the outlet boundary effect will propagate from the outlet towards to the inlet along the core sample. The time for the outlet boundary effect to propagate to the given location is called propagation time. The determination of the critical time at the inlet is based on the accumulative gas flow rate from the upstream gas reservoir to an infinitely long core sample that is proportional to the square root of time for the given boundary and initial conditions (Eq. (7.30)). The critical time at the inlet then corresponds to the time when the above relationship between the flow rate and time, or Eq. (7.30), does not hold any longer. The difference between the inlet and outlet critical times is the propagation time at the inlet; propagation time at the outlet is zero. Given that the governing equation for gas flow in a core is mathematically identical to the nonlinear diffusion equation, the propagation time at a location along the core sample where the pore pressures are monitored (Figure 7.10) is approximately determined using properties of a diffusion process. Note that the propagation time at a given location, for a diffusion process, is proportional to the square of the distance between the location and the downstream boundary. Then the critical time at any location between the inlet and outlet is the summation of the critical time at the outlet and the propagation time at that location. Only the pressure data collected before the critical time are used for estimating the permeability-pore pressure function with Eq. (7.28).

It may be useful to note that the analytical results for the nonlinear diffusive equations have been used for measuring different flow and transport in several applications. Ludwig Boltzmann, the father of statistical physics, put forward the Boltzmann transformation (Eq. (7.24)) to solve nonlinear diffusion equations under the boundary and initial conditions defined by Eq. (7.23). Based on the Boltzmann transformation, Matano (1933) derived an expression for concentration-dependent diffusion coefficient, similar to Eq. (7.28), for solute transport problems. Bruce and Klute (1956) then applied the Boltzmann-Matano method to developing a laboratory technique to estimate soil-water diffusivity that has been widely used in the soil science community. As discussed in this chapter, we recently applied a similar mathematical technique for directly measuring permeability – pore pressure function with a single test run (Liu and Zhang 2020).

7.4.3 Experiments

Three-inch long core samples (with a diameter of 1 in.) from Eagle Ford were used for evaluating the method discussed in Section 7.4.2. These cores were scanned with μ-CT to ensure that they did not contain fractures. Recall that the proposed method is valid only for non-fractured core samples.

During a test with the setup shown in Figure 7.10, the initial pore pressure was set to the atmospheric pressure for two reasons (Liu and Zhang 2020). To do so, the pressure equilibrium process that may take a long time can be avoided. This is also useful to investigate the impact of diffusion that becomes more significant at lower pore pressures. The gas pressure in the upstream gas reservoir was then increased from the atmospheric pressure to 2000 psi and kept constant. When gas (nitrogen herein) flowed from the upstream gas reservoir through the core sample to the downstream gas reservoir, gas pressures as a function of time at three different locations along the core sample (Figure 7.10) were monitored with three pressure transducers. During the tests, a constant confining pressure of 4000 psi was applied to the core samples. To conduct the tests under isothermal conditions, the setup was again placed within an oven with a constant temperature.

As an example, Figure 7.11 shows gas pressures for Sample A as a function of time at the inlet or upstream gas reservoir, outlet or downstream gas reservoir, and from the three pressure transducers along the core sample. Within the test time periods before the critical times, $p(\lambda)$ curves were calculated with Eq. (7.24) and the pressure history data collected from the three pressure transducers. These curves essentially collapse into a single curve, which is consistent with the theory discussed in Section 7.4.2. For gas flow through a core sample under the initial and boundary conditions given in Eq. (7.23), gas pressure is only a function of λ, and $p(\lambda)$ does not depend on the location. The observed consistency indicates that the test setup works appropriately and $p(\lambda)$ can be accurately obtained. Similar consistencies were observed for other two test samples. It is very important to note that Eq. (7.28) used to calculate the permeability-pressure function is the direct result of combining the mass balance with the rational that pressure is only a function of λ. Since the mass balance is always valid and $p(\lambda)$ can be accurately obtained, the permeability function can thus be accurately estimated based on Eq. (7.28). As previously stated, the use of three pressure transducers in the test setup is for the validation purpose discussed above, although only one pressure transducer along the core sample is adequate for practical applications. Because the pressure transducer closest to the sample inlet has the biggest pressure range in which the rock sample can be considered to be infinitely long in terms of gas flow, the pressure data from that transducer was analyzed for permeability estimations in the rest of this section.

The pore pressure dependence of effective permeability determined using the observed $p(\lambda)$ and Eqs. (7.27) and (7.28) for Sample A at the confining stress of 4000 psi are presented in Figure 7.13. The complex behavior shown in Figure 7.13 is because of the combined effects of Knudsen diffusion/slippage flow and mechanical deformation. Within the pore pressure range below 1000 psi, the permeability decreases with increasing pore pressure because of the dominated effect of Knudsen diffusion and slippage flow at low pore pressures. As shown by the theoretical results given in Figure 7.9, the enhancement of effective permeability by Knudsen diffusion and slippage flow becomes less significant at a larger pore pressure. For pore pressures larger than 1000 psi, the permeability-pressure trend reverses and the permeability increases with increasing pore pressure

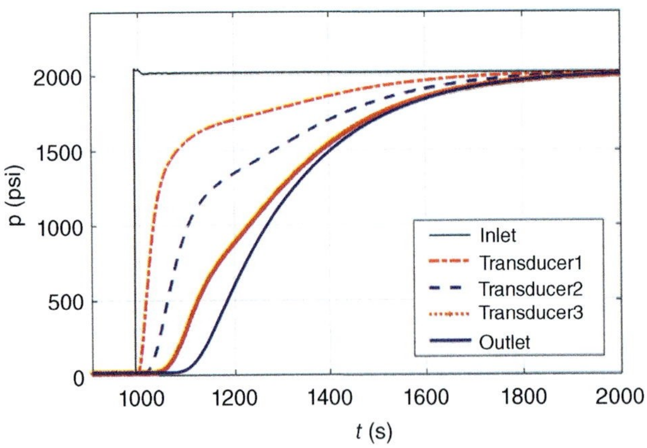

Figure 7.11 The pressure transient curves at inlet, outlet and three pressure transducers for Sample A.

because the mechanical deformation becomes more important than the diffusion effect. At a given confining stress, the increase in pore pressure enlarges pore sizes and consequently the permeability.

The complexity of the permeability – pore pressure function shown in Figure 7.13 is highly relevant to shale gas production in practice because the reservoir is subject to considerable pore pressure changes during production. Our method provides a very efficient way to capture this complexity. For example, the laboratory test for Sample A only took about 30 minutes to determine the permeability curve shown in Figure 7.13. Essentially, the method measures a permeability-pressure curve using the same amount of time as that used to measure a single permeability data point via the "point-by-point" method. If five data points were needed for a reasonable representation of the permeability curve in Figure 7.13, the "point-by-point" method would take five times longer than the current method.

The collapsed pressure curves shown in Figure 7.12, as discussed previously, indicate the validity of the current method. Further comparisons with the other two methods including the steady-state and late-time quasi-steady-state methods shown in Figure 7.13 demonstrate the validity of the proposed method (Liu and Zhang 2020). The steady-state flow method was conducted using a short core sample (0.5 in. long) obtained by cutting Sample A near its inlet, pore pressure of 1000 psi, and confining pressure of 4000 psi. The obtained permeability is slightly higher, but still close enough to the permeability value determined using the new method at the same confining and pore pressures. The slight difference is expected because of the disturbance of rock samples during the cutting and loading/unloading processes in the sample preparation and the test procedures. The second method, called late-time quasi steady-state method in Figure 7.13, is to directly use the late-time test data for estimating a permeability point. At the late time of the test, the steady-state flow is approximately established between the sample inlet and the first pressure transducer. By assuming a steady-state flow there, a permeability at the corresponding pore pressure can be obtained based on Darcy's law, the measured flow rate into the core inlet and pressure gradient between the inlet and the first pressure transducer. The estimated permeability is almost identical to the value from the permeability curve at the given pore pressure.

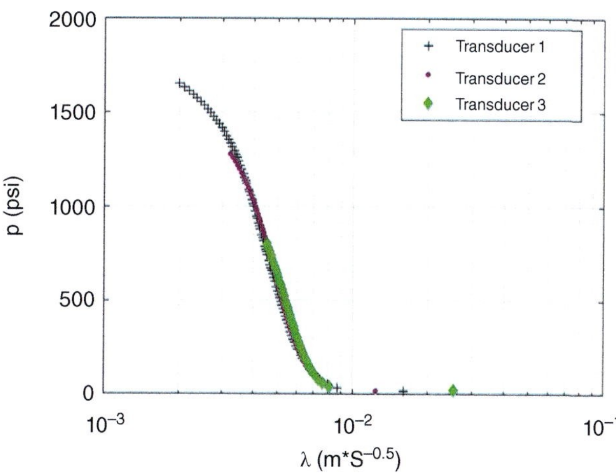

Figure 7.12 Pressure as a function of the transformation variable λ with respect to three pressure transducers for Sample A.

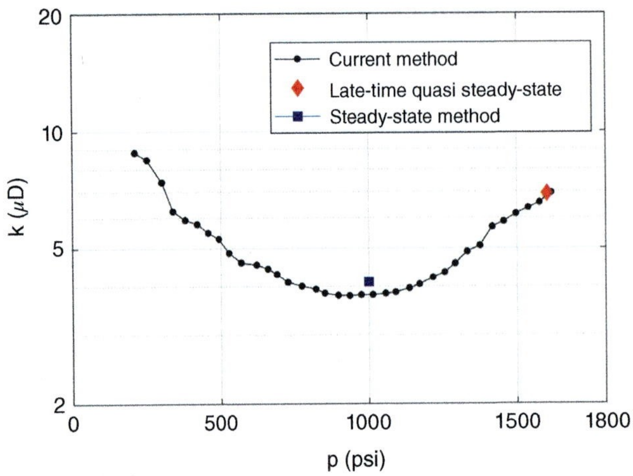

Figure 7.13 Permeability versus pore pressure at the constant confining pressure of 4000 psi for Sample A.

As another example, Figure 7.14 shows the observed permeability as a function of pore pressure for Sample B, which has much lower permeability values on the order of several nD at high pore pressures than Sample A. Recall that Sample A has permeability values on the order of μD. Knudsen diffusion/slippage flow has a much more important effect on permeability at low pore pressures for Sample B than Sample A (Figure 7.13), as evidenced by a more significant permeability enhancement for Sample B. This is consistent with theoretical expectations because the diffusion effect is more pronounced for smaller pore sizes corresponding to smaller permeability values. Interestingly, the permeability for Sample B essentially remains unchanged with pore pressure for pore pressures greater than 800 psi in Figure 7.14. This suggests that the mechanical deformation effect on permeability is insignificant for Sample B under the test conditions (Figure 7.14). This is a result of the fact that a lower permeability, or a smaller porosity, corresponds to a stiffer rock sample (Jaeger et al. 2007).

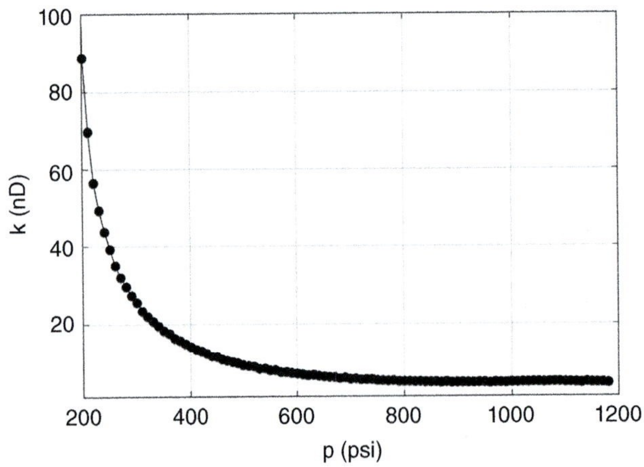

Figure 7.14 Permeability as a function of pore pressure at the constant confining pressure of 4000 psi for Sample B.

7.5 Summary and Conclusions

The laboratory measurement of tight rock permeability is important for several related areas. In this chapter, we briefly reviewed the commonly used laboratory methods for tight rock permeability measurements, including the steady-state flow, PPD, and GRI methods, and then presented the two newly developed laboratory methods.

Motivated by the fact that plug-scale tight rock matrix permeability cannot be measured with the currently available laboratory methods for core samples with artificially induced fractures, we have developed a method to measure both fracture and matrix permeability from them. These induced fractures are common for highly laminated tight rocks mainly as a result of the stress release when the samples are taken from the reservoir to the laboratory. Our method is a generalization of the conventional PPD method and uses the early time data, dominated by fracture flow, to estimate fracture permeability and the late time data, dominated by matrix flow, to estimate matrix permeability.

To deal with the inefficiency of the conventional methods that are based on linear theories for fluid flow in core samples, we have also developed a laboratory method to directly measure, with a single test run, the tight rock permeability as a function of pore pressure from a non-fractured rock sample. It, unlike conventional methods, is based on a nonlinear theory of fluid flow in a core sample and can increase efficiency by at least multiple times for measuring the stress dependency of the permeability.

Both methods were validated with laboratory measurements. The test results indicate that tight rock permeability can be considerably stress-sensitive, and its stress-dependence behavior could be quite complex owing to the combined effects of Knudsen diffusion/slippage flow and mechanical deformation.

References

API (1998). Recommended practices for core analysis. In: *Recommended Practice 40*, 2ee. American Petroleum Institute.

Besnard, M., Buser, M., Fairlie, I. et al. (2019). World nuclear waste report 2019: focus Europe. https://worldnuclearwastereport.org (accessed 16 July 2019).

Beskok, A. and Karniadakis, E.G. (1999). A model for flows in channels, pipes, and ducts at micro and nano scales. *Microscale Thermophysical Engineering* 3 (1): 43–77.

Brace, W.F., Walsh, J.B., and Frangos, W.T. (1968). Permeability of granite under high pressure. *Journal of Geophysical Research* 73: 2225–2236.

Bruce, R.R. and Klute, A. (1956). The measurement of soil moisture diffusivity. *Soil Science Society of America Journal* 20: 458–462.

Chen, H., Liu, H.H., and Chen, J.H. (2018). A method for correcting low permeability laboratory measurements for leaks: theory, methodology and algorithms. *Journal of Natural Gas Science and Engineering* 56: 608–618.

Civan, F. (2010). Effective correlation of apparent gas permeability in tight porous media. *Transport in Porous Media* 82: 375–384.

Cui, X., Bustin, A.M., and Bustin, R.M. (2009). Measurements of gas permeability and diffusivity of tight reservoir rocks: different approaches and their applications. *Geofluids* 9: 208–223.

Dicker, A.I. and Smits, R.M. (1988). A practical approach for determining permeability from laboratory pressure-pulse decay measurements. *SPE International Meeting on Petroleum Engineering* (1–4 November 1988), Tianjin, China. Society of Petroleum Engineers (SPE).

Gan, Z., Griffin, T., Dacy, J. et al. (2018). Fast pressure-decay core permeability measurement for tight rocks. *SPWLA 59th Annual Logging Symposium* (1 October 2018), London, UK. Society of Petrophysicists and Well-Log Analysts (SPWLA).

GRI (1996). Development of laboratory and petrophysical techniques for evaluating shale reservoirs. *GRI Report GRI-95/0496*, Gas Technology Institute.

Hsieh, P.A., Tracy, J.V., Neuzil, C.E. et al. (1981). A transient laboratory method for determining the hydraulic properties of tight rocks – I. Theory. *International Journal of Rock Mechanics and Mining Science and Geomechanics Abstracts* 18: 245–252.

IEAGHG (2011). *Caprock Systems for CO2 Geological Storage*. Cheltenham, UK: IEAGHG.

IPCC (2007). *Climate Change: The Physical Science Basis*. Geneva, Switzerland: IPCC.

IPCC (2014). *Climate Change 2014: Synthesis Report*. Geneva, Switzerland: IPCC.

Jaeger, J.C., Cook, N.G.W., and Zimmerman, R.W. (2007). *Fundamentals of Rock Mechanics*, 4ee. Malden, MA: Blackwell Publishing.

Javadpour, H.S.F., Ettehadtavakkol, A., and Darabi, H. (2014). Nonemperical apparent permeability of shale. *SPE Reservoir Evaluation & Engineering* 17 (3): 414–423. https://doi.org/10.2118/170243-PA.

Jones, S.C. (1997). A Technique for fast pulse-decay permeability measurements in tight rocks. *SPE Reservoir Evaluation & Engineering* 12: 19–25.

Liu, H.H., Rutqvist, J., and Berryman, J.G. (2009). On the relationship between stress and elastic strain for porous and fractured rock. *International Journal of Rock Mechanics and Mining Sciences* 46: 289–296.

Liu, H.H. (2017). *Fluid Flow in the Subsurface: History, Generalization and Applications of Physical Laws*. Switzerland: Springer.

Liu, H.H., Lai, B., Chen, J. et al. (2016). Pressure pulse-decay tests in a dual-continuum medium: late-time behavior. *Journal of Petroleum Science and Engineering* 147: 292–301.

Liu, H.H., Georgi, D., and Chen, J. (2018). Correction of source-rock permeability measurements owing to slip flow and Knudson diffusion: a method and its evaluation. *Petroleum Science* 15 (1): 116–125.

Liu, H.H. and Zhang, J. (2020). An efficient laboratory method to measure the combined effects of Knudsen diffusion and mechanical deformation on shale permeability. *Journal of Contaminant Hydrology* 232: 103652. https://doi.org/10.1016/j.jconhyd.2020.103652.

Luffel, D.L., Hopkins, C.W., and Schettler, P.D. (1993). Matrix permeability measurement of gas productive shales. *SPE Annual Technical Conference and Exhibition* (3–6 October 1993), Houston, Texas. Society of Petroleum Engineers (SPE).

Matano, C. (1933). On the relation between the diffusion-coefficients and concentrations of solid metals (the nickel-copper system). *Japanese Journal of Physics* 8: 109–113.

Mokhtari, M., Alqahtani, A.A., Tutuncu, A.N. et al. (2013). Stress-dependent permeability anisotropy and wettability of shale resources. *Unconventional Resources Technology Conference* (12–14 August 2013), Denver, USA.

NEI (2020). What is nuclear energy. https://www.nei.org/fundamentals/what-is-nuclear-energy (accessed 15 July 2020).

Ning, X. (1992). The measurement of matrix and fracture properties in naturally fractured low permeability cores using a pressure pulse method. Ph.D. Thesis. Texas A&M University.

Ning, X., Fan, J., Holditch, S.A. et al. (1993). The measurement of matrix and fractures properties in naturally fractured cores. *Rocky Mountain Regional/Low Permeability Reservoirs Symposium* (26–28 April 1993), Denver, USA. Society of Petroleum Engineers (SPE).

Nutt, M., Voegele, M., Jové-Colón, C. et al. (2011). Used fuel disposition campaign disposal research and development roadmap. *Used Fuel Disposition Campaign Report*. US Department of Energy.

Ozkan, E., Brown, M., Raghhavan, R. et al. (2011). Comparison of fractured-horizontal-well performance in tight sand and shale reservoirs. *SPE Reservoir Evaluation & Engineering* 14: 248–259.

Profice, S., Hamon, G., and Nicot, B., (2016). Low permeability measurements: insights. *International Symposium of the Society of Core Analysis in St John's Newfoundland and Labrador* (1 February 2016), Canada. Society of Petrophysicists and Well-Log Analysts (SPWLA).

Rutqvist, J., Wu, Y.S., Tsang, C.F. et al. (2002). A modeling approach for analysis of coupled multiphase fluid flow, heat transfer, and deformation in fractured porous rock. *International Journal of Rock Mechanics and Mining Sciences* 39 (4): 429–442.

Sander, R., Pan, Z., and Connell, L.D. (2017). Laboratory measurement of low permeability unconventional gas reservoir rocks: a review of experimental methods. *Journal of Natural Gas Science and Engineering* 37: 248–279.

Sinha, S., Braun, E.M., Determan, M.D. et al. (2013). Steady-state permeability measurements on intact shale samples at reservoir conditions: effect of stress, temperature, pressure, and type of gas. *SPE Middle East Oil and Gas Show and Conference* (10–13 March 2013). Manama, Bahrain. Society of Petroleum Engineers (SPE).

Tinni, A., Fathi, E., Agarwal, R. et al. (2012). Shale permeability measurements on plugs and crushed samples. *SPE Canadian Unconventional Resources Conference* (30 October–1 November 2012). Calgary, Alberta, Canada. Society of Petroleum Engineers (SPE).

Zhang, J., Liu, H.H., and Boudjatit, M. (2020). Matrix permeability measurement from fractured unconventional source-rock samples: method and application. *Journal of Contaminant Hydrology* 233: 103663. https://doi.org/10.1016/j.jconhyd.2020.103663.

Ziarani, S.A. and Aguilera, R. (2012). Knudsen's permeability correction for tight porous media. *Transport in Porous Media* 91 (1): 239–260.

8

Stress-Dependent Matrix Permeability in Unconventional Reservoir Rocks

Athma R. Bhandari[1], *Peter B. Flemings*[1,2], *and Sebastian Ramiro-Ramirez*[1,2]

[1] *Institute for Geophysics, Jackson School of Geosciences, The University of Texas at Austin, Austin, TX, USA*
[2] *Department of Geological Sciences, Jackson School of Geosciences, The University of Texas at Austin, Austin, TX, USA*

8.1 Introduction

The permeability of shales has attracted a considerable amount of research in the last decade. Laboratory measurements that measure permeability as low as ~1 nD using gas or liquid on core plugs under in situ stress conditions are now widely published (for example, Dong et al. 2010; Rosen et al. 2014; Bhandari et al. 2015, 2018, 2019; Chhatre et al. 2015; Ismail and Zoback 2016; Kosanke and Warren 2016; Mckernan et al. 2017). These measurements, made under reservoir stress conditions, remove many drawbacks of the Gas Research Institute (GRI) crushed rock permeability measurement approach (Luffel and Guidry 1992; Luffel et al. 1993). For example, the GRI method does not include the effect of in-situ stress. In addition, there is not a standard analytical approach, which leads to large measurement variability (Passey et al. 2010); furthermore, it is not possible to study sample heterogeneity due to millimeter-scale stratigraphic variation (Sinha et al. 2012). However, laboratory permeability testing on core plugs of clay-rich lithofacies is challenging, and proper care is needed to understand the variable measurements fully.

It is challenging to extract intact core plugs in clay-rich rocks due to disturbances that may occur during both coring and sampling. Most core plugs show multiple bedding plane parallel micro-fractures introduced during core retrieval, core storage, and/or plug extraction from the core. These micro-fractures (Pluymakers et al. 2017) are considered to be related to dehydration of the core when it is disturbed and possibly due to the release of stored internal strain energy due to stress relief (Ougier-Simonin et al. 2016). The width to the length ratio (i.e. the aspect ratio) of such micro-fractures is $<10^{-2}$ and ranges between 10^{-5} and 10^{-3} (Kranz 1983; Ougier-Simonin et al. 2016). Some researchers (Luffel et al. 1993; Suarez-Rivera et al. 2012) argue that these micro-fractures will remain partially open even if reservoir stress conditions are applied again in the laboratory. Hence, they dominate the core plug-based measurements, and the measured permeability may exceed the intact matrix permeability by several orders of magnitude (Luffel et al. 1993; Suarez-Rivera et al. 2012). Even when the applied stress closes the artificial micro-fractures, the critical confining stress required for micro-fracture closure is dependent on the rock facies and sample quality (Suarez-Rivera et al. 2012; Rydzy et al. 2016). Recent results with variable sample quality from the same lithological unit have shown that with confining stress cycling, the micro-fractures

Physics of Fluid Flow and Transport in Unconventional Reservoir Rocks, First Edition.
Edited by Behzad Ghanbarian, Feng Liang, and Hui-Hai Liu.

can be fully closed, and measurements converge to identical permeability if a proper testing protocol is followed (for example, Bhandari et al. 2019). The true stress dependence of permeability in shales, therefore, can be investigated after reversing the sample damage.

The objectives of this chapter are to:

1) show how permeability evolves with confining stress cycling in three different lithofacies and illustrate the general realities of sample damage in clay-rich units,
2) infer matrix permeability behavior, and
3) link our lab-measured permeabilities and their stress dependence to the composition and pore systems in shales.

8.2 Sample Descriptions

We present four samples representing three lithofacies: organic-rich siliceous mudstone (two samples), siltstone, and matrix-supported conglomerate. The tested core plugs (diameter = 38.1 mm; length = 16.9 to 19.2 mm) were drilled with their longitudinal axis oriented parallel to the bedding plane (i.e. horizontal plugs) from preserved vertical cores extracted from the same well. The organic-rich siliceous mudstone (the most dominant lithofacies in the formation) and matrix-supported conglomerate samples are from the Wolfcamp operational unit, and the siltstone sample is from the Bone Spring Formation in the Delaware Basin, West Texas, USA (Ramiro-Ramirez 2022).

Bulk composition analyses of the core plugs' end trimmings using X-ray powder diffraction (XRD) (Figure 8.1) show the organic-rich siliceous mudstone Samples 1 and 2 are composed of 45–52 vol% quartz and 24–33 vol% clays and have Leco TOC of 4 and 3 vol%, respectively. The siltstone has 52 vol% quartz, 13 vol% feldspars, 20 vol% dolomite, ~12 vol% clays, and Leco TOC of 1 vol %. The matrix-supported conglomerate has 43 vol% content in calcite, 29 vol% quartz, 9 vol% clays, and Leco TOC of 2 vol%. The vol% given here is calculated from the measured wt% using the density of the minerals and Leco TOC.

We estimated the total porosity at the "as-received" condition by combining helium porosimetry and nuclear magnetic resonance (NMR) measurements. The helium porosity was 4.4% for the organic-rich siliceous mudstone Sample 1, 6.1% for Sample 2, 7.0% for the siltstone, and 2.0% for the matrix-supported conglomerate. NMR measurements resulted in 6.3%, 8.3%, 1.8%, and 1.7% liquid-filled porosity, respectively.

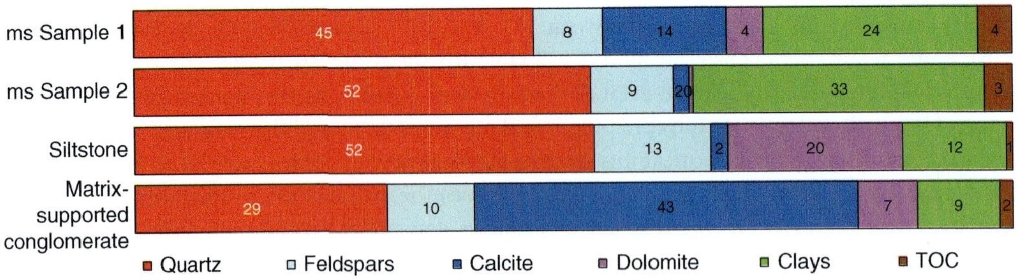

Figure 8.1 Mineralogy and Leco TOC (vol%) for the four samples studied here. "ms" denotes organic-rich siliceous mudstone.

8.3 Permeability Test Program

The permeability test program for the organic-rich siliceous mudstone samples is shown in Figure 8.2. It consists of one gas (argon) permeability measurement using the pulse decay method (Brace et al. 1968; Bhandari et al. 2015) at confining stress of 13.8 MPa and then several liquid (dodecane) permeability measurements using the steady-state method for two confining stress cycles while maintaining the average pore pressure constant at ~7.0 MPa (see further details of the steady-state method in the next paragraph): Cycle 1 (Tests 1–9) and Cycle 2 (Tests 9–21). Due to time constraints, we stopped the experiment at Test 12 in the organic-rich siliceous mudstone Sample 2.

We followed the same confining stress cycling (Figure 8.2) in the siltstone and matrix-supported conglomerate samples. However, we did not conduct the gas permeability measurements and omitted several liquid permeability measurements under some intermediate stress conditions. We directly saturated these samples with dodecane. We measured permeability under stress conditions for Tests 4–7, 9, 11–12, 14–16, 19, and 21 on the siltstone sample and at stress conditions for Tests 6, 9, 12, 16, 19, and 21 on the matrix-supported conglomerate sample. During these tests, we maintained a constant pressure of 6.55–6.81 MPa on the downstream side and first applied a pressure differential typically of 0.69–0.17 MPa across the core plug (therefore, the pressure on the upstream side is higher than the pressure on the downstream side, and this pressure difference drives the flow) to estimate the flow rate. Then, we applied the constant flow rate on the upstream side while maintaining constant pressure on the downstream side. This way, we maintained the average pore pressure (arithmetic average of the upstream pressure and the downstream pressure) constant at

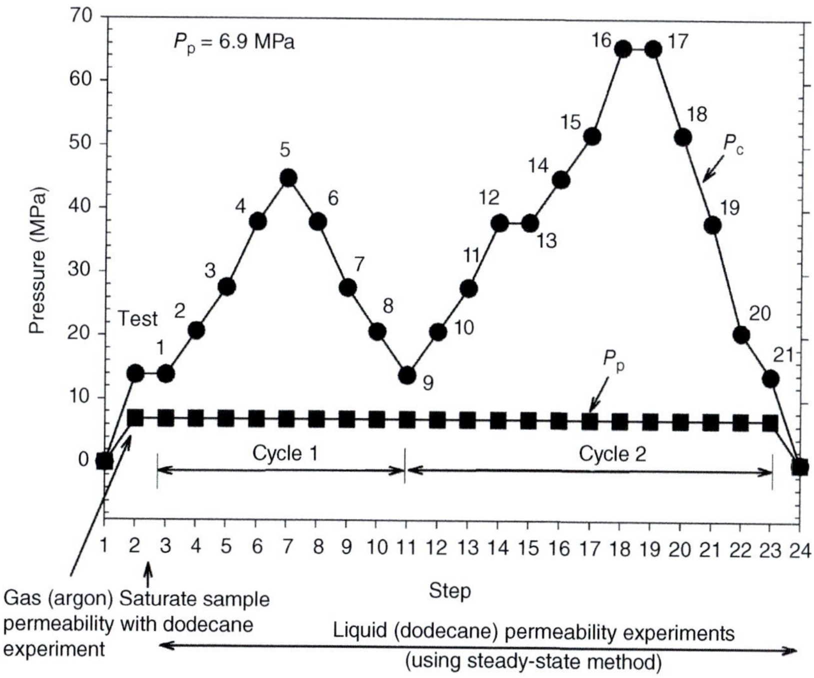

Figure 8.2 Permeability test program *Source:* Bhandari et al. (2019)/With permission of Elsevier. P_c is the confining stress (set equally in the axial and radial directions) and P_p is the average pore pressure. Tests 1–9 correspond to Cycle 1 and Tests 9–21 to Cycle 2.

~7.0 MPa. The total test duration for each sample is ~1 month. The experimental setup and test procedures are described in detail by Bhandari et al. (2019), Ramiro-Ramirez et al. (2020), and Ramiro-Ramirez (2022).

Due to the extremely low fluid flow rates involved, measuring the nano-Darcy range permeability using the steady-state technique in low-permeability rocks is challenging. To accomplish these measurements, we limited our leaks to usually below 10% of the lowest flow rate required (on the order of 1×10^{-5} ml/min) and used a high-precision pump with a flow rate resolution of 0.12 nl/min. In addition, we maintained the temperature of the experimental setup stable at 30.0 °C (max. deviation $<\pm 0.1$ °C; std. deviation = ± 0.04 °C) using a temperature control system.

8.4 Permeability Behavior with Confining Stress Cycling

Figure 8.3 shows the effective permeability to dodecane versus the effective stress $P_c - P_p$ for four samples studied here. The samples behaved differently with the confining stress cycling. The permeability of organic-rich siliceous mudstone Sample 1 exhibited the most stress dependence, and siltstone showed the least stress dependence.

The stress dependence of permeability varied within a narrow range (1.4×10^{-18} m^2 – 1.2×10^{-18} m^2) for the siltstone sample. The permeability decline with increasing effective stress (from 7.0 to 58.5 MPa) is less than 10%. The permeability of organic-rich siliceous mudstone Sample 1 declined by more than three orders of magnitude (from 4.4×10^{-17} m^2 to 2.0×10^{-20} m^2) while increasing the $P_c - P_p$ from 7.0 to 58.5 MPa. It also exhibited significant hysteresis (permeability decrease during loading was not recovered during unloading) in both cycles. The permeability of the organic-rich siliceous mudstone (Sample 2) initially at $P_c - P_p = 7.0$ MPa was 1.6×10^{-19} m^2, which is considerably lower than that of Sample 1 (compare orange triangles to blue circles, Figure 8.3). Sample 2 exhibited a relatively small decline from 1.6×10^{-19} m^2 to 3.3×10^{-20} m^2 during loading and then a slight increase to 5.3×10^{-20} m^2 during unloading in Cycle 1. Next, the permeability decreased slightly to 2.7×10^{-20} m^2 at $P_c - P_p = 30.3$ MPa as we reloaded it in Cycle 2. The matrix-supported conglomerate sample has a permeability of 5.9×10^{-20} m^2 at $P_c - P_p =$ 30.9 MPa, and it increased to 1.0×10^{-19} m^2 at $P_c - P_p = 6.9$ MPa. It decreased slowly to 3.6×10^{-20} m^2 (i.e. a 60% decline) upon loading to $P_c - P_p = 58.5$ MPa, and increased slightly to 6.6×10^{-20} m^2 upon unloading and did not recover fully (in contrast to the siltstone sample).

The initial permeability was 4.4×10^{-17} m^2 for the organic-rich siliceous mudstone Sample 1 and 1.6×10^{-19} m^2 for Sample 2. This difference in measured permeability may be attributed to the sample quality. The μ – CT image of mudstone Sample 1 (Figure 8.4a), taken before permeability testing, shows multiple bedding-parallel micro-fractures, whereas Sample 2 is more intact (Figure 8.4b). The μ – CT images of the samples (voxel size = 24.6 μm) taken after permeability testing (Figure 8.5a,b) show that these micro-fractures were closed substantially by the testing process. Sample 1 had more permeability hysteresis than Sample 2: its permeability declined much more with increasing stress, and the permeability did not increase as much when the sample was unloaded. We interpret that the micro-fractures close during testing and do not reopen to their initial state even after reducing the confining stress to the original condition. The effect is more dramatic with Sample 1 than Sample 2 because it is more damaged initially. These results show that it is important to understand the sample quality before permeability testing to interpret the stress dependence of permeability correctly.

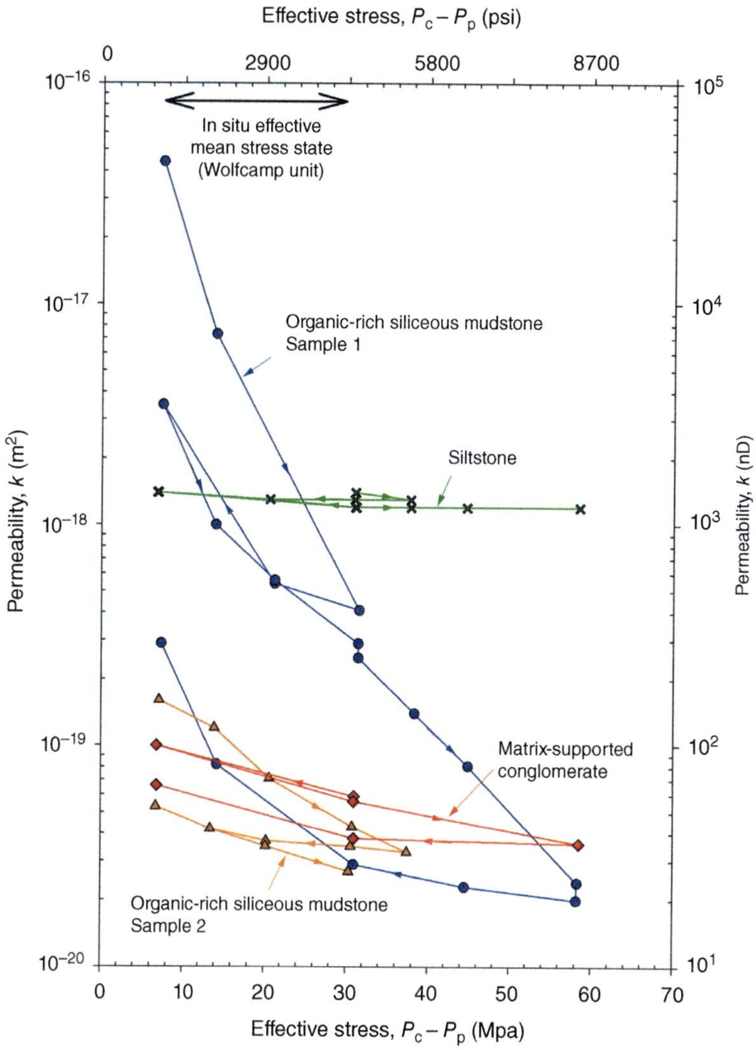

Figure 8.3 Effective permeability to dodecane versus effective stress for siltstone, matrix-supported conglomerate, and organic-rich siliceous mudstone samples. The in situ effective mean stress state at the sampling depth in the Wolfcamp operational unit (based on 0.67–0.90 psi/ft pore pressure gradient, 0.95–1.05 psi/ft lateral stress gradient, and 1.05–1.075 psi/ft vertical stress gradient) is indicated by the double arrow sign.

The siltstone and matrix-supported conglomerate samples look intact with no visible microfractures (Figure 8.4c,d) and appear to be relatively free from sampling artifacts (within the voxel resolution of ~25 μm). These samples have much less permeability hysteresis. Cement, bonding the grains, may make these samples more intact. Ramiro-Ramirez et al. (2020) reported that the dolomitic calcareous sandstone and other carbonate-bearing lithofacies (due to the presence of cement) showed less coring-induced microfractures than organic-rich siliceous mudstone core plugs.

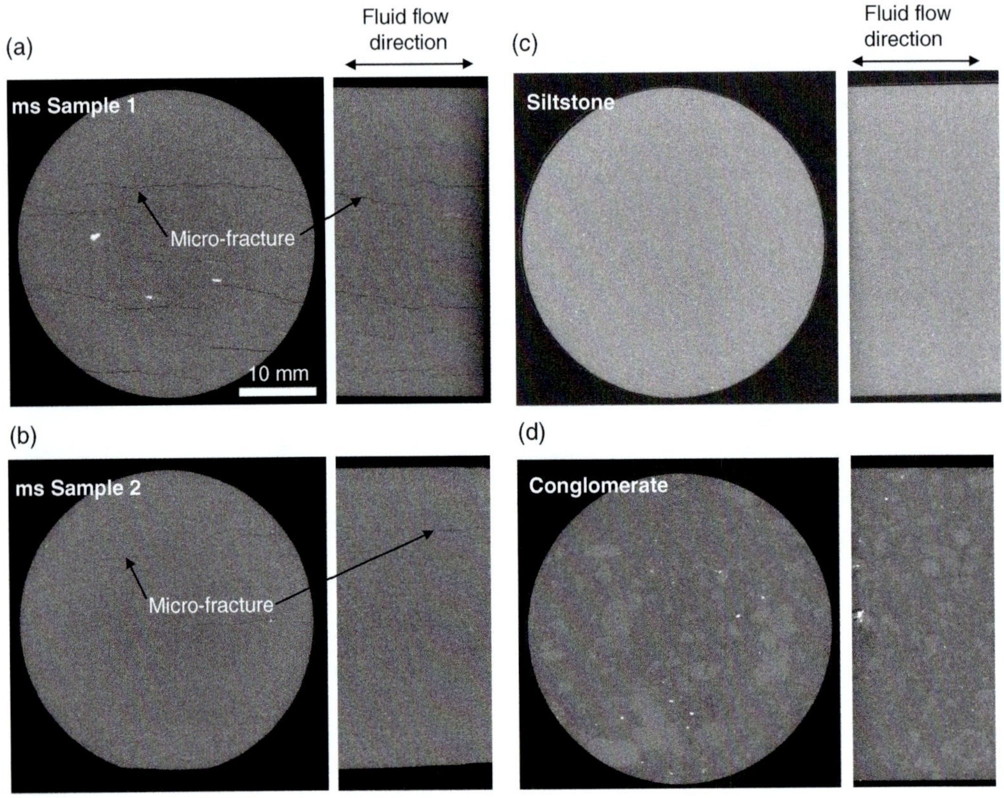

Figure 8.4 Micro-computed tomography (μ-CT) images showing diametrical (left image) and longitudinal (right image) sections of the core plugs taken before permeability testing: (a) ms Sample 1, (b) ms Sample 2, (c) Siltstone, and (d) Conglomerate. The bedding plane is approximately horizontal in these images. *Source:* (a) and (b) Bhandari et al. (2019)/from ELSEVIER. "ms" denotes organic-rich siliceous mudstone.

We interpret that the matrix permeability of the organic-rich siliceous mudstone ranged from $\sim 2.0 \times 10^{-20}\,\text{m}^2$ to $5.0 \times 10^{-20}\,\text{m}^2$ as the effective stress increased from 7.0 to 58.5 MPa (Figure 8.3). For Sample 1, we interpret that the micro-fractures closed after the maximum effective stress of 58.3 MPa was applied in the second cycle. We also concluded that the permeability of $\sim 2.0 \times 10^{-20}\,\text{m}^2$ to $3.0 \times 10^{-20}\,\text{m}^2$ measured during unloading ($P_c - P_p = 58.3 - 31.4$ MPa) characterize the matrix permeability. There may be partial re-opening of micro-fractures below $P_c - P_p = 31.4$ MPa as the permeability increases significantly from the log-linear trend. For Sample 2 (which has a slightly higher clay content, see Figure 8.1), we interpret that the permeability decline during the first cycle was due to the closure of micro-fractures, and we essentially closed the micro-fractures at lower effective stress (37.4 MPa). The permeability measured after reloading the sample ranged from $\sim 5.0 \times 10^{-20}\,\text{m}^2$ to $\sim 3.0 \times 10^{-20}\,\text{m}^2$ as the $P_c - P_p$ increased from 7.0 to 31.4 MPa; these values are likely to characterize the matrix permeability at the in-situ effective stress ($P_c - P_p = 7.0 - 31.4$ MPa). These permeability values are consistent with the matrix permeability estimate of $1.0 \times 10^{-20}\,\text{m}^2$ derived from a dual-time scale pressure dissipation behavior based on history-matching the initial pulse decay experiment data using argon gas in a fractured core model (see Bhandari et al. 2019).

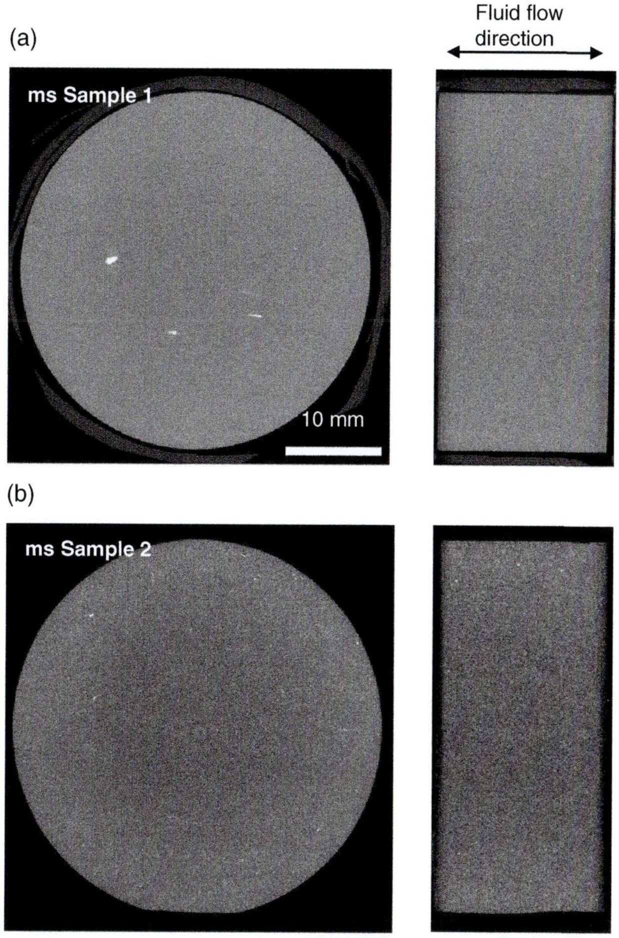

Figure 8.5 Micro-computed tomography (μ-CT) images showing diametrical (left image) and longitudinal (right image) sections of the core plugs taken after permeability testing: (a) ms Sample 1, and (b) ms Sample 2. *Source:* Bhandari et al. (2019)/from ELSEVIER. The bedding plane is approximately horizontal. "ms" denotes organic-rich siliceous mudstone.

We showed that sample damage introduced during core retrieval, core storage, and/or plug extraction from cores can significantly increase permeability under low confining stress conditions, especially in organic-rich siliceous mudstone. However, after cycling the confining stress to in situ effective mean stress conditions ($P_p - P_p = 7.0 - 31.4$ MPa), Figure 8.3) and significantly beyond them (up to $P_c - P_p = 58.5$ MPa), we showed it is possible to reverse such sample damage in mudstones. Through this approach, we demonstrated how to interpret the matrix permeability of the rock based on permeability behavior under various stress conditions. We infer that the clays (due to their higher ductility than other mineral components) play a crucial role in closing these artificial micro-fractures during stress cycling. In the case of Sample 1, our lower-quality sample, effective stresses (up to $P_c - P_p = 58.5$ MPa) significantly higher than the in situ stress ($P_p - P_p = 7.0 -$ 31.4 MPa) were required to completely close micro-fractures. At these stresses, there may be permanent (plastic) compression of pore throats in the matrix in addition to micro-fracture closure.

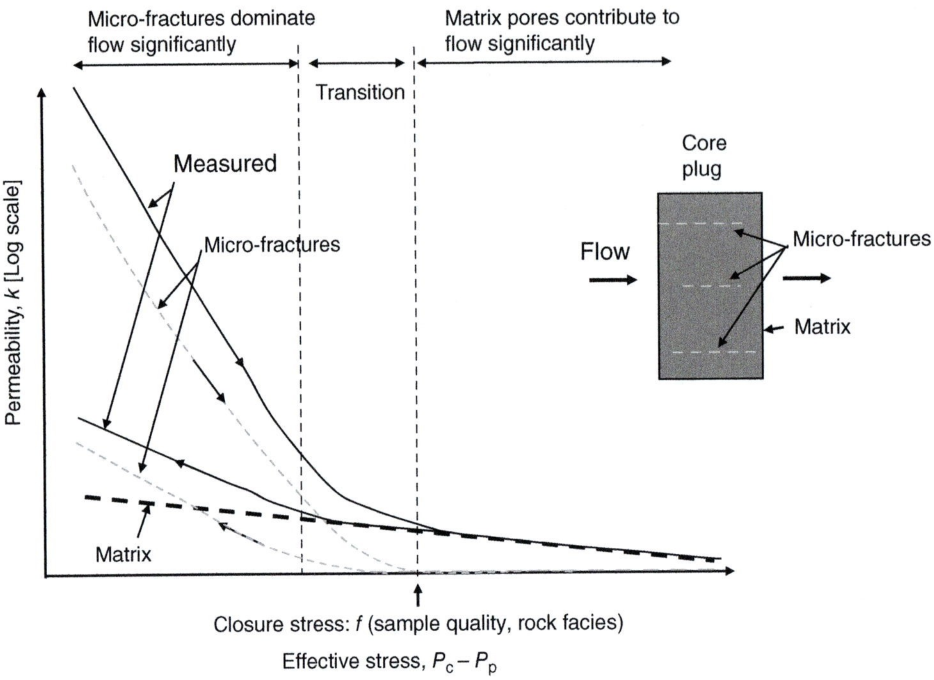

Figure 8.6 Schematic illustration showing how micro-fractures and matrix pores contribute to the measured permeability in mudstones. The grey dashed line represents the permeability due to micro-fractures alone.

It is possible that this may play a role in the observed permeability versus effective stress response. However, since the effect of compressibility of the micro-fractures compared to the matrix is relatively large before micro-fractures closure (Milch et al. 2016; Rydzy et al. 2016), we anticipate the effect of pore throats contraction on the measured overall permeability to be relatively small.

Figure 8.6 illustrates how micro-fractures and matrix pores contribute to the measured permeability in mudstones. The micro-fractures closure stress depends on sample quality and rock facies (composition). Before micro-fractures closure, the micro-fractures dominate flow significantly. After micro-fractures closure, the matrix pores contribute to flow significantly. The micro-fractures do not reopen during unloading until the confining stress is lowered to values much lower than the closure stress.

8.5 Matrix Permeability Behavior

We have shown that it is more challenging to determine permeability in mudstone samples than in siltstone or matrix-supported conglomerate sample because these samples are more likely to suffer disturbance. However, we suggest it is possible to infer the matrix permeability and its dependence on stress, if a comprehensive testing program is conducted over a range of effective stresses ($P_c - P_p = 7.0 - 58.5$ MPa).

Figure 8.7 shows the interpreted matrix permeabilities at different effective stresses for our samples. These permeability values were extracted from the second loading/unloading cycle (Figure 8.3), i.e. after closing micro-fractures present in the rocks. The matrix permeability of

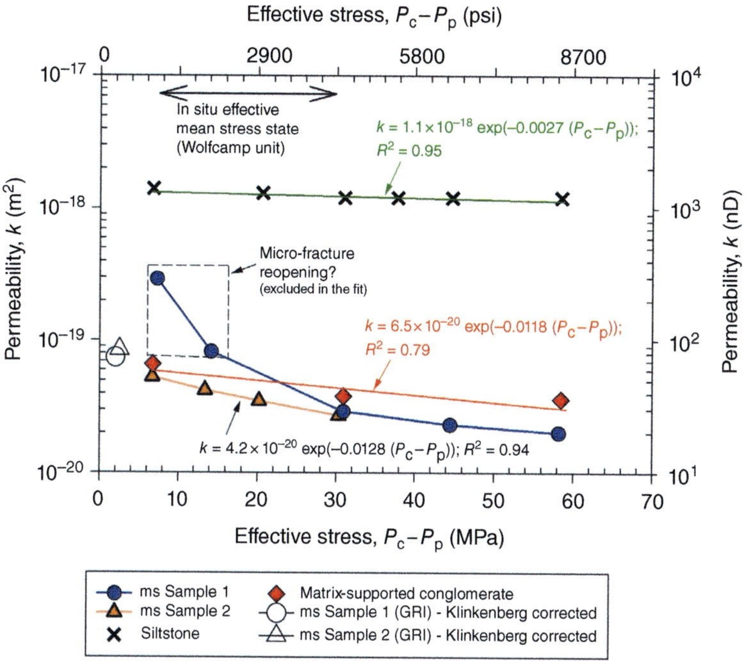

Figure 8.7 Matrix permeability versus effective stress for four plugs. The exponential relationship, $k = k_0 \exp(-\alpha(P_c - P_p))$, was fit to data from the second loading/unloading cycle. k_0 is in m^2, $P_c - P_p$ is in MPa, and α is in MPa^{-1}. The Klinkenberg-corrected permeability using the crushed-rock GRI method for samples from core ms Sample 1 and core ms Sample 2 are shown with open circle and open triangle. The Klinkenberg-uncorrected permeability (not shown) are $4.6 \times 10^{-19}\ m^2$(~462 nD) and $5.2 \times 10^{-19}\ m^2$(~531 nD), respectively. "ms" denotes organic-rich siliceous mudstone. Our estimate of the probable in situ effective mean stress state at the sampling depth in the Wolfcamp operational unit (based on 0.67 – 0.90 psi/ft pore pressure gradient, 0.95 – 1.05 psi/ft lateral stress gradient, and 1.05 – 1.075 psi/ft vertical stress gradient) is indicated by the double arrow sign.

the siltstone is nearly constant with effective stress and around $1.2 \times 10^{-18}\ m^2$($\sim$1200 nD) for 8 MPa $< P_c - P_p <$ 58 MPa. This is almost two orders of magnitude greater than the matrix-supported conglomerate and the organic-rich siliceous mudstone. Our mudstone samples have permeabilities ranging between $2.0 \times 10^{-20}\ m^2$(20 nD) to $5.3 \times 10^{-20}\ m^2$(54 nD), in accord with other core-plug-based measurements (Rafatian and Capsan 2015; Mathur et al. 2016; King et al. 2018) and in situ well testing (Zhan et al. 2016, 2017) reported for Wolfcamp mudstones. We also compared our liquid permeability measurements with the gas permeability measurements using the GRI crushed rock method (Figure 8.7). Interestingly, the equivalent liquid permeability for the organic-rich siliceous mudstone Sample 1 (inverted triangle) is $7.4 \times 10^{-19}\ m^2$($\sim$75 nD) and for the organic-rich siliceous mudstone Sample 2 (circle) is $8.5 \times 10^{-20}\ m^2$($\sim$86 nD). These GRI measurements were conducted under zero confining stress and at a low gas pressure (0.7 – 1.4 MPa or 100 – 200 psi). We followed the same approach as given in Bhandari et al. (2015) to correct the Klinkenberg effect (Klinkenberg 1941). Now we can relate the matrix permeability with the pore space characteristics. This allows us to identify what controls fluid flow at the pore level and explain the differences in permeability among lithofacies.

We fit an exponential relationship, $k = k_0 \exp(-\alpha(P_c - P_p))$ (Best and Katsube 1995), to describe the stress-dependent permeability behavior illustrated in Figure 8.7. α is a measure of pore throat

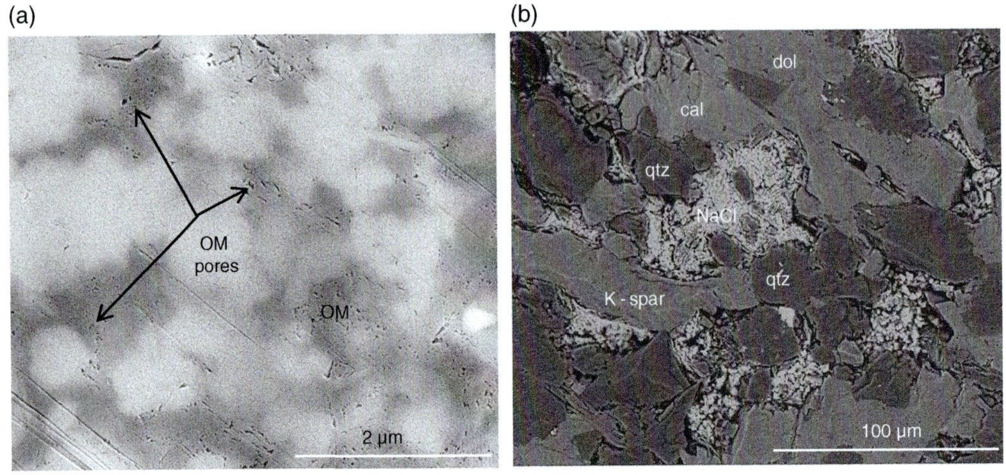

Figure 8.8 Back-scattered SEM image: (a) Organic-rich siliceous mudstone Sample 2. Bhandari et al. (2019)/ from ELSEVIER and (b) siltstone. (Note: SEM image of siltstone taken from twin sample tested using NaCl brine so the pores are filled with precipitated NaCl.) OM is organic matter, qtz denotes quartz, cal represents calcite, dol is dolomite, K-spar indicates feldspar, and NaCl denotes sodium chloride.

compressibility (MPa^{-1}), and k_0 is the permeability (m^2) at zero effective stress. We find α of 0.0027, 0.0118, and 0.0128 MPa^{-1} for the siltstone sample, matrix-supported conglomerate, and organic-rich siliceous mudstone, respectively. The higher pore throat compressibility in mudstone suggests the clay and TOC contents play a crucial role in the stress-dependence of permeability.

The matrix permeability at $P_c - P_p = 30.9$ MPa (Figure 8.7) correlate to pore space characteristics and pore-throat size distribution observed with scanning electron microscopy (SEM) imaging (Figure 8.8) and mercury injection capillary pressure (MICP) measurements (Figure 8.9). The pore system of the organic-rich siliceous samples (for example, see Figure 8.8a for Sample 2) is dominated by nanometer-size interparticle pores between clays, quartz, and carbonate grains and nanometer-size pores within the organic matter. In contrast, the pore system in the siltstone is predominantly composed of micrometer-sized (up to ~15 μm) interparticle pores (Figure 8.8b). The intraparticle pores are primarily in the nanometer-size range but may occasionally be a few micrometers (~2 − 10 μm) in size. The mercury injection capillary pressure measurements (Figure 8.9a) reveal that the peak (i.e. the mode) pore throat radius (Figure 8.9b) is less than 0.01 μm (10 nm) for the mudstone samples ($k = 2.7 - 2.9 \times 10^{-20}\ m^2$), slightly larger than ~0.01 μm for the matrix-supported conglomerate sample ($k = 6.8 \times 10^{-20}\ m^2$), and ~0.1 μm for the siltstone sample ($k = 1.2 \times 10^{-18}\ m^2$).

8.6 Concluding Remarks

This chapter examined the permeability evolution under cycling confining stress conditions (14.0–65.5 MPa) of siltstone, matrix-supported conglomerate, and organic-rich siliceous mudstone samples. We showed that confining stress cycling is crucial to close the micro-fractures in mudstones created from sampling disturbance. In addition, we demonstrated approaches to interpret the matrix permeability behavior and its actual stress dependence.

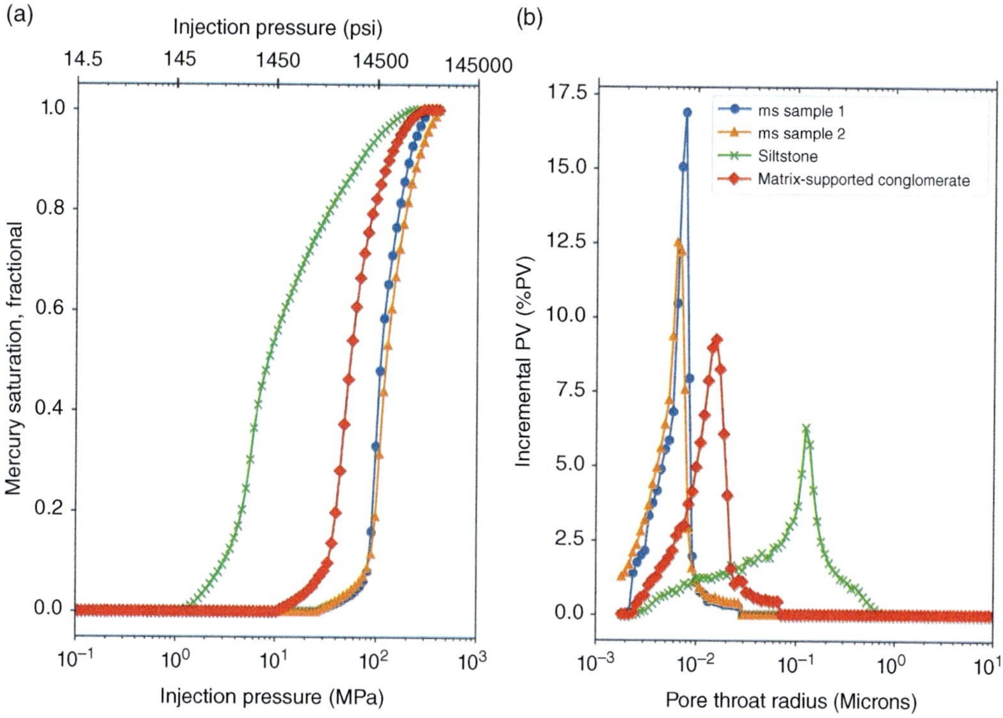

Figure 8.9 (a) Mercury injection capillary pressure versus mercury saturation and (b) pore throat radius versus incremental pore volume injected. Measurements were made on crushed samples of the tested samples. PV indicates pore volume and "ms" denotes organic-rich siliceous mudstone.

We found the bedding parallel matrix permeability in the siltstone sample (~1.2×10^{-18} m^2) is two-orders-of-magnitude greater than the matrix-supported conglomerate and organic-rich siliceous samples (~5.0×10^{-20} m^2 to 2.0×10^{-20} m^2). These measured values agree with the observed pore throat radius distributions and pore systems; the organic-rich siliceous samples have smaller pore throats. In addition, the organic-rich siliceous mudstone matrix permeability exhibited the highest stress dependence. For instance, the pore throat compressibility coefficient in mudstone is 0.0128 MPa^{-1}, five times higher than the siltstone. We attribute this to the greater compressibility of pore throats primarily situated in the clays and organic matter.

The keys to successfully measuring permeability in unconventional reservoir rocks are selecting high-quality samples, completely saturating the core plug with the test fluid, minimizing leaks in the pore fluid lines, precisely controlling temperature, and cycling the confining stress. Micro-CT imaging is a valuable tool to select high-quality samples. Stress cycling is crucial to close the artificial micro-fractures in mudstones, as has been demonstrated previously in other lithologies. Finally, it is becoming possible to reliably examine layer-to-layer permeability variations to explore the stratigraphic heterogeneity and better understand the drainage mechanism in shale reservoirs.

Acknowledgments

The authors are grateful to Shell for funding this research under the Shell-UT Unconventional Research (SUTUR) collaboration.

References

Best, M.E. and Katsube, T.J. (1995). Shale permeability and its significance in hydrocarbon exploration. *The Leading Edge* 14(3): 165–170.

Bhandari, A.R., Flemings, P.B., Hofmann, R., and Polito, P.J. (2018). Stress-dependent in situ gas permeability in the Eagle Ford Shale. *Transport in Porous Media* 123(1): 1–20.

Bhandari, A.R., Flemings, P.B., Polito, P.J. et al. (2015). Anisotropy and stress dependence of permeability in the Barnett Shale. *Transport in Porous Media* 108(2): 393–411.

Bhandari, A.R., Flemings, P.B., Ramiro-Ramirez, S. et al. (2019). Gas and liquid permeability measurements in Wolfcamp samples. *Fuel* 236: 1026–1036.

Brace, W.F., Walsh, J.B., and Frangos, W.T. (1968). Permeability of granite under high pressure. *Journal of Geophysical Research* 73(6): 2225–2236.

Chhatre, S.S., Braun, E.M., Sinha, S. et al. (2015). Steady-state stress-dependent permeability measurements of tight oil-bearing rocks. *Petrophysics* 56(2): 116–124.

Dong, J.-J., Hsu, J.-Y., Wu, W.-J. et al. (2010). Stress-dependence of the permeability and porosity of sandstone and shale from TCDP Hole-A. *International Journal of Rock Mechanics and Mining Sciences* 47(7): 1141–1157.

Ismail, M.I.A. and Zoback, M.D. (2016). Effects of rock mineralogy and pore structure on stress-dependent permeability of shale samples. *Philosophical Transactions of the Royal Society A: Mathematical, Physical and Engineering Sciences* 374(2078): 20150428.

King, H., Sansone, M., Kortunov, P. et al. (2018). Microstructural investigation of stress-dependent permeability in tight-oil rocks. *Petrophysics* 59(1): 35–43.

Klinkenberg, L.J. (1941). The permeability of porous media to liquids and gases. *Drilling and Productions Practices* 2: 200–213.

Kosanke, T. & Warren, A. 2016. Geological Controls on matrix permeability of the Eagle Ford Shale (Cretaceous), South Texas, U.S.A. *The Eagle Ford Shale: A Renaissance in U.S. Oil Production*, Edited by John Breyer, American Association of Petroleum Geologists, 285–300.

Kranz, R.L. (1983). Microcracks in rocks: A review. *Tectonophysics* 100(1–3): 449–480.

Luffel, D.L. and Guidry, F.K. (1992). New core analysis methods for measuring reservoir rock properties of Devonian Shale. *Journal of Petroleum Technology* 44(11): 1184–1190.

Luffel, D.L., Hopkins, C.W., Holditch, S.A., and Schetter, P.D. (1993). Matrix permeability measurement of gas productive shales. *Society of Petroleum Engineers Annual Technical Conference and Exhibition* (3–6 October), Houston, Texas, SPE Paper 26633. 10 p. Society of Petroleum Engineers.

Mathur, A., Sondergeld, C.H., and Rai, C.S. (2016). Comparison of Steady-State and Transient Methods for Measuring Shale Permeability. *SPE Low Perm Symposium*, SPE paper 180259. Denver, Colorado, USA: Society of Petroleum Engineers.

Mckernan, R., Mecklenburgh, J., Rutter, E., and Taylor, K. (2017). Microstructural controls on the pressure-dependent permeability of Whitby mudstone. *Geological Society, London, Special Publications* 454: 39–66.

Milsch, H., Hofmann, H., and Blöcher, G. (2016). An experimental and numerical evaluation of continuous fracture permeability measurements during effective pressure cycles. *International Journal of Rock Mechanics and Mining Sciences* 89: 109–115.

Ougier-Simonin, A., Renard, F., Boehm, C., and Vidal-Gilbert, S. (2016). Microfracturing and microporosity in shales. *Earth-Science Reviews* 162: 198–226.

Passey, Q.R., Bohacs, K., Esch, W.L. et al. (2010). From oil-prone source rock to gas-producing shale reservoir–geologic and petrophysical characterization of unconventional shale-gas reservoirs. *International Oil and Gas Conference and Exhibition in China* (8–10 June). OnePetro. Beijing, China.

Pluymakers, A., Kobchenko, M., and Renard, F. (2017). How microfracture roughness can be used to distinguish between exhumed cracks and in-situ flow paths in shales. *Journal of Structural Geology* 94: 87–97.

Rafatian, N. and Capsan, J. (2015). Petrophysical characterization of the pore space in Permian Wolfcamp rocks. *Petrophysics* 56: 45–57.

Ramiro-Ramirez, S. (2022). Integrated stratigraphic and petrophysical analysis of the Wolfcamp at Delware Basin, West Texas, USA. Ph.D. Thesis. The University of Texas at Austin.

Ramiro-Ramirez, S., Bhandari, A.R., Flemings, P.B., and Reed, R.M. (2020). Porosity and permeability heterogeneity in the Upper Wolfcamp, Delaware Basin, West Texas: implications for production. *SPE/AAPG/SEG Unconventional Resources Technology Conference* (20–22 July). Virtual: Unconventional Resources Technology Conference.

Rosen, R., Mickelson, W., Sharf-Aldin, M. et al. (2014). Impact of experimental studies on unconventional reservoir mechanisms. *SPE Unconventional Resources Conference* (1–3 April 2014), SPE Paper 168965. The Woodlands, Texas, USA.

Rydzy, M.B., Patino, J., Elmetni, N., and Appel, M. (2016). Stressed permeability in shales: effects of matrix compressibility and fractures: a step towards measuring matrix permeability in fractured shale samples. *International Symposium of the Society of Core Analysts*. The Society of Core Analysts, SCA paper SCA2016-027. Snowmass, Colorado, USA (21–26 August 2016).

Sinha, S., Braun, E.M., Passey, Q.R. et al. (2012). Advances in measurement standards and flow properties measurements for tight rocks such as shales. *SPE/EAGE European Unconventional Resources Conference & Exhibition-From Potential to Production* (20–22 March 2012), p. cp-285-00034. European Association of Geoscientists & Engineers.

Suarez-Rivera, R., Chertov, M., Willberg, D.M. et al. (2012). Understanding permeability measurements in tight shales promotes enhanced determination of reservoir quality. *SPE Canadian Unconventional Resources Conference* (30 October–1 November), Calgary, Alberta, Canada: Society of Petroleum Engineers.

Zhan, L., Fair, P., Dombrowski, R. et al. 2016. Using an innovative tool system to estimate in-situ permeability and pressure at multiple targets in a monitoring well in Permian Basin. *SPE Annual Technical Conference and Exhibition* (26–28 September), Dubai, UAE: Society of Petroleum Engineers.

Zhan, L., Fair, P.S., Dombrowski, R.J. et al. 2017. Estimating ultra low permeability at multiple locations using simultaneous impulse tests: a fit for purpose pressure transient solution and its field application. *SPE Annual Technical Conference and Exhibition* (October 9–11), San Antonio, Texas, USA: Society of Petroleum Engineers.

9

Assessment of Shale Wettability from Spontaneous Imbibition Experiments

Zhiye Gao[1,2] *and Qinhong Hu*[3]

[1] *State Key Laboratory of Petroleum Resources and Prospecting, China University of Petroleum, Beijing, China*
[2] *Unconventional Petroleum Research Institute, China University of Petroleum, Beijing, China*
[3] *Department of Earth and Environmental Sciences, University of Texas at Arlington, Arlington, TX, USA*

9.1 Introduction

The exploration and production of hydrocarbons from shale have become a success over the past two decades (Curtis 2002). However, the sustainable and economic development of shale resources is complicated by the wettability characteristics of shales (e.g. Hu et al. 2015). In reservoir engineering, wettability is commonly described as the preference of a solid surface to be in contact with one fluid rather than another. Based on the contact angle value between a fluid droplet and a rock surface, the wettability can be determined as water-wet, oil-wet, and intermediate-wet (Craig 1971; Anderson 1986a). Water- and oil-wet surfaces may co-exist within a single reservoir rock and cause the rock to be both water- and oil-wet, with such rock called to be mixed-wet (Salathiel 1973). Such a mixed-wet characteristic was reported for siliceous shale by many researchers such as Takahashi and Kovscek (2010). Wettability influences the hydrocarbon accumulation in reservoir rocks, and consequently the primary or secondary migration of oil and gas within source or reservoir rocks (Schowalter 1979; Thomas and Clouse 1990). It can significantly affect hydrocarbon recovery efficiency from reservoirs (Morrow 1990). For shale reservoirs, wettability controls the optimal selection of fracturing fluids (Bai et al. 2013; Xu and Dehghanpour 2014). Consequently, it is quite meaningful to accurately determine the shale wettability (Sharifigaliuk et al. 2021).

Conventional reservoir rocks (e.g. carbonate or sandstone) usually have high permeability and relatively homogeneous mineral composition (e.g. carbonates consist mainly of different forms of calcium carbonate). The wettability of conventional reservoir rocks can be determined by the US Bureau of Mines (USBM) test, the Amott test, and contact angle measurement (Amott 1959; Anderson 1986b). In contrast, fine-grained (with grain sizes <62.5 μm) shale has extremely low permeability and contains brittle minerals, ductile clay minerals, and organic matter (OM) (Chalmers et al. 2012; Loucks et al. 2012; Arif et al. 2021). Clean brittle minerals and clay minerals are mostly water-wet, while OM is oil-wet (Odusina et al. 2011; Xu and Dehghanpour 2014). As a result, shale has a mixed wettability at μm-scaled grain sizes. Tight shale requires a very high displacement pressure due to its ultra-low nano-darcy permeability, which invalidates the usage of Amott and USBM tests in determining shale wettability (Xu and Dehghanpour 2014). Furthermore, the interaction between OM and minerals over the long geological history could also change

Physics of Fluid Flow and Transport in Unconventional Reservoir Rocks, First Edition.
Edited by Behzad Ghanbarian, Feng Liang, and Hui-Hai Liu.

shale wettability (Legens et al. 1998). Contact angle measurement, which is widely used in conventional reservoirs, becomes questionable in shale studies because of the issues of surface roughness and contamination (Morrow 1990).

9.2 Spontaneous Imbibition Theory

Spontaneous imbibition (SI) is a useful approach to defining the mixed wettability of reservoir rocks (Anderson 1986b). SI has been extensively applied for shale reservoirs following the successful shale gas exploitation in North America since 2000s.

SI is a capillary-force driven process, during which a wetting fluid displaces a non-wetting one under the influence of capillary suction and sometimes additional gravitational force. Eq. (9.1) was proposed by Handy (1960) to describe the SI process within a water-air system, in which imbibition occurs vertically upward with a negligible gravitational force:

$$Q_{\mathrm{w}}^2 = \left(\frac{2P_{\mathrm{c}}k_{\mathrm{w}}\emptyset A^2 S_{\mathrm{w}}}{\mu_{\mathrm{w}}}\right)t \tag{9.1}$$

where Q_{w} is the volume of water imbibed into the sample, P_{c} is the capillary pressure, k_{w} is the effective permeability to water, $\emptyset$ is porosity, A is the cross-sectional area of sample, S_{w} is fractional water content of pore space, t is imbibition time, and μ_{w} is the viscosity of water.

Handy's equation can be simplified as,

$$H = at^{0.5} \tag{9.2}$$

where H is the cumulative imbibition height (cm) which equals Q_{w}/A and a is sorptivity (cm /s$^{-0.5}$; Philip 1957).

Handy's equation (Eq. 9.1) is based on three assumptions: (i) the wetting front moves in a piston-like manner, (ii) the pressure gradient in the gas phase ahead of the wetting front can be neglected, and (iii) the gravitational forces are much less than capillary force. Handy's equation indicates that the cumulative imbibition height is proportional to the square root of the imbibition time, so that theoretically an imbibition slope of 0.5 should be obtained if the Log cumulative imbibition height is plotted versus Log imbibition time. However, three types of water imbibition slopes were reported in natural rocks by Hu et al. (2012, 2015), which are the theoretical value of 0.5, 0.26, and 0.26 transitioning to 0.5. Percolation theory indicates that the lower water imbibition slope of 0.26 is an indication of low water-wet pore connectivity (Stauffer and Aharony 1994; Hunt et al. 2014).

The imbibition process is influenced by multiple factors such as fluid viscosity and temperature, wettability, boundary conditions (e.g. initial saturation), and pore structure of the rock (Zhang and Austad 2006). The scaling of SI data is deemed to be an effective method to investigate the controlling factors of SI process. Mattax and Kyte (1962) presented a scaling equation, which was later revised by Zhang et al. (1996) and Ma et al. (1997) as Eq. (9.3).

$$t_{\mathrm{D}} = t\sqrt{\frac{k}{\emptyset}}\frac{\sigma}{\sqrt{\mu_{\mathrm{we}}\mu_{\mathrm{nw}}}L_{\mathrm{c}}^2} \tag{9.3}$$

where t_D is dimensionless time, t is imbibition time (s), k is absolute permeability (m^2), σ is interfacial tension (N/m) between wetting and non-wetting phases, μ_{we} is the viscosity of wetting phase (Pa/s), μ_{nw} is the viscosity of non-wetting phase (Pa/s), and L_{c} is the characteristic length (m) to be defined as,

$$L_c = \sqrt{\frac{V_b}{\sum_{i=1}^{n} A_i / l_{A_i}}} \quad (9.4)$$

where V_b is the bulk volume of sample (m^3), n is the total number of surfaces open to imbibition, A_i is the area open to imbibition in the ith direction (m^2), and l_{A_i} is the distance that the wetting front travels from the imbibition face to the non-flow boundary (m).

Imbibition data can be modified to compensate for differences in boundary conditions, interfacial tension, permeability, and porosity by using Eq. (9.3). However, wettability, the remaining factor, is not considered when different fluids are used in SI process. As a result, the difference in imbibition behavior, after scaling the imbibition data using Eq. (9.3), is mainly controlled by sample wettability.

Due to the possible interaction between imbibing fluids and shale components, the imbibition slope of the first-time experimental run with imbibition of either water or *n*-decane (a surrogate oil phase used to probe the oil-wetting characteristics of the samples) is mainly used to evaluate the shale wettability (Gao and Hu 2016).

9.3 Samples and Analytical Methods

9.3.1 SI Experiments

During imbibition experiments, shale samples were cut into rectangular prisms. All sides, except for the top and bottom, were coated with quick-cure epoxy to generate co-current imbibition in the vertical direction. Hu et al. (2001) presented the SI experimental procedure and data processing in detail. Before the shale samples were used in the SI experiments, they were oven-dried at 60 °C for at least 48 hours in order to remove the moisture in the collected pore space to obtain a constant initial water saturation.

The imbibing fluids were water and *n*-decane, and their physical properties are listed in Table 9.1. For either liquid, air is always deemed to be the displaced non-wetting phase in SI experiments. During water imbibition experiments, additional beakers of water were placed inside the imbibition chamber to minimize the evaporation of fluid (water) reservoir. These beakers of water were not used in *n*-decane imbibition experiments. In addition, as the wettability of the shale samples could be changed by imbibed *n*-decane, replicate imbibition tests were not conducted on the same sample during *n*-decane imbibition.

In order to investigate the effect of layered characteristics of shale samples on SI behaviors, we carried out SI experiments in two directions by coating the sample surfaces differently (Figure 9.1); the symbols P or T are used to designate whether the direction of imbibition was either parallel or transverse to the lamination. In addition to the imbibition tests, vapor absorption tests were carried

Table 9.1 Physical properties of related fluids.

Fluid	Density (g/cm^3)	Viscosity (mPa/s)	Interfacial tension with air (mN/m)
Air	—	0.018	—
Water	1.00	1.002	72.0
n-Decane	0.73	0.84	23.9

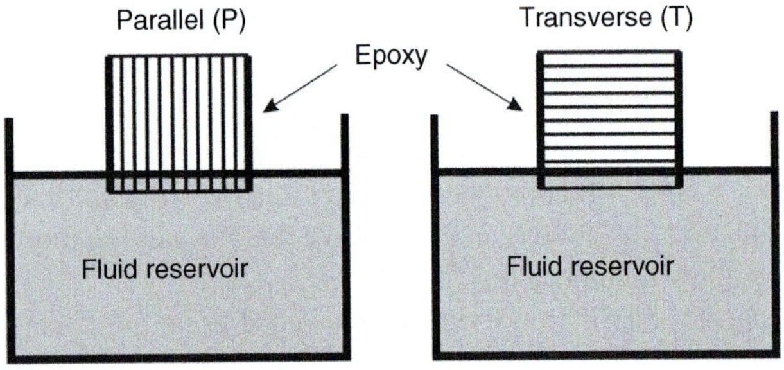

Figure 9.1 Upward imbibition direction at either parallel (P) or transverse (T) to the bedding planes shown as solid lines within the sample; fluid can be either water or *n*-decane.

out for some samples (Table 9.2), by hanging the sample above the fluid reservoir for vapor transport and capillary condensation process.

9.3.2 Barnett Shale from United States

The Barnett Formation is a Mississippian-age marine shelf deposit present in at least eighteen counties covering the Fort Worth Basin in north-central Texas. The Barnett Formation is divided into three members: upper Barnett Shale section, Forestburg Limestone section, and the lower Barnett Shale section (Loucks and Ruppel 2007).

The location of the Blakely #1 (API 497-33041) well was presented by Loucks and Ruppel (2007). The average vitrinite reflectance (R_o) of the Blakely #1 well core samples is about 1.35% to fall into the gas window (Loucks et al. 2009). Four samples from this Blakely #1 well at depths of 7136 ft (2175.0 m), 7169 ft (2185.1 m), 7199 ft (2194.2 m), and 7219 ft (2200.4 m) were used in this study (Table 9.2).

The dimensions of all the Barnett Shale samples used in the SI experiments are listed in Table 9.2. We also conducted mercury intrusion porosimetry (MIP) tests to determine the pore-throat size distribution of these shale samples, following the size-pressure relationship of Washburn (1921). Micromeritics AutoPore IV 9510 was used to perform MIP tests in this study. X-ray diffraction (XRD) analyses were also performed to obtain the mineralogy information of powder-sized samples by using a Rigaku Ultima IV XRD.

9.3.3 Silurian Longmaxi Formation and Triassic Yanchang Formation Shales from China

The Longmaxi Formation contains relatively high TOC content (1.0–4.0 wt.%) and type I kerogen, and is overmature with equivalent vitrinite reflectance (R_{equ}) value of 2.8–3.0% (Liang et al. 2014). Yanchang Formation with lower thermal maturity contains two main shale sections of Chang 7 and Chang 9.

Three Longmaxi Formation Shale samples from different depths in the southeastern Chongqing area were used in this study. In addition, four Yanchang Formation (Chang 7 with a R_o range of 0.83–1.02%) Shale samples from the Ordos Basin were studied (Xiong et al. 2016). The locations of the Ordos Basin and Sichuan Basin are shown in Figure 9.2, while the detailed information of all the Longmaxi and Yanchang Shale samples used in the SI experiments is listed in Table 9.3.

Table 9.2 Properties and dimensions of Barnett samples used in SI experiments.

Sample depth (ft/m)	Porosity	Permeability (m^2)[a]	TOC (%)[b]	BET surface area[c] (m^2/g)	Sample ID[d]	Fluid used[e]	Length (cm)	Width (cm)	Height (cm)
7136/2175.0	0.010	1.14×10^{-21}	1.4	4.94 ± 0.52	B7136 Pa	Water and *n*-Decane	1.7	1.7	1.2
					B7136 Ta	Water	1.4	1.7	1.7
					B7136 Tb	*n*-Decane	1.5	1.7	1.4
7169/2185.1	0.029	2.21×10^{-21}	2.4	9.77 ± 0.34	B7169 Pa	Water and *n*-Decane	1.2	1.8	1.3
					B7169 Ta	Water and *n*-Decane	1.4	1.8	1.8
					B7169 Pb	*n*-Decane vapor	1.7	1.5	1.3
					B7169 Tb	*n*-Decane vapor	1.8	1.4	1.3
7199/2194.2	0.060	4.96×10^{-21}	4.0	15.8	B7199 Pa	Water and *n*-Decane	1.7	1.7	1.2
					B7199 Ta	Water	1.2	1.7	1.7
					B7199 Tb	*n*-Decane	1.7	1.3	3.9
7219/2200.4	0.026	1.78×10^{-21}	3.8	7.93 ± 0.22	B7219 Pa	Water and *n*-Decane	1.7	1.7	1.4
					B7219 Ta	Water and *n*-Decane	1.4	1.7	1.9

[a] Permeability calculated according to Katz and Thompson (1986, 1987) from MIP data.
[b] Total organic carbon (TOC) content from Rowe et al. (2008).
[c] Measured by nitrogen sorption isotherm and Brunauer, Emmett and Teller (BET) theory (average ± standard deviation for triplicate measurements).
[d] a and b are two different cubic samples.
[e] Water imbibition tests were repeated at least three times on the same cubic sample. When both water and *n*-decane SI experiments were performed on the same sample, the *n*-decane experiment was always carried out after the water ones.

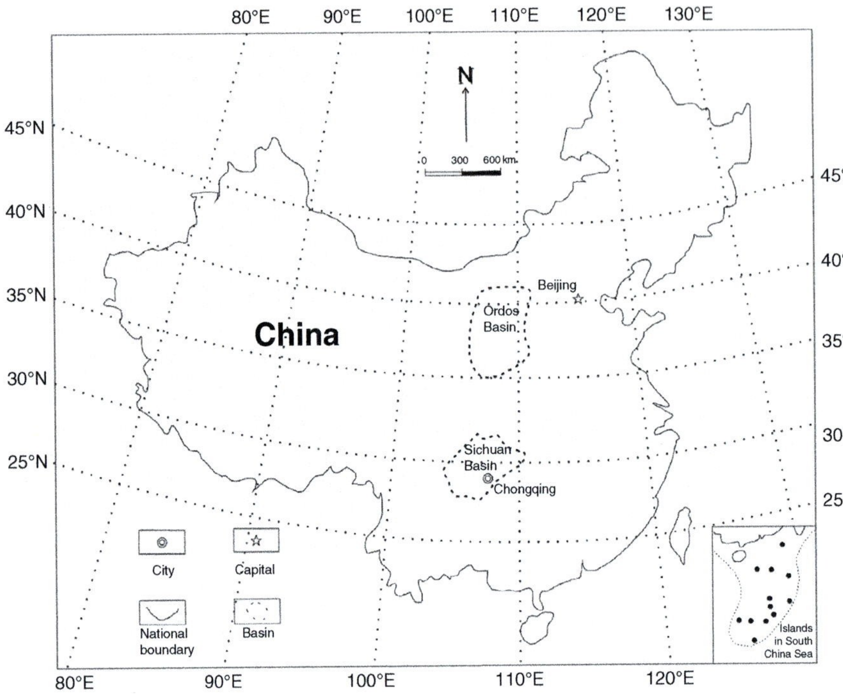

Figure 9.2 The locations of the Sichuan Basin, Ordos Basin, and Chongqing city in China.

The TOC content was measured using a CS-230 carbon and sulfur analyzer (Leco Corp, United States), and the mineral composition was obtained by PANalytical *X*' Pert PRO X-ray diffractometer. Micromeritics AutoPore IV 9510 was used to perform MIP tests to determine pore-throat diameters at macropores (>50 nm) (Hu et al. 2017). Quantachrome Autosorb IQ was employed to perform N_2 physisorption experiments in order to obtain the pore size information of mesopores (2–50 nm) (Sing et al. 1985). Field emission-scanning electron microscopy (FE-SEM) imaging was conducted to directly observe shale pore structure at two-dimensional shale surface, and the three-dimensional (3D) pore structures were also reconstructed using focused ion beam (FIB)-SEM.

9.3.4 Jurassic Ziliujing Formation Shale from China

The Jurassic lacustrine Ziliujing Formation from the northeastern Sichuan Basin has a good potential for shale gas exploitation (Guo 2016). The Da'anzhai Member Shale of the Ziliujing Formation contains relatively high TOC (>1 wt.%) with mainly type II_1–II_2 OM and R_o values from 1.05 to 1.82% (Li et al. 2013).

We selected four Da'anzhai Member Shale samples from X Well in northeastern Sichuan Basin, and Table 9.4 presents detailed information of all the Ziliujing Formation Shale samples used in SI experiments.

As for other shales, an X-ray diffractometer was used to obtain the mineralogy of these samples. The TOC content was measured by a CS-230 carbon and sulfur analyzer. Quantachrome PoreMasterGT 60 was used to perform MIP tests in order to obtain the pore-throat size distribution of these Ziliujing Formation samples. In addition, shale samples were Ar-ion-beam milling processed and then went through FE-SEM tests for pore typing.

Table 9.3 Sample ID and dimension of Longmaxi and Yanchang Formation Shales used in SI experiments.

Formation name	Sample ID	Size (Length × Width × Height) (cm × cm × cm)
Marine Longmaxi Shale	L1 P1	1.06 × 1.08 × 1.01
	L1 T1	1.13 × 1.04 × 1.06
	L1 P2	1.09 × 1.12 × 0.94
	L1 T2	0.97 × 1.05 × 1.00
	L2 P1	1.15 × 1.23 × 1.17
	L2 T1	1.12 × 1.03 × 1.05
	L2 P2	1.08 × 1.06 × 0.98
	L2 T2	1.06 × 1.09 × 1.08
	L3 P1	1.11 × 1.24 × 1.23
	L3 T1	0.96 × 1.17 × 1.18
	L3 P2	1.13 × 1.23 × 1.21
	L3 T2	1.07 × 1.22 × 1.21
Continental Yanchang Shale	Y1 P1	1.07 × 1.29 × 1.24
	Y1 T1	1.23 × 1.23 × 1.30
	Y1 P2	1.16 × 1.28 × 1.15
	Y1 T2	1.21 × 1.13 × 1.27
	Y2 P1	1.25 × 1.27 × 1.27
	Y2 T1	1.24 × 1.29 × 1.32
	Y2 P2	1.23 × 1.23 × 1.12
	Y2 T2	1.21 × 1.26 × 1.23
	Y3 P1	1.22 × 1.15 × 1.17
	Y3 T1	1.25 × 1.29 × 1.28
	Y3 P2	1.31 × 1.31 × 1.23
	Y3 T2	1.25 × 1.14 × 1.27
	Y4 P1	1.18 × 1.16 × 1.09
	Y4 T1	1.12 × 1.16 × 1.08
	Y4 P2	1.09 × 1.11 × 1.13
	Y4 T2	1.04 × 1.08 × 1.10

9.4 Results and Discussion

9.4.1 Complicated Wettability of Barnett Shale Inferred Qualitatively from SI Experiments

A dimensionless weight W_D is proposed in Eq. (9.5) to make a better comparison to account for fluid density, sample porosity, and volume:

$$W_D = \frac{W}{\rho \emptyset V} \tag{9.5}$$

Table 9.4 Sample ID and dimension of Ziliujing Formation Shale used in SI experiments.

Sample ID	Size (Length × Width × Height) (cm × cm × cm)
Z1Pa	1.36 × 1.22 × 1.39
Z1Ta	1.29 × 1.26 × 1.41
Z1Pb	1.36 × 1.04 × 1.36
Z1Tb	1.50 × 1.34 × 1.34
Z2Pa	1.09 × 1.21 × 1.26
Z2Ta	1.10 × 1.19 × 1.28
Z2Pb	1.31 × 1.22 × 1.30
Z2Tb	1.27 × 1.18 × 1.37
Z3Pa	1.10 × 0.89 × 1.26
Z3Ta	1.25 × 1.18 × 1.16
Z3Pb	1.05 × 0.81 × 1.32
Z3Tb	1.31 × 1.18 × 1.37
Z4Pa	0.94 × 1.18 × 1.10
Z4Ta	1.20 × 1.08 × 1.16
Z4Pb	1.24 × 1.12 × 1.24
Z4Tb	1.16 × 1.11 × 1.06

where W is the imbibed weight of fluid (water or *n*-decane, g), ρ is the fluid density (g/cm^3), and V is the sample volume (cm^3). If all the pore spaces are occupied by imbibed fluid, the value of W_D should be 1.

Based on imbibition plots, water imbibition behavior can be divided into four stages with their own representative slopes (Becker 2019; Hu et al. 2019). Stage I lasts a few seconds and shows the highest rate of change among all four stages but can have fluctuations resulting from the initial sample settling. Once the sample has settled, it comes to Stage II, when the sample first imbibes fluid. Stage II shows the second highest rate of change among all four stages due to the edge-accessible pores and microfractures being invaded first. Stage III starts when fluid has filled the edge-accessible pores and begins to enter the inner pore network. Stage III with a nearly linear and constant slope is used to represent the connectivity of pore networks for the sample. If present, Stage IV is the last stage when fluid reaches the sample top. The imbibition slope of Stage III, together with dimensionless imbibed weight, is used to characterize the wettability of our shale samples.

9.4.1.1 Wettability of Barnett Shale

More oil-wet samples have a higher *n*-decane imbibition slope than water imbibition, for both P and T samples (the symbols P or T are used to designate imbibition direction to be parallel, or transverse, to the lamination). Similar *n*-decane imbibition behavior was observed for the P and T samples of B7136 Forestburg Limestone (Figure 9.3a). Both B7136 Pa and B7136 Tb shale samples began to imbibe *n*-decane as soon as they contacted *n*-decane. The imbibing direction had little impact

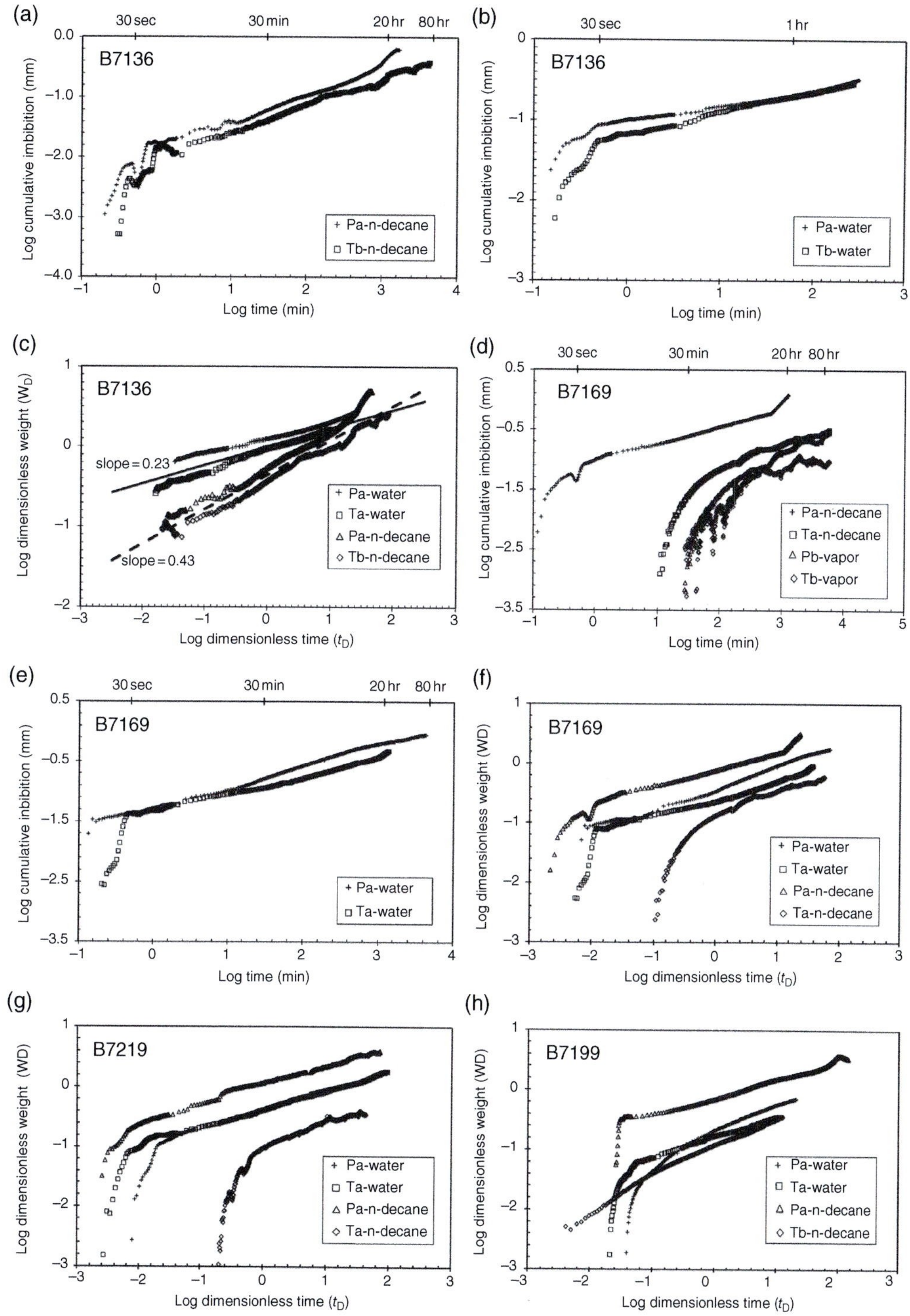

Figure 9.3 Imbibition behavior of Barnett samples. (a) The *n*-decane imbibition test for B7136 sample; (b) The water imbibition test for B7136 sample; (c) Log dimensionless weight (W_D) versus log dimensionless time (t_D) for B7136 sample; (d) The n-decane imbibition test for B7169 sample; (e) The water imbibition test for B7169 sample; (f) Log dimensionless weight (W_D) versus log dimensionless time (t_D) for B7169 sample; (g) Log dimensionless weight (W_D) versus log dimensionless time (t_D) for B7219 sample; (h) Log dimensionless weight (W_D) versus log dimensionless time (t_D) for B7199 sample.

on the *n*-decane imbibition, which indicated that *n*-decane could move more freely either along or across the lamination for its better wettability and connectivity. The *n*-decane and water imbibition curves of B7136 samples are shown in Figure 9.3a,b, respectively. The water and *n*-decane imbibition curves are compared in Figure 9.3c. The slope of *n*-decane experiments is about 0.43 nearly twice as large as the slope of water experiments on the same sample, which indicated its more oil-wet characteristics.

For more water-wet shale samples, P and T samples, which were different samples and from the same rock, always show different *n*-decane imbibition behaviors. During the first 10 minutes, almost no *n*-decane entered B7169 Ta but a large amount of *n*-decane imbibed into B7169 Pa (Figure 9.3d). The results of *n*-decane vapor absorption experiments for B7169 samples are also shown in Figure 9.3d, which show a shape similar to the *n*-decane imbibition curve of B7169 Ta. Because of the more water-wet characteristics of the sample, *n*-decane did not enter the T samples during the first 30 minutes. However, during this period, *n*-decane vapor may fill the pore spaces to change the sample wettability to a more oil-wet condition. Therefore this would cause the increased *n*-decane imbibition in these shale samples. The experimental results of P samples, however, suggest different controlling factors. The *n*-decane imbibition curve of B7169 Pa appears to be much closer to a more oil-wet imbibition plot, which is probably caused by *n*-decane migration within some cracks parallel to the lamination. There was no significant difference between P and T tests of water imbibition (Figure 9.3e). Wetting fluid can migrate freely along or across the lamination. The water and *n*-decane imbibition plots of B7169 and B7219 samples are presented in Figure 9.3f,g, respectively.

B7199 samples showed intermediate imbibition behavior (Figure 9.3h). The water imbibition of B7199 P and T samples showed different behaviors, and the imbibition slope of P sample was greater, which was different from more water-wet samples. The P sample imbibed much more *n*-decane than the T sample at the beginning of the experiment, similar to the more water-wet samples. However, B7199 Tb began to imbibe *n*-decane in a linear fashion at the very beginning (at about two minutes), which was similar to the more oil-wet sample B7136. The *n*-decane imbibition of T samples has a higher slope than the water imbibition, which indicates the oil-wet tendency of B7199 samples. Consequently, B7199 is considered as mixed-wet.

9.4.1.2 Properties of Barnett Samples and Their Correlation to Wettability

Results of MIP tests are shown in Figure 9.4. The dominant pore-throat sizes of Barnett samples derived from MIP analyses are less than 0.1 μm (at 70–80%). An appreciable amount of micron-sized pores were also observed in MIP analyses (Figure 9.4). All of our samples have a similar volume controlled by pore-throat diameters at larger than 60 μm (Figure 9.4). These large "pores" are probably related to micro-cracks. Hu et al. (2014) presented the SEM images of B7219 sample, showing the existence of micro-cracks with widths of several micrometers.

The mineralogical composition of Barnett samples is presented in Table 9.5. The B7136 samples have much higher carbonate content than other siliceous mudstones. Although the pure calcite mineral is water-wet, the reaction between carbonates and organic fluids from the maturation of OM-rich shale could lead to a less hydrophilic wettability (Legens et al. 1998), which could be used to interpret the more oil-wet characteristics of B7136 samples.

For siliceous samples, OM plays an important role in affecting the wettability. Generally, OM-hosted pores are oil-wet, and the inorganic pores are water-wet (Odusina et al. 2011). The porosities of these Barnett samples are positively related with TOC values. The volume fraction of <10 nm pores increases with an increasing TOC content: 18%, 50%, 53%, and 61% for B7136, B7169,

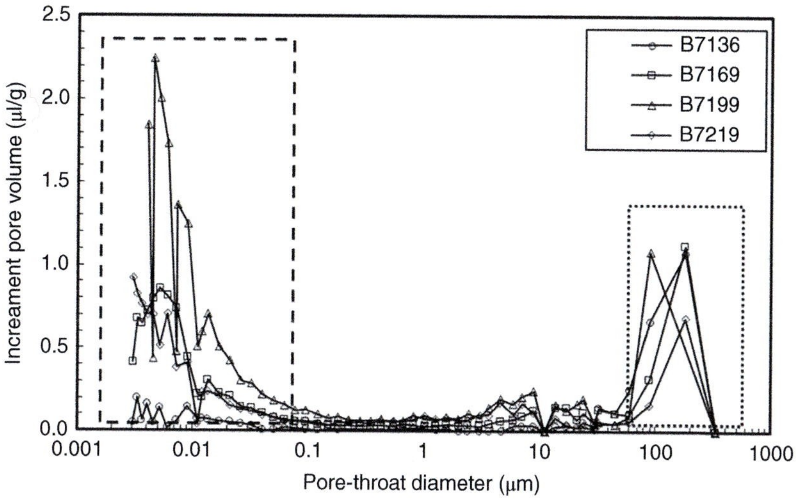

Figure 9.4 Incremental pore volume versus pore-throat diameter for Barnett samples from MIP tests.

Table 9.5 Mineralogical composition of Barnett samples from XRD analyses.

	Mineral weight (%)							
Sample ID	**Quartz**	**Calcite**	**Dolomite**	**Pyrite**	**Illite/mica**	**Mixed layer (I/S)**	**Chlorite**	**Others**
B7136	14.7	62.2	10.3	0.6	10.4	0	0	1.8[a]
B7169	29.4	20.6	14.7	2	16.8	9.3	7.2	0
B7199	56.9	2.6	0	2.9	0	37.6	0	0
B7219	29.6	24.8	6.4	1.7	9.8	0	0	27.7[b]

[a] Including 1.8% albite.
[b] Including 22.6% fluorapatite and 5.0% albite.

B7219, and B7199, respectively (Figure 9.5). Consequently, B7199 sample with the highest TOC content could contain more OM-hosted pores than B7169 and B7219 samples, which contributes to their wettability differences. For more water-wet B7169 and B7219 samples, the water-wet pores are better connected, compared with oil-wet OM-hosted pores, and easier to form continuous water-wet pathways. While for mixed-wet B7199 sample, OM-hosted nanopores become abundant and are better connected, which could form continuous (though sparsely) oil-wet pathways in addition to water-wet pathways.

9.4.1.3 Low Pore Connectivity to Water of Barnett Samples

Our results reported in Table 9.6 show that the water imbibition slopes (around 0.28) of Barnett samples are much lower than the theoretical 0.5 value, which shows the low pore connectivity to water for these Barnett samples.

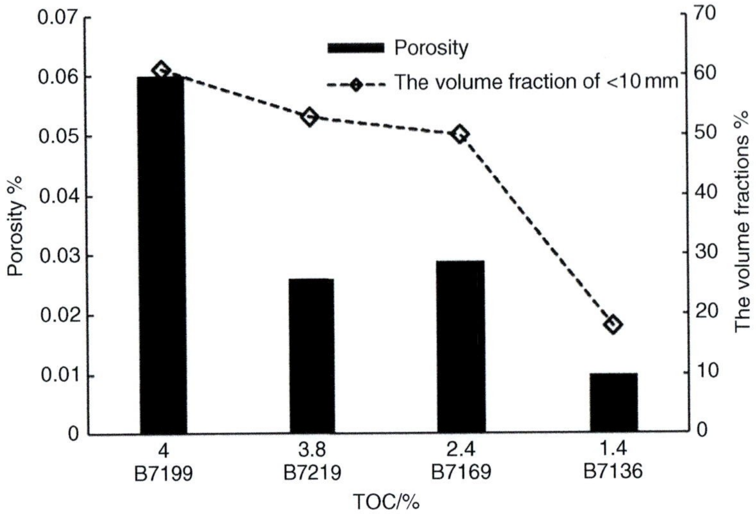

Figure 9.5 Relationship between TOC and porosity and pore volume fraction controlled by pore-throat diameters of <10 nm.

Table 9.6 Water imbibition slope for Barnett samples.

Sample ID	Water imbibition slope
B7136	0.269 ± 0.005
B7169	0.273 ± 0.050
B7199	0.284 ± 0.062
B7219	0.313 ± 0.019

Results for only T samples presented to eliminate the effect of cracks within samples.
Imbibition slopes as average ± standard deviation for at least triplicate measurements on the same sample.

9.4.2 More Oil-Wet Longmaxi Formation Shale and More Water-Wet Yanchang Formation Shale

9.4.2.1 TOC and Mineralogy

The main compositional differences between Longmaxi and Yanchang Shale formations are the TOC and clay minerals (Table 9.7). The TOC contents of Yanchang Shale are much higher (5–8 wt.%) than Longmaxi Shale (1–4 wt.%). The clay minerals of Yanchang Shale samples are mainly mixed layer (illite/smectite) clay mineral, while illite is the dominant clay mineral in the Longmaxi Shale.

9.4.2.2 Pore Structure Difference Between Longmaxi and Yanchang Samples

The pore structure derived from MIP tests is shown in Figure 9.6a. The macropore volume of Yanchang samples, except for sample Y4, is much greater than the Longmaxi samples. The mesopore size distribution obtained from N_2 physisorption is shown in Figure 9.6b, and the mesopore volume of the Longmaxi samples is greater than that of the Yanchang samples.

Table 9.7 TOC and mineralogy of Longmaxi and Yanchang Formation Shale samples.

Sample ID	TOC (wt.%)	Quartz	Feldspar	Calcite	Pyrite	Total clays	Mixed layer (illite-smectite) (%)	Illite (I) (%)	Kaolinite (%)	Chlorite (%)
L1	3.8	34.3	7.4	5.8	5.5	47.0	2.8	36.7	2.4	5.2
L2	1.0	32.8	14.3	9.2	1.9	41.8	1.3	33.4	0.0	7.1
L3	1.4	93.8	NA	NA	2.8	3.4	1.5	1.5	0.4	0.0
Y1	5.4	20.3	12.2	NA	NA	67.5	29.7	30.4	0.0	7.4
Y2	5.9	46.4	33.4	NA	9.2	10.8	8.7	1.6	0.4	0.0
Y3	5.1	30.6	13.7	NA	NA	55.7	44.0	8.4	0.6	2.8
Y4	8.5	17.1	14.2	NA	25.3	43.4	34.7	6.5	0.0	2.2

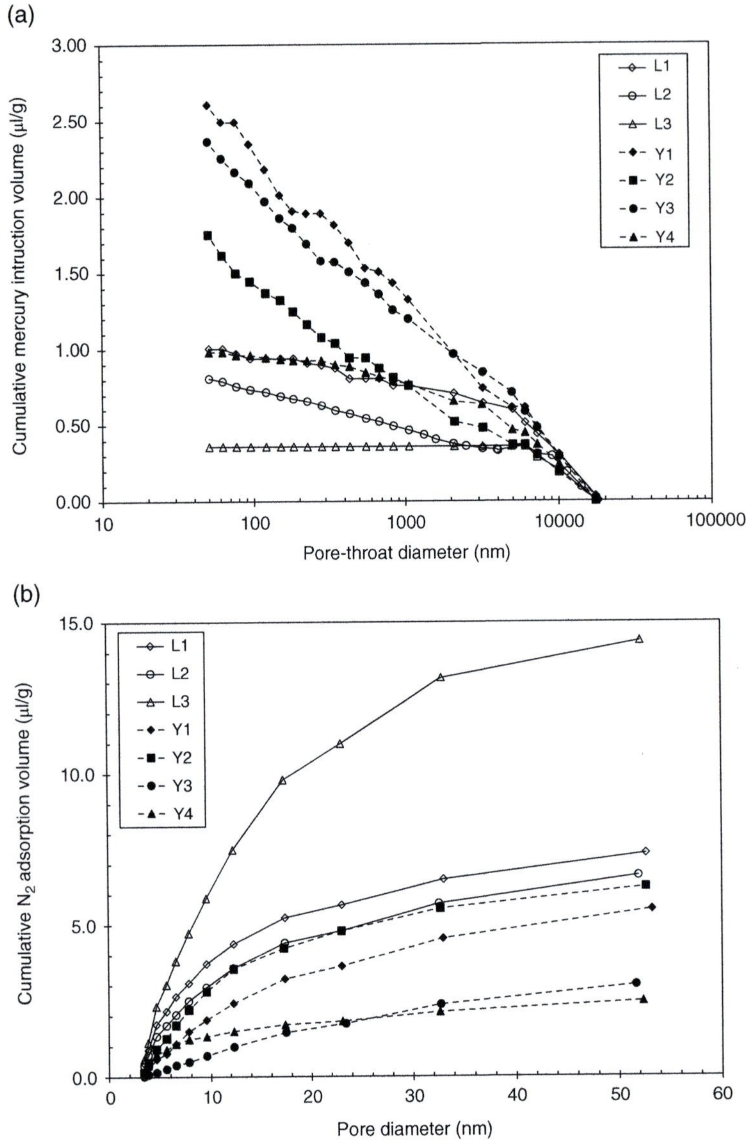

Figure 9.6 Pore size characteristics of Longmaxi and Yanchang samples: (a) Pore-throat size distribution of macropores obtained by MIP tests; (b) Pore size distribution of mesopores derived from N_2 physisorption tests.

Table 9.8 Presents the calculated MIP porosity, and the total porosities of Yanchang Formation Shale samples are mostly greater than those of the Longmaxi Formation shale samples studied here. Longmaxi Shale samples have smaller pore sizes and larger surface areas compared with the Yanchang Shale samples.

The representative FE-SEM images are presented in Figur 9.7 for respectfully L1, Y1, and Y2 samples. It can be observed that OM pores (Figure 9.7a), inter-particle pores around calcite (Figure 9.7b), and intercrystalline pores within pyrite (Figure 9.7c) are the main pore types within L1 sample. The pores within clay minerals are nearly eliminated because the clay minerals are strongly compacted during the diagenesis of L1 sample, as shown by Figure 9.7d. However, for two Yanchang Shale samples (Y1 and Y2), inter-particle pores among different brittle minerals

Table 9.8 MIP porosity, BET surface area, and average pore diameter, as well as their averaged values, of Longmaxi (L1 to L3) and Yanchang Formation (Y1 to Y4) Shale samples.

Sample ID	MIP porosity (%)	Average	Multi-pointBET surface area (m^2/g)	Average	Average pore diameter from N_2 physisorption (nm)	Average
L1	0.65	0.73	12.54	12.24	4.40	5.03
L2	0.36		9.01		5.15	
L3	1.18		15.18		5.53	
Y1	1.94	2.26	2.27	2.30	13.06	13.11
Y2	3.13		2.86		11.01	
Y3	2.55		0.81		22.87	
Y4	1.43		3.27		5.50	

(Figure 9.7f,j) and pores within clay minerals (Figure 9.7g,k) are the two dominant pore types, but the OM pores are poorly developed (Figure 9.7e,i). OM pores with circular shapes could be occasionally found in Y1 sample (Figure 9.7h). Furthermore, some micro-fractures are observed in Y2 sample (Figure 9.7l).

The reconstructed 3D pore network of L1 sample and Y3 sample by FIB-SEM are shown in Figure 9.8a–c and d–f, respectively. The pore size information extracted from 3D images of L1 and Y3 samples are presented in Figure 9.8g,h. The median pore size (pore diameter corresponding to 50% of total cumulative frequency) of Y3 sample is about 90 nm, while it is only about 35 nm for L1 sample.

The average porosity of Yanchang Shale samples is 2.26%, which is greater than 0.73% of the average porosity of Longmaxi samples (Table 9.8). However, the average BET surface area of Longmaxi samples is 12.24 m^2/g much greater than the average BET surface area of Yanchang samples, which is 2.30 m^2/g.

9.4.2.3 Water and Oil Imbibition Experiments

The triplicate imbibition curves of samples L1 T1 and Y1 T1 are presented in Figure 9.9a,b. The water imbibition slopes of all T samples are listed in Table 9.9. As can be seen, the reported slopes are less than the theoretical value of 0.5, which indicates the low water-wet pore connectivity of our samples. The relationship between the 1st water imbibition slope of T samples and MIP porosity is shown in Figure 9.9c. Although there is no unique relationship in characterizing the imbibition slope and porosity for these seven samples, linear relationships with high correlations exist in Longmaxi and Yanchang samples. The first water imbibition slope is positively related to the porosity, which indicates that the higher the porosity is, the better its water-wet pore connectivity is. As to the reproducibility and stability of water imbibition experiments, Samples L1 T1 and Y1 T1 with the highest clay contents exhibit the poorest reproducibility among Longmaxi and Yanchang samples (Table 9.9). The strong interaction between water and clay minerals changes the pore structure and finally affects the imbibition behaviors of both Longmaxi and Yanchang Shales.

For convenience, the first water or *n*-decane imbibition slope is listed together with sample information in Table 9.10. The imbibition slopes of three samples (L1 P2, Y2 T2, and Y4 T2) are not available (too low values) due to the negligible weight of imbibed *n*-decane in imbibition experiments.

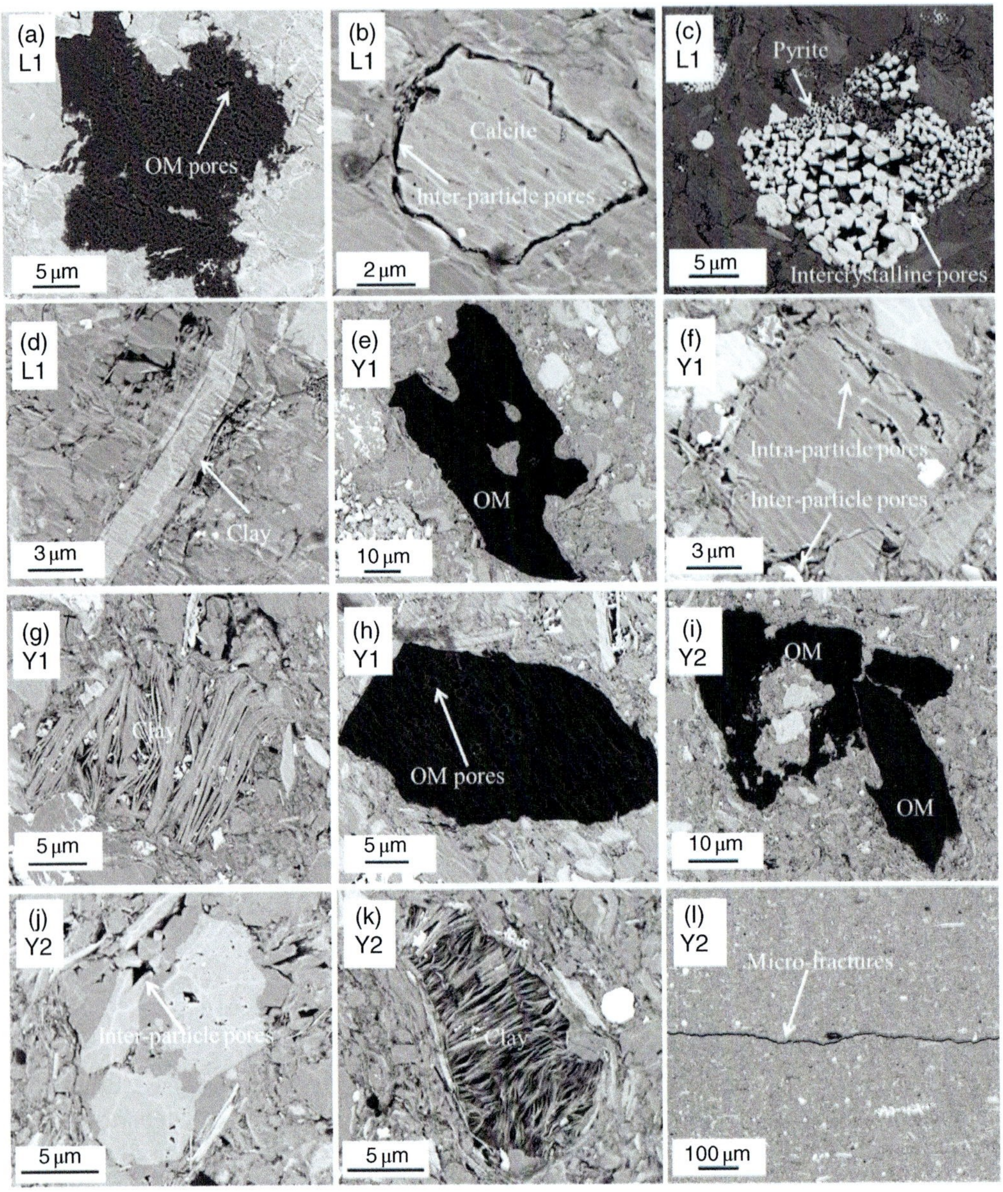

Figure 9.7 FE-SEM photos of Longmaxi and Yanchang Formation Shale samples. (a) Organic matter (OM) pores in sample L1; (b) interparticle pores around calcite in sample L1; (c) intercrystalline pores within pyrite in sample L1; (d) pores within clay minerals in sample L1; (e) Organic matter (OM) pores in sample Y1; (f) interparticle pores among different minerals and intraparticle pores within some mineral grains in sample Y1; (g) pores within clay minerals in sample Y1; (h) OM pores with regular circular shape in sample Y1; (i) Organic matter (OM) pores in sample Y2; (j) interparticle pores among different minerals and intraparticle pores within some mineral grains in sample Y2; (k) pores within clay minerals in sample Y2; (l) microfractures parallel to the bedding planes in sample Y2.

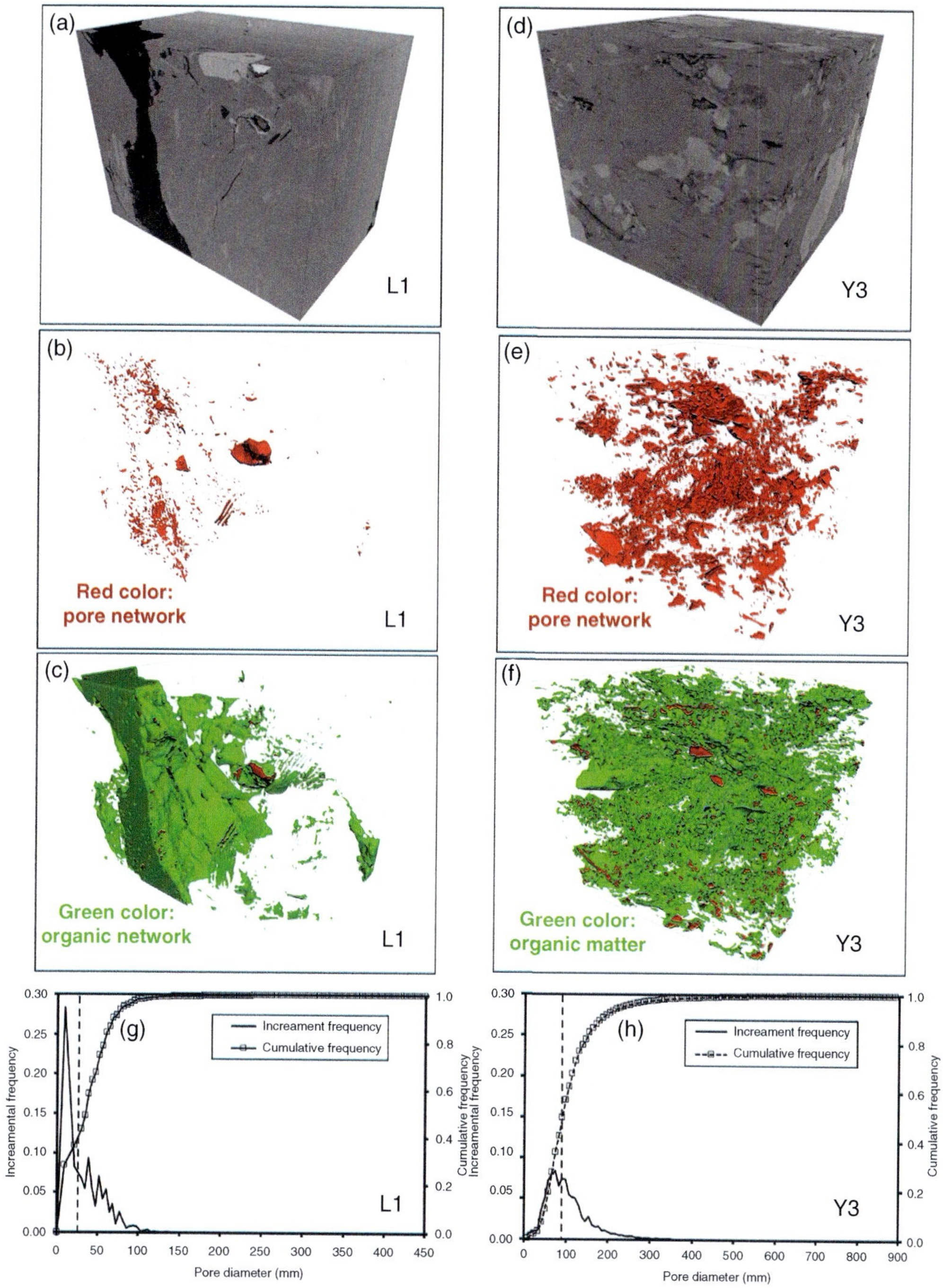

Figure 9.8 FIB-SEM images and pore size distribution information of Samples L1 and Y3. (L1:12.34 × 10.17 × 8.00 μm; Y3:10.00 × 10.00 × 10.00 μm). (a) The reconstructed three-dimensional photo of sample L1; (b) pore network of sample L1; (c) distribution of organic matter in sample L1; (d) The reconstructed three-dimensional photo of sample Y3; (e) pore network of sample Y3; (f) distribution of organic matter in sample Y3; (g) The pore-size distribution information extracted from focused ion beam–scanning electron microscopy photos of sample L1; (h) The pore-size distribution information extracted from focused ion beam–scanning electron microscopy photos of sample Y3.

Based on the oil imbibition slopes of T samples, Longmaxi samples L2 and L3 show higher oil-wet pore connectivity, while Yanchang samples Y2 and Y4 show lower oil-wet pore connectivity with almost no *n*-decane imbibed. For other samples, the oil imbibition slope of T samples is around 0.26, which indicates the low oil-wet pore connectivity of these samples.

Unlike the high slope of T samples, the water imbibition slope of P samples indeed shows a weakly positive relationship with porosity (Figure 9.9d), which means that the higher porosity is also good for the formation of connected water-wet pathway along the bedding planes.

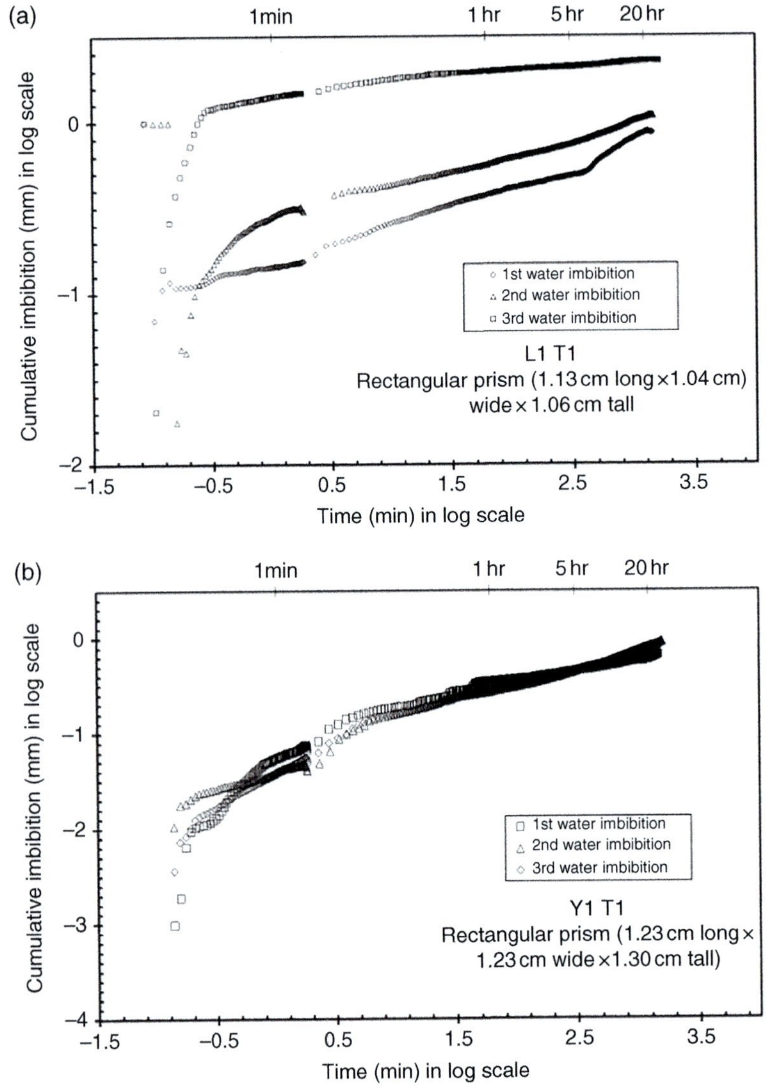

Figure 9.9 Imbibition behavior of Longmaxi and Yanchang samples and its relationship with MIP porosity. (a) Cumulative water imbibition versus time for sample L1 T1; (b) Cumulative water imbibition versus time for sample Y1 T1; (c) The relationship between first water imbibition slope of transverse (T) samples and mercury intrusion porosimetry (MIP) porosity for the Longmaxi and Yanchang samples; (d) The relationship between first water imbibition slope of parallel (P) samples and mercury intrusion porosimetry (MIP) porosity for the Longmaxi and Yanchang samples.

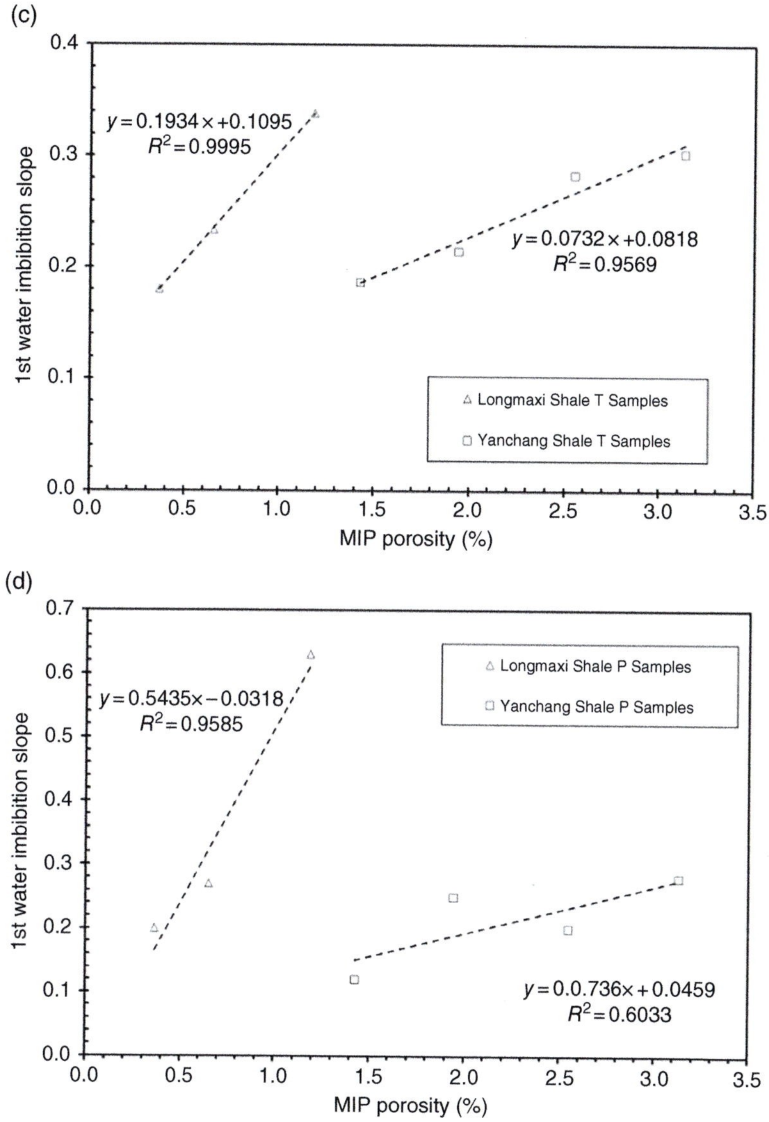

Figure 9.9 (Continued)

The oil imbibition behaviors are quite different between the Longmaxi and Yanchang samples. The *n*-decane imbibition slopes of Yanchang T samples are no more than 0.27 and two samples (Y2 and Y4) show almost no *n*-decane imbibition (Table 9.10). However, the *n*-decane imbibition slopes of Yanchang P samples are greater than 0.3 and those reported for T samples. The preferential pathway of oil migration in the Yanchang samples is probably caused by the developed bedding planes and more water-wet wettability.

Compared with the Yanchang samples, the Longmaxi Shale T samples show greater *n*-decane imbibition slopes (Table 9.10), which is related to the more developed OM-hosted pores in these Longmaxi samples.

Table 9.9 Water imbibition slopes of all transverse (T) samples of Longmaxi and Yanchang Formation Shales.

Sample ID	Water imbibition slope			Average imbibition slope
	1st	2nd	3rd	
L1 T1	0.234	0.161	0.046	0.147 ± 0.095
L2 T1	0.181	0.063	0.118	0.121 ± 0.059
L3 T1	0.339	0.318	0.407	0.355 ± 0.047
Y1 T1	0.215	0.330	0.382	0.309 ± 0.085
Y2 T1	0.303	0.319	0.235	0.286 ± 0.045
Y3 T1	0.284	0.314	0.240	0.279 ± 0.037
Y4 T1	0.187	0.248	0.183	0.206 ± 0.036

Table 9.10 SI results of Longmaxi and Yanchang Formation Shale samples.

Formation name	Sample ID	Fluid used	Imbibition slope
Marine Longmaxi Shale	L1 P1	Water	0.27
	L1 T1	Water	0.23
	L1 P2	*n*-Decane	0.00
	L1 T2	*n*-Decane	0.26
	L2 P1	Water	0.20
	L2 T1	Water	0.18
	L2 P2	*n*-Decane	0.34
	L2 T2	*n*-Decane	0.46
	L3 P1	Water	0.63
	L3 T1	Water	0.34
	L3 P2	*n*-Decane	0.47
	L3 T2	*n*-Decane	0.42
Continental Yanchang Shale	Y1 P1	Water	0.25
	Y1 T1	Water	0.22
	Y1 P2	*n*-Decane	0.37
	Y1 T2	*n*-Decane	0.27
	Y2 P1	Water	0.28
	Y2 T1	Water	0.30
	Y2 P2	*n*-Decane	0.33
	Y2 T2	*n*-Decane	0.00
	Y3 P1	Water	0.20
	Y3 T1	Water	0.28
	Y3 P2	*n*-Decane	0.51
	Y3 T2	*n*-Decane	0.25
	Y4 P1	Water	0.12
	Y4 T1	Water	0.19
	Y4 P2	*n*-Decane	0.67
	Y4 T2	*n*-Decane	0.00

9.4.2.4 Wettability of Longmaxi and Yanchang Shale Samples Deduced from SI Experiments

For the Longmaxi samples, the water imbibition slope of P samples is greater than that of the T samples, and the *n*-decane imbibition slope of T samples is much greater than the water imbibition slope of T samples (Table 9.10), both of which indicate the more oil-wet property of these shale samples. In such samples, the migration of water across the bedding planes is restricted, while the oil phase could move across the bedding planes more easily. The more oil-wet wettability of the Longmaxi samples largely depends on their pore structure, which is significantly controlled by its high thermal maturity. The well-developed OM-related pores are easier to form a connected oil-wet pathway within these Longmaxi samples, as shown in Figure 9.7a,b.

For the Yanchang samples, the transport of oil phase across the bedding planes is more difficult than its transport along the bedding planes. However, the migration of water across bedding planes is unblocked. The wettability of the Yanchang samples used here is more water-wet, which is related to the more developed inorganic pores in Yanchang samples (Figure 9.7f,g).

9.4.3 Complicated Wettability of Ziliujing Formation Shale

9.4.3.1 TOC and Mineralogy

Table 9.11 shows the composition of four shale samples from the Ziliujing formation. The TOC content ranges from 0.97 to 1.72 wt.% with an average value of 1.41 wt.%.

As to the mineralogy of our shale samples, one main characteristic is the high clay content (from 41.3 to 54.0 wt.% with an average value of 46.4 wt.%). The clay minerals of those shale samples mainly consist of mixed-layer (illite/smectite), illite and chlorite.

9.4.3.2 Pore Structure

The pore-throat size information obtained from MIP is presented in Figure 9.10. All four samples show bimodal distributions with two peaks around 10 nm and 1 μm. Sample Z1 has the lowest porosity and smallest median pore diameter (D_{50}), while Sample Z4 has the highest porosity and the largest D_{50} (Table 9.12); D_{50} is the pore-throat diameter corresponding to 50% of total cumulative mercury intrusion. Low specific surface area of Sample Z1 is probably because of the higher contribution (>40%) of large pores (>100 nm) to the total pore space (Figure 9.10a).

Very few pores are developed in Sample Z1 (Figure 9.11), which is consistent with its lowest MIP porosity. Due to the lowest TOC content of Sample Z1, OM pores are rarely observed, and inorganic pores related to clay minerals are the main pore type. OM pores as well as OM-clay interparticle pores are commonly observed in Samples Z2 and Z4, and inorganic pores are also well developed. OM pores and OM-clay interparticle pores are the main pore types in Sample Z3, while inorganic pores are poorly developed.

Table 9.11 TOC and mineralogy (wt.%) of Ziliujing Formation Shale samples used in this study.

Sample ID	TOC	Plagioclase	Pyrite	Quartz	Calcite	Total clays	I/S	Illite	Kaolinite	Chlorite
Z1	0.97	5.2	1.2	36.4	10.1	47.1	16.0	12.2	7.1	11.8
Z2	1.32	4.5	2.3	35.3	3.9	54.0	22.7	14.6	6.5	10.3
Z3	1.72	3.6	4.0	25.5	23.9	43.0	18.9	12.0	4.3	7.7
Z4	1.63	6.2	0.6	41.0	10.9	41.3	15.7	9.9	7.0	8.7

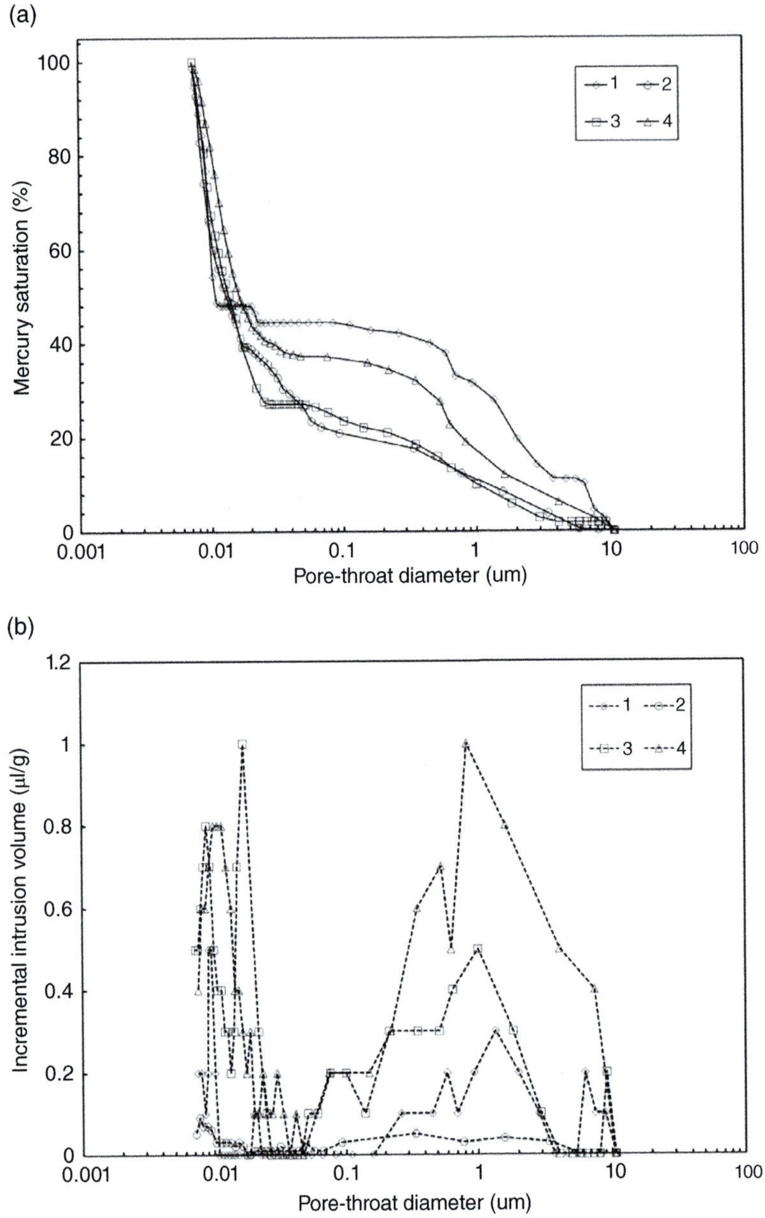

Figure 9.10 Pore-throat diameter distribution of Ziliujing Formation Shale samples by MIP tests as a function of (a): Mercury saturation and (b): Incremental intrusion volume.

Table 9.12 Pore structure information of Ziliujing Formation Shale samples obtained from MIP tests.

Sample ID	Porosity (%)	Median pore diameter D_{50} (nm)	Specific surface area (m^2/g)
Z1	0.95	10.6	1.01
Z2	2.25	12.8	2.39
Z3	2.49	13.5	3.05
Z4	3.49	16.2	2.97

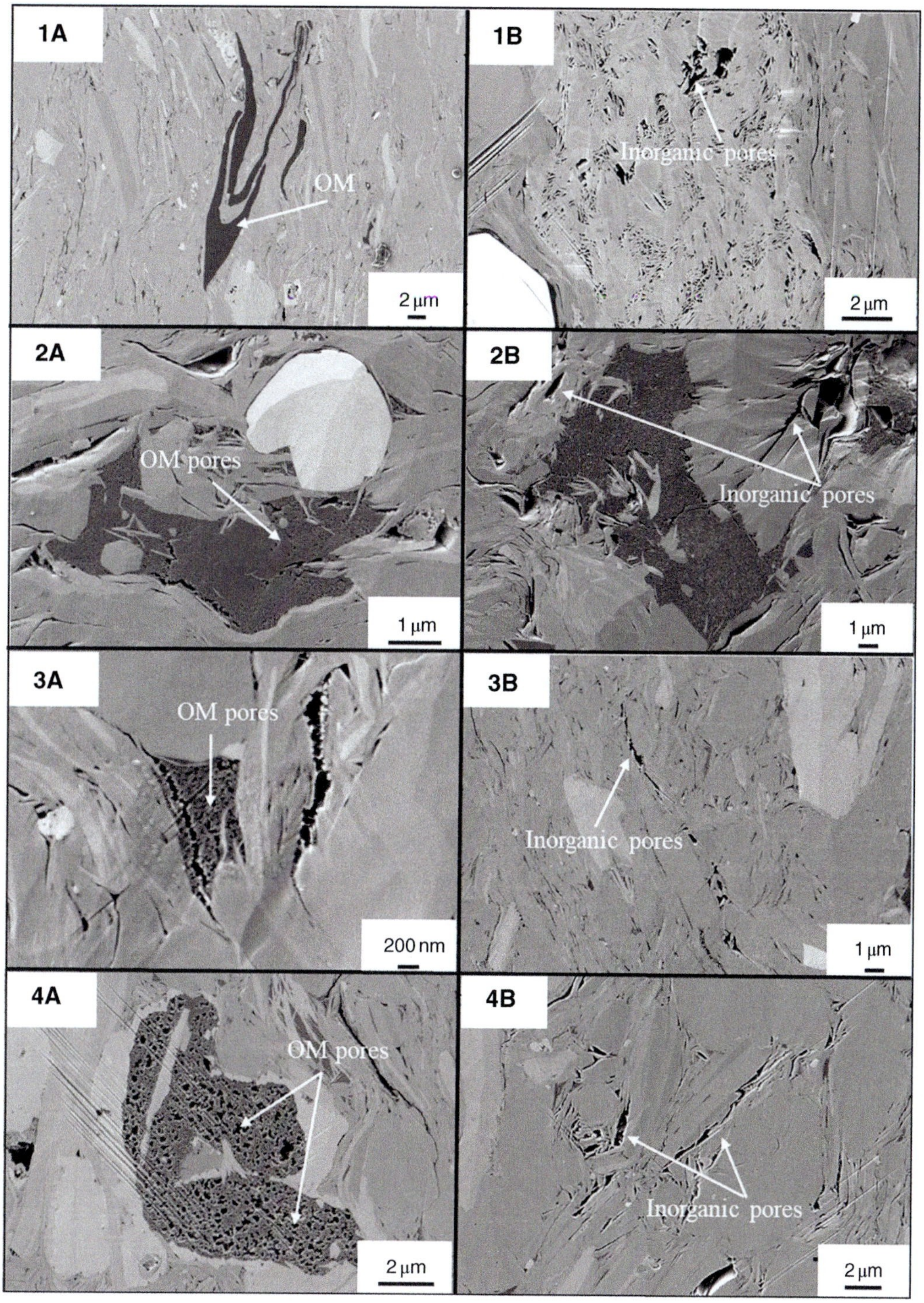

Figure 9.11 FE-SEM images of Ziliujing Formation Shale samples. *Source:* Gao et al. (2018)/from American Association of Petroleum Geologists.

9.4.3.3 Water and Oil Imbibition Experiments

The water imbibition slopes are listed in Table 9.13. All the water imbibition slopes are much less than the theoretical value of 0.5, indicating the low water-wet pore connectivity of these Ziliujing Shale samples. The relationship between the 1st water imbibition slope and MIP porosity was also investigated for T samples. Except for Sample Z3Ta with the lowest imbibition slope of 0.05, the first water imbibition slopes are positively correlated with the MIP porosity (Figure 9.12).

Table 9.13 Spontaneous imbibition results of Ziliujing Formation Shale samples.

Sample ID	Fluid used	1st	2nd	3rd	Average
Z1Pa	Water	0.245	0.339	0.321	0.302 ± 0.050
Z1Ta	Water	0.153	0.318	0.200	0.224 ± 0.085
Z1Pb	*n*-Decane	0.328			
Z1Tb	*n*-Decane	0.188			
Z2Pa	Water	0.256	0.241	0.251	0.249 ± 0.008
Z2Ta	Water	0.277	0.259	0.154	0.230 ± 0.066
Z2Pb	*n*-Decane	0.394			
Z2Tb	*n*-Decane	0.493			
Z3Pa	Water	0.359	0.383	0.172	0.305 ± 0.116
Z3Ta	Water	0.050	0.249	0.113	0.137 ± 0.102
Z3Pb	*n*-Decane	0.187			
Z3Tb	*n*-Decane	0.466			
Z4Pa	Water	0.309	0.337	0.289	0.312 ± 0.024
Z4Ta	Water	0.372	0.154	0.184	0.237 ± 0.118
Z4Pb	*n*-Decane	0.457			
Z4Tb	*n*-Decane	0.312			

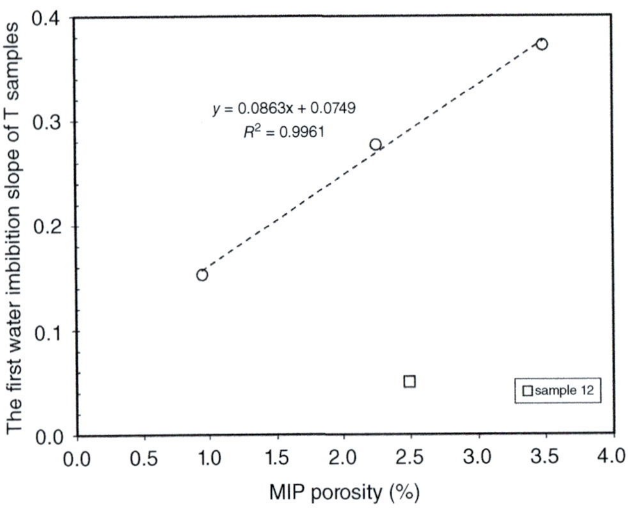

Figure 9.12 1st water imbibition slope of T samples versus MIP porosity (%) for Ziliujing Formation Shale samples.

9.4.3.4 Wettability of Ziliujing Formation Shale Indicated from SI Experiments and its Correlation to Shale Pore Structure and Composition

The water and *n*-decane imbibition slopes of P samples are much higher than those of T samples. The very low imbibition slopes of T samples indicate that it is difficult to form a connected pathway for both oil and water across the bedding planes, which is closely related with the less developed pore spaces (1A and 1B of Figure 9.11). Consequently, the bedding planes work as barriers to limit the migration of both oil and water within Sample Z1. The wettability of the sample is determined as slightly more oil-wet because of its higher *n*-decane imbibition slopes compared with water imbibition slopes for both P and T samples.

For both P and T samples of Sample Z2, the *n*-decane imbibition slopes are much higher than water imbibition slopes, which shows a more oil-wet property. Another SI characteristic of Sample Z2 is the greater water and *n*-decane imbibition slopes of T samples compared with P samples, which means that the effect of bedding planes on the fluid migration for this sample is insignificant. Water and *n*-decane could transport across the bedding planes more freely due to the better-connected water-wet and oil-wet pathways in this direction. As shown in Figure 9.11 (2A and 2B), the dominant OM-related pores and widely distributed microfractures are the two main pore types, which form the better oil-wet and good water-wet pathways, respectively.

The water and *n*-decane imbibition curves of Sample Z3 are quite different. Water could move along the bedding planes but could not penetrate the bedding planes easily, which is caused by the less developed water-wet inorganic pores in Sample Z3 (Figure 9.11). However, the migration of *n*-decane within Sample Z3 is less affected by the blockage effect of bedding planes. The abundant OM pores and OM-clay interparticle pores could build a good oil-wet pore network.

The more water-wet wettability of Sample Z4 is determined by the higher water imbibition slope of Sample Z4Ta compared with the *n*-decane imbibition slope of Sample Z4Tb. Unlike Sample Z3, the bedding planes block the migration of *n*-decane but not water for Sample Z4. The better-connected water-wet pathway benefits from the protection of the high quartz content of Sample Z4. The *n*-decane imbibition slope of Sample Z4Ta is much higher than that of Sample Z1Tb, which is closely related to the well-developed oil-wet OM pores and OM-clay interparticle pores within Sample Z4 (Figure 9.11).

9.4.4 Shale Wettability Evolution Model

The co-existence of oil-wet OM pores and water-wet inorganic pores at nm-μm scales in shale complicates its wettability at mm-cm scales. The pore structure controls shale wettability. Simplified pore network model is established for more oil-wet, more water-wet, mixed-wet, and intermediate-wet shale samples, respectively (Figure 9.13) (Gao et al. 2019). For more oil-wet (or more water-wet) shale samples, the oil-wet (or water-wet) pores are more developed and better connected to form an oil-wet (or water-wet) pathway. Both oil and water-wet pores are well-connected in mixed-wet shale samples, while neither oil-wet pores nor water-wet pores are well-connected in the intermediate-wet shale samples.

Thermal maturity significantly controls shale pore structure (Mastalerz et al. 2013) and thus affects shale wettability. The wettability of different shale samples inferred from SI experiments and thermal maturities are listed in Table 9.14. A new shale wettability evolution model is shown in Figure 9.14 (Gao et al. 2019) with four stages:

Stage I with thermal maturity $0.5\% < R_o < 1.3\%$: Organic pores begin to form due to the primary cracking of OM but are not well developed. The primary inorganic pores are eliminated by

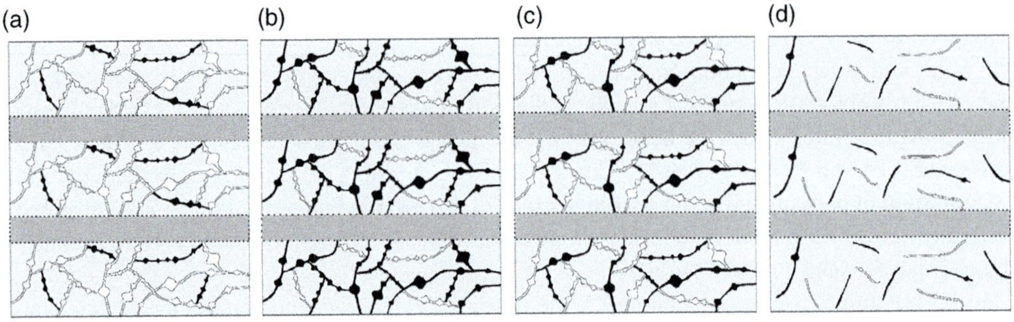

Figure 9.13 Simplified pore network models for shale samples with different wettabilities (black lines: oil-wet pore networks; white lines: water-wet pore networks; dotted lines: mixed-wet pore spaces with larger width among bedding planes; gray region: grains). (a) more water-wet shale sample; (b) more oil-wet shale sample; (c) mixed-wet shale sample; (d) intermediate-wet shale sample.

Table 9.14 Thermal maturity and wettability inferred from SI experiments for shale samples from various formations reported in the literature.

Age	Formation	Location	R_o or R_{equ} (%)	Wettability
Triassic	Yanchang	Ordos Basin, China	0.83–1.02 Xiong et al. (2016)	More water-wet Gao and Hu (2018)
Silurian	Longmaxi	Sichuan Basin, China	2.84–3.05 Tian et al. (2013)	More oil-wet Gao and Hu (2018)
Mississippian	Barnett	Fort Worth Basin, Texas, United States	1.35 Loucks et al. (2009)	Three types of wettability Gao and Hu (2016)
Jurassic	Ziliujing	Sichuan Basin, China	1.05–1.82 Li et al. (2013)	Three types of wettability Gao et al. (2018)
Triassic	Montney	British Columbia and Alberta, Canada	2.06 Chalmers and Bustin (2012)	More oil-wet Lan et al. (2015)
Devonian	Horn River	Horn River Basin, Canada	1.6–2.5 Ross and Bustin (2008, 2009)	More water-wet Lan et al. (2015)
Silurian	Longmaxi Wufeng	Sichuan Basin, China	2.2–3.06 Guo and Zhang (2014)	More oil-wet Yang et al. (2017)
Silurian	Longmaxi	Upper Yangzi Platform, China	2.17 Sun et al. (2017)	Mixed-wet Sun et al. (2017)
Permian	Longtan		2.82 Sun et al. (2017)	More water-wet Sun et al. (2017)
Cambrian	Niutitang		3.22 Sun et al. (2017)	Intermediately wet Sun et al. (2017)

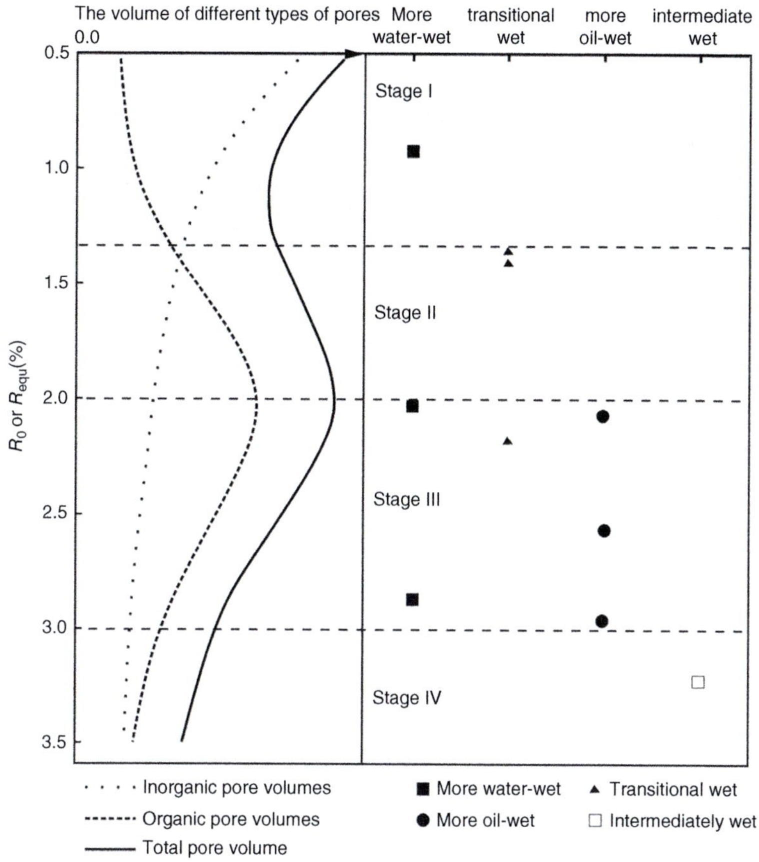

Figure 9.14 Wettability evolution and pore structure evolution models; for convenience, the term of transitionally wet is short for variable wettability including more water-wet, mixed-wet, and more oil-wet. *Source:* Gao et al. (2019)/With permission of Elsevier.

compaction and oil filling, however, still as the dominant pore type to build a well-connected water-wet pathway.

Stage II with thermal maturity 1.3% < R_o < 2.0%: The inorganic pores are reduced more due to the enhanced compaction, while many organic pores are generated by secondary cracking of OM. However, the development level of inorganic pores and organic pores is complicated, and shale samples exhibit variable wettabilities (Figure 9.14).

Stage III with thermal maturity 2.0% < R_o < 3.0%: Organic pores are much more developed by secondary cracking of OM and become the dominant pore type, while inorganic pores are reduced significantly due to the strong compaction, which leads to the more oil-wet wettability.

Stage IV with thermal maturity R_o > 3.0%: Organic pores are decreased due to the weakened secondary cracking of OM and the strong compaction, which causes the intermediate-wet characteristics.

9.5 Conclusions

The SI experiments were used to determine the overall shale wettability of different formations from China and United States, which showed the wide applicability of SI experiments in shale wettability evaluation. The effect of pore structure on shale wettability was also investigated in this study. The co-existence of oil-wet and water-wet pore networks in shale samples was confirmed, which led to the mixed wettability of shale samples. The dominant oil-wet (or water-wet) pore network determined the more oil-wet shale wettability. In order to show the close relationship between shale wettability and pore structure, four simplified pore network models with different wettability were proposed. Also, a new shale wettability evolution model was proposed to show the wettability change corresponding to the pore structure alteration during thermal maturation process. It should be noted that further studies are needed to verify and improve the proposed shale wettability evolution model.

In summary, SI experiment is a useful way to determine shale wettability, which also reveals the close relationship between shale wettability and pore connectivity. However, SI experiments are mainly used to qualitatively characterize shale wettability. A more representative and reliable quantitative wettability index is still not established and requires further study; recently Wang et al. (2021) established a wettability classification scheme by considering both contact angle measurement and SI experiments. Finally, the *in-situ* conditions of shale wettability should be evaluated by SI experiments in the future.

Acknowledgments

The funding for this work was provided by the National Natural Science Foundation of China (Nos. 41972145, 41690134, and 41830431).

References

Amott, E. (1959). Observations relating to the wettability of porous rock. *Transactions of AIME* 216: 156–162.

Anderson, W. (1986a). Wettability literature survey-part 1: rock/oil/brine interactions and the effects of core handling on wettability. *Journal of Petroleum Technology* 38 (10): 1125–1144.

Anderson, W. (1986b). Wettability literature survey-part 2: wettability measurement. *Journal of Petroleum Technology* 38 (11): 1246–1262.

Arif, M., Zhang, Y., and Iglauer, S. (2021). Shale wettability: data sets, challenges, and outlook. *Energy & Fuels* 35 (4): 2965–2980.

Bai, B., Elgmati, M., Zhang, H., and Wei, M. (2013). Rock characterization of Fayetteville shale gas plays. *Fuel* 105: 645–652.

Becker, S.J. (2019). Laboratory-scale petrophysical evaluation of lithofacies effect on reservoir & source quality and core-calibrated well log analysis in Pennsylvanian-Permian Wolfcamp-Spraberry intervals, Midland Basin, Texas, USA. MS Thesis, The University of Texas at Arlington, 191 pp.

Chalmers, G.R. and Bustin, R.M. (2012). Geological evaluation of Halfway–Doig–Montney hybrid gas shale–tight gas reservoir, northeastern British Columbia. *Marine and Petroleum Geology* 38 (1): 53–72.

Chalmers, G.R., Bustin, R.M., and Power, I.M. (2012). Characterization of gas shale pore systems by porosimetry, pycnometry, surface area, and field emission scanning electron microscopy/transmission electron microscopy image analyses: examples from the Barnett, Woodford, Haynesville, Marcellus, and Doig units. *AAPG Bulletin* 96 (6): 1099–1119.

Craig, F.F. Jr. (1971). The reservoir engineering aspects of waterflooding. *Richardson, Texas, SPE Monograph Series* 3: 12–44.

Curtis, J.B. (2002). Fractured shale-gas systems. *AAPG Bulletin* 86 (11): 1921–1938.

Gao, Z. and Hu, Q. (2016). Wettability of Mississippian Barnett Shale samples at different depths Investigations from directional spontaneous imbibition. *AAPG Bulletin* 100 (1): 101–114.

Gao, Z. and Hu, Q. (2018). Pore structure and spontaneous imbibition characteristics of marine and continental shales in China. *AAPG Bulletin* 102 (10): 1941–1961.

Gao, Z., Yang, S., Jiang, Z. et al. (2018). Investigating the spontaneous imbibition characteristics of continental Jurassic Ziliujing Formation shale from the northeastern Sichuan Basin and correlations to pore structure and composition. *Marine and Petroleum Geology* 98: 697–705.

Gao, Z., Fan, Y., Hu, Q. et al. (2019). A review of shale wettability characterization using spontaneous imbibition experiments. *Marine and Petroleum Geology* 109: 330–338.

Guo, T. (2016). Key geological issues and main controls on accumulation and enrichment of Chinese shale gas. *Petroleum Exploration and Development* 43 (3): 349–359.

Guo, T. and Zhang, H. (2014). Formation and enrichment mode of Jiaoshiba shale gas field, Sichuan Basin. *Petroleum Exploration and Development* 41: 31–40.

Handy, L.L. (1960). Determination of effective capillary pressures for porous media from imbibition data. *Transactions of AIME* 219: 75–80.

Hu, Q., Persoff, P., and Wang, J. (2001). Laboratory measurement of water imbibition into low-permeability welded tuff. *Journal of Hydrology* 242: 64–78.

Hu, Q. H., R. P. Ewing, and S. Dultz, 2012, Low pore connectivity in natural rock: *Journal of Contaminant Hydrology*, v. 133, p. 76–83, doi:https://doi.org/10.1016/j.jconhyd.2012.03.006.

Hu, Q. H., X. Gao, Z. Gao, et al., 2014. Pore accessibility and connectivity of mineral and kerogen phases in shales. *SPE, AAPG, SEG Unconventional Resources Technology Conference* (25–27 August 2014), Denver, Colorado. http://archives.datapages.com/data/urtec/2014/1922943.htm?q=%2BtitleStrip%3A%22pore+accessibility+connectivity+mineral+kerogen+phases+shales%22 (accessed 11 March 2015).

Hu, Q., Ewing, R.P., and Rowe, H.D. (2015). Low nanopore connectivity limits gas production in Barnett formation. *Journal of Geophysical Research – Solid Earth* 120 (12): 8073–8087.

Hu, Q., Zhang, Y., Meng, X. et al. (2017). Characterization of multiple micro-nano pore networks in shale oil reservoirs of Paleogene Shahejie Formation in Dongying Sag of Bohai Bay Basin East China. *Petroleum Exploration and Development* 44 (5): 720–730.

Hu, Q., Mann, G., and Zhao, J. (2019). Pore structure and fluid uptake of the Yeso, Abo and Cisco formations in the Permian Basin in southeast New Mexico, U.S.A. *Interpretation* 7: SK1–SK17.

Hunt, A.G., Ewing, R.P., and Ghanbarian, B. (2014). *Percolation Theory for Flow in Porous Media*, 3ee), Lect. Notes Phys. 880. Heidelberg: Springer.

Katz, A.J. and Thompson, A.H. (1986). A quantitative prediction of permeability in porous rock. *Physical Review B* 34 (11): 8179–8181.

Katz, A.J. and Thompson, A.H. (1987). Prediction of rock electrical conductivity from mercury injection measurements. *Journal of Geophysical Research – Solid Earth* 92 (B1): 599–607.

Lan, Q., Xu, M., Binazadeh, M. et al. (2015). A comparative investigation of shale wettability: the significance of pore connectivity. *Journal of Natural Gas Science and Engineering* 27: 1174–1188.

Legens, C., Palermo, T., Toulhoat, H. et al. (1998). Carbonate rock wettability changes induced by organic compound adsorption. *Journal of Petroleum Science and Engineering* 20: 277–282.

Li, Y., Feng, Y., Liu, H. et al. (2013). Geological characteristics and resource potential of lacustrine shale gas in the Sichuan Basin SW China. *Petroleum Exploration and Development* 40 (4): 454–460.

Liang, C., Jiang, Z., Zhang, C. et al. (2014). The shale characteristics and shale gas exploration prospects of the Lower Silurian Longmaxi shale, Sichuan Basin, South China. *Journal of Natural Gas Science and Engineering* 21: 636–648.

Loucks, R.G. and Ruppel, S.C. (2007). Mississippian Barnett Shale: lithofacies and depositional setting of a deep-water shale-gas succession in the Fort Worth Basin, Texas. *AAPG Bulletin* 91 (4): 579–601.

Loucks, R.G., Reed, R.M., Ruppel, S.C., and Jarvie, D.M. (2009). Morphology, genesis, and distribution of nanometer-scale pores in siliceous mudstones of the Mississippian Barnett Shale. *Journal of Sedimentary Research* 79: 848–861.

Loucks, R.G., Reed, R.M., Ruppel, S.C., and Hammes, U. (2012). Spectrum of pore types and networks in mudrocks and a descriptive classification for matrix-related mudrock pores. *AAPG Bulletin* 96 (6): 1071–1098.

Ma, S., Morrow, N.R., and Zhang, X. (1997). Generalized scaling of spontaneous imbibition data for strongly water-wet systems. *Journal of Petroleum Science and Engineering* 18: 165–178.

Mastalerz, M., Schimmelmann, A., Drobniak, A., and Chen, Y. (2013). Porosity of Devonian and Mississippian New Albany Shale across a maturation gradient: insights from organic petrology, gas adsorption, and mercury intrusion. *AAPG Bulletin* 97 (10): 1621–1643.

Mattax, C.C. and Kyte, J.R. (1962). Imbibition oil recovery from fractured, water-drive reservoir. *SPE Journal* 2 (2): 177–184.

Morrow, N.R. (1990). Wettability and its effect on oil recovery. *Journal of Petroleum Technology* 42 (12): 1476–1484.

Odusina, E.O., Sondergeld, C.H., and Rai, C.S. (2011). NMR study of shale wettability. *Canadian Unconventional Resources Conference*, Alberta, Canada (15–17 November 2011).

Philip, J.R. (1957). The theory of infiltration: 4. Sorptivity and algebraic infiltration equations. *Soil Science* 84: 257–265.

Ross, D.J.K. and Bustin, R.M. (2008). Characterizing the shale gas resource potential of Devonian–Mississippian strata in the western Canada sedimentary basin: application of an integrated formation evaluation. *AAPG Bulletin* 92: 87–125.

Ross, D.J.K. and Bustin, R.M. (2009). The importance of shale composition and pore structure upon gas storage potential of shale gas reservoirs. *Marine and Petroleum Geology* 26: 916–927.

Rowe, H.D., Loucks, R.G., Ruppel, S.C., and Rimmer, S.M. (2008). Mississippian Barnett Formation, Fort Worth Basin, Texas: bulk geochemical inferences and Mo–TOC constraints on the severity of hydrographic restriction. *Chemical Geology* 257: 16–25.

Salathiel, R. (1973). Oil recovery by surface film drainage in mixed-wettability rocks. *Journal of Petroleum Technology* 25: 1216–1224.

Schowalter, T.T. (1979). Mechanics of secondary hydrocarbon migration and entrapment. *AAPG Bulletin* 63 (5): 723–760.

Sharifigaliuk, H., Mahmood, S.M., Al-Bazzaz, W., and Khosravi, V. (2021). Complexities driving wettability evaluation of shales toward unconventional approaches: a comprehensive review. *Energy & Fuels* 35 (2): 1011–1023.

Sing, K.S., Everett, D.H., Haul, R.A.W. et al. (1985). Reporting physisorption data for gas/solid systems with special reference to the determination of surface area and porosity. *Pure and Applied Chemistry* 57 (4): 603–619. https://doi.org/10.1351/pac198557040603.

Stauffer, D. and Aharony, A. (1994). *Introduction to Percolation Theory*, 2ee. London: Taylor and Francis.

Sun, M., Yu, B., Hu, Q. et al. (2017). Pore connectivity and tracer migration of typical shales in south China. *Fuel* 203: 32–46.

Takahashi, S. and Kovscek, A.R. (2010). Spontaneous countercurrent imbibition and forced displacement characteristics of low-permeability, siliceous shale rocks. *Journal of Petroleum Science and Engineering* 71: 47–55.

Tian, H., Pan, L., Xiao, X. et al. (2013). A preliminary study on the pore characterization of Lower Silurian black shales in the Chuandong Thrust Fold Belt, southwestern China using low pressure N_2 adsorption and FE-SEM methods. *Marine and Petroleum Geology* 48: 8–19.

Thomas, M.M. and Clouse, J.A. (1990). Primary migration by diffusion through kerogen: II. Hydrocarbon diffusivities in kerogen. *Geochimica et Cosmochimica Acta* 54 (10): 2781–2792.

Wang, Q., Zhou, W., Hu, H. et al. (2021). Pore geometry characteristics and fluid-rock interaction in the Haynesville Shale, East Texas, United States. *Energy & Fuels* 35 (1): 237–250.

Washburn, E.W. (1921). Note on a method of determining the distribution of pore sizes in a porous material. *Proceedings of the National Academy of Sciences* 7: 115–116.

Xiong, F., Jiang, Z., Chen, J. et al. (2016). The role of the residual bitumen in the gas storage capacity of mature lacustrine shale: a case study of the Triassic Yanchang shale, Ordos Basin China. *Marine and Petroleum Geology* 69: 205–215.

Xu, M. and Dehghanpour, H. (2014). Advances in understanding wettability of gas shales. *Energy & Fuels* 28 (7): 4362–4375.

Yang, R., Hao, F., He, S. et al. (2017). Experimental investigations on the geometry and connectivity of pore space in organic-rich Wufeng and Longmaxi shales. *Marine and Petroleum Geology* 84: 225–242.

Zhang, P. and Austad, T. (2006). Wettability and oil recovery from carbonates: effects of temperature and potential determining ions. *Colloids and Surfaces A: Physicochemical and Engineering Aspects* 279: 179–187.

Zhang, X., Morrow, N.R., and Ma, S. (1996). Experimental verification of a modified scaling group for spontaneous imbibition. *SPE Reservoir Engineering* 11: 280–285.

10

Permeability Enhancement in Shale Induced by Desorption

Brandon Schwartz and Derek Elsworth

Department of Energy and Mineral Engineering, EMS Energy Institute and G3 Center, Penn State University
University Park, PA, USA

10.1 Introduction

The role of ultra-low permeability of shales in staunching fluid flow has confounded attempts to recover hydrocarbons but been a boon to their use for the sequestration of energy wastes. Recent advances in reservoir completions have improved the economics of shale applications – the most important two being horizontal drilling and multistage hydraulic fracturing. These two techniques are used in tandem to reduce drainage path lengths from the shale matrix to the high permeability pathways created by the stimulation treatment. The large initial production rates achieved for new wells result from rapid drainage to the newly created fracture surface area. The rate-limiting step for long-term production remains the low matrix permeability – apparent only after well production extends significantly beyond the initial production peak. Similarly, injection rates into depleted shale reservoirs for enhanced gas production or for carbon sequestration must remain low in order to control pressure buildup. This chapter highlights key factors that control permeability evolution in shale and methods that can be used to promote permeability enhancement through desorption.

10.1.1 Shale Mineralogical Characteristics

Shale is a sedimentary rock composed primarily of clays, silicates, carbonates, and organic matter (Sone and Zoback 2013). Within the matrix, individual mineral components form thin laminae separated by fracture planes which allow for greater fluid flow parallel to bedding (Crook et al. 2002; Bandyopadhyay 2009; Bonnelye et al. 2017) than perpendicular to it. These bedding features are observed to cause brittle failure planes when coring (Lyu et al. 2015; Heng et al. 2020). Porosity is concentrated between laminae as micro-fissures that create a set of preferential flow paths throughout the shale (Keller et al. 2011; Yan et al. 2015). Imaging of kerogen in shales shows a spongelike network of nanopores that create high surface area for sorptive storage (Kang et al. 2011). This arrangement leads to nanoscale diffusion out of the kerogen into micro-fissures and bedding-parallel fractures during fluid flow (Wu et al. 2016; Fan and Ettehadtavakkol 2017; Liu et al. 2019b).

Physics of Fluid Flow and Transport in Unconventional Reservoir Rocks, First Edition.
Edited by Behzad Ghanbarian, Feng Liang, and Hui-Hai Liu.

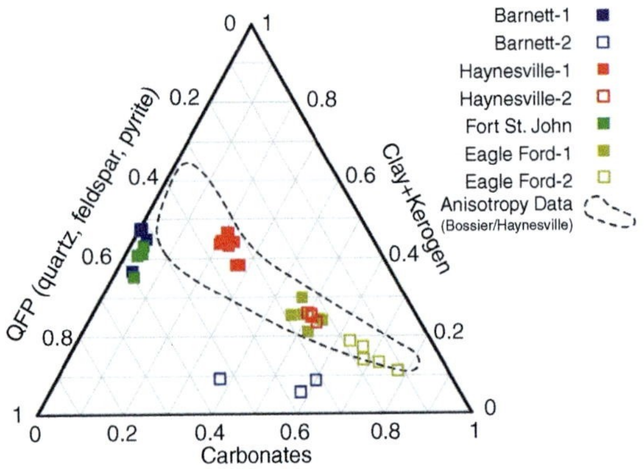

Figure 10.1 Ternary diagram of shale mineralogy. *Source:* Sone (2012) (Fig. 2.1)/ Stanford University.

Figure 10.1 shows a ternary diagram representing the mineralogy of several shales. Clays and kerogen are generally the predominant mineral components, with carbonates, quartz, feldspar, and pyrite comprising the remainder. Hydrocarbons are predominantly stored within the clays and organic matter. Organic matter is strongly sorptive, while clays are weakly sorptive (Ross and Bustin 2009). The other mineralogical components of shale are non-sorptive. The disparate mineralogy and mineral distribution in shales allow for the creation of porosity via shrinkage of sorptive components (Wang et al. 2012). Desorptive shrinkage relaxes the solid matrix locally surrounding flow paths that were originally constricted due to adsorptive swelling of clays and organic matter. This desorption occurs naturally in shales during pressure depletion, which causes an increase in permeability that competes with permeability reduction due to increased effective stress (Jia et al. 2018). While flow paths are concentrated in the clays and kerogen, shales with higher silicate or carbonate contents are also stiffer, such that they lose less permeability with increasing net effective stress than softer shales (Schwartz et al. 2019a). Understanding controls on permeability evolution due to changes in pore pressure, net effective stress, and adsorption capacity is an area of active research.

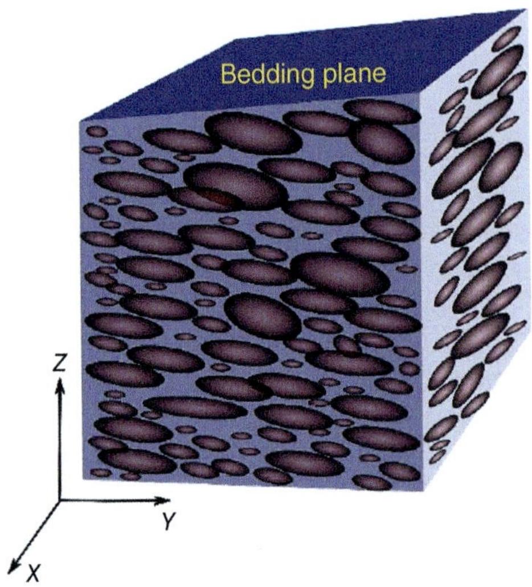

Figure 10.2 Macropores aligned parallel to bedding in shale. *Source:* Adapted from Gu and Mildner (2016) (Fig. 1).

10.1.2 Flow Network

Shale is typified by strongly anisotropic transport and mechanical properties (Sone and Zoback 2013; Schwartz et al. 2019b). Permeability in shale can be 10–100 times lower in the bedding-perpendicular direction than in the bedding-parallel direction (Bolton et al. 2000; Kwon et al. 2004; Pan et al. 2015). This permeability anisotropy is due to the concentration of porosity between clay-mineral laminae parallel to bedding (Ulm and Abousleiman 2006; Horne 2013). Flow paths parallel to bedding are characterized as macropores and long micro-fractures that create a preferential flow network through the shale (Marschall et al. 2005; Keller et al. 2011; Daigle et al. 2017). Figure 10.2 shows a schematic of bedding-parallel ellipsoidal macropores, which are concentrated between laminae in the bedding-parallel direction. Inside each lamina, fluid transport originates in nanopores and is driven by diffusion (Javadpour et al. 2007). Both parallel- and perpendicular-to-bedding, storage, and flow occur predominantly in the kerogen and clay components (Sondergeld et al. 2010; Curtis et al. 2012).

10.1.2.1 Bedding-Parallel Flow Network

While it remains challenging to image nanopores, the pore volume of the preferential flow network is sufficiently large that it is readily imaged using numerous techniques (Gu and Mildner 2016; Curtis et al. 2019; Goral et al. 2019). In all cases, imaging techniques reveal planar micro-fractures and macropores separating laminae. Figures 10.3–10.5 give characteristic examples of the bedding-parallel flow network from the literature. In all cases, flow paths can be characterized as long, flat cracks oriented parallel to bedding and between laminae.

Flow paths comprised of an ensemble flow network of component fractures may be individually represented by the cubic law to model permeability (Witherspoon et al. 1980; Bai and Elsworth

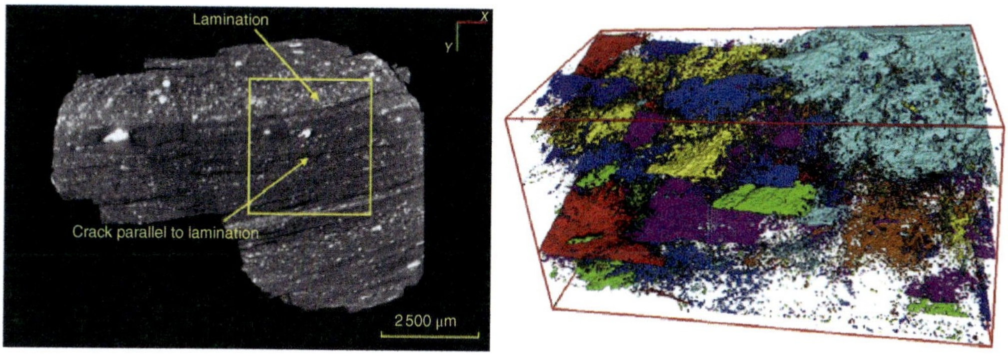

Figure 10.3 Three-dimensional characterization of micro-fractures in shale. *Source:* Qi et al. (2018)/from Elsevier.

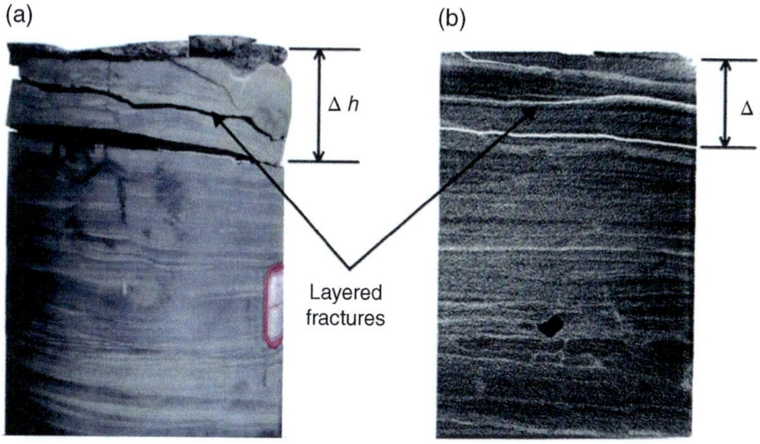

Figure 10.4 Planes of weakness parallel to laminae in shale. (a) is the optical image and (b) is the CT image. *Source:* Huang and Hu (2018)/Royal Society of Chemistry.

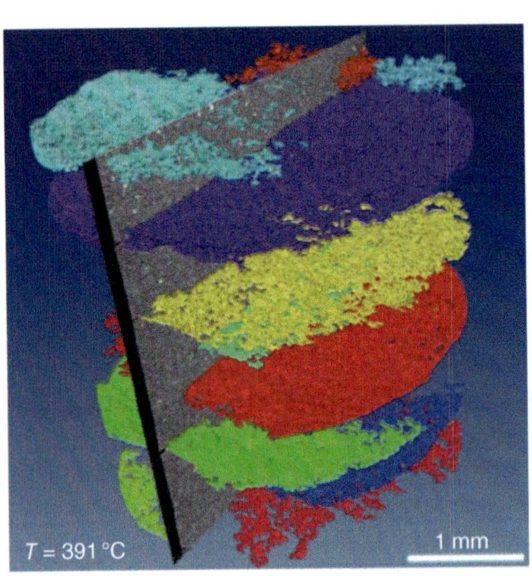

Figure 10.5 Imaging of bedding-parallel fractures in shale expanding during heating. *Source:* Kobchenko et al. (2011).

2000; Izadi et al. 2011). The cubic law may overestimate the permeability of single fractures due to the neglect of variations in fracture roughness and tortuosity (Wang et al. 2015). Permeability evolution in fractured, sorptive media is typically cast as an ensemble response to evolving strains (Liu et al. 2011; Wang et al. 2012; Schwartz et al. 2019c), and these strains are often competitive with each other (Liu and Harpalani 2013). For example, thermally induced permeability evolution in coals is modeled as an ensemble response to thermal and sorptive strains and dependent on applied boundary conditions (Qu et al. 2012).

10.1.2.2 Bedding-Perpendicular Flow Paths

Flow paths in the bedding-perpendicular direction are typified as tortuous nanotubes connected along grain boundaries. Figure 10.6 shows several such pathways in a small section of a shale sample from Chen et al. (2015). These openings can be so small that Darcian flow (driven by pressure gradient) is no longer an appropriate representation (Javadpour 2009; Civan et al. 2012). This happens because of gas slippage at the walls of nano-sized fractures or pores – in contrast to no-slip boundary conditions for liquid flow as presumed in the cubic or Pouseuille law quantifying Darcian flow. Conversely, in the case of flow along grain boundaries in the bedding-perpendicular direction, the diameter of openings approaches the mean free path of gas molecules such that Darcian flow breaks down. The flow is no longer driven by pressure gradients but by concentration gradients. This requires flow models based on diffusion and Fick's law to properly predict fluid transport (Izadi et al. 2011; Chen et al. 2015).

Because gas flow in shale originates in nanopores and within laminae along clay platelet boundaries, gas permeability models must account for the Knudsen diffusion-driven flow that occurs in these spaces (Cao et al. 2017). As the gas approaches the bedding-parallel flow paths, fluid transport begins to transition to the Darcian flow regime. Figure 10.7 shows part of the preferential flow network in a shale, which comprises a series of nanotubes connected to larger macropores. Such systems require dual permeability models that can couple Fickian and Darcian flow when considering field-scale response (Li and Elsworth 2015). Figure 10.8 shows the multiple scales that may need to be considered in field applications – in particular, moving from nanopores within organic storage, nested within the matrix, and draining to adjacent hydraulic fractures. In the laboratory, however,

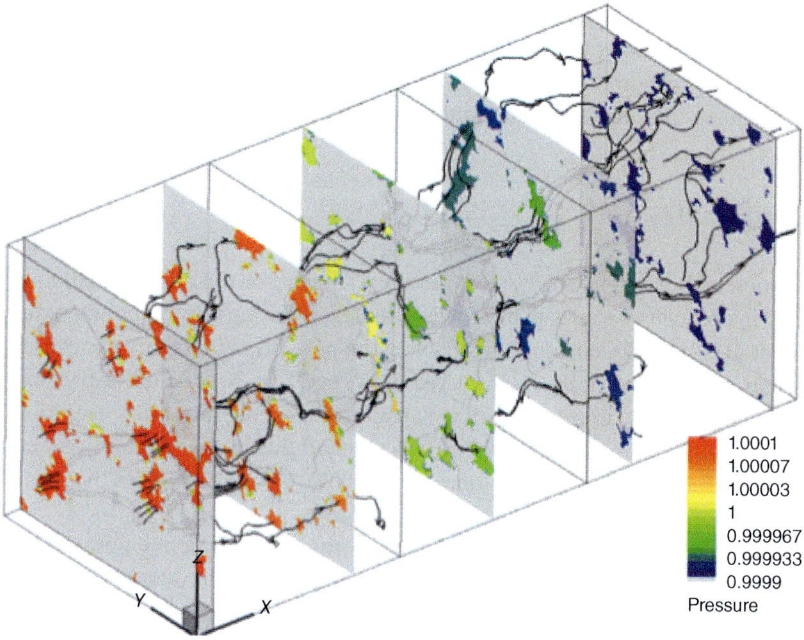

Figure 10.6 Fluid transport along grain boundaries within shale. *Source:* Chen et al. (2015) (Fig. 3)/Springer Nature.

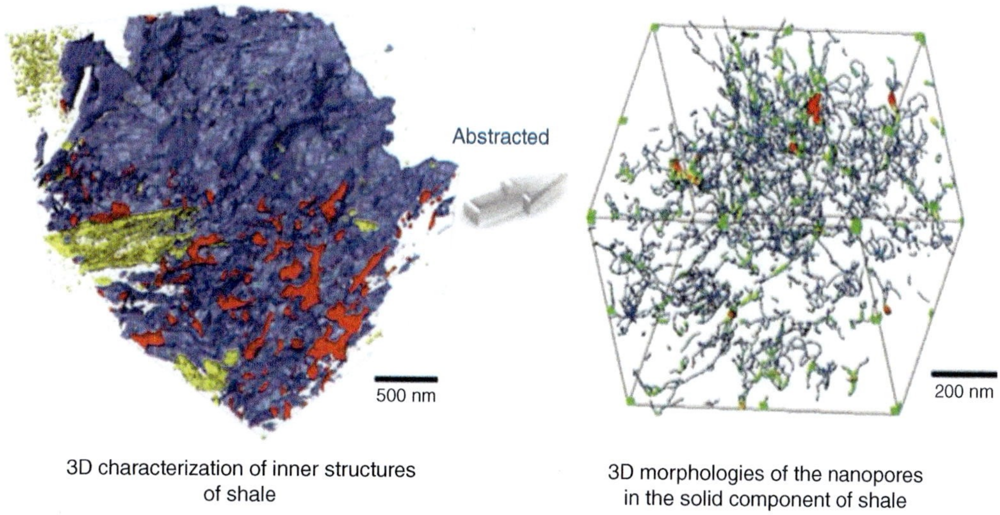

Figure 10.7 Preferential flow network within shale. *Source:* Ma et al. (2017) (Fig. 10) / With permission of Geological Society of London.

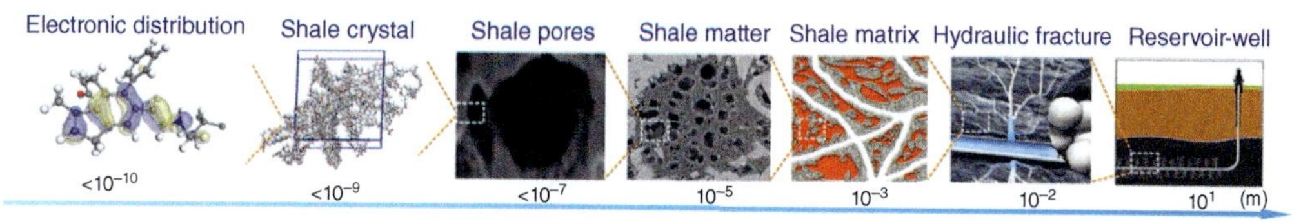

Figure 10.8 Multiscale modeling of fluid transport through shale. *Source:* Wang et al. (2020)/from ELSEVIER.

shale gas permeability is generally measured individually in either bedding-parallel or -perpendicular directions using pressure pulse tests (Brace et al. 1968). As a result, measurements taken in the bedding-parallel direction are measuring the permeability of the system of planar micro-fractures and macropores that make up the preferential flow network.

10.2 Adsorption in Shales

Shales experience adsorption in the presence of sorptive gases, such as carbon dioxide and methane. Adsorption in the organic matter causes shales to internally swell, constricting flow paths and reducing permeability (Izadi et al. 2011). Concurrently, increasing the pore pressure of any gas causes poromechanical expansion, leading permeability to increase (Wang et al. 2012). Bulk expansion due to sorptive swelling is additive to poromechanical expansion, whereas permeability evolution is competitive due to these strains (Harpalani and Chen 1997; Cui and Bustin 2005; Robertson and Christiansen 2007; Mazumder and Wolf 2008; Siriwardane et al. 2009). Figure 10.9 shows the range of permeability responses to sorptive gas at increasing pore pressures.

Sorptive swelling occurs along the surface of the adsorbent. In the presence of any appreciable amount of gas, swelling is essentially isotropic (Robertson and Christiansen 2006). However, because shale is a mechanically transversely isotropic material each of the three principal components of the sorptive strain primarily impacts only one of the two (or three) directional permeabilities – the permeability effect is directionally decoupled. Conversely, the orthogonal impact of each component of the sorptive strain is captured through the Poisson effect – and directionally coupled. As such, three-dimensional models of permeability evolution in shale can be reduced to two dimensions without loss of model fidelity (Schwartz et al. 2019c).

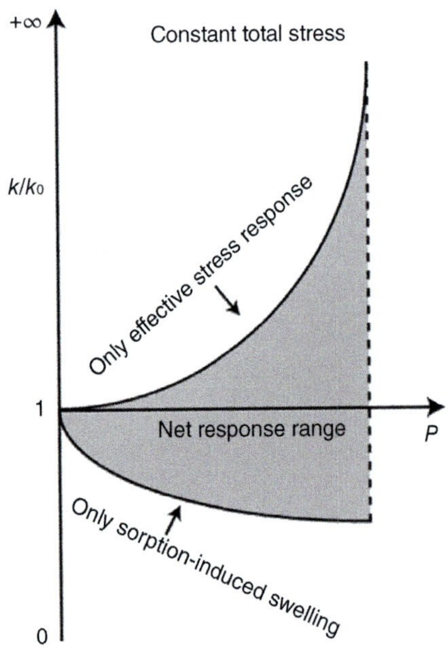

Figure 10.9 U-shaped permeability response with increasing pore pressure for a sorptive gas. *Source:* After Wang et al. (2012) (Fig. 5)/John Wiley & Sons intermediate between effective stress-only response and sorption-only response.

10.2.1 Langmuir Theory

Sorptive strain has been shown to follow a Langmuir curve that is proportional to the fractional coverage (Robertson 2005). Fractional coverage of adsorption sites ranges from 0 to 100% and can be cast as

$$\theta = \frac{P}{P + P_L} \tag{10.1}$$

or

$$\theta = \frac{bP}{1 + bP} \tag{10.2}$$

where P is the pressure, P_L is the Langmuir pressure, and b is the affinity constant. P_L corresponds to 50% fractional coverage and is readily estimated from laboratory data. Inspection of Eq. (10.1) and Eq. (10.2) also shows that b is equal to $1/P_L$. It is also equal to

$$b = b_{\infty} e^{Q/RT} \tag{10.3}$$

where b_∞ is the affinity constant at infinite temperature, Q is the heat of adsorption, R is the universal gas constant, and T is temperature. The value of b_∞ can be extrapolated from laboratory data and is equal to

$$b_\infty = \frac{\alpha}{k_{\mathrm{d}_\infty}\sqrt{2\pi MRT}} \tag{10.4}$$

where α is a sticking (attachment) coefficient, $k_{\mathrm{d}\infty}$ is the rate constant for desorption at infinite temperature, and M is the molecular weight of the gas. For a given gas and solid, the only variable in Eq. (10.4) is T (Do 1998), such that b_∞ varies with $T^{-1/2}$ linearly as:

$$b_\infty = b_{\infty 0}\left(\frac{T}{T_0}\right)^{-\frac{1}{2}}. \tag{10.5}$$

where $b_{\infty 0}$ is the affinity constant at the reference temperature T_0. Sorptive strain is linear with fractional coverage (Robertson 2005). With the maximum strain occurring at 100% coverage, sorptive strain is defined as:

$$\varepsilon_\mathrm{s} = \varepsilon_\mathrm{L}\frac{b\mathrm{P}}{1 + b\mathrm{P}} \tag{10.6}$$

where ε_L is the Langmuir strain at 100% coverage of sorption sites. The Langmuir strain corresponds to the maximum possible sorptive strain for a given gas and solid and is typically estimated by curve fitting. Figure 10.10b shows standard isotherms for sorptive strain. For a given ε_L the sorptive strain can be cast as a function of varying pressure and temperature:

$$\varepsilon_\mathrm{s} = \varepsilon_\mathrm{L}\frac{b_\infty e^{Q/RT}P}{1 + b_\infty e^{Q/RT}P}. \tag{10.7}$$

While Figure 10.10a shows the variation of sorptive strain isotherms with pressure, Figure 10.10b shows sorptive strain along isobars with varying temperatures. Both figures show that desorption occurs with decreasing pressure (or partial pressure) and increasing temperature.

10.2.2 Competing Strains in Permeability Evolution

There are several strains that arise during desorptive processes. Of particular note are mechanical strains that occur with changing pore pressures and thermal strains that develop as a result of changing temperatures. These strains are competitive with sorptive strain, and all three strains must be accounted for in permeability models (Liu et al. 2010; Chen et al. 2012).

10.2.2.1 Poro-Sorptive Strain

Figure 10.11 illustrates that fissure-like elliptical-section pores in shale are concentrated between bedding planes and can be characterized using pore spacing s, pore aperture b, and pore length a. Figure 10.11a shows an idealized pore surrounded by matrix rock including organic matter (OM). With increasing pore pressure, the pore expands due to the traction along the pore boundary (Figure 10.11b). Increasing pore size also causes bulk expansion of the shale. If the gas species is sorptive, then an additional deformation occurs as the surrounding organic matter swells (Figure 10.11c). Because the deformation originates outside of pores, it creates a separate bulk expansion that concurrently constricts flow paths by causing pore closure. Combining these mechanisms, increasing pore pressure with a sorptive species causes pore diameter to increase due to poromechanical expansion and to simultaneously decrease due to sorptive swelling

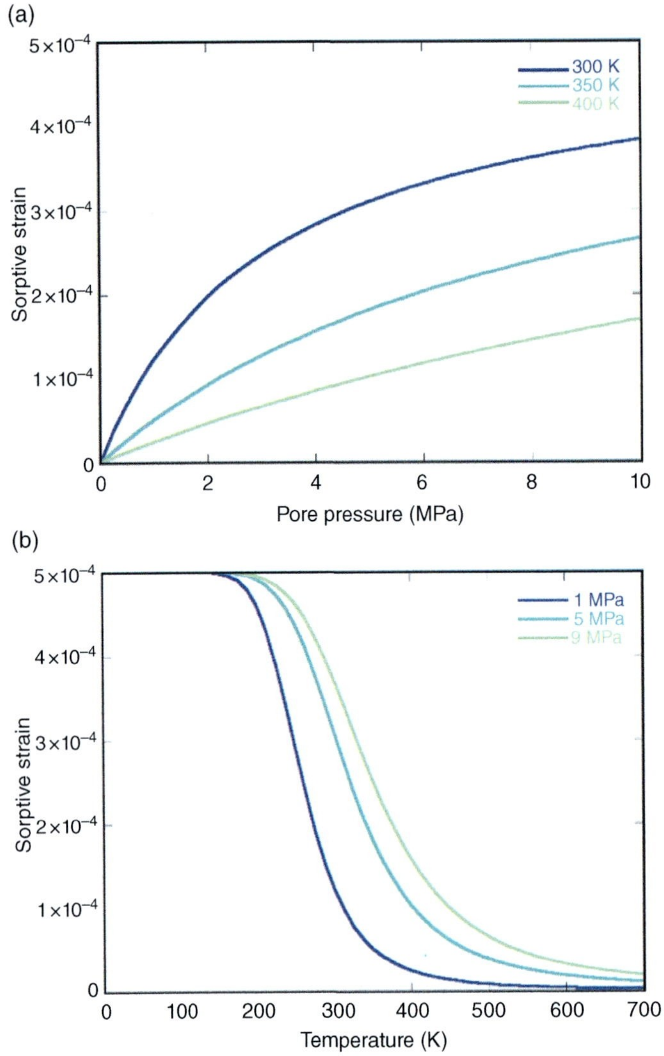

Figure 10.10 Behavior of sorptive gases at constant temperature or pressure. (a) Langmuir isotherms for sorptive strain, and (b) isobars for sorptive strain with varying temperatures.

(Fig. 10.11d). If the injection pressure of a sorptive gas is higher than the equilibrium pore pressure, poromechanical expansion of the pore space competes with the permeability reduction caused by sorptive swelling (Liu et al. 2010; Kumar et al. 2016). At any given pore pressure, the bulk expansion from a sorptive gas is larger than that from a non-sorptive gas because the sorptive swelling causes an additional deformation. The permeability response results from the net change in the pore aperture.

This ensemble response to sorptive gases can be further illustrated in Figure 10.12. In Figure 10.12a, strain is plotted against increasing pore pressure with expansion shown as positive strain. The sorptive strain follows a Langmuir-type curve that grows rapidly at low pore pressures and slows as it approaches a maximum strain at higher pore pressures. After 100% of the available adsorption sites are filled, there is no further sorptive strain. Instead, additional pore pressure only

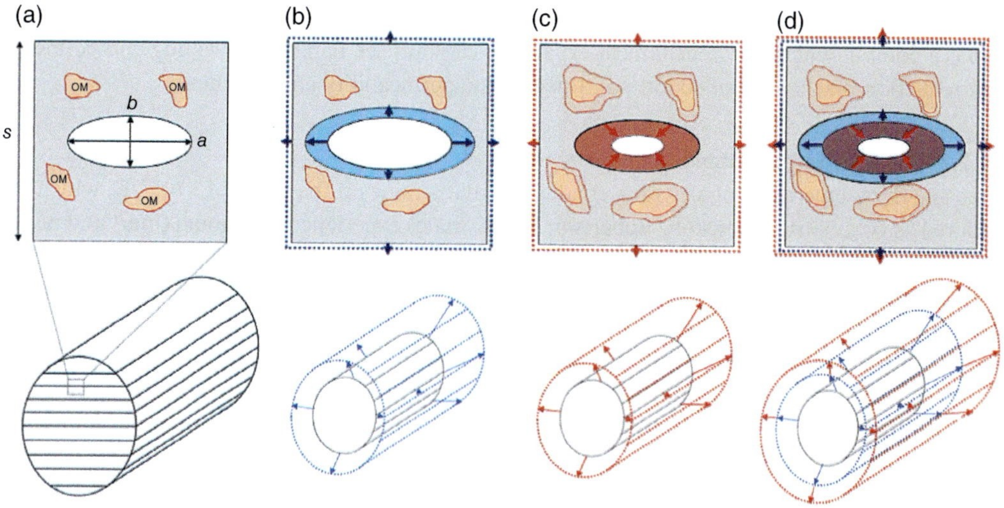

Figure 10.11 (a) Diagram of parallel processes occurring in fissure-like elliptical-section pores in shale core. (b) Shows poromechanical expansion, (c) displays sorptive swelling, and (d) the combined effect.

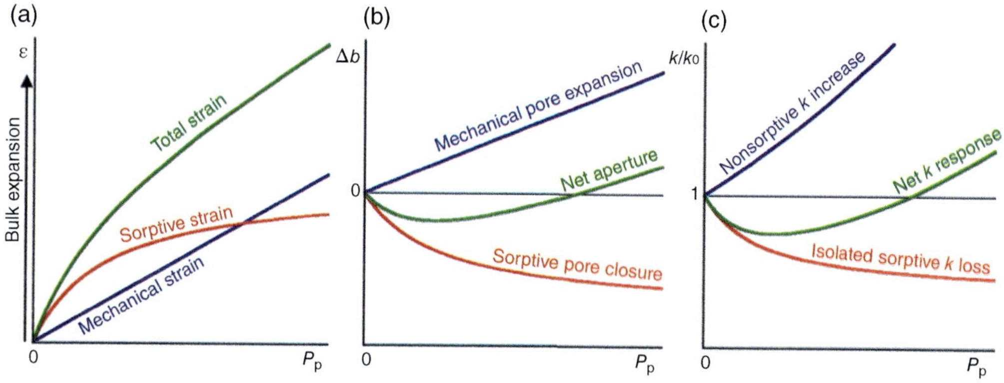

Figure 10.12 Competitive responses in the deformation of pores due to sorptive and mechanical strain. (a) Both strains cause positive bulk expansion and are additive components of the total strain. (b) Mechanical strain causes pore dilation, while sorptive strain reduces the aperture. (c) The net permeability response will be due to the net aperture change. P_p denotes the pore pressure.

causes additional pore expansion (Wang et al. 2012). The total strain curve would be the sum of the sorptive strain and the poromechanical strain. Figure 10.12b illustrates that the net pore response would be positive or negative depending on whether sorptive pore closure is larger than mechanical pore expansion. At lower pore pressures, the pore closure caused by sorptive swelling is larger than the pore expansion caused by increased pore pressure. At larger pore pressures, incremental poromechanical expansion outpaces incremental sorptive swelling, and the net pore response is expansion. Figure 10.12c translates the net pore response to the net permeability response. At low pore pressures, permeability decreases due to the dominant effect of sorptive swelling. As the rate of permeability reduction decreases due to Langmuir swelling, the rate of permeability enhancement increases due to poromechanical expansion. At higher pressures, permeability can increase beyond the initial permeability as poromechanical expansion becomes the dominant effect. Therefore, there is a point at which poromechanical expansion can outpace Langmuir swelling and permeability reduction reaches an inflection point to become permeability enhancement.

10.2.2.2 Thermal-Sorptive Strain

Increasing the temperature of any rock creates thermal strains on the order of ~10^{-5}/K, which is measured as volumetric expansion (Robertson 1988). Shales experience traditional (inert and non-reactive) thermal expansion in the range 300–400 K typical of reservoirs (Gilliam and Morgan 1987), but at ~470 K, they begin to collapse as kerogen undergoes pyrolysis and illitic clays experience a reduction in spacing between platelets. Thermal expansion of shales comprising multi-mineralic aggregates is lower than that of its individual mineral components (Robertson 1988) as pores and cracks reduce skeletal stiffness and reduce aggregate thermal expansion coefficient to be less than that of the weighted average of its mineral composition (Cooper and Simmons 1977). This difference is due to mineral expansion being distributed partly to bulk volume expansion and partly to pore closure, indicating that thermal strains potentially reduce permeability. The magnitude of permeability loss depends on the percentage of mineral expansion that contributes to pore volume reduction. Thermal strain significantly reduces permeability for constant volume reservoirs (Qu et al. 2010) – one end member behavior for the spectrum of boundary conditions. However, increasing temperature also causes desorption in sorptive rocks that can both lead to permeability enhancement (Sakurovs et al. 2008; Li et al. 2009) or permeability loss (Sinha et al. 2013) depending on the comparative influence of the competitive thermal and sorptive strains.

The Langmuir model predicts that desorption also occurs with increasing temperature at constant pore pressure. This leads to counterintuitive mechanical behavior in sorptive rocks with increasing temperature. Rock stiffness in shales tends to decrease upon heating and increase upon cooling (Closmann and Bradley 1979; Zhu and Arson 2014). Indeed, laboratory tests on European black shales showed that rock stiffness decreases with increasing temperature in single-stage uniaxial compression tests (Rybacki et al. 2016). Increasing temperature also leads to a softening of coals, indicated by a decrease in bulk modulus (Khan and Jenkins 1984). Pore closure causes the dynamic bulk modulus to increase as the shale becomes stiffer due to decreased porosity (Schwartz et al. 2019b). The net result is that thermally induced desorption can lead to permeability enhancement if the change in sorptive strain is larger than the change in thermal strain.

10.3 Permeability Models for Sorptive Media

Shale and coal are both fractured, sorptive media and their responses to sorptive gases are similar (Kumar et al. 2016). Many permeability models have been developed to capture the effect of sorption in coals (Seidle et al. 1992; Palmer and Mansoori 1996; Cui and Bustin 2005; Shi and Durucan 2005; Palmer 2009), and permeability evolution in fractured, sorptive media is typically cast in terms of evolving strains (Liu et al. 2011). These strains occur due to organic material swelling within the matrix and sorption within flow channels (Li et al. 2017a). In flow channels composed of sorptive minerals, the layer of adsorbed gas against the flow channel wall directly reduces the flow channel aperture (Sakhaee-Pour and Bryant 2012). Conversely, the shrinkage of organic matter during desorption increases pore aperture (Levine 1996; Liu and Rutqvist 2010). The magnitude of permeability reduction caused by sorptive swelling represents the potential permeability enhancement in the event of desorption. In coals, methane desorption has been shown to increase permeability two- to fivefold (Harpalani and Schraufnagel 1990). This has been directly observed in situ, where permeability increased during depletion as organic shrinkage expanded fractures, which had a larger effect than the additional compression caused by the overburden stress (Mavor and Vaughn 1998).

10.3.1 Strain Based Models

Permeability evolution – either permeability loss or permeability enhancement – is due to a change in the conductivity of the flow network. These changes are often cast as evolving strains that can originate internal or external to the shale volume and result in a change in the pore structure. Considering all three strains that can typically arise in a shale reservoir – alternately resulting from changes in pore pressure, temperature, or fractional coverage of adsorption sites – changes to the pore structure arise from the net impact of these three strains as:

$$\Delta b = \lambda s\left(\Delta\varepsilon_{\mathrm{p}} + \Delta\varepsilon_{\mathrm{s}} + \Delta\varepsilon_{\mathrm{th}}\right) \tag{10.8}$$

where b is the aperture of a macropore or micro-fissure, s is the spacing between pores, and λ is a boundary condition ranging from 0 for free volume expansion to 1 for fixed bulk volume (no boundary expansion). The three strains – poromechanical, sorptive, and thermal – are shown as positive values; however, the sorptive strain would be of the opposite sense to, and sign of, both the poromechanical and thermal strains. Using the cubic law, the sorptive component of permeability evolution can be isolated as:

$$\left(\frac{k}{k_0}\right)_{\mathrm{s}} = \left(1 + \lambda\frac{s}{b_0}\Delta\varepsilon_{\mathrm{s}}\right)^3. \tag{10.9}$$

Combining Eq. (10.9) with Eq. (10.7) gives sorptive permeability evolution as:

$$\left(\frac{k}{k_0}\right)_{\mathrm{s}} = \left(1 + \lambda\frac{s}{b_0}\left(\varepsilon_{\mathrm{L}}\frac{b_{\infty 0}\left(\frac{T}{T_0}\right)^{-\frac{1}{2}}\cdot e^{\frac{Q}{RT}}\cdot P}{1 + b_{\infty 0}\left(\frac{T}{T_0}\right)^{-\frac{1}{2}}\cdot e^{\frac{Q}{RT}}\cdot P} - \varepsilon_{\mathrm{s0}}\right)\right)^3 \tag{10.10}$$

where we note that the change in permeability from some reference permeability k_0 is dependent on the reference sorptive strain $\varepsilon_{\mathrm{s0}}$ that exists at k_0. It can be readily shown that the sorptive component of permeability evolution increases when (i) temperature increases or (ii) pressure decreases. This is also apparent from an inspection of Langmuir isotherms: decreasing pressure or increasing temperature leads to desorption, which dilates flow paths previously constricted by nearby adsorptive swelling. Therefore, the total permeability evolution in terms of evolving thermal, poromechanical, and sorptive strains is:

$$\frac{k}{k_0} = \left(1 + \lambda\frac{s}{b_0}\left(\Delta\varepsilon_{\mathrm{p}} + \Delta\varepsilon_{\mathrm{s}} + \Delta\varepsilon_{\mathrm{th}}\right)\right)^3 \tag{10.11}$$

In terms of each individual strain, Eq. (10.11) can be expanded to

$$\frac{k}{k_0} = \left(1 + \lambda\frac{s}{b_0}\left(\varepsilon_{\mathrm{L}}\frac{b_{\infty 0}\left(\frac{T}{T_0}\right)^{-\frac{1}{2}}\cdot e^{\frac{Q}{RT}}\cdot P}{1 + b_{\infty 0}\left(\frac{T}{T_0}\right)^{-\frac{1}{2}}\cdot e^{\frac{Q}{RT}}\cdot P} - \varepsilon_{\mathrm{s0}} - \alpha_{\mathrm{th}}\Delta T + \frac{\Delta P_{\mathrm{p}}}{K}\right)\right)^3 \tag{10.12}$$

where α_{th} is the coefficient of thermal expansion of the shale and K is the bulk modulus. While Eq. (10.12) can become unwieldy to use, the key insight is that the initial sorptive strain is dependent on the initial pore pressure and initial temperature. Also, if pressure or temperature is constant, the equation reduces to a simpler form.

10.4 Competing Processes during Permeability Evolution

From the foregoing, it is clear that competing impacts of changes in pressure, effective stress, temperature, and sorption are manifest, in a complex manner, on strains. Here, we resolve the impact of competing strains on the evolution of permeability observed within an isothermal experiment. Then, we isolate the sorption component of permeability evolution to understand the permeability loss associated with adsorption, whether thermally induced or pore pressure induced.

10.4.1 Resolving Competing Strains

We measured the bulk strain of the shale core when injected with either methane or helium gas. We plotted strain versus pore pressure for both helium and methane in Figure 10.13. The resolution of the LVDT is 0.1 μm with a smaller displacement rounded to the nearest 0.1 μm. Because the lengths of our samples are on the order of millimeters, a strain of 10^{-4} is the minimum resolvable increment. Strains less than 0.1 μm are automatically bit-limited by the LVDT. We included boundaries in Figure 10.13 to show the possible range of strains that could be obscured by the limits of resolution of the LVDT. In the case of methane injection, the measured strain is the sum of the poromechanical expansion and the sorptive swelling:

$$\varepsilon_{tot} = \varepsilon_p + \varepsilon_s \tag{10.13}$$

such that

$$\varepsilon_s = \varepsilon_{tot} - \varepsilon_p \tag{10.14}$$

where

$$\varepsilon_s = \varepsilon_L \frac{P}{P + P_L}. \tag{10.15}$$

The strain caused by methane injection includes both ε_p and ε_s, meaning it represents ε_{tot} with increasing pore pressure. The helium strain curve represents poromechanical expansion only and is equal to ε_p. In order to find ε_s, we subtracted the helium-induced strain from the methane-induced strain at each pressure and fit the resulting data to a Langmuir-type strain curve according to Eq. (10.4). From the plot below, the sorptive strain at 6 MPa is 1.15×10^{-4}, which can be used to calculate P_L and ε_L.

The Langmuir strain is isolated using Eqs. (10.14) and (10.15). The strain data in Figure 10.13 were used as input into Eq. (10.14) and a Langmuir strain curve was fit to the result using Eq. (10.15). Results for the isolated sorptive strain are plotted in Figure 10.14. At low pressures, the lower-than-expected measurements of strain for methane suggest that the core may not have been fully saturated, although the strain data are within the margin of error for the LVDT. In Eq. (10.15), a Langmuir pressure P_L of 7 MPa and a Langmuir strain ε_L of 2.75×10^{-4} provided an excellent fit to Eq. (10.14).

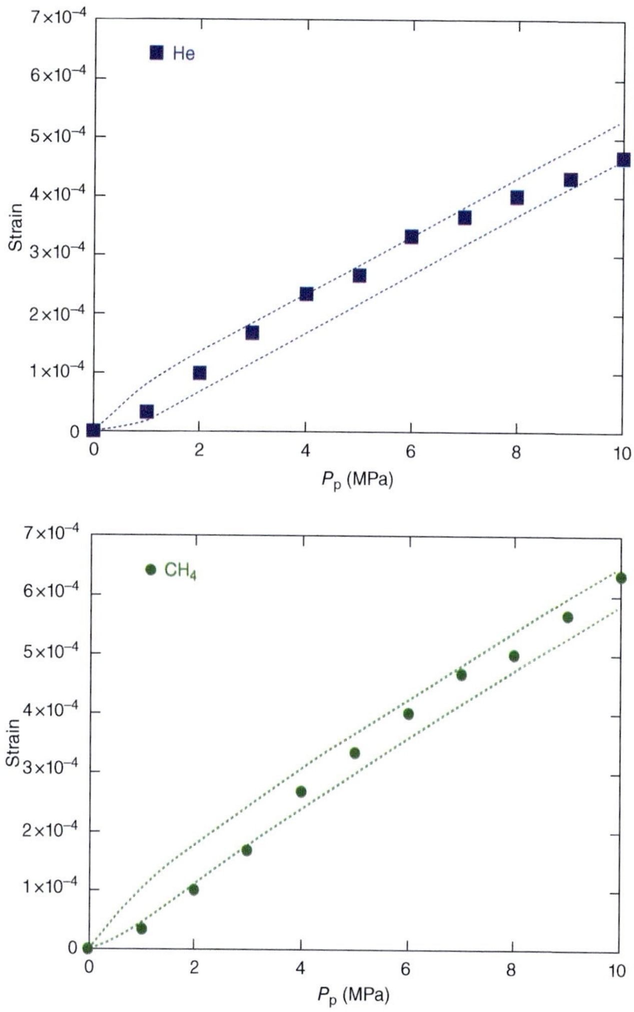

Figure 10.13 Strain vs. pore pressure for helium (top) and methane (bottom). The blue and green dotted lines show measurements ±0.1 μm of a linear fit to the strain data in order to illustrate the range of strain values that could be obscured by the resolution of the LVDT.

10.4.2 Solving for Sorption-Induced Permeability Evolution

When considering permeability evolution to methane with increasing pore pressure, the poromechanical and sorptive strains act in parallel to create a net response at each location in the matrix. The permeability evolution can be represented as processes in parallel as (Wang et al. 2012):

$$\frac{k}{k_0} = \left(1 + \left(\frac{s}{b_0} + \frac{a(1-\nu)}{b_0}\right)(\varepsilon_p + \varepsilon_s)\right)^3 \tag{10.16}$$

We group the variables in Eq. (10.16) into sorptive and non-sorptive components:

$$\frac{k}{k_0} = (1 + A - B)^3 \tag{10.17}$$

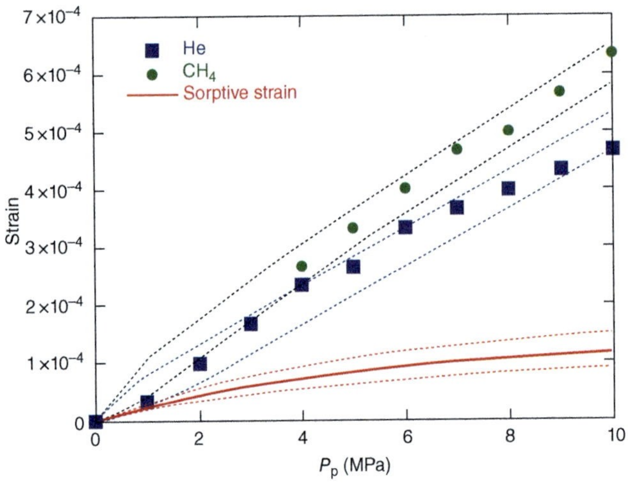

Figure 10.14 Isolating the sorptive strain using Eqs. (10.14) and (10.15). Subtracting the helium-induced strain data from the methane-induced strain data and fitting a Langmuir-type strain curve yields a Langmuir pressure P_L of 7 MPa and a Langmuir strain ε_L of 2.75 × 10^{-4} (mm/mm).

$$A = \frac{\Delta b_p}{b_0} = \left(\frac{s}{b_0} + \frac{a(1-\nu)}{b_0}\right)\varepsilon_p = \left(\frac{s}{b_0} + \frac{a(1-\nu)}{b_0}\right)\frac{\Delta\sigma'}{E} \tag{10.18}$$

$$B = \frac{\Delta b_s}{b_0} = \left(\frac{s}{b_0} + \frac{a(1-\nu)}{b_0}\right)\varepsilon_s = \left(\frac{s}{b_0} + \frac{a(1-\nu)}{b_0}\right)\varepsilon_L\frac{P}{P+P_L} \tag{10.19}$$

where k is the new permeability, k_0 is the original permeability, b_0/a represents an average aspect ratio of pores (b_0/a can be estimated by history matching field data, for example, Bandyopadhyay (2009) estimated b_0/a to be 0.10 in Bakken shale), A represents the change in the aperture size due to poromechanical expansion, and B represents the reduction in the aperture size due to sorptive swelling. Equation (10.17) can then be rearranged as follows:

$$\left(\frac{k}{k_0}\right)^{\frac{1}{3}} = 1 + A - B \tag{10.20}$$

$$\left(\frac{k}{k_0}\right)^{\frac{1}{3}} + 1 = \sqrt[3]{(1+A)^3} + (1-B) \tag{10.21}$$

$$(1-B)^3 = \left[\left(\frac{k}{k_0}\right)^{\frac{1}{3}} + 1 - \sqrt[3]{(1+A)^3}\right]^3. \tag{10.22}$$

We note that $(1 - B)^3$ is the permeability evolution in the absence of poromechanical expansion and $(1 + A)^3$ is the permeability evolution in the absence of Langmuir swelling. In Eq. (10.11), k/k_0 is the ensemble permeability response measured when using methane. In terms of the gases used in the experiments, we could cast Eq. (10.22) as follows:

$$\left(\frac{k}{k_0}\right)_{\text{sorptive}} = \left[\left(\frac{k}{k_0}\right)^{\frac{1}{3}}_{CH_4} + 1 - \left(\frac{k}{k_0}\right)^{\frac{1}{3}}_{He}\right]^3. \tag{10.23}$$

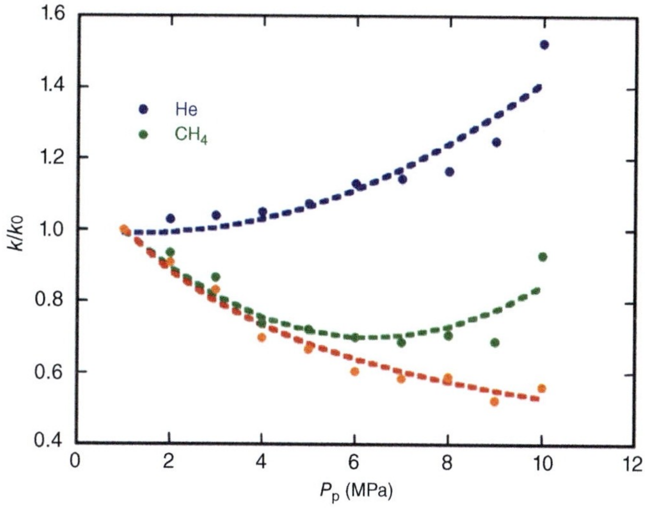

Figure 10.15 Solving for sorptive permeability evolution for processes in parallel. Permeability measurements were made sequentially with helium and then methane. The isolated sorptive permeability from Eq. (10.23) is in orange. The red regression curve is from Eq. (10.19).

We plotted the permeability data measured via helium and methane against the pore pressure in Figure 10.15. The data were then used as input into Eq. (10.23) to show the signature of the sorption component of permeability evolution. We input the sorption data from Eq. (10.15) into Eq. (10.19) and found the line of best fit for the data from Eq. (10.23) by using s/b_0 equal to 1500 and b/a equal to 7×10^{-3}(where b/a is further constrained by published values for aspect ratios in shale). In Figure 10.15, the red curve represents the permeability reduction due to the sorption predicted from Eq. (10.19) and the orange-filled circles are from Eq. (10.23). We found an excellent fit and good agreement between these two independent methods of solving for sorptive strain and sorptive permeability evolution.

10.5 Desorption Processes Yielding Permeability Enhancement

There are four methods by which native hydrocarbons can be desorbed from organic matter and clays in shale matrix: (i) lowering the pore pressure through depletion, (ii) lowering the partial pressure of methane by injecting a non-sorptive or sorptive gas, (iii) injecting a sorptive gas with higher adsorption affinity, or (iv) increasing the temperature of the shale. The first two are examples of desorption caused by changing pressure (ΔP), the third causes desorption by changing gas composition (ΔC), and the fourth by changing temperature (ΔT). We discuss each of these in turn.

10.5.1 Pressure Depletion

During fluid production, the pore pressure of the reservoir reduces, and methane desorbs to remain in equilibrium with the solid surface. This process has been shown to follow a Langmuir adsorption model, where fractional coverage can vary between 0 and 100% based on the gas pressure according to Eq. (10.1).As a reservoir undergoes pressure depletion, Eq. (10.1) shows that fractional coverage decreases. This is due to the drop in energy associated with the decrease in the pressure of the gas

(Do 1998). It has been noted in shale and coal reservoirs that permeability can increase almost 20-fold during production (Harpalani and Chen 1997), which is counterintuitive as the effective stress is typically increasing. In these instances, the permeability gained via desorption is evidently greater than the permeability lost from increasing effective stress (Li et al. 2017b).

10.5.2 Lowering Partial Pressure

When a second gas species is injected into pore space originally occupied by adsorbed methane, a binary mixture forms with a net pressure and two partial pressures of the individual species. In simplest terms, the net pressure is the weighted average of the partial pressures and molar concentrations of each gas species:

$$Pn = X_1 n_1 + X_2 n_2 \tag{10.24}$$

where X_1 and X_2 are the partial pressures of each gas species. Injecting a non-sorptive gas – commonly chosen to be nitrogen in the field and helium or argon in the laboratory – has the benefit of not interacting with the solid surface. In the case of non-sorptive gas injection, the only mechanism causing desorption is the lowering of the partial pressure of methane (Schwartz and Elsworth 2019). This can be achieved by increasing the pore pressure through gas injection or by injecting a non-sorptive gas at constant pore pressure. As methane is produced downstream, the non-sorptive gas maintains the pressure in the pore space. As nitrogen is abundant – air is composed of approximately 78% N_2 – and moderately inexpensive, it is the favored choice for field applications. Figure 10.16 shows the permeability enhancement possible from lowering the partial pressure of methane during a nitrogen flood. The partial pressure can also be lowered by injecting a sorptive gas species, with the additional overprinted effect from the injected gas interacting with the solid surface.

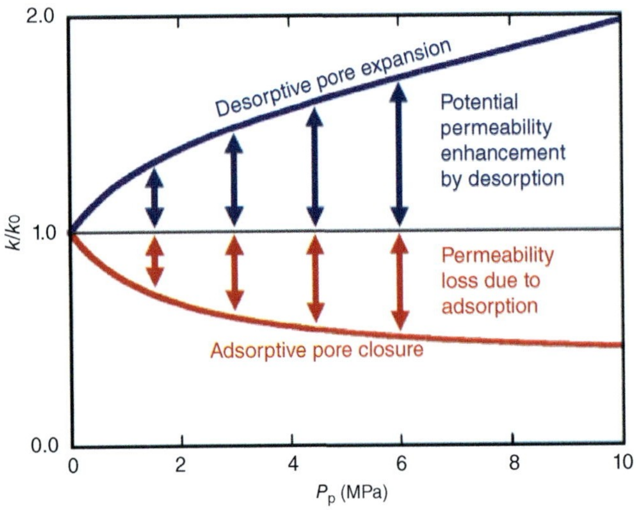

Figure 10.16 Schematic of permeability enhancement expected for a nitrogen flood at constant pore pressure.

10.5.3 Sorptive Gas Injection

Aside from changing the pore pressure, desorption can occur with changing the gas composition. Each gas species of a mixture has its own adsorption affinity, and changing the mixture composition results in changing adsorbed volumes for each species. The most common sorptive gas considered for injection is carbon dioxide, as it is highly sorptive to shales and coals and is also a greenhouse gas. Depending on the mineral composition of the shale, CO_2 can have anywhere from 5 to 20 times higher adsorption affinity to shale than CH_4 (Liu et al. 2019b). Injecting CO_2 into a pore where CH_4 is present creates a binary mixture that lowers the partial pressure of the CH_4 in the same way as with non-sorptive gas injection. However, because CO_2 has a much higher affinity to adsorption than CH_4, sorption sites begin to fill with CO_2 instead of CH_4 as the system moves toward a new equilibrium. Therefore, CH_4 desorbs and is replaced by CO_2 on the sorptive surface. This creates an opportunity for the now gaseous CH_4 to be produced from the pore space and to traverse through the larger flow network within the shale matrix (Luo et al. 2015; Du et al. 2019).

As CO_2 has a higher sorption affinity, it creates an additional sorptive strain over the strain caused by CH_4. Practically, this additional swelling is observed as an additional permeability loss associated with CO_2. Whether or not the permeability loss associated with CO_2 injection is compensated by the extra CH_4 production is an area of active research. Some have proposed injecting a mixture of CO_2–N_2, where optimized production can be achieved at some ratio of CO_2 to N_2 (Li and Elsworth 2019; Fan et al. 2019).

10.5.4 Desorption with Increased Temperature

The Langmuir equation shows that equilibrium sorption changes with changing pressure or temperature. Figure 10.10b shows that whereas desorption occurs during pressure decrease, it also occurs during temperature increase. As the gas gains kinetic energy with increased temperature, it has additional energy to escape from the sorptive surface. While reservoirs are typically considered as isothermal during production, temperature can be increased in the lab and potentially in the field. The permeability response to heat is dependent on the magnitude of desorptive shrinkage compared to the thermal expansion of minerals. Thermal expansion in shale is linear within the range of temperatures typically experienced in reservoirs: 300–400 K (Robertson 1988). However, thermally induced desorption is not linear but based on fractional coverage. The competition between desorption and mineral expansion creates an inverted U-shaped permeability response with increasing temperature. Laboratory observations suggest that some shales may experience substantial permeability increase in fracture sets and matrix rock, if subjected to temperatures of 300–350 K (Schwartz and Elsworth 2021).

10.6 Permeability Enhancement Due to Nitrogen Flooding

Nitrogen is an inexpensive and ubiquitously available gas often used in enhanced oil and gas recovery. As a non-sorptive gas, nitrogen enhances permeability by inducing desorption of methane in clays and organic matter. As nitrogen infiltrates the sorptive pore space, the partial pressure of methane is reduced. The equilibrium between adsorbed and free gas is shifted toward the gaseous phase, and methane can be produced from the pore space together with nitrogen. As desorption occurs, organic matter loses volume and flow paths dilate. The magnitude of permeability enhancement depends on the sorptive affinity of the solids and the density of pores affected. For a shale with

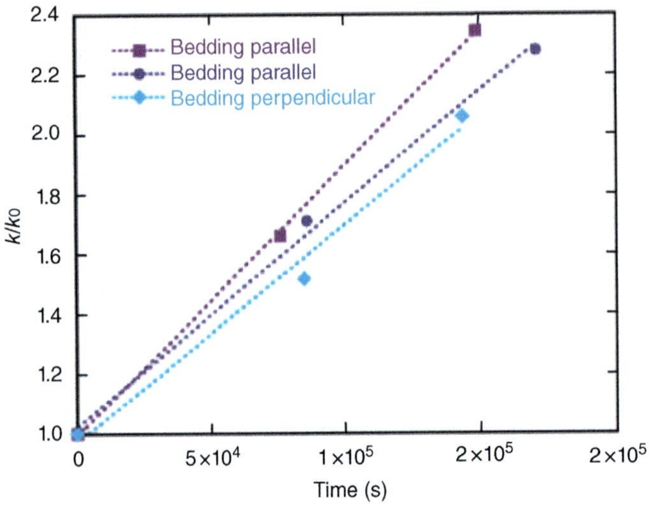

Figure 10.17 Normalized permeability evolution k/k_0. Permeability increased 206% in the bedding-parallel direction and 234% in the bedding-perpendicular direction. *Source:* Adapted from Schwartz and Elsworth (2019).

high pore density, permeability enhancement would be small compared to a shale with low pore density. Therefore, a general conjecture is that shales with lower initial permeability should experience greater permeability enhancement under the same sorptive conditions, although this remains to be verified. In laboratory measurements of permeability enhancement for samples of Marcellus shale, injecting nitrogen gas at near-constant pore pressure caused matrix permeability to more than double in both the bedding-parallel and bedding-perpendicular directions (Schwartz and Elsworth 2019). Figure 10.17 shows permeability enhancement during nitrogen flooding for two shale samples in the bedding parallel direction and one sample in the bedding perpendicular direction. In all cases, permeability more than doubled over the course of two days.

10.7 Discussion

Of the four methods discussed in Section 5 that induce desorption, pressure depletion is the only one that occurs naturally during production operations. The relationship between desorption-induced permeability enhancement and concurrent stress-induced permeability reduction during depletion remains an area of active research. As pore density, pore geometry, mineral distribution, and sorptive capacity differ between shales, the stress-permeability relationship would also vary (Zheng et al. 2015; Schwartz et al. 2019a). Of the other methods discussed, thermal flooding has recently been demonstrated in the laboratory (Schwartz and Elsworth 2021 but has not yet been adopted by the industry. Several challenges exist with thermal flooding in shales – in particular, the energy input required to saturate a shale reservoir would be extensive. While steam flooding has been successful in heavy oils as a mechanism to reduce fluid viscosity, the low permeability of shales prohibitively reduces the effectiveness of such techniques.

The other two methods of desorption include lowering the partial pressure of the natural gas via non-sorptive gas injection and preferential desorption of natural gas via CO_2 injection. Since the beginning of the current shale gas boom, CO_2 injection has been proposed as a form of secondary recovery (Kim et al. 2017; Jin et al. 2017; Zeng et al. 2020). CO_2 preferentially adsorbs to organic matter allowing methane to be produced via desorption, and this technique has garnered interest in the oil and gas industry. The recent expansion of tax credits available for carbon sequestration through Section 45Q offers additional incentives to operators to attempt CO_2 flooding in US gas shales. However, several challenges exist that must be overcome before CO_2 flooding for enhanced gas recovery (EGR) achieves widescale adoption. The main challenge associated with EGR in gas shales is that each horizontal well is hydraulically isolated. Since well pads tend to have three to four laterals in each direction (for a total of six to eight laterals per well pad), some have suggested refracturing the middle laterals to conduct a horizontal sweep from the leftmost lateral to the rightmost. This would impact an equivalent volume of two wells for three laterals or three wells for four laterals (Eshkalak et al. 2014). Alternatively, the huff-n-puff method of soaking a hydraulically isolated well and allowing a mixture of reservoir fluids and injectate to be produced could be a more effective design. The main drawbacks of the huff-n-puff method are that soaking times can be long and that CO_2 is produced and mixed with hydrocarbons. Additionally, CO_2 can reduce the permeability of the shale matrix by several orders of magnitude due to swelling. It has been proposed that mixing N_2 with CO_2 might allow for successful EGR with permeability loss correspondingly limited or removed.

The permeability response of shales to CO_2-N_2 mixtures is an area of active research. CO_2 has a higher affinity to adsorb within the organic pore space of coals and shales, leading to more storage, larger sorptive strains, and greater permeability loss (Chareonsuppanimit et al. 2012; Li and Elsworth 2015). Comparatively, N_2 flooding has been shown to double the permeability of shales (Schwartz and Elsworth 2019) and may compensate for the permeability loss due to CO_2 adsorption. Injection of N_2 reduces permeability loss via poromechanical expansion and by lowering the partial pressure of methane. The Langmuir strain is independent of temperature for each gas species but does change based on the sorption affinity of each species present (Robertson and Christiansen 2007; Singh and Cai 2018). Whereas carbon dioxide has a stronger affinity to adsorb than methane, nitrogen is a very weakly sorptive gas. Indeed, it can be treated as a non-sorptive gas when compared to methane or CO_2 adsorption in shales. This creates a complex response as the CO_2–N_2 injectate mixes with native CH_4. Whereas the adsorption of CO_2 lowers permeability by increasing methane desorption (Pan and Connell 2012; Liu et al. 2017), the permeability evolution due to the strains induced as a function of changing CO_2–N_2–CH_4 composition remains undefined. We suggest further study comparing these three gases to determine which of these competing mechanisms has a larger impact on ultimate recovery.

Another application related to CO_2 injection in shales is permanent geologic carbon sequestration (GCS), with the shale acting as a caprock for saline aquifer injection or as a reservoir in the case of depleted shale fields. Whereas hydrocarbon recovery during EGR is significantly impacted by the loss of permeability associated with CO_2 injection, permanent GCS does not require that hydrocarbons be produced as a result of the injection. In the case of saline aquifer injection, there are typically no hydrocarbons present in the first place. Injecting CO_2 into depleted shale reservoirs has several advantages over saline aquifers, including adsorptive trapping and sorptive storage. Whereas leakage and induced seismicity can be large risks in GCS projects (Pawar et al. 2015, 2016), shale reservoirs are supposed to be hydraulically isolated from the surrounding intervals. Depleted shale fields may also have larger tolerance for pressure buildup, as they are already below discovery pressure. In these ways, they may represent a reduced level of risk compared to saline aquifer injection. Moreover, depleted shale fields have the benefit of production data from thousands of wells, making the local subsurface well characterized. This may further reduce the uncertainty when compared with fluid injection in greenfields.

Shales represent a unique class of subsurface reservoirs, both for hydrocarbon production and for the sequestration of energy wastes. While much work has been done to understand these sorptive media, there are several important challenges that future work should address. First, the transition from Fickian flow through nanopores to Darcian flow within bedding-parallel micro-fissures – and the roles of each during fluid storage or production – should be better understood. While micro-fissures make up the predominant flow paths in the preferential pore network of shale, the majority of storage occurs in the nanoporous, sorptive material between laminae (Sondergeld et al. 2010; Curtis et al. 2012). The comparative influence of bedding-parallel and bedding-perpendicular permeability evolution during desorption should be better constrained. Second, some shales have been reported to accommodate multilayer adsorption, leading to larger sorptive storage volumes (Yu et al. 2014, 2016; Zhou et al. 2018). It should be determined if additional layers of adsorbed gas would further impact permeability by creating an additional sorptive strain or by limiting the mean free path in transport. Third, the engineering design for gas flooding in shales needs to be optimized. It may be that the huff-n-puff method is the best available option for EGR. However, if that is the case, then the future of EGR in shale may be limited. Gas injection across parallel laterals reduces the reservoir volume stimulated by EGR but may have the benefit of higher sweep efficiency. Hydraulically connecting parallel wells through refracturing could provide the injectivity necessary for large-scale CO_2 sequestration via EGR or GCS. Lastly, understanding the complex response of shale to CO_2–N_2–CH_4 mixtures will be critical to any efforts to convert depleted shale fields into EGR or GCS assets. While CO_2 could be injected alone, the beneficial permeability enhancement when comingled with N_2 is apparent. It has been shown that larger volumes of CO_2 can be sequestered in shale when mixed with 25% nitrogen than when CO_2 is injected alone (Li and Elsworth 2019). This is also critical for developing flue gas (or minimally separated flue gas) injection methods for GCS, which could forego the significant costs associated with stripping CO_2 out of flue gas using amine separation or some other method. Each of these challenges is directly tied to permeability evolution in shale by desorption processes, and it promises to be an important field in the decades to come.

10.8 Conclusion

Internal swelling is caused by the sorptive strain that develops during adsorption, which increases with increasing pore pressure, decreasing temperature, or increasing concentration of a sorptive gas species. Conversely, decreasing the pore pressure, increasing the temperature, or increasing the concentration of a non-sorptive gas species will lead to desorption and a reduction of the sorptive strain. The magnitude of permeability enhancement will depend on the distribution of sorptive mineral components relative to flow paths, the geometry (aspect ratio) of the flow paths, and the initial permeability.

Desorption processes in shale represent a critical pathway to permeability enhancement for oil and gas development, carbon sequestration, and hydrogen storage. The volume of adsorbed gas is controlled by the pressure, temperature, and composition of fluids in the pore space of a shale. Clays make up a much larger mass fraction of shales compared to organic carbon but are only weakly adsorptive compared to the total organic carbon. As flow paths such as macropores and micro-fractures are concentrated near clays, and organic carbon such as kerogen, quantifying the permeability evolution due to desorptive processes is paramount to understanding flow through porous, sorptive media.

References

Bai, M. and Elsworth, D. (2000). *Coupled Processes in Subsurface Deformation, Flow, and Transport*. American Society of Civil Engineers.

Bandyopadhyay, K. (2009). *Seismic Anisotropy: Geological Causes and Its Implications to Reservoir Geophysics*. Stanford University.

Bolton, A.J., Maltman, A.J., and Fisher, Q. (2000). Anisotropic permeability and bimodal pore-size distributions of fine-grained marine sediments. *Marine and Petroleum Geology* 17 (6): 657–672. https://doi.org/10.1016/S0264-8172(00)00019-2.

Bonnelye, A., Schubnel, A., David, C. et al. (2017). Elastic wave velocity evolution of shales deformed under uppermost crustal conditions. *Journal of Geophysical Research, Solid Earth* 122 (1): 130–141. https://doi.org/10.1002/2016JB013540.

Brace, W., Walsh, J.B., and Frangos, W.T. (1968). Permeability of granite under high pressure. *Journal of Geophysical Research* 73 (6): 2225–2236. https://doi.org/10.1029/JB073i006p02225.

Cao, P., Liu, J., and Leong, Y.K. (2017). A multiscale-multiphase simulation model for the evaluation of shale gas recovery coupled the effect of water flowback. *Fuel* 199: 191–205. https://doi.org/10.1016/j.fuel.2017.02.078.

Chareonsuppanimit, P., Mohammad, S.A., Robinson, R.L. Jr., and Gasem, K.A. (2012). High-pressure adsorption of gases on shales: measurements and modeling. *International Journal of Coal Geology* 95: 34–46. https://doi.org/10.1016/j.coal.2012.02.005.

Chen, Z., Liu, J., Pan, Z. et al. (2012). Influence of the effective stress coefficient and sorption-induced strain on the evolution of coal permeability: model development and analysis. *International Journal of Greenhouse Gas Control* 8: 101–110. https://doi.org/10.1016/j.ijggc.2012.01.015.

Chen, L., Zhang, L., Kang, Q. et al. (2015). Nanoscale simulation of shale transport properties using the lattice Boltzmann method: permeability and diffusivity. *Nature Scientific Reports* 5 (1): 1–8.

Civan, F., Rai, C.S., and Sondergeld, C.H. (2012). Determining shale permeability to gas by simultaneous analysis of various pressure tests. *SPE Journal* 17 (3): 717–726. https://doi.org/10.2118/144253-PA.

Closmann, P.J. and Bradley, W.B. (1979). The effect of temperature on tensile and compressive strengths and Young's modulus of oil shale. *Society of Petroleum Engineers Journal* 19 (5): 301–312. https://doi.org/10.2118/6734-PA.

Cooper, H.W. and Simmons, G. (1977). The effect of cracks on the thermal expansion of rocks. *Earth and Planetary Science Letters* 36 (3): 404–412. https://doi.org/10.1016/0012-821X(77)90065-6.

Crook, A.J., Yu, J.G., and Willson, S M. (2002). Development of an orthotropic 3D elastoplastic material model for shale. *SPE/ISRM Rock Mechanics Conference*. Society of Petroleum Engineers. https://doi.org/10.2118/78238-MS.

Cui, X. and Bustin, R.M. (2005). Volumetric strain associated with methane desorption and its impact on coalbed gas production from deep coal seams. *AAPG Bulletin* 89 (9): 1181–1202. https://doi.org/10.1306/05110504114.

Curtis, M.E., Sondergeld, C.H., Ambrose, R.J., and Rai, C.S. (2012). Microstructural investigation of gas shales in two and three dimensions using nanometer-scale resolution imaging microstructure of gas shales. *AAPG Bulletin* 96 (4): 665–677. https://doi.org/10.1306/08151110188.

Curtis, M. E., Sondergeld, C. H., and Rai, C. S. (2019 October). Visualization of pore connectivity using mercury injection capillary pressure measurements, micro X-ray computed tomography, and cryo scanning electron microscopy. *URTeC*, pp. 3143–3154, Denver, Colorado (22–24 July 2019). (URTeC); Society of Exploration Geophysicists. https://doi.org/10.15530/urtec-2019-637.

Daigle, H., Hayman, N.W., Kelly, E.D. et al. (2017). Fracture capture of organic pores in shales. *Geophysical Research Letters* 44 (5): 2167–2176. https://doi.org/10.1002/2016GL072165.

Do, D.D. (1998). *Adsorption Analysis: Equilibria and Kinetics*, vol. 2. World Scientific.

Du, X., Gu, M., Liu, Z. et al. (2019). Enhanced shale gas recovery by the injections of CO_2, N_2, and CO_2/N_2 mixture gases. *Energy & Fuels* 33 (6): 5091–5101. https://doi.org/10.1021/acs.energyfuels.9b00822.

Eshkalak, M.O., Al-shalabi, E.W., Sanaei, A. et al. (2014). Enhanced gas recovery by CO_2 sequestration versus re-fracturing treatment in unconventional shale gas reservoirs. *Abu Dhabi International Petroleum Exhibition and Conference*. Society of Petroleum Engineers. https://doi.org/10.2118/172083-MS.

Fan, D. and Ettehadtavakkol, A. (2017). Analytical model of gas transport in heterogeneous hydraulically-fractured organic-rich shale media. *Fuel* 207: 625–640. https://doi.org/10.1016/j.fuel.2017.06.105.

Fan, C., Elsworth, D., Li, S. et al. (2019). Modelling and optimization of enhanced coalbed methane recovery using CO_2/N_2 mixtures. *Fuel* 253: 1114–1129. https://doi.org/10.1016/j.fuel.2019.04.158.

Gilliam, T.M. and Morgan, I.L. (1987). *Shale: Measurement of Thermal Properties (No. ORNL/TM-10499)*. Oak Ridge, TN: Oak Ridge National Laboratory https://doi.org/10.2172/6163318.

Goral, J., Walton, I., Andrew, M., and Deo, M. (2019). Pore system characterization of organic-rich shales using nanoscale-resolution 3D imaging. *Fuel* 258: 116049. https://doi.org/10.1016/j.fuel.2019.116049.

Gu, X. and Mildner, D.F.R. (2016). Ultra-small-angle neutron scattering with azimuthal asymmetry. *Journal of Applied Crystallography* 49 (3): 934–943. https://doi.org/10.1107/S1600576716005586.

Harpalani, S. and Chen, G. (1997). Influence of gas production induced volumetric strain on permeability of coal. *Geotechnical and Geological Engineering* 15 (4): 303–325. https://doi.org/10.1007/BF00880711.

Harpalani, S. and Schraufnagel, R.A. (1990). Influence of matrix shrinkage and compressibility on gas production from coalbed methane reservoirs'. *SPE Annual Technical Conference and Exhibition*. Society of Petroleum Engineers. https://doi.org/10.2118/20729-MS.

Heng, S., Li, X., Liu, X., and Chen, Y. (2020). Experimental study on the mechanical properties of bedding planes in shale. *Journal of Natural Gas Science and Engineering* 76: 103161. https://doi.org/10.1016/j.jngse.2020.103161.

Horne, S.A. (2013). A statistical review of mudrock elastic anisotropy. *Geophysical Prospecting* 61 (4): 817–826. https://doi.org/10.1111/1365-2478.12036.

Huang, F. and Hu, B. (2018). Macro/microbehavior of shale rock under the dynamic impingement of a high-pressure supercritical carbon dioxide jet. *RSC Advances* 8 (66): 38065–38074. https://doi.org/10.1039/C8RA07480A.

Izadi, G., Wang, S., Elsworth, D. et al. (2011). Permeability evolution of fluid-infiltrated coal containing discrete fractures. *International Journal of Coal Geology* 85 (2): 202–211. https://doi.org/10.1016/j.coal.2010.10.006.

Javadpour, F. (2009). Nanopores and apparent permeability of gas flow in mudrocks (shales and siltstone). *Journal of Canadian Petroleum Technology* 48 (8): 16–21. https://doi.org/10.2118/09-08-16-DA.

Javadpour, F., Fisher, D., and Unsworth, M. (2007). Nanoscale gas flow in shale gas sediments. *Journal of Canadian Petroleum Technology* 46 (10): https://doi.org/10.2118/07-10-06.

Jia, B., Tsau, J.S., and Barati, R. (2018). A workflow to estimate shale gas permeability variations during the production process. *Fuel* 220: 879–889. https://doi.org/10.1016/j.fuel.2017.11.087.

Jin, L., Hawthorne, S., Sorensen, J. et al. (2017). Advancing CO_2 enhanced oil recovery and storage in unconventional oil play – experimental studies on Bakken shales. *Applied Energy* 208: 171–183. https://doi.org/10.1016/j.apenergy.2017.10.054.

Kang, S.M., Fathi, E., Ambrose, R.J. et al. (2011). Carbon dioxide storage capacity of organic-rich shales. *SPE Journal* 16 (4): 842–855. https://doi.org/10.2118/134583-PA.

Keller, L.M., Holzer, L., Wepf, R., and Gasser, P. (2011). 3D geometry and topology of pore pathways in Opalinus clay: implications for mass transport. *Applied Clay Science* 52 (1–2): 85–95. https://doi.org/10.1016/j.clay.2011.02.003.

Khan, M.R. and Jenkins, R.G. (1984). Thermoplastic properties of coal at elevated pressures: 1. Evaluation of a high-pressure microdilatometer. *Fuel* 63 (1): 109–115. https://doi.org/10.1016/0016-2361(84)90263-1.

Kim, T.H., Cho, J., and Lee, K.S. (2017). Evaluation of CO_2 injection in shale gas reservoirs with multi-component transport and geomechanical effects. *Applied Energy* 190: 1195–1206. https://doi.org/10.1016/j.apenergy.2017.01.047.

Kobchenko, M., Panahi, H., Renard, F. et al. (2011). 4D imaging of fracturing in organic-rich shales during heating. *Journal of Geophysical Research, Solid Earth* 116 (B12): https://doi.org/10.1029/2011JB008565.

Kumar, H., Elsworth, D., Mathews, J.P., and Marone, C. (2016). Permeability evolution in sorbing media: analogies between organic-rich shale and coal. *Geofluids* 16 (1): 43–55. https://doi.org/10.1111/gfl.12135.

Kwon, O., Kronenberg, A.K., Gangi, A.F. et al. (2004). Permeability of illite-bearing shale: 1. Anisotropy and effects of clay content and loading. *Journal of Geophysical Research, Solid Earth* 109 (B10): https://doi.org/10.1029/2004JB003055.

Levine, J.R. (1996). Model study of the influence of matrix shrinkage on absolute permeability of coal bed reservoirs. *Geological Society, London, Special Publications* 109 (1): 197–212. https://doi.org/10.1144/GSL.SP.1996.109.01.14.

Li, X. and Elsworth, D. (2015). Geomechanics of CO_2 enhanced shale gas recovery. *Journal of Natural Gas Science and Engineering* 26: 1607–1619. https://doi.org/10.1016/j.jngse.2014.08.010.

Li, Z. and Elsworth, D. (2019). Controls of CO_2-N_2 gas flood ratios on enhanced shale gas recovery and ultimate CO_2 sequestration. *Journal of Petroleum Science and Engineering* 179: 1037–1045. https://doi.org/10.1016/j.petrol.2019.04.098.

Li, Z.Q., Xian, X.F., and Long, Q.M. (2009). Experiment study of coal permeability under different temperature and stress. *Journal of China University of Mining and Technology* 38 (4): 523–527.

Li, J., Yu, T., Liang, X. et al. (2017a). Insights on the gas permeability change in porous shale. *Advances in Geo-Energy Research* 1 (2): 69–73. https://doi.org/10.26804/ager.2017.02.01.

Li, X., Feng, Z., Han, G. et al. (2017b). Permeability evolution of propped artificial fractures in Green River shale. *Rock Mechanics and Rock Engineering* 50 (6): 1473–1485. https://doi.org/10.1007/s00603-017-1186-2.

Liu, S. and Harpalani, S. (2013). A new theoretical approach to model sorption-induced coal shrinkage or swelling. *AAPG Bulletin* 97 (7): 1033–1049. https://doi.org/10.1306/12181212061.

Liu, H.H. and Rutqvist, J. (2010). A new coal-permeability model: internal swelling stress and fracture–matrix interaction. *Transport in Porous Media* 82 (1): 157–171. https://doi.org/10.1007/s11242-009-9442-x.

Liu, J., Chen, Z., Elsworth, D. et al. (2010). Linking gas-sorption induced changes in coal permeability to directional strains through a modulus reduction ratio. *International Journal of Coal Geology* 83 (1): 21–30. https://doi.org/10.1016/j.coal.2010.04.006.

Liu, J., Chen, Z., Elsworth, D. et al. (2011). Interactions of multiple processes during CBM extraction: a critical review. *International Journal of Coal Geology* 87 (3–4): 175–189. https://doi.org/10.1016/j.coal.2011.06.004.

Liu, J., Yao, Y., Liu, D., and Elsworth, D. (2017). Experimental evaluation of CO_2 enhanced recovery of adsorbed-gas from shale. *International Journal of Coal Geology* 179: 211–218. https://doi.org/10.1016/j.coal.2017.06.006.

Liu, B., Qi, C., Mai, T. et al. (2019b). Competitive adsorption and diffusion of CH_4/CO_2 binary mixture within shale organic nanochannels. *Journal of Natural Gas Science and Engineering* 53: 329–336. https://doi.org/10.1016/j.jngse.2018.02.034.

Luo, X., Wang, S., Wang, Z. et al. (2015). Adsorption of methane, carbon dioxide and their binary mixtures on Jurassic shale from the Qaidam Basin in China. *International Journal of Coal Geology* 150: 210–223. https://doi.org/10.1016/j.coal.2015.09.004.

Lyu, Q., Ranjith, P.G., Long, X. et al. (2015). Effects of coring directions on the mechanical properties of Chinese shale. *Arabian Journal of Geosciences* 8 (12): 10289–10299. https://doi.org/10.1007/s12517-015-1977-2.

Ma, L., Fauchille, A.L., Dowey, P.J. et al. (2017). Correlative multi-scale imaging of shales: a review and future perspectives. *Geological Society, London, Special Publications* 454 (1): 175–199. https://doi.org/10.1144/SP454.11.

Marschall, P., Horseman, S., and Gimmi, T. (2005). Characterisation of gas transport properties of the Opalinus Clay, a potential host rock formation for radioactive waste disposal. *Oil & Gas Science and Technology* 60 (1): 121–139. https://doi.org/10.2516/ogst:2005008.

Mavor, M.J. and Vaughn, J.E. (1998). Increasing coal absolute permeability in the San Juan Basin fruitland formation. *SPE Reservoir Evaluation and Engineering* 1 (3): 201–206. https://doi.org/10.2118/39105-PA.

Mazumder, S. and Wolf, K.H. (2008). Differential swelling and permeability change of coal in response to CO_2 injection for ECBM. *International Journal of Coal Geology* 74 (2): 123–138. https://doi.org/10.1016/j.coal.2007.11.001.

Palmer, I. (2009). Permeability changes in coal: analytical modeling. *International Journal of Coal Geology* 77 (1–2): 119–126. https://doi.org/10.1016/j.coal.2008.09.006.

Palmer, I. and Mansoori, J. (1996). How permeability depends on stress and pore pressure in coalbeds: a new model. *SPE Annual Technical Conference and Exhibition*. Society of Petroleum Engineers. https://doi.org/10.2118/36737-MS.

Pan, Z. and Connell, L.D. (2012). Modelling permeability for coal reservoirs: a review of analytical models and testing data. *International Journal of Coal Geology* 92: 1–44. https://doi.org/10.1016/j.coal.2011.12.009.

Pan, Z., Ma, Y., Connell, L.D. et al. (2015). Measuring anisotropic permeability using a cubic shale sample in a triaxial cell. *Journal of Natural Gas Science and Engineering* 26: 336–344. https://doi.org/10.1016/j.jngse.2015.05.036.

Pawar, R.J., Bromhal, G.S., Carey, J.W. et al. (2015). Recent advances in risk assessment and risk management of geologic CO_2 storage. *International Journal of Greenhouse Gas Control* 40: 292–311. https://doi.org/10.1016/j.ijggc.2015.06.014.

Pawar, R.J., Bromhal, G.S., Chu, S. et al. (2016). The National Risk Assessment Partnership's integrated assessment model for carbon storage: a tool to support decision making amidst uncertainty. *International Journal of Greenhouse Gas Control* 52: 175–189. https://doi.org/10.1016/j.ijggc.2016.06.015.

Qi, C., Wang, X., Wang, W. et al. (2018). Three-dimensional characterization of micro-fractures in shale reservoir rocks. *Petroleum Research* 3 (3): 259–268. https://doi.org/10.1016/j.ptlrs.2018.08.003.

Qu, H.Y., Liu, J.S., Pan, Z.J., and Connell, L. (2010 June). Impact of thermal processes on CO2 injectivity into a coal seam. *IOP Conference Series: Materials Science and Engineering* (vol. 10, no. 1, p. 012090). IOP Publishing. https://doi.org/10.1088/1757-899X/10/1/012090.

Qu, H., Liu, J., Chen, Z. et al. (2012). Complex evolution of coal permeability during CO2 injection under variable temperatures. *International Journal of Greenhouse Gas Control* 9: 281–293. https://doi.org/10.1016/j.ijggc.2012.04.003.

Robertson, E.C. (1988). *Thermal Properties of Rocks*. United States Department of the Interior Geological Survey https://doi.org/10.3133/OFR88441.

Robertson, E.P. (2005). *Measurement and Modeling of Sorption-Induced Strain and Permeability Changes in Coal (No. INL/EXT-06-11832)*. Idaho National Laboratory (INL) https://doi.org/10.2172/911830.

Robertson, E.P. and Christiansen, R.L. (2006). *A Permeability Model for Coal and Other Fractured, Sorptive-Elastic Media (No. INL/CON-06-11830)*. Idaho National Laboratory (INL) https://doi.org/10.2118/104380-MS.

Robertson, E.P. and Christiansen, R.L. (2007). Modeling laboratory permeability in coal using sorption-induced strain data. *SPE Reservoir Evaluation and Engineering* 10 (3): 260–269. https://doi.org/10.2118/97068-PA.

Ross, D.J. and Bustin, R.M. (2009). The importance of shale composition and pore structure upon gas storage potential of shale gas reservoirs. *Marine and Petroleum Geology* 26 (6): 916–927. https://doi.org/10.1016/j.marpetgeo.2008.06.004.

Rybacki, E., Meier, T., and Dresen, G. (2016). What controls the mechanical properties of shale rocks? – part II: brittleness. *Journal of Petroleum Science and Engineering* 144: 39–58. https://doi.org/10.1016/j.petrol.2016.02.022.

Sakhaee-Pour, A. and Bryant, S. (2012). Gas permeability of shale. *SPE Reservoir Evaluation and Engineering* 15 (4): 401–409. https://doi.org/10.2118/146944-PA.

Sakurovs, R., Day, S., Weir, S., and Duffy, G. (2008). Temperature dependence of sorption of gases by coals and charcoals. *International Journal of Coal Geology* 73 (3–4): 250–258. https://doi.org/10.1016/j.coal.2007.05.001.

Schwartz, B. and Elsworth, D. (2019). Permeability enhancement in gas shale due to nitrogen flooding. *Unconventional Resources Technology Conference*, pp. 2745–2756, Denver, Colorado (22–24 July 2019). Unconventional Resources Technology Conference (URTeC); Society of Exploration Geophysicists. https://doi.org/10.15530/urtec-2019-106.

Schwartz, B. and Elsworth, D. (2021). Inverted U-shaped permeability enhancement due to thermally induced desorption determined from strain-based analysis of experiments on shale at constant pore pressure. *Fuel* 302: 121178.

Schwartz, B., Huffman, K., Thornton, D., and Elsworth, D. (2019a). The effects of mineral distribution, pore geometry, and pore density on permeability evolution in gas shales. *Fuel* 257: 116005. https://doi.org/10.1016/j.fuel.2019.116005.

Schwartz, B., Elsworth, D., and Marone, C. (2019b). Relationships between mechanical and transport properties in Marcellus shale. *International Journal of Rock Mechanics and Mining Sciences* 119: 205–210. https://doi.org/10.1016/j.ijrmms.2019.04.020.

Schwartz, B., Huffman, K., Thornton, D., and Elsworth, D. (2019c). A strain based approach to calculate disparities in pore structure between shale basins during permeability evolution. *Journal of Natural Gas Science and Engineering* 68: 102893. https://doi.org/10.1016/j.jngse.2019.05.006.

Seidle, J.P., Jeansonne, M.W., and Erickson, D.J. (1992). Application of matchstick geometry to stress dependent permeability in coals. *SPE Rocky Mountain Regional Meeting*. Society of Petroleum Engineers. https://doi.org/10.2118/24361-MS.

Shi, J.Q. and Durucan, S. (2005). A model for changes in coalbed permeability during primary and enhanced methane recovery. *SPE Reservoir Evaluation and Engineering* 8 (4): 291–299. https://doi.org/10.2118/87230-PA.

Singh, H. and Cai, J. (2018). A mechanistic model for multi-scale sorption dynamics in shale. *Fuel* 234: 996–1014. https://doi.org/10.1016/j.fuel.2018.07.104.

Sinha, S., Braun, E.M., Determan, M.D. et al. (2013, March). Steady-state permeability measurements on intact shale samples at reservoir conditions-effect of stress, temperature, pressure, and type of gas. *SPE Middle East Oil and Gas Show and Conference*. Society of Petroleum Engineers. https://doi.org/10.2118/164263-MS.

Siriwardane, H., Haljasmaa, I., McLendon, R. et al. (2009). Influence of carbon dioxide on coal permeability determined by pressure transient methods. *International Journal of Coal Geology* 77 (1–2): 109–118. https://doi.org/10.1016/j.coal.2008.08.006.

Sondergeld, C.H., Newsham, K.E., Comisky, J.T. et al. (2010 January). Petrophysical considerations in evaluating and producing shale gas resources. *SPE Unconventional Gas Conference*. Society of Petroleum Engineers. https://doi.org/10.2118/131768-MS.

Sone, H. (2012). *Mechanical Properties of Shale Gas Reservoir Rocks, and Its Relation to the In-Situ Stress Variation Observed in Shale Gas Reservoirs*. Stanford University.

Sone, H. and Zoback, M.D. (2013). Mechanical properties of shale-gas reservoir rocks – part 1: static and dynamic elastic properties and anisotropy. *Geophysics* 78 (5): D381–D392. https://doi.org/10.1190/geo2013-0050.1.

Ulm, F.J. and Abousleiman, Y. (2006). The nanogranular nature of shale. *Acta Geotechnica* 1 (2): 77–88. https://doi.org/10.1007/s11440-006-0009-5.

Wang, S., Elsworth, D., and Liu, J. (2012). A mechanistic model for permeability evolution in fractured sorbing media. *Journal of Geophysical Research, Solid Earth* 117 (B6): https://doi.org/10.1029/2011JB008855.

Wang, L., Cardenas, M.B., Slottke, D.T. et al. (2015). Modification of the local cubic law of fracture flow for weak inertia, tortuosity, and roughness. *Water Resources Research* 51 (4): 2064–2080. https://doi.org/10.1002/2014WR015815.

Wang, H., Chen, L., Qu, Z. et al. (2020). Modeling of multi-scale transport phenomena in shale gas production – a critical review. *Applied Energy* 262: 114575. https://doi.org/10.1016/j.apenergy.2020.114575.

Witherspoon, P.A., Wang, J.S., Iwai, K., and Gale, J.E. (1980). Validity of cubic law for fluid flow in a deformable rock fracture. *Water Resources Research* 16 (6): 1016–1024. https://doi.org/10.1029/WR016i006p01016.

Wu, K., Chen, Z., Li, X. et al. (2016). A model for multiple transport mechanisms through nanopores of shale gas reservoirs with real gas effect–adsorption-mechanic coupling. *International Journal of Heat and Mass Transfer* 93: 408–426. https://doi.org/10.1016/j.ijheatmasstransfer.2015.10.003.

Yan, C., Deng, J., Hu, L. et al. (2015). Brittle failure of shale under uniaxial compression. *Arabian Journal of Geosciences* 8 (5): 2467–2475. https://doi.org/10.1007/s12517-014-1373-3.

Yu, W., Sepehrnoori, K., and Patzek, T.W. (2014). Evaluation of gas adsorption in Marcellus shale. *SPE Annual Technical Conference and Exhibition*. Society of Petroleum Engineers. https://doi.org/10.2118/170801-MS.

Yu, W., Sepehrnoori, K., and Patzek, T.W. (2016). Modeling gas adsorption in Marcellus shale with Langmuir and bet isotherms. *SPE Journal* 21 (2): 589–600. https://doi.org/10.2118/170801-PA.

Zeng, T., Miller, C.S., and Mohanty, K.K. (2020). Combination of a chemical blend with CO_2 huff-n-puff for enhanced oil recovery in oil shales. *Journal of Petroleum Science and Engineering* 194: 107546. https://doi.org/10.1016/j.petrol.2020.107546.

Zheng, J., Zheng, L., Liu, H.H., and Ju, Y. (2015). Relationships between permeability, porosity and effective stress for low-permeability sedimentary rock. *International Journal of Rock Mechanics and Mining Sciences* 78: 304–318. https://doi.org/10.1016/j.ijrmms.2015.04.025.

Zhou, S., Xue, H., Ning, Y. et al. (2018). Experimental study of supercritical methane adsorption in Longmaxi shale: insights into the density of adsorbed methane. *Fuel* 211: 140–148. https://doi.org/10.1016/j.fuel.2017.09.065.

Zhu, C. and Arson, C. (2014). A thermo-mechanical damage model for rock stiffness during anisotropic crack opening and closure. *Acta Geotechnica* 9 (5): 847–867. https://doi.org/10.1007/s11440-013-0281-0.

11

Multiscale Experimental Study on Interactions Between Imbibed Stimulation Fluids and Tight Carbonate Source Rocks

Feng Liang, Hui-Hai Liu, and Jilin Zhang

Aramco Americas: Aramco Research Center–Houston, Houston, TX, USA

11.1 Introduction

Fractured carbonates contain 60% of the remaining oil and more than 40% of gas reserves worldwide (Beydoun 1998). Acid fracturing and matrix acidizing are the most common ways to stimulate conventional reservoirs due to the reactivity between carbonates and acids (McLeod 1984, 1989; Olsen and Karr 1986; Bybee 2004). However, acids are not as effective as in stimulating carbonate-rich tight gas or shale formation due to their significantly low permeabilities (ranging from nano to micro Darcies) as well as their heterogeneous pore structures (Nelson 2009; Sondergeld et al. 2010; Loucks et al. 2012). The combination of horizontal drilling and proppant fracturing has been proved to be the most effective stimulation technique to produce natural gas from tight carbonate reservoirs. Because the low permeability of rock matrix bottlenecks the flow of hydrocarbon fluids, creating complex fractures in the rock matrix is highly desirable to improve and maintain conductive flow paths for enhancing the production. Baihly et al. (2010) observed that the gas production rate declines rapidly within the first few months of production from horizontal wells in different shale gas basins (Barnett Shale, Fayetteville Shale, Woodford Shale, Haynesville Shale, and Eagle Ford Shale play). They also noted that the early production rate is affected by the wellbore cleanup conditions since the gas is produced along with water and the residual fracturing fluids. Wang (2017) conducted a comprehensive study on factors that control the shale gas production and production-decline trend in fractured systems. These factors include the petrophysical properties from the reservoir itself, such as existence of adsorption gas, matrix permeability, rock mechanical properties (e.g. creep and viscoelastic behaviors), clay content, organic content and its maturity, evolution of shale matrix apparent permeability, and parameters related to the stimulation process such as simulated reservoir volume, the density, conductivity, and connectivity of fracture networks. He concluded that regardless of the matrix permeability, denser fracture networks and larger fracture-surface area inside the stimulated reservoir volume always leads to higher production.

On the other hand, during hydraulic fracturing, massive amounts of fracturing fluids are pumped to crack-open the formation, and only a small percentage of the fluids are recovered during the flow-back process (Cheng 2012; King 2012; Soeder 2017). Despite extensive previous studies of the interaction between carbonate rocks and fracturing fluid on the core-scale level (Akrad et al. 2011; Lai et al. 2016), there have been very few microscopic studies documented, including

Physics of Fluid Flow and Transport in Unconventional Reservoir Rocks, First Edition.
Edited by Behzad Ghanbarian, Feng Liang, and Hui-Hai Liu.

fluid flow pathways and mineralogical changes of the fluid and rock interaction. Given the complexity of the tight organic-rich carbonate source rocks and their high heterogeneity, microscopic studies on the texture, and the effects of the imbibed fluid on the mechanical, morphological, and flow properties of tight organic-rich carbonate rocks are very important. Also, since various components were exposed to the fracturing fluids under the reservoir conditions, it is important to investigate whether different laminae are equally exposed to the fracturing fluids and whether fracturing fluid travels with any preference to certain laminae of the rock formation in situ. Will the flow path have any preference when microfractures are present? If there is a preferred pathway, how much difference is expected among the different flow pathways? What does the fluid do to the formation? Will mechanical properties change after fluid exposure? Are the morphology and mineralogy of the formation rock going to change? What changes will there be in terms of flow property? It is critical to have some answers to the above questions and understand the dual continuum flow phenomena in tight carbonate rocks in order to design more efficient simulation techniques for hydraulic fracturing.

In this study, we have designed a few sets of experiments, which were conducted at multiscale levels including the microscale and core scale to address the above questions. Two sets of tight organic-rich carbonate rock samples were used. One sample set was a highly laminated lime mudstone from a basin in a Middle Eastern field that was used in microscopic studies, such as fluid uptake pathways, morphological, mineralogic, and micromechanical aspects. The second sample set was from an outcrop of the Eagle Ford Shale considered to be analogous to the set from the Middle Eastern field in terms of mineralogy. The sample set was used in core-scale studies on mechanical and flow properties including permeability and porosity. Correlations between the observations from multiscale experiment results have been discussed to figure out the factors that will affect the production rate and production-decline rate.

11.2 Fluid Uptake Pathways

The fluid uptake study was conducted by dynamically capturing the fluid pathway through a highly laminated tight organic-rich carbonate thin-section sample from the Middle Eastern field using X-ray radiographic images. The selected sample was composed of two distinguished zones, an organic matter-rich zone (OMZ) and a calcite-rich zone (Cal-Z). Microfractures were presented in both zones in the thin section prepared from the as-received sample. Potassium iodide (KI, 5 wt% in DI water) was added to the treatment fluid to enhance the contrast of the images. The spontaneous imbibition test was conducted for 100 minutes with one radiograph captured at every 1-minute interval. The photomicrographs were analyzed and the fluid front movement in both zones was captured (Liang et al. 2020).

11.2.1 Experimental Methods

11.2.1.1 Materials

11.2.1.1.1 Rock Sample A thin section of the rock sample with dimension of 22 × 11 × 3.3 mm (length × width × thickness) was used in this study. The total organic carbon (TOC) of the selected tight organic-rich carbonate rock sample was around 8.7 wt%. The mineral content, determined by X-ray diffraction (XRD), showed that it was mainly composed of 94.5 wt% calcite, 0.5 wt% dolomite, 2.5 wt% quartz, 0.5 wt% pyrite, 0.6 wt% gypsum, and a minor amount of clay (0.6 wt% kaolinite and 0.8 wt% illitic clay). The rock surface was further fine-trimmed by a target surfacing trimmer and polished by a broad triple ion-beam mill.

11.2.1.2 Experimental Procedure

11.2.1.2.1 3D Microscale Visualization of Thin-Section Rock Sample in As-Received State A Nikon HMXST225 Micro-Computed Tomography (microCT) system was used to image the tight organic-rich carbonate rock sample at two different resolutions, 5.48 and 2.79 μm, respectively. The X-ray used in the imaging process was 180 kV with a 2200 × 2200 pixel detector. 3000 projections were collected and reconstructed with 16-bit precision.

11.2.1.2.2 Dynamic Study of Spontaneous Imbibition Test Ghanbari and Dehghanpour (2015) reported that liquid imbibition and ion diffusion rates parallel to the lamination are higher than those in the perpendicular direction. Similarly, our imbibition study was conducted using a 5 wt% KI solution along the maximum liquid imbibition rate direction, which is parallel to the lamination direction. This is consistent with the fracturing fluid flow direction during hydraulic fracturing. The gravity effects on the fluid movement rate along the lamination direction were minimized by placing the thin-section sample with lamination in the horizontal direction and the treatment fluid (5 wt% KI solution) on the left side of the sample. In this way, the fluid imbibition direction was in the horizontal direction. Then, the sample setup was transferred to an X-ray chamber within 3 minutes and set up for radiographic imaging. The entire spontaneous imbibition test was conducted for 100 minutes with one X-ray radiographic image taken at 1-minute intervals. During the acquisition of the images, the sample remained stationary without rotation. The fluid front of the collected radiographic images was identified by importing them into an in-house MATLAB program to examine and measure the sudden change of attenuation along the fluid imbibition direction. Then the distance of the fluid front was measured/calculated between the sample edge and the fluid front.

11.2.2 Results and Discussion

11.2.2.1 Surface Characterization

Computed X-ray tomography (CT) of the prepared thin-section sample prior to the imbibition test was performed using 5.48 and 2.79 μm resolution, respectively. The 3D image for each resolution was reconstructed (Figure 11.1). It was observed that this area was composed of two distinguished zones, an OMZ, shown as a dark-grey band in the middle of the image, and Cal-Z, shown as a light-grey color present above and below the OMZ layer in Figure 11.1. The thickness of the OMZ layer in this sample was approximately 1.2 millimeters (mm).

As documented in our earlier studies (Liang et al. 2018, 2019) with tight carbonate samples from similar field, Cal-Z zones contain mainly calcite, while OMZs contain less calcite, more aluminosilicate minerals, such as quartz, feldspar, and clay minerals, and more organic matter (OM) in comparison to the Cal-Z zones. As shown in Figure 11.1a, at the resolution of 5.48 μm, multiple microfractures were present in both the OMZs and Cal-Z zones on slices of the reconstructed image. Microfractures are occasionally seen in Cal-Z zones, and they appear to be parallel to the lamination. The length of the microfractures can be up to a few millimeters. In Figure 11.1b at the higher resolution of 2.79 μm, a higher density of microfractures was observed in the OMZs with orientations in different directions. Some of them were parallel to the lamination direction, while others were perpendicular to or at high angles with the lamination. Some large irregular shaped microfractures were also observed in the OMZ. It is very difficult to distinguish from images whether

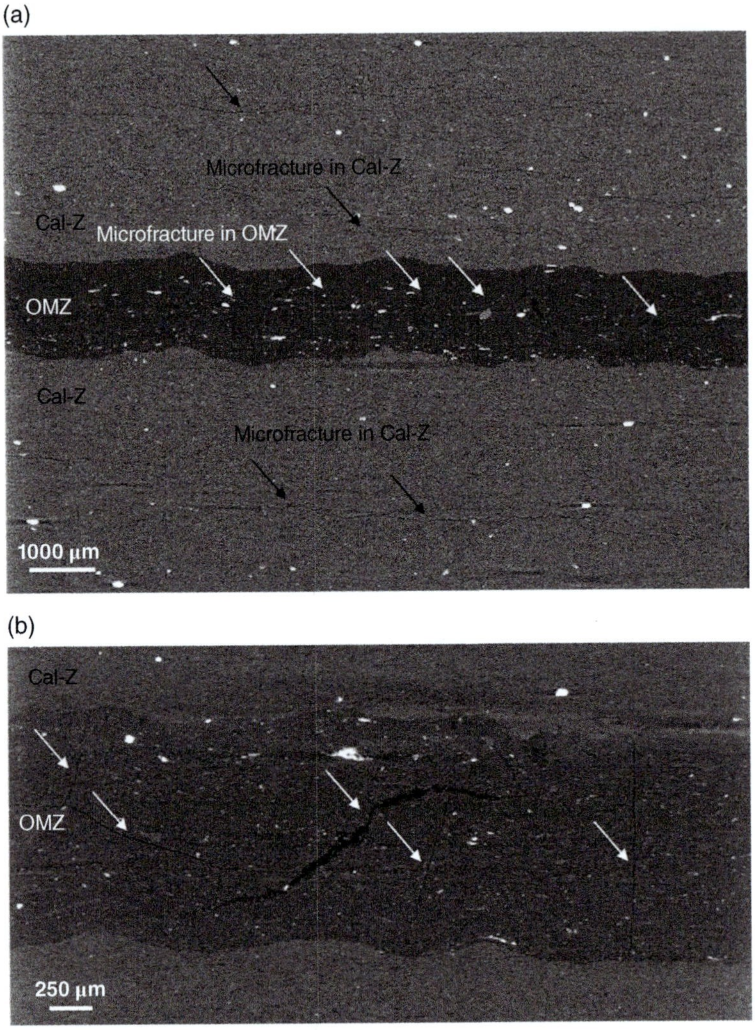

Figure 11.1 MicroCT image at resolution of (a) 5.48 μm and (b) 2.79 μm for one section with highly laminated areas for the as-received sample.

these microfractures were naturally occurring, or induced by the sample preparation processes, or from desiccation when the rock sample was exposed to air, though extra cautionary steps were considered/taken in order to minimize the damage during sample preparation. In fact, there is no guarantee that there was no induced damage to the prepared sample. Since large numbers of microfractures are generated during the hydraulic fracturing process, especially with slickwater treatment, this sample could be treated as the initial state of the formation sample in the far field during hydraulic fracturing process for the fluid uptake pathway study.

The set of microCT images at the resolution of 2.79 μm was reconstructed and further processed. In order to overcome the challenges with the large dataset size and the microstructure complexity, supervised artificial intelligence (AI) was used during the micrograph analysis for fracture segmentation (Howard et al. 2019; Zhang et al. 2019). Some features and patterns, which might fail to be

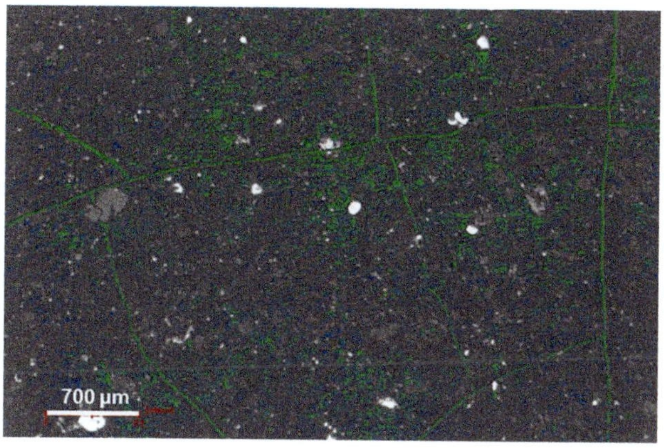

Figure 11.2 Color-mapped fracture and porosity network. Green: connected fractures and porosity network; Blue: disconnected porosity.

recognized with the conventional algorithms, can be detected with supervised AI. Once the recognized features are validated on a smaller scale, they can be deployed to process massive amounts of similar data automatically. In this study, fractures and matrix from identified date sets were used in the training data sets; then automation for recognizing fractures and matrix was conducted for the rest of the data sets. The analyzed data suggested that there were connected microfracture networks in the OMZ. Figure 11.2 shows the analyzed results for one area in the OMZ, in which green color represents the connected fractures and porosity network while blue color represents the disconnected porosity.

11.2.2.2 Spontaneous Imbibition Tests

The imbibition study was conducted by placing the thin-section sample with layers in a horizontal direction and flowing the treatment fluid from left to right. In this way, the fluid flow was along the maximum fluid imbibition rate direction, i.e. the direction of sample lamination was parallel to the fluid-imbibition direction. A 5 wt% KI solution was used as the treatment fluid. Due to the high attenuation of the KI fluid, the zones that contain KI fluid appear to be relatively brighter on the radiographs highlighting the fluid flow pathways. The first X-ray radiograph was taken at 4 minutes after the start of the experiment and the entire spontaneous imbibition test was conducted for 100 minutes with one image taken at 1-minute intervals. Figure 11.3a–d show the radiographs of the fluid-imbibed conditions at the selected experimental times of 4, 10, 30, and 100 minutes. There are a couple of interesting observations in this series of images. First, the OMZ is relatively brighter than Cal-Z zone in Figure 11.3, which is completely opposite to the contrast as observed in the original state (Figure 11.1a). This indicates that KI fluid transported very fast in the microfracture abundant OMZ and even at the 4 minutes of the imbibition test, the fluid front had already traveled from the left side of the sample to the right side of the sample. Secondly, some bright strings were observed in Figure 11.3a, which indicates that the abundant microfractures in Cal-Z zone were filled with KI fluid in less than 4 minutes. This means that the fluid moved very fast in microfractures containing Cal-Z as well. Different from the microfracture-dominated areas, the fluid front progressed very slowly in the Cal-Z matrix. The fluid front is marked in Figure 11.3d at 100 minutes into the imbibition test.

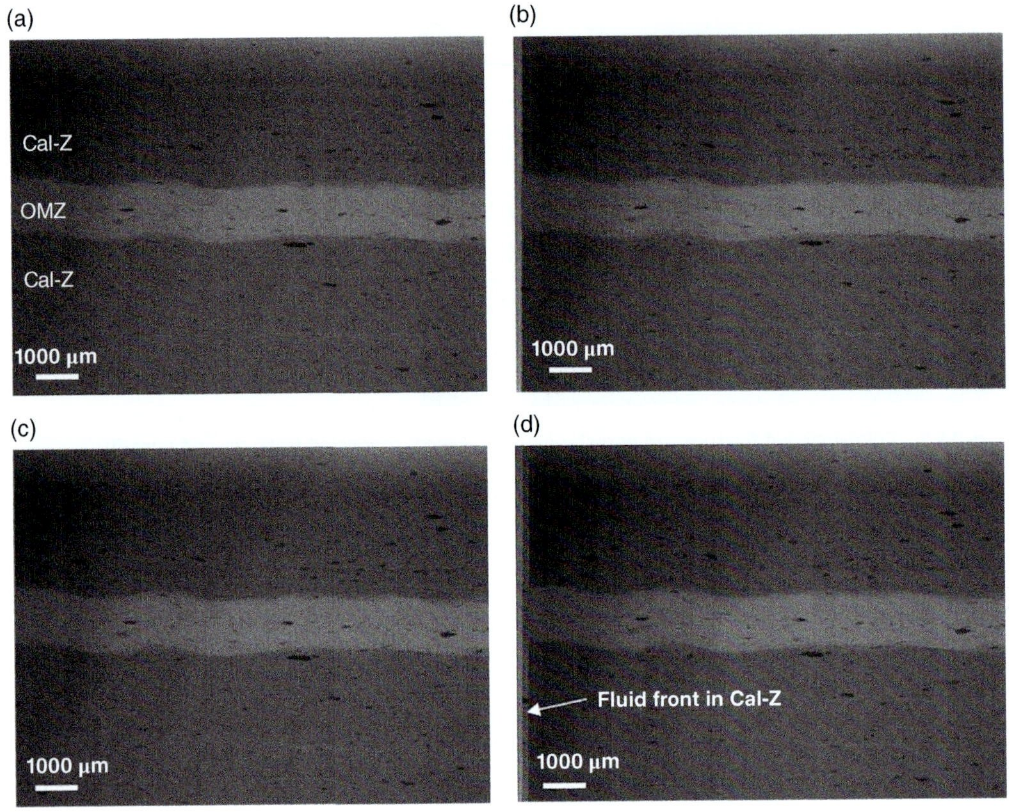

Figure 11.3 Dynamic study of microscale imaging at different imbibition time: full field of view. (a) $t = 4$ minutes; (b) $t = 10$ minutes; (c) $t = 30$ minutes; (d) $t = 100$ minutes.

Figure 11.4 shows the zoomed-in section in the fluid contact side to accurately measure the distance of fluid front movement in the matrix of the Cal-Z zone at the selected experimental time, in which 120 μm was determined to be the travel distance at 100 minutes of the imbibition time. The distance of the fluid front vs. time was plotted in Figure 11.5. Except at the very beginning of the imbibition test, the fluid front movement rate in the matrix of the Cal-Z zone was calculated to be at about 72 μm/h.

In summary, the test results indicate that the fluid flow pathways are dominated by the connected fracture networks. The fluid flow in the matrix is much slower than that in the connected fractures.

11.3 Mechanical Property Change After Fluid Exposure

The structural complexity of these laminated source shales occurs at all scales from macro to micro, and nanometer highlighting the need for tools that can characterize the fluid flow of this inhomogeneous anisotropic porous material on multiple scales when possible (Zeszotarski et al. 2004). In this section, two sets of experiments were designed and conducted to evaluate the rock property evolution after fluid exposure, one at the cylindrical core-scale level and the other one at the microscale level. The outcrop from an Eagle Ford Shale was used in this study.

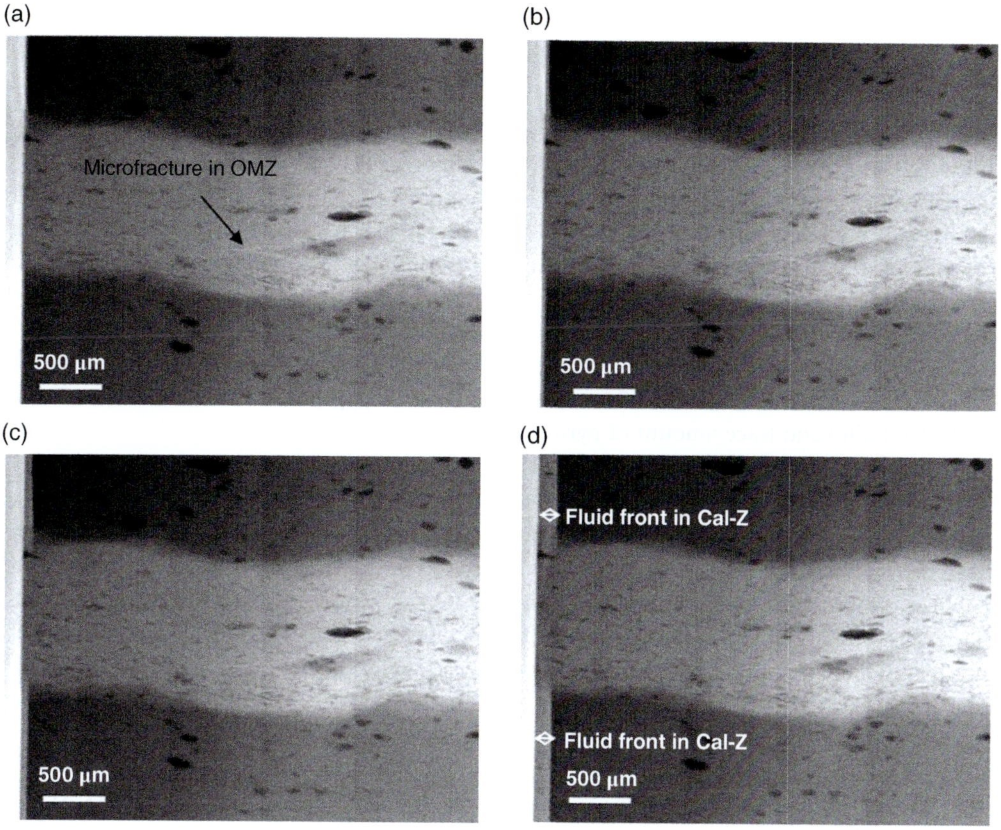

Figure 11.4 Enlarged view of microscale imaging at different imbibition time: zoomed-in section in the fluid contact side. (a) $t = 4$ minutes; (b) $t = 10$ minutes; (c) $t = 30$ minutes; (d) $t = 100$ minutes.

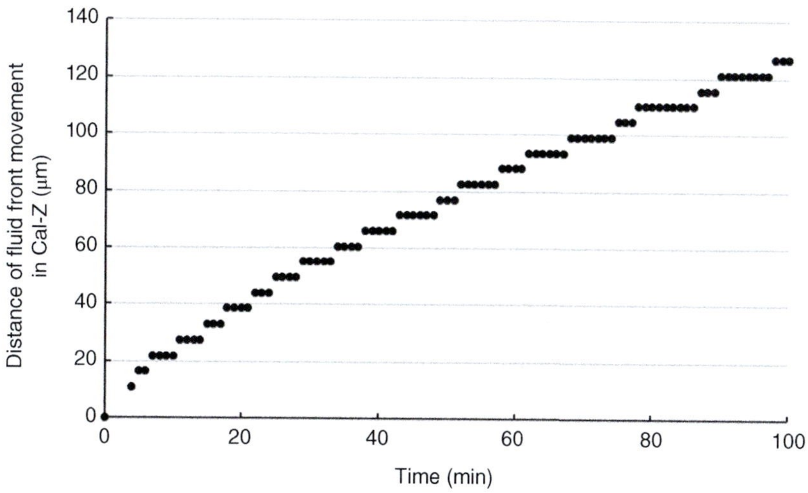

Figure 11.5 Distance of fluid front with time in the matrix of Cal-Z zone determined from dynamic radiograph images.

11.3.1 Experimental Methods

11.3.1.1 Materials

11.3.1.1.1 Rock Sample Adjacent cylindrical Eagle Ford core samples were prepared perpendicular to the bedding plan for mechanical testing both at the core scale and microscale. All the core plugs were 1-in. in diameter, while the lengths were different depending on the type of test. The length of the core plugs used for the unconfined compressive strength (UCS) test was 2-in. The length of the core plugs used for the Brazilian tensile strength test was 0.4-in. The length of the core plugs used for the microindentation test was 0.5-in. in length. Core plug surfaces were then fine-trimmed using a target surface trimmer and polished using an Allied High Tech Multiprep surface polisher with abrasive discs. The TOC of these outcrop Eagle Ford samples was around 2–4.5 wt%. The mineral content, determined by XRD, shows that it is mainly composed of 76.3 wt% calcite, 1.5 wt% dolomite, 17.5 wt% quartz, 1.8 wt% gypsum, a minor amount of clay (1.7 wt% kaolinite) and trace amount of pyrite.

11.3.1.1.2 Treatment Fluids for Cylindrical Core Plugs Three fluids were prepared as the treatment fluids for the cylindrical core plugs mechanical testing. Two different forms of friction reducer were used in this set of experiments. The friction reducer used in this study was an acrylamide-based terpolymer composed of acrylamide, acrylic acid, and 2-acrylamido-2-methylpropanesulfonic acid. Fluid 1 was 2 wt% KI; Fluid 2 was 0.015 wt% friction reducer (CELB-217-063-2, 80% active, product from ChemEOR, in a solid form) in 2%/KCl; Fluid 3 was 0.015 wt% friction reducer (DP/EM 5015, 30% active, product from SNF, in an emulsion form) in 2% KCl.

11.3.1.2 Experimental Procedure

11.3.1.2.1 UCS and Brazilian Tensile Strength Test The UCS test procedure was based on ASTM D7012-14 (2014). Two sets of strain gauges were attached to the center of the sample to measure deformation in both the axial and radial directions. The stress–strain curve was generated during the UCS test. Young's modulus was calculated from the linear portion of the stress–strain curve. The Brazilian tensile strength test was performed to determine the tensile strength of specimens. The testing procedure was based on ASTM D3967-08 (2008).

11.3.1.2.2 Fluid Treatment for Cylindrical Core Plugs In order to quantify the effects of fluids on shale mechanical properties, UCS and Brazilian tests were conducted on both "untreated" and "treated" sample specimens. Untreated specimens refer to cylindrical samples that were tested in as-received state, i.e. before the fluid treatment. Treated specimens refer to core plugs being soaked in three days (72 hours) under room temperature.

11.3.1.2.3 Microindentation Testing and Its Mapping Procedure The microindentation instrument used in this study was a Nanovea PB1000. A core plug was first labeled with 2 cm tape at the bottom and marked the point 3 mm from the left on the tape as $x = 0$. Then, the core plug was fixed on the stage to make sure that the straight line of the label was parallel with the y direction of the stage. We outlined the measure area with the indenter tip (red dots in Figure 11.6). Figure 11.6 shows the mapping protocol for microindentation testing. The microindentation of as-received outcrop Eagle Ford shale was conducted first with the mapped 25 points (0.1 mm apart) labeled as "untreated." To shorten the fluid treatment time, deionized (DI) water was used as the treatment fluid. The entire shale sample was submerged in DI water for 2 hours. The microindentation test of

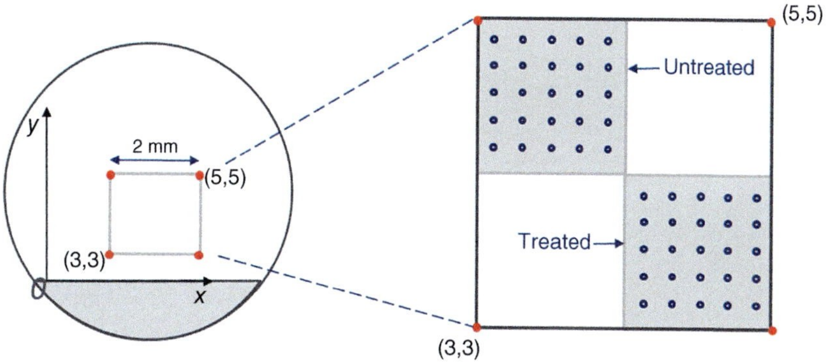

Figure 11.6 Illustration of mapping protocol for microindentation testing.

the DI water-treated rock sample was then conducted again with the mapped 25 points labeled as "treated." For the single-load microindentation, the maximum load is 0.5 N, the loading and unloading rate are both 1 N/min, when it reaches the max load, the creep time is 7 seconds.

11.3.2 Results and Discussion

11.3.2.1 UCS and Brazilian Test on Cylindrical Core Plugs

The UCS, Brazilian tensile strength, and Young's modulus of the outcrop Eagle Ford samples, untreated and treated with three different fluids, are shown in Figure 11.7. The general trend observed is that the value of all three parameters decreased to various degrees when compared with the untreated sample. In the case of UCS, the degree of reduction for the treated samples ranged from 50 to 60%. Although the Brazilian tensile strength of the three fluid-treated samples remained in a similar range, the failure patterns for Fluids 2 and 3 treated samples were different. Figure 11.8 shows the fracture patterns for these two-fluids treated samples using industrial Computed Tomography (CT) scan with a resolution of 20 μm. There was a primary fracture cross-through for the sample treated with Fluid 2 (Figure 11.8a), which behaved more like a brittle failure. While for the sample treated with the Fluid 3, there was no primary fracture observed during the failure process (Figure 11.8b). This suggests that the sample became more ductile.

11.3.2.2 Microindentation Test

Microindentation and nanoindentation have proven to be great tools to mechanically characterize geological porous material and, more lately, source rocks (Barsoum et al. 2004; Ulm and Abousleiman 2006; Broitman 2017). Instrumented indentation technique (including both micro- and nanoindentation) is achieved by a high-resolution system that can drive an indenter tip made of diamond into and withdraw it from a sample material. During the process, the loads and displacements of the tip are recorded by positional sensors from which the mechanical properties can be derived. Microindentation and nanoindentation testing differ in a couple of ways, but the results are expected to be able to compare, if the tip geometries used can provide similar strains in the material when testing, such as the Vickers and Berkovich shaped tips. Both Berkovich and Vickers shaped tips are commonly used on microindentation, while the Berkovich shaped indenter tip is preferred in nanoindentation. The difference between the tip shapes is that the Vickers tip is a four-sided pyramid, while the Berkovich tip is a three-sided pyramid. The Berkovich tip is sharper than the

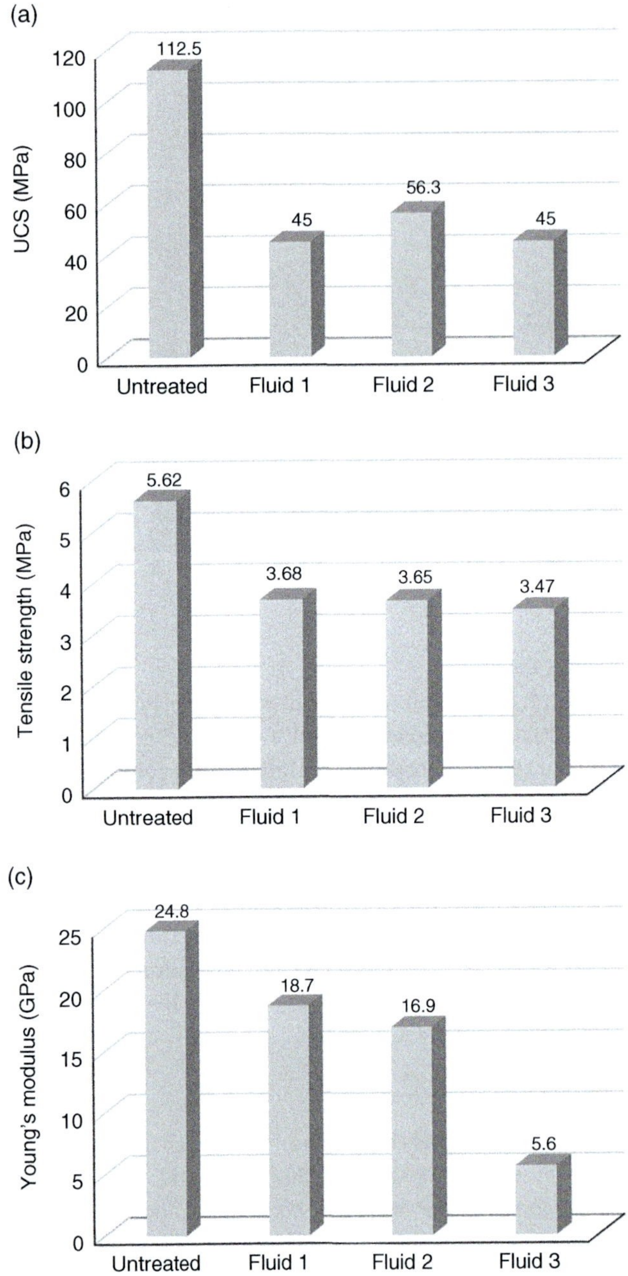

Figure 11.7 UCS (a), Brazilian tensile strength (b), and Young's modulus (c) of outcrop Eagle Ford samples for untreated and treated with three different fluids. Fluid 1 was 2 wt% KI; Fluid 2 was 0.015 wt% friction reducer (CELB-217-063-2, 80% active, from ChemEOR in a solid form) in 2 wt% KCl; Fluid 3 was 0.015 wt% friction reducer (DP/EM 5015, 30% active from SNF in an emulsion form) in 2 wt% KCl.

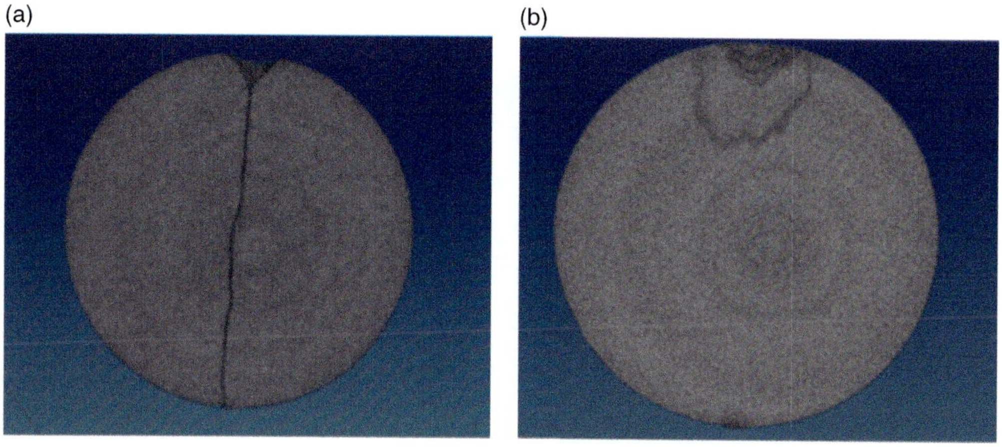

Figure 11.8 Failure patterns of Brazilian tensile strength test. (a) Treated with Fluid 2; (b) Treated with Fluid 3.

Vickers tip, but it is more fragile. The Vickers indenter is a safer option for deep surface penetration when high load indentation (usually a few Newtons or more) is required. Even though the hardness values are calculated as the peak force divided by the area of contact (Eq. 11.1) for both microindentation and nanoindentation, the definition of the contact area differs between these two techniques.

$$\text{Hardness} = \frac{(\text{Peak Force})}{(\text{Contact Area})} \tag{11.1}$$

For microindentation, the contact area in Eq. (11.1) is the surface area of the tip-faces that are in contact with the sample, while for nanoindentation the contact area is defined as the projected area of the contact between the sample and the tip.

Figure 11.9 shows the load-depth curve of the as-received outcrop Eagle Ford sample untreated (a) and treated (b) with DI water. The depth at the load of 0.5 N ranges from 2.9 to 6.6 μm for the untreated sample; the depth ranges from 4.4 to 10.4 μm for the DI water-treated sample. The depth is deeper in general after DI water treatment, which indicates the softening effect by DI water treatment.

Table 11.1 lists the parameter comparison for the outcrop Eagle Ford sample before and after DI water treatment. Under the same peak force, the maximum depth for the treated samples is deeper and the degree of hardness reduction was nearly 30% for the treated sample, which indicates that the rock sample was softened after DI water treatment.

11.4 Morphology and Minerology Changes After Fluid Exposure

The study on this part of the research is to quantify the morphological and mineralogical changes in OM rich zones when organic-rich tight carbonate rock samples are exposed to three different types of aqueous-based treatment fluid.

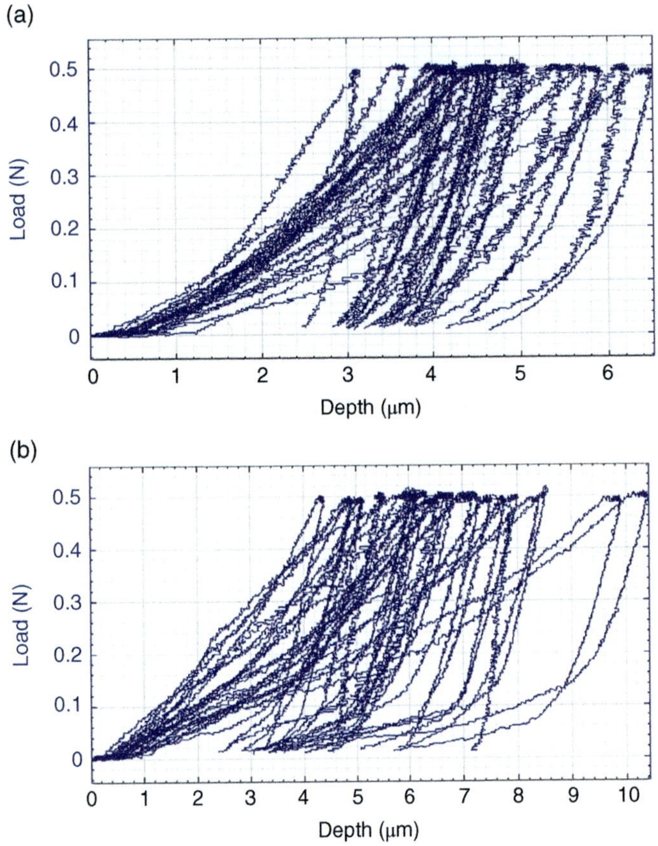

Figure 11.9 Load-Depth curve of microindentation testing of the outcrop Eagle Ford cylindrical core plugs. (a) untreated; (b) treated with DI water.

Table 11.1 Summary of the microindentation test for outcrop eagle ford sample.

		Max depth (nm)	Hardness (MPa)
Untreated	Average	4600	11.36
	Standard deviation	762	44.83
Treated	Average	6554	7.96
	Standard deviation	1427	17.34

11.4.1 Experimental Methods

11.4.1.1 Materials

11.4.1.1.1 Rock Samples A highly laminated lime mudstone sample from the same basin in the Middle Eastern Field used for the microCT study was used for the morphological and mineralogic studies. Three thin-section samples with dimensions 7.5 × 7.5 × 2.5 mm (length × width × thickness) were prepared and glued to a piece of fiberglass. Each rock surface was then fine-trimmed using a Leica EM TXP trimmer and milled using a Leica EM TIC 3X broad ion beam mill. The TOC of the bulk sample was around 8.2 wt%. The mineral content of the thin-section sample determined by XRD showed that it was mainly composed of 92 wt% calcite, 2 wt% dolomite, 3 wt% quartz, and minor amounts of pyrite (less than 1 wt%), kaolinite (less than 1 wt%), and illitic clay (less than 1 wt%).

11.4.1.1.2 Treatment Fluids Three aqueous-based treatment fluids (2 wt% KCl, 0.015 wt% friction reducer, and synthetic seawater, hereinafter referred as Fluid 4, 5, 6, respectively) were selected in this study to investigate the interactions between the fluids and the tight organic-rich carbonate source rocks. Fluid 4 was 2 wt% KCl (pH = 7.1), which does not contain any polymers. Fluid 5 was 0.015 wt% friction reducer (DP/EM 5015, 30% active, product from SNF, in an emulsion form) in 2 wt% KCl. This fluid was selected because of the common practice of using slickwater fracturing in stimulating the unconventional fields. Fluid 6 was synthetic seawater. Seawater was selected because it has been considered as a preferred base fluid for preparing stimulation fluid for offshore stimulation jobs or some areas where there is a freshwater shortage issue. Concerns associated with seawater usage mainly derive from its potential negative impacts to the reservoir formation and downhole tools such as mineral precipitation, corrosion, and scale formation. Thus, a comprehensive study on the impact of seawater on the formation from microscopic imaging and flow characteristics standpoint is needed. The composition of the synthetic seawater used in our study was based on the water analysis of the representative seawater sample from one Middle Eastern region (Table 11.2). The total dissolved solids (TDS) are around 57,000 mg/l.

Table 11.2 Water analysis of the representative seawater sample from one middle eastern region.

Ions	Concentration (mg/l)
Calcium	618
Magnesium	2108
Potassium	595
Sodium	18,451
Strontium	11
Chloride	30,694
Sulfate	4142
Carbonate	<1
Bicarbonate	150
TDS	56,800

11.4.1.2 Experimental Procedure

11.4.1.2.1 Surface Characterization using SEM Coupled with EDS The texture and microstructure of the thin-section rock samples were analyzed using scanning electron microscope (SEM) images performed by a Zeiss Sigma HPVP Scanning Electron Microscope. SEM images were taken at the same locations before and after fluid imbibition to capture the sample's morphological changes due to the fluid treatment. The energy dispersive X-ray spectroscopy (EDS) mapping performed by a coupled Bruker XFlash 6130 Energy Dispersive Spectroscopy detector was used to investigate the elemental changes on the thin-section surface exposed to the electron beam before and after fluid treatment.

11.4.1.2.2 Fluid Treatment Due to the fact that the amount of mineral dissolution is related to the mass ratio of the treatment fluid to the rock sample, the ratio in this study was chosen to be the same at 10 : 1 for all three fluid treatments. Each thin-section rock sample was submerged in a treatment fluid for 48 hours at ambient temperature and then vacuum-dried at 80 °C for 24 hours.

11.4.1.2.3 Quantification of Dissolved Ions in the Treatment Fluids The three treatment fluids immediately after preparation were first analyzed as controls using inductively coupled plasma (ICP) with optical emission spectrometer (OES) technique. After the sample imbibition test, all three treatment fluids were analyzed to quantify the dissolved ions such as Mg^{2+}, Ca^{2+}, and Sr^{2+}, with the results being expressed in mg/l (or ppm).

11.4.2 Results and Discussion

11.4.2.1 SEM and EDS Mapping of Thin-Section Surface before Fluid Treatment

11.4.2.1.1 Morphology Characterization Before the fluid treatment, multiple SEM images and EDS maps were acquired at different locations for each of the polished thin-section surfaces at the initial condition. Figure 11.10 shows one set of SEM images of the representative area on the polished surface (Figure 11.10a) and their magnified images from the sequenced blue-framed areas (Figure 11.10b–d). Similar to what we see from the microCT image in Figure 11.1, two distinguished zones were observed in Figure 11.10a. A Cal-Z, mainly composed of calcite crystals, was shown as light grey color in image. The OMZ, ranging from 400 to 500 μm in width in this sample, was shown as a darker gray band in the middle of the image. The texture of the OMZ could be seen in more details from the magnified images in Figure 11.10b–d. The OM rich areas appeared as black stringers in layered structures, which was also intermixed with other minerals such as calcite, dolomite, pyrite, etc. Nanopore structures within the OM were observed in the higher magnified image (Figure 11.10d). The intra-organic pore was the dominant pore type present in this sample. Hakami et al. (2016) have reported similar porous OM in a similar basin. Pommer and Milliken (2015) also reported that the nanopore generated within OM was the dominant pore type in Eagle Ford shale play in North America.

11.4.2.1.2 Mineral Identification Another representative OMZ from the same thin-section sample was characterized by SEM image and the elemental mapping from EDS analysis shown in Figure 11.11. Figure 11.11a was the secondary electron SEM image; Figure 11.11b–d were EDS maps of Ca, Si, and Al, respectively. The mineral compositions could be semi-quantified using the pseudo composite mapping, which combines all detectable elements from calcium (Ca), silicon (Si), aluminum (Al), magnesium (Mg), etc. For example, in Figure 11.11e, the areas where both Fe

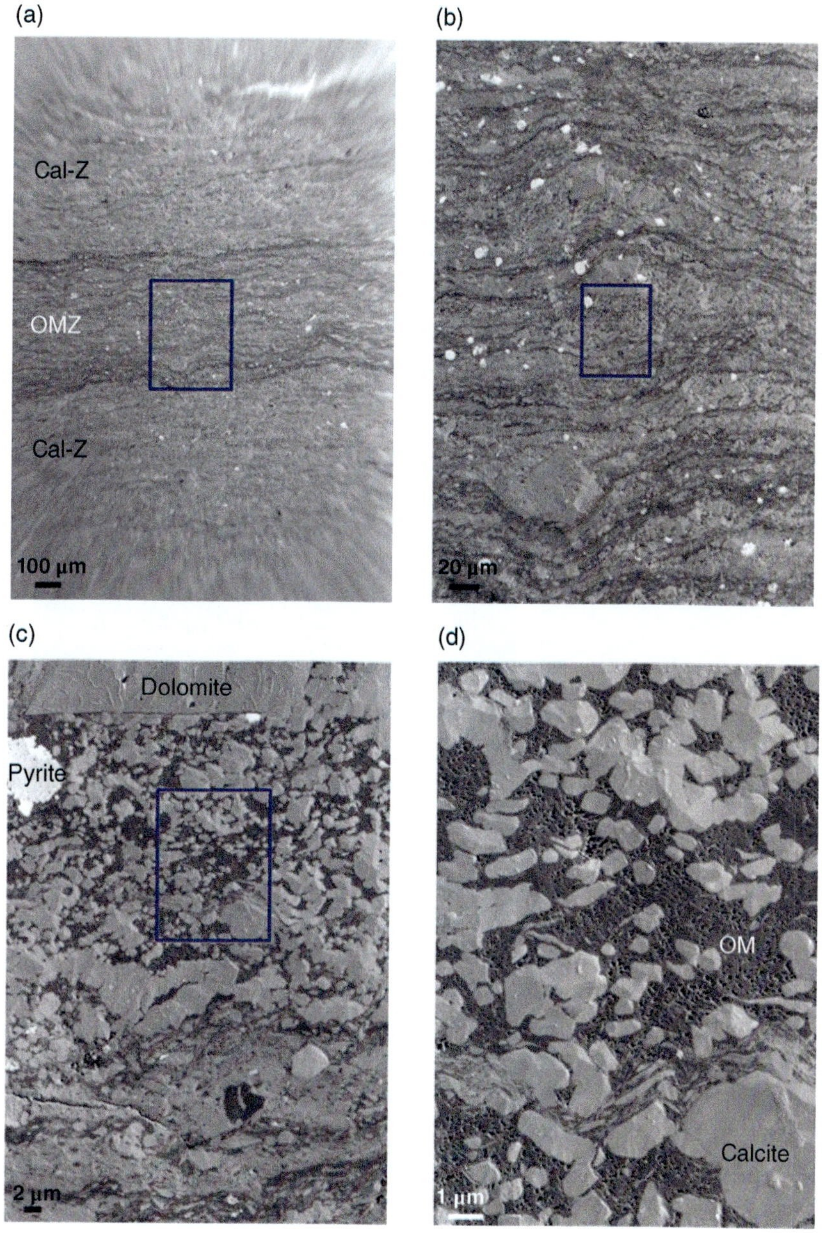

Figure 11.10 SEM images of the thin-section sample prepared from the tight organic-rich carbonate rock sample from the Middle Eastern Field (a) and their sequential magnified images from blue-framed areas (b–d).

and S show high amounts are identified as pyrite; the areas where both Ca and Mg occur are identified as dolomite; the areas where Ca is high but Mg is undetectable are identified as calcite; the areas where high Si yet no Al are identified as quartz; and the area where both high Si and Al are high are identified as clay minerals. This is the way we registered different minerals from SEM images. In the pseudo-color composite mapping Figure 11.11e, calcite was shown as

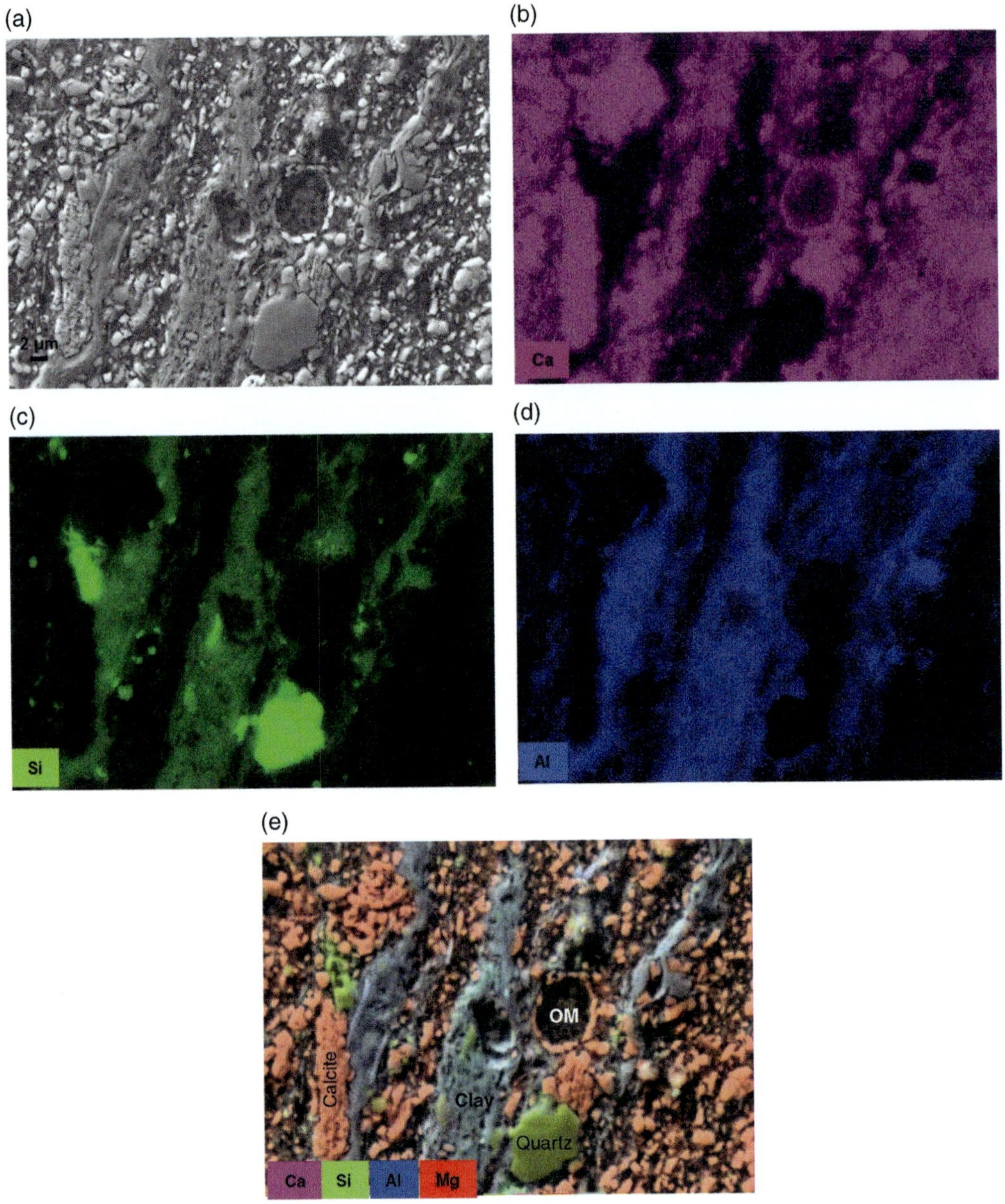

Figure 11.11 SEM images and elemental mapping from the EDS analysis for the thin-section sample prepared from the tight carbonate rock sample from the Middle East field. (a) secondary electron SEM image; (b) EDS mapping of Ca; (c) EDS mapping of Si; (d) EDS mapping of Al; (e) pseudo-color composite map from EDS analysis.

orange, while quartz was shown as green, clay minerals were shown as bluish-grey, and the dark grey areas were OM.

Figure 11.12 shows another example of mineral identification using the SEM coupled with the EDS mapping. A different color scheme was chosen to highlight the minerals of interest in this photomicrograph. This set of images was taken in the OMZ. In Figure 11.12a, two clusters of

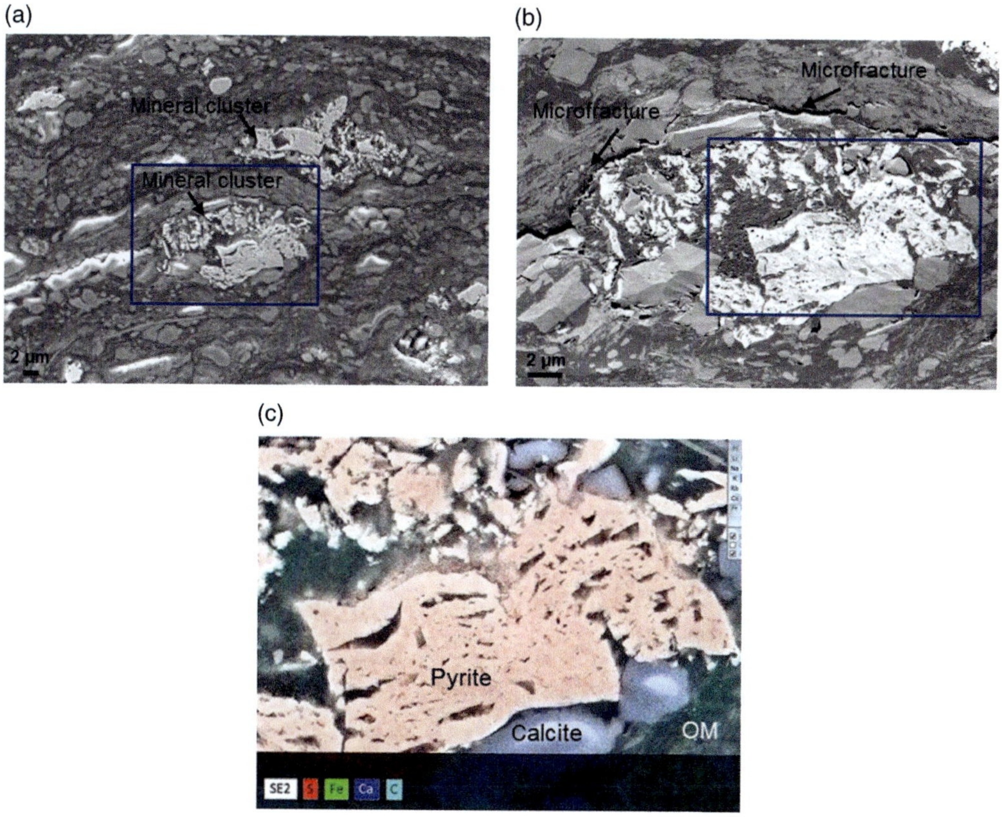

Figure 11.12 SEM image and elemental mapping from the EDS analysis. (a) secondary electron SEM image; (b) magnified area of the blue-framed area in a; (c) pseudo-color composite map from the EDS analysis of S, Fe, Ca, and C for blue-framed area in b.

mineral crystals (size around 20 × 10 μm) were observed inside the OM rich area. Figure 11.12b is a magnified portion of the mineral clusters in the blue-framed area in Figure 11.12a. Due to the laminated nature of the OMZs, microfractures that were parallel to the laminations were observed (Figure 11.12b). Figure 11.12c shows the pseudo-color composite mapping from the EDS analysis of S, Fe, Ca, and C for the blue-framed area in Figure 11.12b, confirming that the mineral crystals in Figure 11.12b were mainly pyrite. In this case, calcite crystals showed as bright light blue and OM showed in dark green (Figure 11.12c).

11.4.2.2 SEM and EDS Mapping of Thin-Section Surface after Fluid Treatment

The SEM and EDS mapping of the thin-section sample surface were performed at the same locations after the three aqueous-based fluid treatment fluids. Again, Fluid 4 was 2 wt% KCl (pH = 7.1). Fluid 5 was 0.015 wt% friction reducer (DP/EM 5015, 30% active from SNF in an emulsion form) in 2% KCl. Fluid 6 was synthetic seawater.

11.4.2.2.1 Treatment with Fluid 4 (2 wt% KCl) Figure 11.13 shows two pairs of SEM images in the OMZ before and after 2 wt% KCl treatment. The pair images (Figure 11.13a,b) were conducted at a lower resolution than the pair Figure 11.13c,d. In its original state shown in the first pair

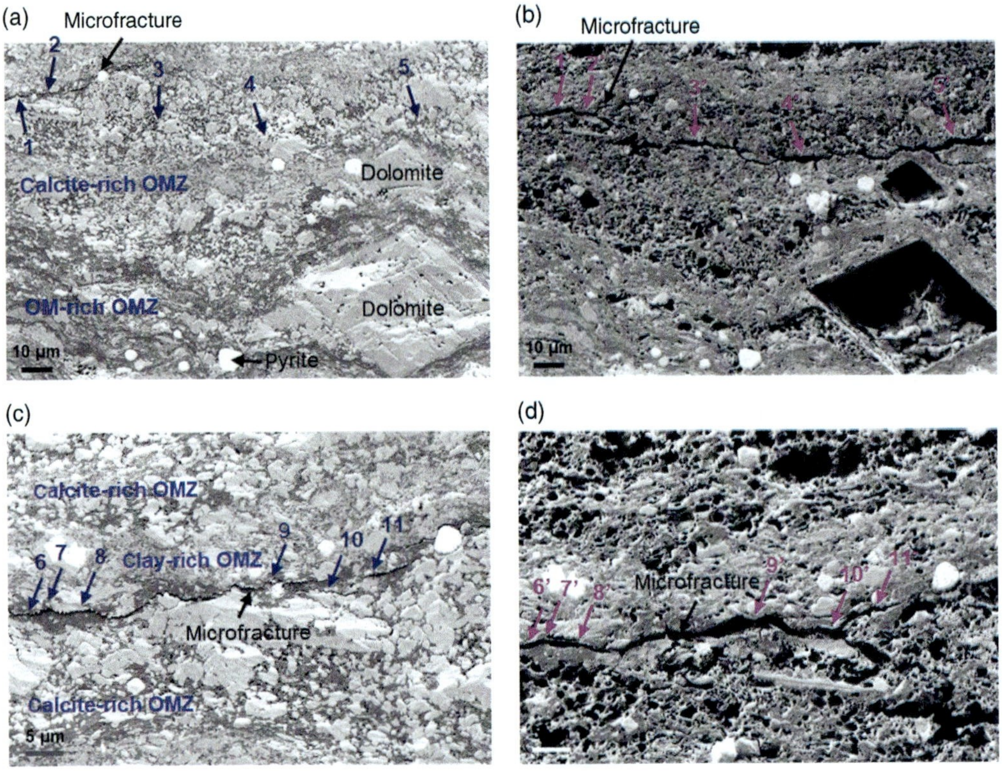

Figure 11.13 Two pairs of SEM images from the OMZ of source rock section sample before (a and c) and after 2 wt% KCl fluid treatment (b and d). Small calcite crystals were dissolved.

(Figure 11.13a), calcite crystals appeared as light gray and pyrite crystals appeared as bright nearly circular-shaped crystal clusters, called framboids. The upper portion of the image Figure 11.13a has more calcite present, which was labeled as *calcite-rich OMZ*. The lower portion was dominated with stringing-featured, OM-rich gray area, which was labeled as *OM-rich OMZ*. Two euhedral rhomb-shaped dolomite crystal showed on this image, and they were also surrounded by OM-rich stringers. In the upper side of calcite-rich OMZ in Figure 11.13b, the surface of fluid treated sample was relatively rough (high relief) and appeared more porous. This might be due to the dissolution of calcite crystals in 2 wt% KCl solution, leaving cavities on the rock surface. In the lower portion of OM-rich OMZ in Figure 11.13b where silicate and aluminosilicate minerals are present, fewer cavities were observed except that two dolomite crystals were detached from the surface. This was probably because less calcite is present in this area and also the mineral dissolution initiated around the interfaces between the dolomite crystals and OMZ, then somehow the dolomite crystal was dislodged from its original location. There was no OM dissolution observed in this pair of images.

In images Figure 11.13c,d, one clay-rich area was located in the center of the OMZ, which was labeled as *clay-rich OMZ*. Both top and bottom zones were *calcite-rich OMZ*. There was one microfracture cross the entire image area in nearly horizontal direction (Figure 11.13c). After the 2 wt% KCl fluid treatment, the calcite dissolution was more pronounced in calcite-rich OMZ, while relatively less dissolution was observed in the clay-rich OMZ (Figure 11.13d).

Table 11.3 Aperture size change due to 2 wt% KCl fluid treatment in Figure 11.13.

	Before 2 wt% KCl fluid treatment		After 2 wt% KCl fluid treatment	
	n	Aperture (μm)	*n*′	Aperture (μm)
Figure 11.13a,b	1	0.24	1′	1.64
	2	0.22	2′	0.59
	3	0 (mineral-filled)	3′	1.55
	4	0 (mineral-filled)	4′	1.93
	5	0 (mineral-filled)	5′	1.87
Figure 11.13c,d	6	0.22	6′	0.68
	7	0.23	7′	0.40
	8	0.22	8′	0.60
	9	0.16	9′	1.39
	10	0.13	10′	1.37
	11	0.36	11′	0.55

Another feature observed with 2 wt% KCl fluid treatment from the two pairs of SEM images in Figure 11.13 was the creation and enlargement of microfractures and/or reopening filled fractures. After 2 wt% KCl fluid treatment, there was one pronounced long fracture crossing the entire image shown in both Figure 11.13b,d. In their initial state (Figure 11.13a,c), these two microfractures were not quite as pronounced. This might be due to mineral dissolution from the pre-filled fractures with the fluid treatment. Since the microfracture had an irregular shape, the apertures of the microfracture in the paired images were measured at multiple points with the direction perpendicular to the fracture (Table 11.3). Each individual point n' (1′ to 11′) in Figure 11.13b,d corresponded to the point n (1–11) in images Figure 11.13a,c. In general, for the pre-existing microfractures, the apertures in the treated rock sample were measured to be 2–7 times larger than the ones in its original state. For the newly generated microfracture across points 3′, 4′, and 5′ due to fluid treatment, the aperture size ranged from 1 to 2 μm.

11.4.2.2.2 Treatment with Fluid 5 (0.015 wt% Friction Reducer) Figure 11.14 shows three pairs of SEM images in the OMZ before and after 0.015 wt% friction reducer (in 2 wt% KCl) treatment. The pair images (Figure 11.14a,b) were conducted at lower resolution than the other two pairs of images. At its initial condition (Figure 11.14a), the sample surface appeared to be tight and pores were not visible under this image resolution. There was a partially filled fracture across the image in the diagonal direction (the direction of lamination). After the friction reducer fluid treatment, multiple empty spots were shown on the image Figure 11.14b, which indicated that the rock surface became more porous after fluid treatment.

Figure 11.14c,d are the second pair of SEM images in a calcite-rich OMZ and they were acquired at higher magnification than Figure 11.14a,b before and after the friction reducer fluid treatment. At its initial condition (Figure 11.14c), the surface was shown with abundant calcite crystals with different sizes. There was also one porous OM-rich mass (about 8 μm in diameter) shown on the top left side of the image surrounded by calcite crystals. After slickwater fluid treatment, some relatively small calcite crystals (less than 5 μm), which were on at the initial state disappeared after

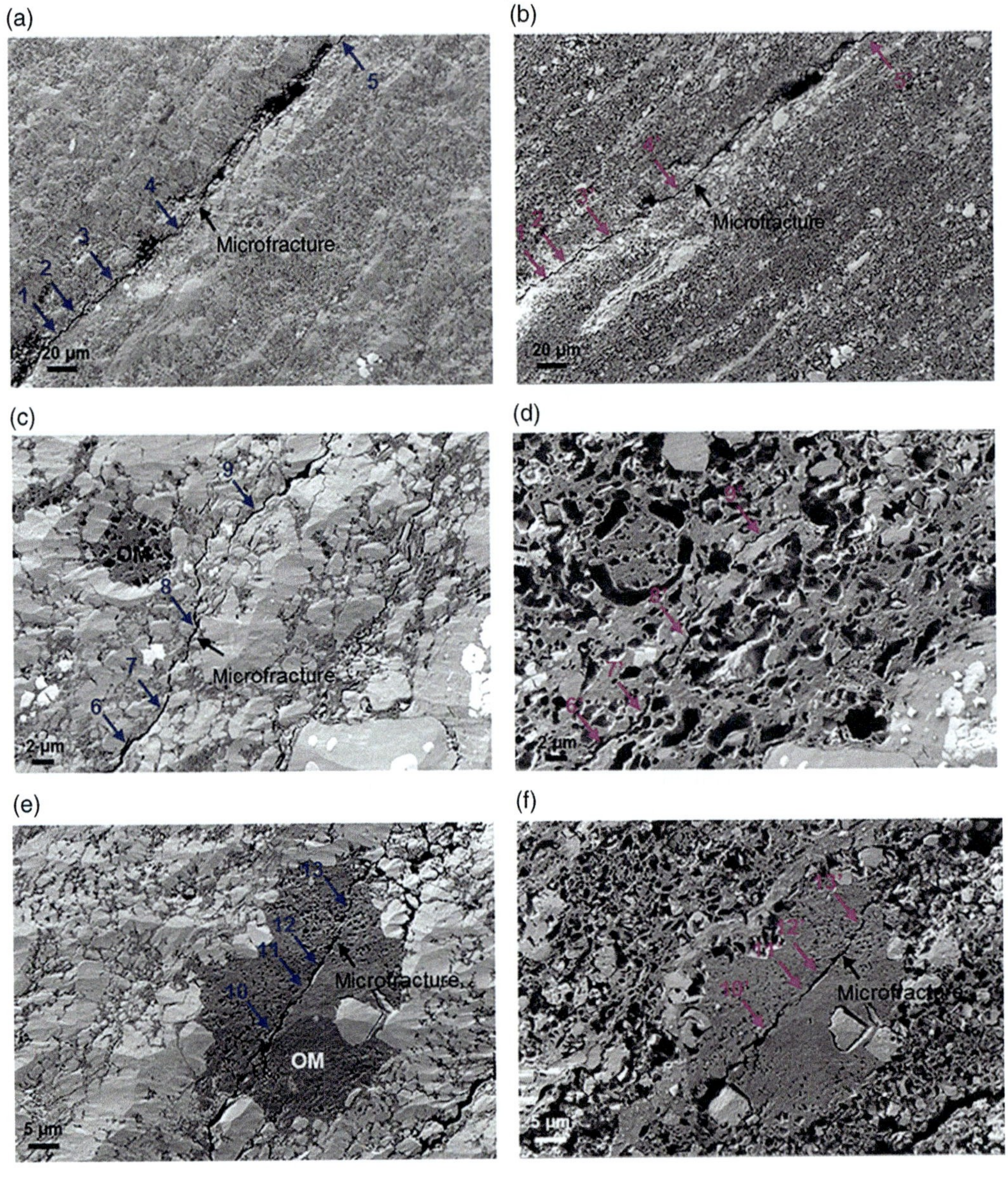

Figure 11.14 Three pairs of SEM images from OMZ of source rock section sample before (a, c, and e) and after 0.015 wt% friction reducer in 2 wt% KCl fluid treatment (b, d, and f). Small calcite crystals were dissolved.

fluid treatment (Figure 11.14d), which could be due to dissolution by the fluid treatment. The porous OM-rich area and the large-sized calcite crystals (bigger than 5 μm) seemed to remain unchanged.

Figure 11.14e,f are the third pair of high-resolution SEM images for an area in the OMZ before and after friction reducer fluid treatment. At its initial condition (Figure 11.14e), there was a large OM mass (approximately 40 × 25 μm) shown on the surface, which had a microfracture across it. Some intra- nanopores were observed on the left side of the OM, while relatively fewer were shown on the right side of the microfracture. Similar to Figure 11.14c, this OM mass was also surrounded

Table 11.4 Aperture size Change due to 0.015 wt% friction reducer (in 2 wt% KCl) fluid treatment in Figure 11.14.

	Before 0.015 wt% friction reducer Fluid Treatment		After 0.015 wt% friction reducer Fluid Treatment	
	n	Aperture (μm)	*n′*	Aperture (μm)
Figure 11.14a,b	1	1.13	1′	1.04
	2	1.43	2′	1.62
	3	1.09	3′	1.04
	4	0.25	4′	0.44
	5	0.72	5′	1.56
Figure 11.14c,d	6	0.49	6′	0.45
	7	0.18	7′	0.25
	8	0.22	8′	0.32
	9	0.16	9′	0.22
Figure 11.14e,f	10	0.38	10′	0.44
	11	0.18	11′	0.38
	12	0.67	12′	0.73
	13	0.19	13′	0.23

by calcite crystals. After friction reducer fluid treatment, the relatively smaller-sized calcite crystals (less than 5 μm) were completed dissolved by the fluid treatment (Figure 11.14f), while for relatively large-sized calcite or dolomite particles (bigger than 5 μm), the mineral dissolution was only observed on the edge and the surface. The other components such as silicate minerals and OM remain unchanged after friction reducer fluid exposure.

In order to quantify the changes of the microfracture due to the fluid treatment, the apertures at multiple points were measured with the direction perpendicular to the fracture for the paired images (Table 11.4). The aperture of the microfracture remains in a similar range after friction reducer fluid treatment.

11.4.2.2.3 Treatment with Fluid 6 (Synthetic Seawater) Figure 11.15 shows three pairs of SEM images in the OMZ before and after the synthetic seawater treatment. The pair images (Figure 11.15a,b) were conducted at lower resolution than the other two pairs of images. At its initial condition (Figure 11.15a), there was one open microfracture, which was parallel to the lamination direction observed. Different from 2 wt% KCl and 0.015 wt% friction reducer fluid treatments, very minimal mineral dissolution was observed since the rock surface after seawater fluid treatment did not contain many newly generated empty spots. Another observation was that significant amount of mineral precipitation mainly composed of NaCl was observed covering the surface of the pre-existing microfracture.

Two more pairs of SEM images (Figure 11.15c,d and e,f) with higher magnification in two other areas in the OMZ before and after the synthetic water treatment were presented. Similar to what was observed in the pair images of Figure 11.15a,b, the rock surface after seawater fluid treatment does not contain many newly generated empty spots, indicating that the calcite dissolution is very little. A small amount of mineral precipitation was also observed on the rock surface. The aperture of the

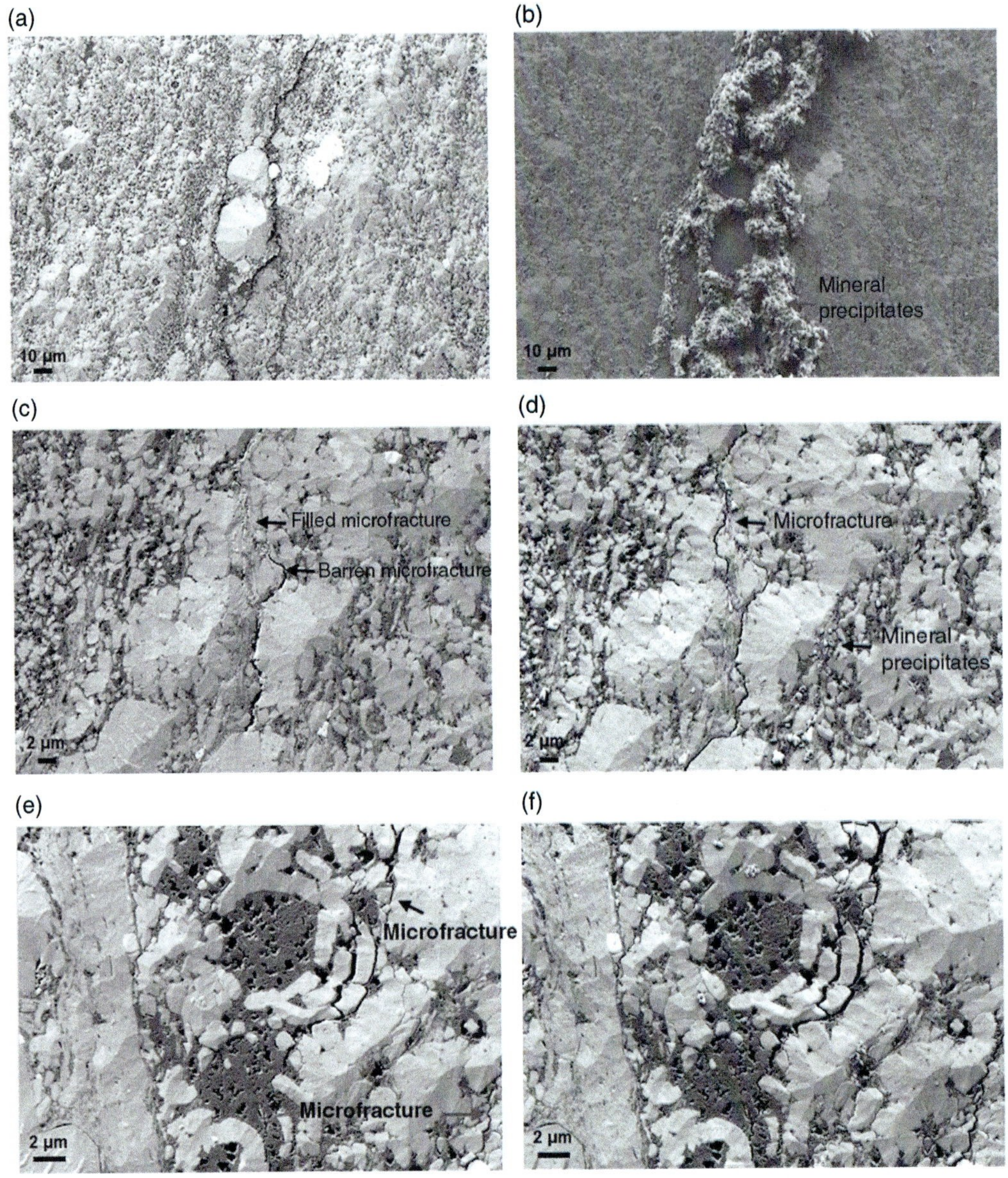

Figure 11.15 Three pairs of SEM images from the OMZ of source rock section sample before (a, c, and e) and after the synthetic seawater fluid treatment (b, d, and f).

microfracture remains in a similar range after synthetic water fluid treatment as well. Thus, when we design the treatment fluid using seawater, precautions to prevent scale deposition are very important.

11.4.2.3 Quantification of Dissolved Ions in the Treatment Fluids

From the mineralogy analysis, the main minerals in the Middle Eastern sample were calcite, dolomite, quartz, and minor amounts of pyrite, kaolinite, and illitic clay. Among these minerals, calcite has the highest solubility in water than other minerals.

Table 11.5 Quantification of cations dissolution in treatment fluids.

		Cations dissolution (normalized as μg/g of rock)		
		2 wt% KCl	0.015 wt% friction reducer	Synthetic seawater
Cations	Mg^{2+}	35.8	34.2	0.0
	Ca^{2+}	700.9	522.7	111.2
	Sr^{2+}	13.6	11.6	9.6

Calcite dissolution on the rock surface was observed in all three treatment fluids in the morphology study using the SEM coupled with the EDS. A more significant amount of calcite dissolution was observed for the 2 wt% KCl and 0.015 wt% friction reducer fluids treated sample than the synthetic seawater treated sample. This might be due to the solution difference of these three fluids. Since seawater contains relatively high concentrations of Ca^{2+} and Mg^{2+}, lowering the solubility of calcite. Dissolution or dislocation of dolomite were also observed in 2% KCl fluid treated rock sample. Other minerals such as quartz and majority of OM remain undissolved in all three aqueous-based fluids.

The expected majority of dissolved ions in the treatment fluids were Ca^{2+}, Mg^{2+} along with other commonly observed cations in carbonate rocks, such as Sr^{2+}. To quantify the changes of the cations, the concentrations of these three cations for the prepared treatment fluids were analyzed by ICP-OES as the control. Then the fluids after the rock treatment were analyzed as well. The difference of these three cations for all three fluids after rock treatment were presented as the μg dissolution per gram of rock sample for a better comparison (Table 11.5).

The ICP analysis for the treatment fluids further confirms the mineral dissolution observed from the surface morphology studies. It also quantifies the difference of dissolution capacities among these three fluids. As shown in Table 11.5, the Ca^{2+} dissolution was higher for the 2 wt% KCl treatment than that for the friction reducer treatment fluid. 700.9 μg of Ca^{2+} ion per gram of rock was dissolved in 2 wt% KCl and 522.7 μg of Ca^{2+} ion per gram of rock was dissolved in 0.015 wt% friction reducer treatment fluids. Around 35 μg of Mg^{2+} ion per gram of rock was dissolved in these two fluids. The increase of ion concentrations in Ca^{2+} and Mg^{2+} verifies the calcite and dolomite crystals dissolution into the treatment fluids. The dissolution of calcite in seawater was lower than that in 2 wt% KCl and 0.015 wt% friction reducer, which is consistent with our observation from the surface morphology study. Small amounts of Sr^{2+} ion increases (less than 14 μg per gram of rock) were also observed in all three treatment fluids, possibly resulting from the presence of minor or trace amounts of strontium carbonate or sulfate minerals in this suite of rocks.

11.5 Flow Property Change After Fluid Exposure

To further assess the impacts of morphological change due to aqueous-based fluids treatment in terms of mineral dissolution and precipitation on flow parameters of rock samples, permeability measurements with fluid treatment experiments for core plugs were conducted. Due to the lack of the whole core sample from the Middle Eastern field used in the SEM studies, the analogous outcrop Eagle Ford core plugs were used to measure the absolute permeability value.

11.5.1 Experimental Methods

11.5.1.1 Materials

11.5.1.1.1 Rock Sample The TOC of the outcrop Eagle Ford sample was around 5 wt%. The mineral content determined by XRD showed that it was mainly composed of 66 wt% calcite, 26 wt% quartz, 1 wt% dolomite, 2 wt% gypsum, small amount of pyrite (less than 1 wt%), kaolinite (3 wt%), and illitic clay (1 wt%).

11.5.1.1.2 Treatment Fluids The three aqueous-based treatment fluids used in the flow property study were the same as the morphological studies in the precious section. Fluid 4 was 2 wt% KCl (pH = 7.1), which does not contain any polymers. Fluid 5 was 0.015 wt% friction reducer (DP/EM 5015, 30% active from SNF in an emulsion form) in 2 wt% KCl. Fluid 6 was synthetic seawater, which has the same composition as that shown in Table 11.2.

11.5.1.2 Experimental Procedure

11.5.1.2.1 Fluid Treatment and Flow Characteristics Assessment for Core Plugs The size of the outcrop Eagle Ford core plugs used in flow properties study was 1.0 in. in diameter and 1.0 in. in length. Each core plug sample was submerged in the treatment fluid for 72 hours at ambient temperature and then vacuum-dried at 80 °C for 24 hours. The absolute permeability for these core plugs was measured before and after fluid treatment. The permeability was measured using an AP-608 Automated Permeameter–Porosimeter from Coretest Systems and the methodology was based on the pulse decay method proposed by Jones (1997).

11.5.2 Results and Discussion

11.5.2.1 Changes in Flow Characteristics

The absolute permeability measurement under different effective stresses for three Eagle Ford outcrop core plugs before and after fluid treatment are summarized in Table 11.6. After fluid treatment, the absolute permeability under different stress conditions was significantly enhanced. The permeability improvement was calculated and listed in the table as well.

The observed permeability increase after fluid–rock interaction is aligned with our expectation that the mineral dissolution, new microfractures generation, or sealed fractures reopening could result in the permeability increase. Thirty-five times of absolute permeability increase was observed under our testing conditions with 2 wt% KCl fluid treatment. Note that the observed absolute permeability enhancement considerably depends on the test conditions, such as rock and fluid chemical compositions, temperature, fluid pressure, volumetric solid–fluid ratio and others, while the test conditions for this work were carefully documented. Caution needs to be taken when applying the observed values for permeability enhancement to different conditions.

Interestingly, the absolute permeability resulting from mineral dissolution and enlargement of the microfractures becomes more sensitive to the stress. The stress sensitivity is characterized by the slope of permeability-effective stress curves shown in Figure 11.16. Using 2 wt% KCl fluid treatment as an example, the permeability after fluid treatment has a sensitivity factor of 4×10^{-4} (psi^{-1}), whereas the permeability before fluid treatment shows a sensitivity factor of 1×10^{-4} (psi^{-1}). The sensitivity factor increased 4 times after fluid treatment in this case.

This observed stress sensitivity is consistent with what has been reported in the literature, in which permeability of open microfractures and the alike, rather than sealed/cemented microfractures are relatively stress sensitive (Khan and Teufel 2000; Laubach et al. 2004; Muralidharan et al. 2004; Abass et al. 2007; Tian 2014; Lavrov 2017).

Table 11.6 Absolute permeability measurement under different effect stress for three Eagle Ford outcrop core plugs before and after fluid treatment.

		Absolute permeability (md)		
Treatment fluid	**Effective stress (psi)**	**Before fluid treatment (Initial)**	**After fluid treatment (Treated)**	**Permeability improvement (Treated/Initial)**
2 wt% KCl	500	0.024	0.923	38
	1000	0.02	0.773	39
	2000	0.019	0.502	26
0.015 wt% friction reducer	500	0.016	0.363	23
	1000	0.012	0.276	23
	2000	0.01	0.21	21
synthetic seawater	500	0.195	2.306	12
	1000	0.159	1.48	9
	2000	0.122	0.731	6

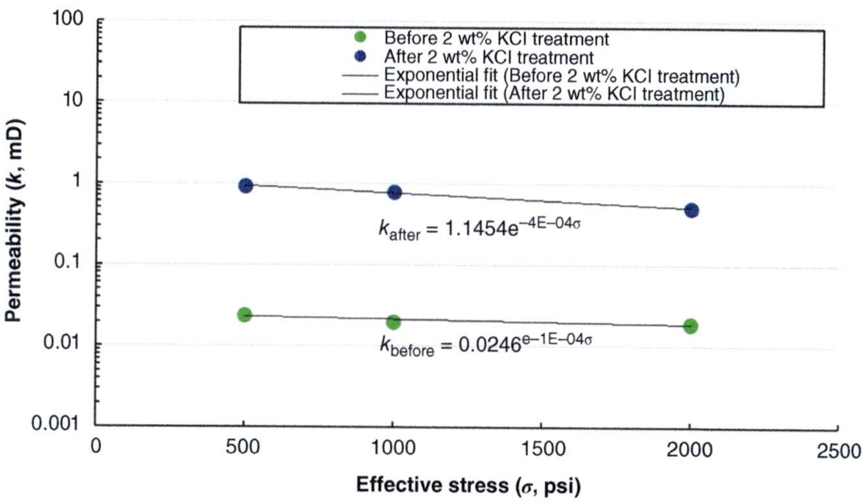

Figure 11.16 Absolute permeability measurement under different confining pressures for three Eagle Ford outcrop core plugs before and after fluid treatment.

11.6 Conclusions

In this study, we have conducted a multiscale investigation, including the microscale and core scale tests to study the fluid intake pathways, and their impact on the organic-rich tight carbonate formation rocks in terms of mechanical properties, morphology change, minerology change, and flow property change. Two sets of organic-rich tight carbonate rock samples were used. One sample set was a highly laminated lime mudstone from a basin in a Middle Eastern Field that was used in microscopic studies such as fluid uptake pathways, morphological, and mineralogic aspects. The second sample set was from an outcrop of the Eagle Ford Shale considered to be analogous to the set from the Middle Eastern Field in terms of mineralogy. The sample set was used in core scale level studies on rock mechanical and flow properties such as permeability. Below are what we find from our studies:

- The results from the spontaneous imbibition tests revealed that the fluid traveled very fast and reached to the other end of the sample in less than 4 minutes in microfractures within both the OMZs and Cal-Z zones. The water-front progressed very slowly in the Cal-Z matrix. This indicates that the main fluid flow pathways for fracturing fluid could be within the connected fracture network at different scales, instead of the matrix initially.
- Aqueous-based stimulation fluids significantly lower the mechanical properties including both the unconfined compressive stress and Brazilian tensile strength of the organic-rich tight carbonate rocks. Microindentation test shows the softening effect by the fluid treatment as well.
- The SEM and EDS results for the fluid-treated source rock samples show that significant calcite dissolution has been observed in 2 wt% KCl and friction reducer fluid treated sample. In the cases associated with 2 wt% KCl and friction reducer fluid, reopening of some mineral-filled natural fractures was observed. The enlargement of microfractures was slightly more noticeable for samples treated with 2 wt% KCl (2–7 times) than that for fraction reducer fluid treatment at the micron-scale level. Some NaCl salt precipitation and new microfractures generation were observed for samples treated with seawater.
- The absolute permeability increases significantly after 72 hours treated in 2 wt% KCl and friction reducer fluid treatment than synthetic seawater treatment. While more studies are needed under different test conditions, this study provides experimental evidences that aqueous-based fracturing fluid may potentially have positive effect on gas production from organic-rich carbonate source rock.

These findings shed a light on the development of an improved hydraulic fracturing treatment by incorporating the factors that positively affect the production rate while minimizing the factors that will likely lead to the production rate decline. Below are the factors affecting the production rate in a positive matter that we should keep in mind when developing new stimulation technologies:

- *Enhance connectivity of natural or induced microfractures and keep them open during production.* From the microscopic scale experimental investigation, we have seen the importance of the fractures and microfractures generated by stimulation process as the main fluid flow pathways. Improving and maintaining the connectivity between the natural and induced microfractures in the far-field and the primary fracture networks are essential to enhancing the well production rate in the unconventional fields. These microfractures, besides the primary fracture networks generated by hydraulic fracturing, should be an important contributor to the high production rate at the initial production stage. The commonly observed quick production rate decline at least partially results from the closure of these microfractures. During hydrocarbon production, the pore pressure declines and accordingly the effective stress increases, resulting in the closure of these microfractures without proppants during the production. Enhancing connectivity of natural or induced microfractures and keep them open during production will derive a positive factor in production.
- *Improve dissolution factor and reduce the scale/precipitate formation.* Considerable amounts of dissolutions have been seen with the neutral pH fluid. Enhancing the dissolution factor by acid fluids for carbonate formation will improve the flow path in a positive way. Addition of the scale inhibitors or pre-treat seawater could prevent the precipitates or scale formation.

- *Minimize softening effects of source rocks.* Additives that could be used along with the stimulation fluid to reduce or minimize the softening effects of the source rocks can potentially reduce the production rate decline.

References

Abass, H.H., Ortiz, I., Khan, M.R. et al. (2007). Understanding stress dependent permeability of matrix, natural fractures, and hydraulic fractures in carbonate formations. *SPE Saudi Arabia Section Technical Symposium*, Dhahran, Saudi Arabia (7–8 May 2007). SPE-110973-MS. https://doi.org/10.2118/110973-MS.

Akrad, O., Miskimins, J., and Prasad, M. (2011). The effects of fracturing fluids on shale rock mechanical properties and proppant embedment. *SPE Annual Technical Conference*, Denver, Colorado (30 October–2 November, 2011). SPE-146658-MS. https://doi.org/10.2118/146658-MS.

ASTM D3967-08 (2008). *Standard Test Method for Splitting Tensile Strength of Intact Rock Core Specimens*. West Conshohocken, Pennsylvania: ASTM International www.astm.org.

ASTM D7012-14 (2014). *Standard Test Methods for Compressive Strength and Elastic Moduli of Intact Rock Core Specimens Under Varying States of Stress and Temperatures*. Pennsylvania: ASTM International, West Conshohocken www.astm.org.

Baihly, J., Altman, R., Malpani, R. et al. (2010). Shale gas production decline trend comparison over time and basins. *SPE Annual Technical Conference and Exhibition*, Florence, Italy (19–22 September 2010). SPE-135555-MS. https://doi.org/10.2118/135555-MS.

Barsoum, M.W., Murugaiah, A., Kalidindi, S.R. et al. (2004). Kinkling nonlinear elastic solids, nanoindentations, and geology. *Physical Review Letters* 92 (25): 255508. https://doi.org/10.1103/PhysRevLett.92.255508.

Beydoun, Z.R. (1998). Arabian plate oil and gas: why so rich and so prolific? *Episodes* 21 (2): 74–81. https://doi.org/10.18814/epiiugs/1998/v21i2/001.

Broitman, E. (2017). Indentation hardness measurements at macro-, micro-, and nanoscale: a critical overview. *Tribology Letters* 65: 23. https://doi.org/10.1007/s11249-016-0805-5.

Bybee, K. (2004). Acid fracturing a carbonate reservoir. *Journal of Petroleum Technology* 56 (7): 49–52. SPE-0704-0049-JPT. https://doi.org/10.2118/0704-0049-JPT.

Cheng, Y. (2012). Impact of water dynamics in fractures on the performance of hydraulically fractured wells in gas-shale reservoirs. *Journal of Canadian Petroleum Technology* 51 (2): 143–151. SPE-127863-PA. https://doi.org/10.2118/127863-PA.

Ghanbari, E. and Dehghanpour, H. (2015). Impact of rock fabric on water imbibition and salt diffusion in gas shales. *International Journal of Coal Geology* 138: 55–67. doi: 10.1016/j.coal.2014.11.003.

Hakami, A., Al-Mubarak, A., Al-Ramadan, K. et al. (2016). Characterization of carbonate mudrocks of the Jurassic Tuwaiq Mountain formation, Jafurah basin, Saudi Arabia: implications for unconventional reservoir potential evaluation. *Journal of Natural Gas Science and Engineering* 33: 1149–1168. https://doi.org/10.1016/j.jngse.2016.04.009.

Howard, J., Lin, S., and Zhang, S. (2019). Uncertainty quantification in image segmentation for image-based rock physics in a shaly sandstone. *Petrophysics* 60 (2): 240–254. https://doi.org/10.30632/PJV60N2-2019a2.

Jones, S.C. (1997). A technique for faster pulse-decay permeability measurements in tight rocks. *SPE Formation Evaluation* 12 (1): 19–25. SPE-28450-PA. https://doi.org/10.2118/28450-PA.

Khan, M. and Teufel, L.W. (2000). The effect of geological and geomechanical parameters on reservoir stress path and its importance in studying permeability anisotropy. *SPE Reservoir Evaluation & Engineering* 3 (5): 394–400. https://doi.org/10.2118/66184-PA.

King, G. (2012). Hydraulic fracturing 101: what every representative, environmentalist, regulator, investor, university research neighbor and engineer should know about estimating frac risk. *Journal of Petroleum Technology* 64 (4): 34–42. SPE-0412-0034-JPT. https://doi.org/10.2118/0412-0034-JPT.

Lai, B., Liang, F., Zhang, J. et al. (2016). Fracturing fluids effects on mechanical properties of organic rich shale. *50th US Rock Mechanics/Geomechanics Symposium*, Houston, TX (26–29 June 2016). ARMA-2016-180.

Laubach, S.E., Olson, J.E., and Gale, J.F.W. (2004). Are open fractures necessarily aligned with maximum horizontal stress? *Earth and Planetary Science Letters* 222: 191–195. https://doi.org/10.1016/j.epsl.2004.02.019.

Lavrov, A. (2017). Fracture permeability under normal stress: a fully computational approach. *Journal of Petroleum Exploration and Production Technology* 7: 181–194. https://doi.org/10.1007/s13202-016-0254-6.

Liang, F., Lai, B., Zhang, J. et al. (2018). An experimental study on interactions between imbibed fracturing fluid and organic-rich tight carbonate source rocks. *SPE Journal* 23 (6): 2133–2146. SPE-188338-PA. https://doi.org/10.2118/188338-PA.

Liang, F., Zhang, J., Liu, H.-H. et al. (2019). Multiscale experimental studies on interactions between aqueous-based fracturing fluids and tight organic-rich carbonate source rocks. *SPE Reservoir Evaluation & Engineering* 22 (2): 402–417. SPE-192411-PA. https://doi.org/10.2118/192411-PA.

Liang, F., Zhang, J., Liu, H.-H. et al. (2020). Microscopic scale experimental investigation into fracturing fluid uptake pathways for tight carbonate source rocks. *International Petroleum Technology Conference*, Dhahran, Saudi Arabia (13–15 January 2020). IPTC-20093-MS. https://doi.org/10.2523/IPTC-20093-MS.

Loucks, R.G., Reed, R.M., Ruppel, S.C. et al. (2012). Spectrum of pore types and networks in mudrocks and a descriptive classification for matrix-related mudrock pores. *AAPG Bulletin* 96 (6): 1071–1098. https://doi.org/10.1306/08171111061.

McLeod, H.O. (1984). Matrix acidizing. *SPE Journal* 36 (12): 2055–2069. SPE-13752-PA. https://doi.org/10.2118/13752-PA.

McLeod, H.O. (1989). Significant factors for successful matrix acidizing. *SPE Centennial Symposium at New Mexico Tech*, Socorro, New Mexico, USA (16–19 October 1989). SPE-20155-MS. https://doi.org/10.2118/20155-MS.

Muralidharan, V., Putra, E., and Schechter, D.S. (2004). Experimental and simulation analysis of fractured reservoir experiencing different stress conditions. *Canadian International Petroleum Conference*, Calgary, Alberta (8–10 June 2004). Paper 2004-229. https://doi.org/10.2118/2004-229.

Nelson, P.H. (2009). Pore-throat sizes in sandstones, tight sandstones, and shales. *AAPG Bulletin* 93 (3): 329–340. https://doi.org/10.1306/10240808059.

Olsen, T.N. and Karr, G.K. IV. (1986). Treatment optimization of acid fracturing in carbonate formations. *SPE Rocky Mountain Regional Meeting*, Billings, Montana, USA (19–21 May 1986). SPE-15165-MS. https://doi.org/10.2118/15165-MS.

Pommer, M. and Milliken, K. (2015). Pore types and pore-size distribution across thermal maturity, Eagle Ford formation, Southern Texas. *AAPG Bulletin* 99 (9): 1713–1744. https://doi.org/10.1306/03051514151.

Soeder, D.J. (2017). *Unconventional: The Development of Natural Gas from the Marcellus Shale*, 1ee. Boulder, Colorado: Geological Society of America.

Sondergeld, C.H., Ambrose, R.J., Rai, C.S. et al. (2010). Micro-structural studies of gas shales. *SPE Unconventional Gas Conference*, Pittsburgh, Pennsylvania, USA (23–25 February 2010). SPE-131771-MS. https://doi.org/10.2118/131771-MS.

Tian, Y. (2014). Experimental study on stress sensitivity of naturally fractured reservoirs. *SPE International Student Paper Contest at the SPE Annual Technical Conference and Exhibition*, Amsterdam, The Netherlands (27–29 October 2014). SPE-173463-STU. https://doi.org/10.2118/173463-STU.

Ulm, F.-J. and Abousleiman, Y. (2006). The nanogrannular nature of shale. *Acta Geotechnica* 1 (2): 77–88.

Wang, H. (2017). What factors control shale-gas production and production-decline trend in fractured systems: a comprehensive analysis and investigation. *SPE Journal* 22 (2): 562–581. SPE-179967-PA. https://doi.org/10.2118/179967-PA.

Zeszotarski, J.C., Chromik, R.R., Vinci, R.P. et al. (2004). Imaging and mechanical property measurements of kerogen via nanoindentation. *Geochimca et Cosmochimica Acta* 68: 4113–4119. https://doi.org/10.1016/j.gca.2003.11.031.

Zhang, S., Byrnes, A.P., Jankovic, J. et al. (2019). Management, analysis, and simulation of micrographs with cloud computing. *Microscopy Today* 29 (2): 26–33.

Part III

Large-Scale Petrophysics

12

Effective Permeability in Fractured Reservoirs

Percolation-Based Effective-Medium Theory

Behzad Ghanbarian

Porous Media Research Lab, Department of Geology, Kansas State University, Manhattan, KS, USA

12.1 Introduction

Geologic formations are multi-scale and multi-physical systems. Shale and tight reservoirs, as new resources of hydrocarbon and energy supplies, are among notable examples (Wang et al. 2017). However, their low porosity and permeability values make them distinct from other conventional reservoirs, such as carbonates and sandstones. With recent progress in hydraulic fracturing and horizontal drilling in the past decade (Friehauf et al. 2009; Malhotra et al. 2014; Saini et al. 2020), unconventional reservoirs became one of the main energy suppliers in the US. However, modeling fluid flow and transport in such multi-scale systems still requires investigations due to their complex structures. This is because naturally or hydraulically fractured reservoirs are composed of matrix and fracture networks. The matrix in shale reservoirs consists of pores with sizes of a few to several hundred nanometers (Nelson 2009; Rijfkogel et al. 2019), while the fracture network may contain fractures of aperture with sizes of a few microns to several hundred microns (Gale et al. 2007; Olson et al. 2009).

Unconventional reservoir rocks are spatially and temporally heterogeneous at both discrete and continuum scales (Hyman et al. 2016). At the discrete scale, heterogeneity is due to spatial variation in pore size, pore connectivity, surface roughness, and mineralogy. At the continuum scale, the position and orientation of fractures and faults cause spatial heterogeneity. Such small- and large-scale heterogeneities make the flow and transport modeling challenging in unconventional reservoirs. In the literature, various approaches, including theoretical (Ebigbo et al. 2016; Fokker 2001; Hunt et al. 2014; Sævik et al. 2013) and numerical (Bogdanov et al. 2007; Chen et al. 2019; Hyman et al. 2018; Lang et al. 2014) methods, have been developed to model fluid flow and transport in naturally fractured geologic formations or hydraulically fractured reservoirs. In what follows, we briefly describe two theoretic approaches used to determine effective permeability in porous media.

12.1.1 Percolation Theory

Percolation theory is a theoretical method from statistical physics to study the effect of heterogeneity and connectivity on flow and transport in porous media. Within percolation theory, there exist two frameworks: insulator-conductor and conductor-superconductor (Stauffer and Aharony

Physics of Fluid Flow and Transport in Unconventional Reservoir Rocks, First Edition.
Edited by Behzad Ghanbarian, Feng Liang, and Hui-Hai Liu.

1994). In the insulator-conductor framework, a system is composed of insulating and conducting components with zero and finite conductivities, respectively. In conductor-superconductor framework, however, a system consists of finite-conductivity and infinite-conductivity components.

In percolation theory, a key feature is the presence of a percolation threshold at which an infinite spanning cluster emerges along with many finite clusters. Below the threshold, all clusters are finite in size, and there is no spanning (infinite) cluster, while above it, there are still finite clusters along with the infinite cluster. Above but near the percolation threshold, transport modes, such as effective permeability, conform to the following power-law scaling

$$k_{\mathrm{eff}} \propto (p - p_{\mathrm{c}})^{t}, \quad p > p_{\mathrm{c}} \tag{12.1}$$

where p is the occupation probability and p_{c} is the percolation threshold. In very homogeneous media with negligible local fluctuation in conductivity, the scaling exponent t would be independent of the structure and geometry of a system, a feature called universality (Hunt et al. 2014). The universal value of t is 1.3 in two-dimensional and 2 in three-dimensional systems constructed of one insulating component and one conducting component (Stauffer and Aharony 1994). The value $t = 2$ in three dimensions incorporates the effects of connectivity and tortuosity (Ghanbarian et al. 2013). In heterogeneous systems with broad conductance distributions, one should expect $t > 2$ (Feng et al. 1987). However, in correlated systems, Sahimi (1994) and Sahimi and Mukhopadhyay (1996) showed that the value of t could be less than 2.

As stated earlier, Eq. (12.1) is theoretically valid in the region above and near the percolation threshold. However, evidence from experiments and numerical simulations indicate that it is valid over a much broader region above the threshold (Hunt et al. 2014). To apply Eq. (12.1) to porous media, such as rocks and fracture networks, one needs to replace p by volumetric fraction of pores or fractures and p_{c} by critical volume of pores or fractures (Balberg 1987; Hunt et al. 2014).

Within the conductor-superconductor framework, the effective permeability follows a power law similar to Eq. (12.1). However, its exponent is denoted by s, and its value in three dimensions is 0.76 (Bergman and Stroud 1992). However, a range from 0.37 to 1.30 has been reported in the literature (McLachlan et al. 2007; Youngs 2002).

It is difficult to address the effective permeability of fractured reservoirs within the context of percolation theory, either in the insulator-conductor or the conductor-superconductor framework. In what follows, we briefly review concepts of effective-medium theory, another technique from statistical physics, that can be used to scale up effective permeability in porous media.

12.1.2 Effective-Medium Theory

Effective-medium theory (EMT) is another technique from statistical physics in which a heterogeneous formation is replaced by a homogeneous one whose effective permeability is the same as the actual permeability of the heterogeneous formation (David et al. 1990; Doyen 1988; Koplik et al. 1984). Replacing the heterogeneous system with an equivalent homogeneous one causes local perturbations because the local permeability is different than the effective permeability. The effective permeability of a formation can accordingly be calculated by forcing the average of these local perturbations to be equal to zero (Ghanbarian and Daigle 2016; Kirkpatrick 1973):

$$\int \frac{k - k_{\mathrm{eff}}}{k + \left[\frac{Z}{2} - 1\right] k_{\mathrm{eff}}} f(k) \mathrm{d}k = 0 \tag{12.2}$$

where k is the local permeability, $f(k)$ is the permeability distribution, and Z is the average coordination number. Eq. (12.2) with $Z = 4$ in two dimensions and 6 in three dimensions reduces to the self-consistent approximation.

In a formation composed of two components, e.g. matrix and fracture, Eq. (12.2) reduces to

$$f_{\mathrm{m}}\frac{k_{\mathrm{m}}-k_{\mathrm{eff}}}{k_{\mathrm{m}}+\left[\frac{Z}{2}-1\right]k_{\mathrm{eff}}}+(1-f_{\mathrm{m}})\frac{k_{\mathrm{f}}-k_{\mathrm{eff}}}{k_{\mathrm{f}}+\left[\frac{Z}{2}-1\right]k_{\mathrm{eff}}}=0 \tag{12.3}$$

where f_{m} and $(1 - f_{\mathrm{m}})$ represent the matrix and fracture fractions, respectively, k_{m} is the matrix permeability, and k_{f} is the fracture permeability.

Results of effective-medium theory are reliable in systems with small variances (Adler and Berkowitz 2000; Ghanbarian and Daigle 2016; Ghanbarian and Javadpour 2017; Ghanbarian et al. 2016; Sahimi 2011). Therefore, effective-medium theory is not applicable to highly heterogeneous formations with large local permeability variations.

12.2 Objectives

Percolation theory provides a general theoretic framework to calculate the effective permeability in homogeneous and highly heterogeneous media where the ratio of one component conductivity to another one is infinitely large (e.g. insulator-conductor or conductor-superconductor systems). Although effective-medium theory is applicable to systems where more than one permeable component exists, it is restricted to homogeneous and relatively heterogeneous media with small local fluctuations. In addition, it is well documented in the literature that power laws from percolation theory are theoretically valid above and near the percolation threshold (Hunt et al. 2014; Stauffer and Aharony 1994), whereas effective-medium theory results are accurate far above the threshold (Sahimi 2003, 2011). The main objectives of this chapter are to (i) present applications of percolation-based effective-medium theory to model effective permeability in matrix-fracture systems and (ii) compare theoretical estimations of k_{eff} with two- and three-dimensional numerical simulations.

12.3 Percolation-Based Effective-Medium Theory

By combining concepts of percolation theory, Eq. (12.1), and the two-component model of effective-medium theory, Eq. (12.3), McLachlan (1987) proposed percolation-based effective-medium theory (P-EMT). Adopting his approach for fractured reservoirs composed of matrix and fractures results in

$$(1-\rho)\frac{k_{\mathrm{m}}^{1/t}-k_{\mathrm{eff}}^{1/t}}{k_{\mathrm{m}}^{1/t}+\left[\frac{1-\rho_{\mathrm{c}}}{\rho_{\mathrm{c}}}\right]k_{\mathrm{eff}}^{1/t}}+\rho\frac{k_{\mathrm{f}}^{1/t}-k_{\mathrm{eff}}^{1/t}}{k_{\mathrm{f}}^{1/t}+\left[\frac{1-\rho_{\mathrm{c}}}{\rho_{\mathrm{c}}}\right]k_{\mathrm{eff}}^{1/t}}=0 \tag{12.4}$$

where ρ is the fracture density, and t is the scaling exponent from percolation theory. ρ_{c} is the critical fracture density below which k_{eff} is mainly controlled by the matrix and its permeability k_{m}. Above ρ_{c}, however, it is the fracture network that controls k_{eff}. Note that if $k_{\mathrm{m}} = k_{\mathrm{f}}$, then one should expect $k_{\mathrm{eff}} = k_{\mathrm{m}} = k_{\mathrm{f}}$. This criterion is well met by Eq. (12.4) because in such a case the terms $k_{\mathrm{m}}^{1/t} - k_{\mathrm{eff}}^{1/t}$ and $k_{\mathrm{f}}^{1/t} - k_{\mathrm{eff}}^{1/t}$ would be equal to zero.

In Eq. (12.4), k_{eff} is implicitly explained in terms of fracture density ρ. Rewriting Eq. (12.4) yields

$$\rho = \frac{\left(k_{\mathrm{eff}}^{1/t} - k_{\mathrm{m}}^{1/t}\right)\left(\rho_{\mathrm{c}} k_{\mathrm{f}}^{1/t} + (1-\rho_{\mathrm{c}}) k_{\mathrm{eff}}^{1/t}\right)}{\left(k_{\mathrm{f}}^{1/t} - k_{\mathrm{m}}^{1/t}\right) k_{\mathrm{eff}}^{1/t}} \tag{12.5}$$

Recently, Sadeghi et al. (2018) proposed an explicit form of the percolation-based effective-medium theory as follows:

$$k_{\mathrm{eff}} = \left[b_1 + b_2\rho + b_2\sqrt{b_3 + 2b_1 b_2^{-1}\rho + \rho^2}\right]^t \tag{12.6a}$$

in which

$$b_1 = -\frac{\rho_{\mathrm{c}} k_{\mathrm{f}}^{1/t} + (1-\rho_{\mathrm{c}}) k_{\mathrm{m}}^{1/t}}{2(1-\rho_{\mathrm{c}})} \tag{12.6b}$$

$$b_2 = \frac{k_{\mathrm{f}}^{1/t} - k_{\mathrm{m}}^{1/t}}{2(1-\rho_{\mathrm{c}})} \tag{12.6c}$$

$$b_3 = \frac{\left[\rho_{\mathrm{c}} k_{\mathrm{f}}^{1/t} - (1-\rho_{\mathrm{c}}) k_{\mathrm{m}}^{1/t}\right]^2 + 4(1-\rho_{\mathrm{c}}) k_{\mathrm{f}}^{1/t} k_{\mathrm{m}}^{1/t}}{k_{\mathrm{f}}^{1/t} - k_{\mathrm{m}}^{1/t}} \tag{12.6d}$$

To show the sensitivity of percolation-based effective-medium theory to parameters ρ_{c} and t, we presumed a formation with $k_{\mathrm{f}} = 10^{-17}\ \mathrm{m}^2$ (10 μD) and $k_{\mathrm{m}} = 10^{-20}\ \mathrm{m}^2$ (10 nD) and varied the values of ρ_{c} and t. Figure 12.1 shows the effective permeability against the fracture density on both semi-log and log–log scales for two cases: (i) the critical fracture density ρ_{c} (Figures 12.1a and b) and (ii) 12.2 the scaling exponent t (Figures 12.1c and d).

In Figures 12.1a and b, we set $t = 1$ but varied ρ_{c} from 0 to 0.1 ($\rho_{\mathrm{c}} = 0$, 0.01, 0.05, and 0.1). Figure 12.1b clearly shows that the trend between k_{eff} and ρ on the logarithmic scale seems to be power law at ρ values near 1. Figure 12.1c presents the results for $\rho_{\mathrm{c}} = 0.2$ and different t values ranged from 0.5 to 2 ($t = 0.5$, 1, 1.5, and 2). As can be seen, the shape of the effective permeability and the slope of the $\rho_{\mathrm{c}} - k_{\mathrm{eff}}$ curve over the entire range of fracture density are controlled by the value of t (Figure 12.1d). The smaller the exponent t, the sharper the variation in k_{eff}.

12.4 Comparison with Simulations

In this section, we compare estimations by P-EMT with simulations reported in the literature as well as those performed in this study. In the following, we first compare theoretical estimations with two- and three-dimensional simulations of Chen et al. (2019). Next, estimations by P-EMT are compared with our three-dimensional numerical simulations.

12.4.1 Chen et al. (2019)

In the Chen et al. (2019) study, discrete fracture-matrix networks were composed of lines and elliptical fractures, respectively, in two and three dimensions. Chen et al. (2019) adopted a generalized lattice Boltzmann (LB) method to simulate fluid flow and to determine the effective permeability k_{eff} in such networks. In what follows, we briefly describe their simulations in two and three dimensions.

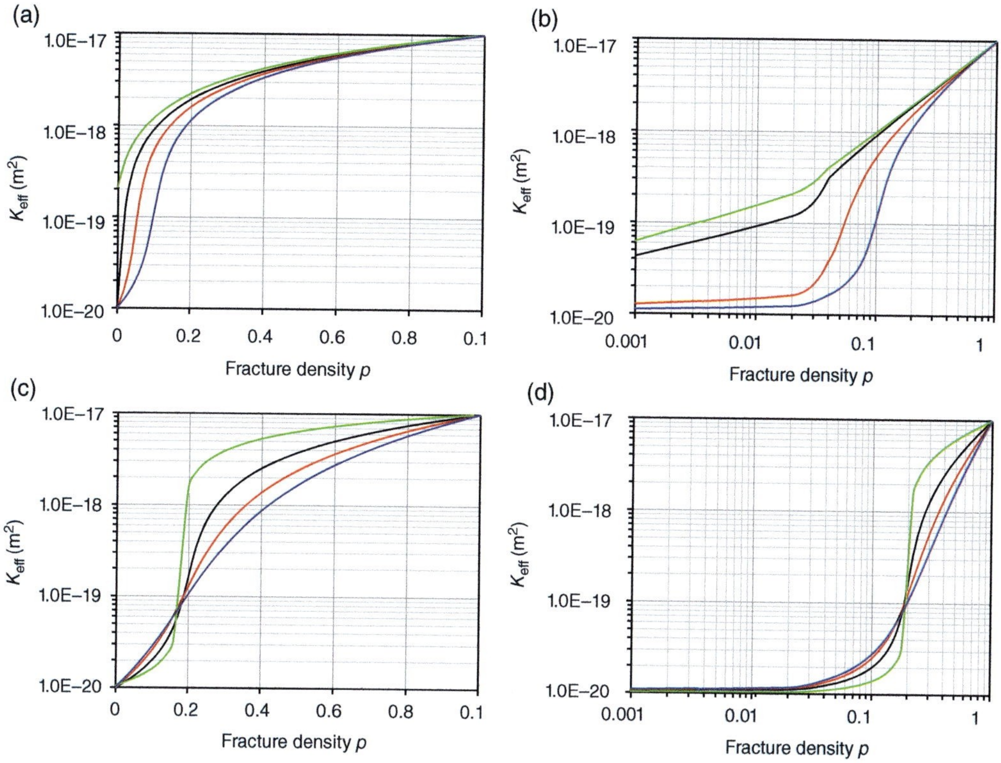

Figure 12.1 Effective permeability k_{eff} versus fracture density ρ determined by percolation-based effective-medium theory (P-EMT). (a) Eq. (12.6) with $k_f = 10^{-17}$ m², $k_m = 10^{-20}$ m², $t = 1$ and $\rho_c = 0, 0.01, 0.05, 0.1$ respectively represented by green, black, red, and blue, (b) the same as (a) but on log–log scale, (c) Eq. (12.6) with $k_f = 10^{-17}$ m², $k_m = 10^{-20}$ m², $\rho_c = 0.2$ and $t = 0.5, 1, 1.5$, and 2 respectively represented by green, black, red, and blue, and (d) the same as (c) but on log–log scale.

12.4.1.1 Two-Dimensional Simulations

Chen et al. (2019) represented a two-dimensional matrix-fracture system by a square domain of size L (= 200 lattices or 4 cm). The domain was initially composed of a matrix only with permeability $k_m = 10^{-20}$ m² (10 nD). First, two random numbers were generated based on a uniform distribution to determine the coordinates of two points in the domain. Next, a fracture with the width of one lattice unit and permeability of 3.33×10^{-9} m² (3.33×10^3 D; based on the cubic law) was created by connecting the two points via a straight line. Using the same approach, fractures were added to the domain until the desired fracture density was reached. In their study, the fracture density was defined as the ratio of the total fracture length to the entire area of the domain

$$\rho = \frac{L_f}{L^2} \tag{12.7}$$

where L is the domain size, and L_f is the total fracture length.

The matrix permeability was set equal to 10^{-20} m² (10 nD), while the fracture permeability k_f varied between 1×10^{-17} m² (10 μD) and 1×10^{-14} (10 mD). The generalized LB method of Guo and Zhao (2002), in which the generalized Navier–Stokes equation is solved, was applied to simulate fluid flow along one direction using a pressure drop of 2×10^{-6} (lattice units).

The implemented pressure drop was low enough to guarantee laminar flow. Chen et al. (2019) applied the periodic boundary conditions and the Darcy equation to determine k_{eff}.

In Figure 12.2, we show two-dimensional simulations from Chen et al. (2019) for four different matrix-fracture systems with $\log(k_f/k_m) = 3$, 4, 5, and 6. The theoretical estimations of k_{eff} by P-EMT are also presented in this figure. We found that Eq. (12.6) with $t = 1$ and $\rho_c = 0$ accurately estimates the k_{eff} at different fracture densities. The average relative error is less than 23% and the calculated RMSLE values are smaller than 0.099 indicating well agreement between the theoretical estimations and the numerical simulations in two dimensions.

The scaling exponent $t = 1$ is in well agreement with the exponents 0.79 (for $\log(k_f/k_m) = 3$), 0.92 (for $\log(k_f/k_m) = 4$), 0.94 (for $\log(k_f/k_m) = 5$), and 0.95 (for $\log(k_f/k_m) = 6$) derived by Chen et al. (2019) who fitted a power law to their $\rho_c - k_{eff}$ simulations ($\rho > 0.01$). We should point out that the

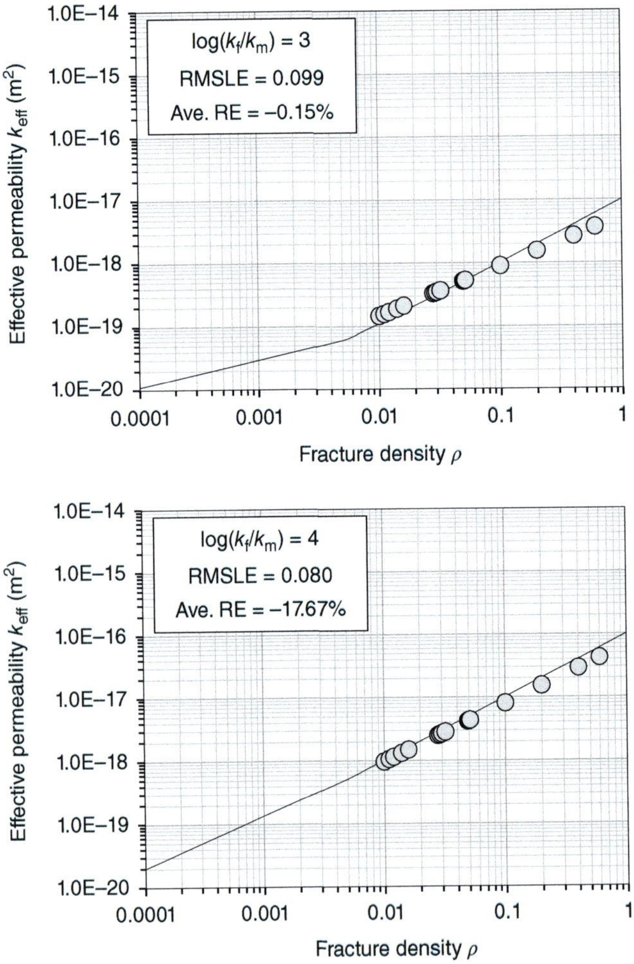

Figure 12.2 Effective permeability simulated by the lattice-Boltzmann method (filled circles) and estimated by the percolation-based effective-medium theory (solid line) versus fracture density. Simulations were carried out on two-dimensional domains with $\log(k_f/k_m) = 3$, 4, 5, and 6. For theoretical estimations by percolation-based effective-medium theory and Eq. (12.6), we used $t = 1$, $\rho_c = 0$, $k_m = 10^{-20}$ m^2, and $k_f = 10^{-17}$, 10^{-16}, 10^{-15}, and 10^{-14} m^2. *Source:* Data adapted from Chen et al. (2019).

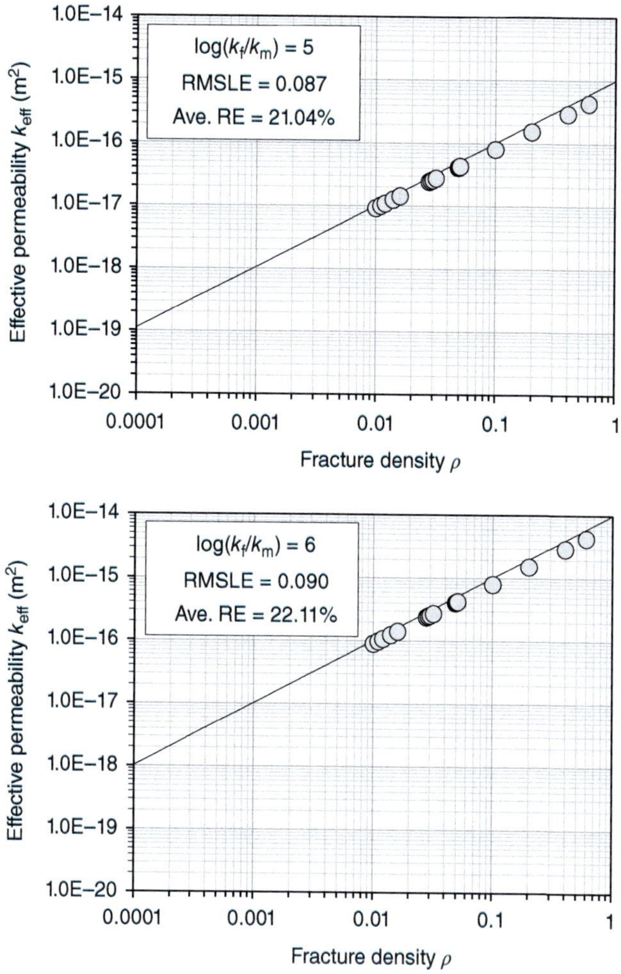

Figure 12.2 (Continued)

universal scaling exponent t = 1.3 in two dimensions from percolation theory (Stauffer and Aharony 1994) resulted in less accurate estimations of k_{eff} than $t = 1$ (results not shown). Note that P-EMT (Eq. 12.4) with t = 1 and $\rho_c = 2/Z$ reduces to the effective-medium theory (Eq. 12.3). Although having $t = 1$ is consistent with the effective-medium theory, $\rho_c = 0$ means $Z = \infty$ within the EMT framework, which is not physically supported. The value $\rho_c = 0$ is in accord with the simulations since Chen et al. (2019) included a boundary region around the fractures, so they did not percolate through the domain.

12.4.1.2 Three-Dimensional Simulations

Chen et al. (2019) generated the matrix-fracture system in three dimensions using the three-dimensional DFN (Hyman et al. 2014, 2015) based on which individual fractures are represented as planar polygons. In the DFN approach, fractures are assigned a shape, location, width, and orientation. In the study of Chen et al. (2019), the fracture network was composed of 855 elliptical fractures. The radius of fractures followed a truncated power-law distribution with lower and upper

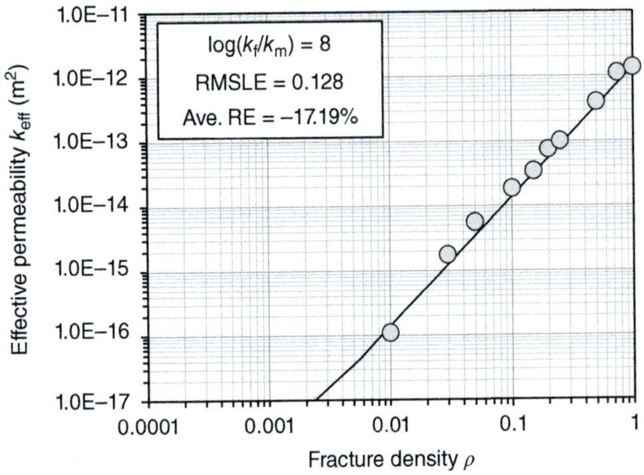

Figure 12.3 Effective permeability simulated by the lattice-Boltzmann method (filled circles) and estimated by the percolation-based effective-medium theory (solid line) versus fracture density. Simulations were carried out on three-dimensional domains with $\log(k_f/k_m) = 8$. For theoretical estimations by percolation-based effective-medium theory and Eq. (12.6), we used $t = 2$, $\rho_c = 0$, $k_f = 1.35 \times 10^{-12}$ m^2, and $k_m = 1.35 \times 10^{-20}$ m^2. *Source:* Data adapted from Chen et al. (2019).

cutoffs $r_{min} = 0.1$ μm and $r_{max} = 1$ μm and exponent 2.28. The fracture orientation was drawn from a Fisher distribution, and the fracture location was uniformly distributed through the domain. Chen et al. (2019) defined fracture density in three dimensions as the ratio of the total fracture area to the entire matrix-fracture system volume. After generating the three-dimensional matrix-fracture systems using the stochastic discrete fracture-matrix networks, the generalized LB method of Guo and Zhao (2002) was used to simulate fluid flow in the matrix-fracture systems.

Figure 12.3 shows the simulated effective permeability against the fracture density. We also show the k_{eff} estimated by P-EMT. As can be seen, Eq. (12.6) with $t = 2$, $\rho_c = 0$, $k_f = 1.35 \times 10^{-12}$ m^2, and $k_m = 1.35 \times 10^{-20}$ m^2 accurately estimates the effective permeability with RMSLE = 0.128 and average relative error = −17.19%. Although the negative average relative error indicates that, on average, P-EMT underestimates the k_{eff}, there is good agreement between the theory and the simulations (Figure 12.3). The exponent $t = 2$ is consistent with the universal scaling exponent in three dimensions from percolation theory (Stauffer and Aharony 1994). The critical fracture density $\rho_c = 0$ is also in accord with the two-dimensional estimations by Eq. (12.6) shown in Figure 12.2.

12.4.2 New Three-Dimensional Simulations

In this study, we generated three three-dimensional matrix-fracture systems using cubic domains of size of 2 cm and the DFNWORKS software package (Hyman et al. 2015). Within the DFN framework, individual fractures were represented by a single family of planar discs with a constant radius of 5 m within a 50 m cubic domain, which is different from the three-dimensional simulations of Chen et al. (2019). The orientation of fractures was sampled from a Fisher distribution with an intensity parameter of 0.1 that resulted in a uniform covering of the unit sphere and equiprobable random orientations. We considered networks composed of 50, 100, 200, 267, 300, 400, and 500 fractures. At a count of 267, the fracture network percolated through the domain.

After generating the three-dimensional fracture networks, we created an octree-refined volume mesh using the upscaled discrete fracture matrix (UDFM) method presented by Sweeney et al. (2020). In the UDFM, the mesh surrounding fractures was refined for a better representation of the network and fewer false connections between fractures. The spatially variable mesh was composed of tetrahedron, but we computed the dual mesh, which is composed of three-dimensional Voronoi control volumes. If a fracture intersected a control volume, then it was considered as a fracture cell. If not, then it was tagged as a matrix cell. The value of matrix permeability varied from one system to another ($k_m = 10^{-12}$, 10^{-14}, and 10^{-16} m^2), while the fracture permeability was fixed at 10^{-10} m^2. The outlet flow rate was calculated using the flow solver PFLOTRAN (Lichtner et al. 2015), which solves the pressure Poisson equation. Pressures 2 and 1 MPa were applied as boundary conditions at the two sides of the domains. The effective permeability for each network was determined using the Darcy equation and the numerical simulations under steady-state conditions.

Figure 12.4 shows the simulated effective permeability against the fracture density for the three matrix-fracture systems studied here. We also display the estimated values by P-EMT, Eq. (12.6) with $t = 2$, $\rho_c = 0.1$, $k_f = 10^{-10}$ m^2, and $k_m = 10^{-12}$, 10^{-14}, and 10^{-16} m^2. As can be seen, there exists a good agreement between the simulations and the theory. The average relative error ranged from -50% for $\log(k_f/k_m) = 6$ to 3% for $\log(k_f/k_m) = 2$. We found RMSLE = 0.025, 0.234, and 0.526, respectively, for systems with $\log(k_f/k_m) = 2$, 4, and 6. Although the value of RMSLE increases from $\log(k_f/k_m) = 2$ to 6, the obtained results are promising. Our k_{eff} estimations are based on first principles from percolation-based effective-medium theory without any ad hoc parameters.

The value $t = 2$ is in accord with the universal scaling exponent in three dimensions from percolation theory (Stauffer and Aharony 1994). It is also consistent with our results obtained from three-dimensional simulations of Chen et al. (2019) presented in Figure 12.3. Regarding the critical fracture density, the value $\rho_c = 0.1$ is very close to the critical volume fraction of 0.114 reported by Clark et al. (2020), who investigated the effective permeability in irregular composite porous media composed of two distinct materials with two different permeability values.

12.5 Conclusion

We proposed applications of percolation-based effective-medium theory (P-EMT) to model the effective permeability in fractured reservoirs where the matrix and fracture network contribute to fluid flow. The proposed model estimates the k_{eff} from matrix permeability, k_m, fracture permeability, k_f, fracture density, ρ, critical fracture density, ρ_c, and scaling exponent, t. To evaluate P-EMT, theoretical estimations were compared with two- and three-dimensional numerical simulations. Results showed good agreement between the theory and the simulations. We found that the scaling exponent $t = 1$ resulted in accurate estimations of k_{eff} in two dimensions, while $t = 2$ led to good agreement between the theory and simulations in three dimensions. In contrast to other empirical and semi-empirical models, our proposed k_{eff} model based on percolation-based effective-medium theory has no ad hoc parameters and is predictive once the parameters t, ρ_c, k_m, and k_f are known for fractured reservoirs. Further investigations are still required to better understand how the scaling exponent t and critical fracture density ρ_c may vary from one geologic formation to another.

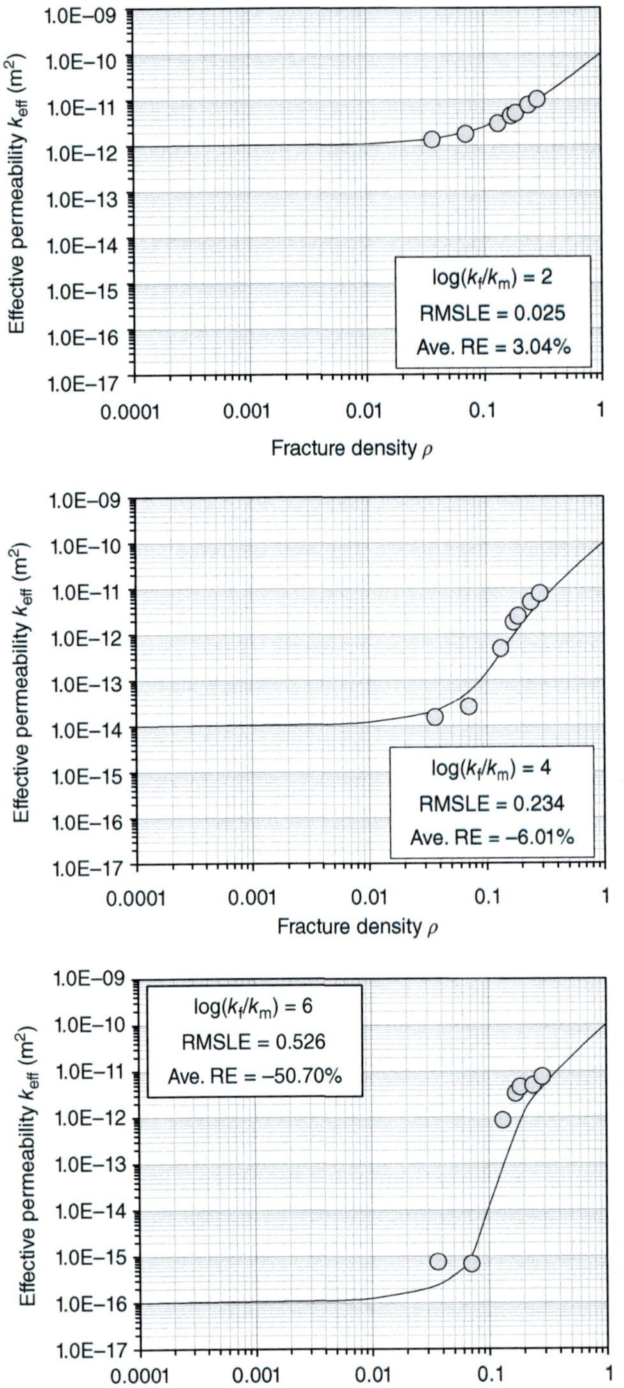

Figure 12.4 Effective permeability simulated by the numerical method (filled circles) and estimated by the percolation-based effective-medium theory (solid line) versus fracture density. Simulations were carried out on three-dimensional domains with $\log(k_f/k_m) = 2, 4$, and 6. For theoretical estimations by percolation-based effective-medium theory and Eq. (12.6), we used $t = 2$, $\rho_c = 0.1$, $k_f = 10^{-10}$ m^2, and $k_m = 10^{-12}$, 10^{-14}, and 10^{-16} m^2.

Acknowledgment

The author is grateful to Jeffrey D. Hyman, Los Alamos National Laboratory, for providing the numerical simulations used in this study, and to Kansas State University for the support through the faculty start-up fund.

References

Adler, P.M. and Berkowitz, B. (2000). Effective medium analysis of random lattices. *Transport in Porous Media* 40 (2): 145–151. https://doi.org/10.1023/A:1006611011600.

Balberg, I. (1987). Recent developments in continuum percolation. *Philosophical Magazine B: Physics of Condensed Matter: Statistical Mechanics, Electronic, Optical and Magnetic Properties* 56 (6): 991–1003. https://doi.org/10.1080/13642818708215336.

Bergman, D.J. and Stroud, D. (1992). Physical properties of macroscopically inhomogeneous media. *Solid State Physics* 46: 147–269.

Bogdanov, I.I., Mourzenko, V.V., and Thovert, J. (2007). Effective permeability of fractured porous media with power-law distribution of fracture sizes. *Physical Review E* 76: 1–15. https://doi.org/10.1103/PhysRevE.76.036309.

Chen, L., Hyman, J.D., Lei, Z. et al. (2019). Effect of fracture density on effective permeability of matrix-fracture system in shale formations. In: *Geological Carbon Storage: Subsurface Seals and Caprock Integrity*, vol. 238 (ed. S. Vialle, J. Ajo-Franklin and J.W. Carey), 137–146. AGU Geophysical Monograph.

Clark, C.L., Winter, C.L., and Corley, T. (2020). Effects of percolation on the effective conductivity of irregular composite porous media. *Advances in Water Resources* 137: 103507.

David, C., Guéguen, Y., and Pampoukis, G. (1990). Effective medium theory and network theory applied to the transport properties of rock. *Journal of Geophysical Research, Solid Earth* 95 (B5): 6993–7005.

Doyen, P.M. (1988). Permeability, conductivity, and pore geometry of sandstone. *Journal of Geophysical Research* 93 (B7): 7729. https://doi.org/10.1029/JB093iB07p07729.

Ebigbo, A., Lang, P.S., Paluszny, A., and Zimmerman, R.W. (2016). Inclusion-based effective medium models for the permeability of a 3D fractured rock mass. *Transport in Porous Media* 113 (1): 137–158. https://doi.org/10.1007/s11242-016-0685-z.

Feng, S., Halperin, B.I., and Sen, P.N. (1987). Transport properties of continuum systems near the percolation threshold. *Physical Review B* 35 (1): 197–214. https://doi.org/10.1103/PhysRevB.35.197.

Fokker, P.A. (2001). General anisotropic effective medium theory for the effective permeability of heterogeneous reservoirs. *Transport in Porous Media* 44 (2): 205–218. https://doi.org/10.1023/A:1010770623874.

Friehauf, K.E., Sharma, M.M., and Sullivan, R.B. (2009). Application of a new compositional model for hydraulic fracturing with energized fluids: a south Texas case study. Paper presented at the SPE Hydraulic Fracturing Technology Conference, The Woodlands, Texas (19 January), SPE-119265-MS. Society of Petroleum Engineers.

Gale, J.F., Reed, R.M., and Holder, J. (2007). Natural fractures in the Barnett Shale and their importance for hydraulic fracture treatments. *AAPG Bulletin* 91: 603–622.

Ghanbarian, B. and Daigle, H. (2016). Permeability in two-component porous media: effective-medium approximation compared with lattice-Boltzmann simulations. *Vadose Zone Journal* 15 (2): https://doi.org/10.2136/vzj2015.05.0071.

Ghanbarian, B. and Javadpour, F. (2017). Upscaling pore pressure-dependent gas permeability in shales. *Journal of Geophysical Research, Solid Earth* 122 (4): 2541–2552. https://doi.org/10.1002/2016JB013846.

Ghanbarian, B., Hunt, A.G., Sahimi, M. et al. (2013). Percolation theory generates a physically based description of tortuosity in saturated and unsaturated porous media. *Soil Science Society of America Journal* 77 (6): 1920. https://doi.org/10.2136/sssaj2013.01.0089.

Ghanbarian, B., Sahimi, M., and Daigle, H. (2016). Modeling relative permeability of water in soil: application of effective-medium approximation and percolation theory. *Water Resources Research* 52 (7): 5025–5040. https://doi.org/10.1002/2015WR017903.

Guo, Z. and Zhao, T.S. (2002). Lattice Boltzmann model for incompressible flows through porous media. *Physical Review E* 66: 036304.

Hunt, A., Ewing, R., and Ghanbarian, B. (2014). *Percolation Theory for Flow in Porous Media*. Springer.

Hyman, J.D., Gable, C.W., Painter, S.L., and Makedonska, N. (2014). Conforming Delaunay triangulation of stochastically generated three dimensional discrete fracture networks: a feature rejection algorithm for meshing strategy. *SIAM Journal on Scientific Computing* 36: A1871–A1894.

Hyman, J.D., Karra, S., Makedonska, N. et al. (2015). DFNWorks: a discrete fracture network framework for modeling subsurface flow and transport. *Computers & Geosciences* 84: 10–19.

Hyman, J.D., Jiménez-Martínez, J., Viswanathan, H.S. et al. (2016). Understanding hydraulic fracturing: a multi-scale problem. *Philosophical Transactions of the Royal Society A: Mathematical, Physical and Engineering Sciences* 374: 20150426.

Hyman, J.D., Karra, S., Carey, J.W. et al. (2018). Discontinuities in effective permeability due to fracture percolation. *Mechanics of Materials* 119: 25–33.

Kirkpatrick, S. (1973). Percolation and conduction. *Reviews of Modern Physics* 45: 574–588.

Koplik, J., Lin, C., and Vermette, M. (1984). Conductivity and permeability from microgeometry. *Journal of Applied Physics* 56 (11): 3127–3131. https://doi.org/10.1063/1.333872.

Lang, P.S., Paluszny, A., and Zimmerman, R.W. (2014). Permeability tensor of three-dimensional fractured porous rock and a comparison to trace map predictions. *Journal of Geophysical Research, Solid Earth* 119: 6288–6307.

Lichtner, P., Hammond, G., Lu, C. et al. (2015). PFLOTRAN User Manual: A Massively Parallel Reactive Flow and Transport Model for Describing Surface and Subsurface Processes. *Technical Report (Report No.: LA-UR-15-20403).* Los Alamos National Laboratory.

Malhotra, S., Lehman, E.R., and Sharma, M.M. (2014). Proppant placement using alternate-slug fracturing. *SPE Journal* 19: 974–985.

McLachlan, D.S. (1987). An equation for the conductivity of binary mixtures with anisotropic grain structures. *Journal of Physics C: Solid State Physics* 20 (7): 865–877. https://doi.org/10.1088/0022-3719/20/7/004.

McLachlan, D.S., Sauti, G., and Chiteme, C. (2007). Static dielectric function and scaling of the ac conductivity for universal and nonuniversal percolation systems. *Physical Review B* 76: 014201.

Nelson, P.H. (2009). Pore-throat sizes in sandstones, tight sandstones, and shales. *AAPG Bulletin* 93 (3): 329–340. https://doi.org/10.1306/10240808059.

Olson, J.E., Laubach, S.E., and Lander, R.H. (2009). Natural fracture characterization in tight gas sandstones: integrating mechanics and diagenesis. *AAPG Bulletin* 93: 1535–1549.

Rijfkogel, L.S., Ghanbarian, B., Hu, Q., and Liu, H.H. (2019). Clarifying pore diameter, pore width, and their relationship through pressure measurements: a critical study. *Marine and Petroleum Geology* 107: 142–148.

Sadeghi, M., Ghanbarian, B., and Horton, R. (2018). Derivation of an explicit form of the percolation-based effective-medium approximation for thermal conductivity of partially saturated soils. *Water Resources Research* https://doi.org/10.1002/2017WR021714.

Sævik, P.N., Berre, I., Jakobsen, M., and Lien, M. (2013). A 3D computational study of effective medium methods applied to fractured media. *Transport in Porous Media* 100: 115–142.

Sahimi, M. (1994). Long-range correlated percolation and flow and transport in heterogeneous porous media. *Journal de Physique I* 4: 1263–1268.

Sahimi, M. (2003). *Heterogeneous Materials I: Linear Transport and Optical Properties*. New York: Springer-Verlag.

Sahimi, M. (2011). *Flow and Transport in Porous Media and Fractured Rock: From Classical Methods to Modern Approaches*, 2e. Weinheim, Germany: Wiley-VCH.

Sahimi, M. and Mukhopadhyay, S. (1996). Scaling properties of a percolation model with long-range correlations. *Physical Review. E, Statistical Physics, Plasmas, Fluids, and Related Interdisciplinary Topics* 54 (4): 3870–3880. https://doi.org/10.1103/PhysRevE.54.3870.

Saini, G.S., Hender, D., James, C. et al. (2020). An automated physics-based workflow for identification and classification of drilling dysfunctions drives drilling efficiency and transparency for completion design. Paper presented at the SPE Canada Unconventional Resources Conference, Virtual (24 September), SPE-200006-MS. Society of Petroleum Engineers.

Stauffer, D. and Aharony, A. (1994). *Introduction to Percolation Theory*. London: Taylor and Francis.

Sweeney, M.R., Gable, C.W., Karra, S. et al. (2020). Upscaled discrete fracture matrix model (UDFM): an octree-refined continuum representation of fractured porous media. *Computational Geosciences* 24: 293–310.

Wang, L., Wang, S., Zhang, R. et al. (2017). Review of multi-scale and multi-physical simulation technologies for shale and tight gas reservoirs. *Journal of Natural Gas Science and Engineering* 37: 560–578. https://doi.org/10.1016/j.jngse.2016.11.051.

Youngs, I.J. (2002). Exploring the universal nature of electrical percolation exponents by genetic algorithm fitting with general effective medium theory. *Journal of Physics D: Applied Physics* 35 (23): 3127–3137. https://doi.org/10.1088/0022-3727/35/23/314.

13

Modeling of Fluid Flow in Complex Fracture Networks for Shale Reservoirs

Hongbing Xie[1], Xiaona Cui[1], Wei Yu[1,2], Chuxi Liu[2], Jijun Miao[1], and Kamy Sepehrnoori[2]

[1] *SimTech LLC, Katy, TX, USA*
[2] *The University of Texas at Austin, Hildebrand Department of Petroleum and Geosystems Engineering, Austin, TX, USA*

13.1 Shale Reservoirs with Complex Fracture Networks

Shale oil and gas reservoirs with extremely low porosity and permeability and natural fractures are located all over the world, as displayed in Figure 13.1. Recently, advances in horizontal drilling and multi-stage hydraulic fracturing have enabled rapid economic developments of shale gas and oil reservoirs. It has been reported that shale gas production in 2018 accounted for 70% of total US natural gas production (US Energy Information Administration 2019). During the hydraulic fracturing process, the pre-existing natural fractures in shale might be activated, and a very complex fracture network might be formed, as shown in Figures 13.2 and 13.3. The created complex fracture networks will greatly enhance the contact surface area between shale and fractures and help the extraction of fluid flow from shale rocks (Maxwell et al. 2002; Fisher et al. 2004; Gale et al. 2007; Warpinski et al. 2009).

Generally, there is high uncertainty in natural fracture distribution and hydraulic fracture geometry, which plays an important role in fluid flow in unconventional reservoirs. Many fracture diagnostic techniques, such as microseismic, distributed temperature sensing, and distributed acoustic sensing, and many fracture propagation models have been widely applied to characterize and predict the fracture geometries (Olson 2008; Vermylen and Zoback 2011; Weng et al. 2011; Fisher and Warpinski, 2012; Wu and Olson 2014, 2015; Raterman et al. 2017; Gale et al. 2018; Maity 2018). Although the understanding of complex fracture networks has been improved, the impacts of natural fracture on well performance through reservoir simulation are still required to be ascertained.

13.2 Complex Fracture Reservoir Simulation

Accurate reservoir simulation with 3D complex hydraulic and natural fractures is vital to help evaluate the impact of fracture complexity on well performance. However, it is very challenging to simulate hydraulically fractured reservoirs due to the complexity of fracture orientation and geometry. Three traditional numerical methods are often used, including dual porosity dual permeability (DPDK) (Warren and Root 1963), local grid refinement (LGR) (Cipolla et al. 2009), and unstructured grids (Wang 2015). However, the DPDK method is hard to capture large-scale

Physics of Fluid Flow and Transport in Unconventional Reservoir Rocks, First Edition.
Edited by Behzad Ghanbarian, Feng Liang, and Hui-Hai Liu.

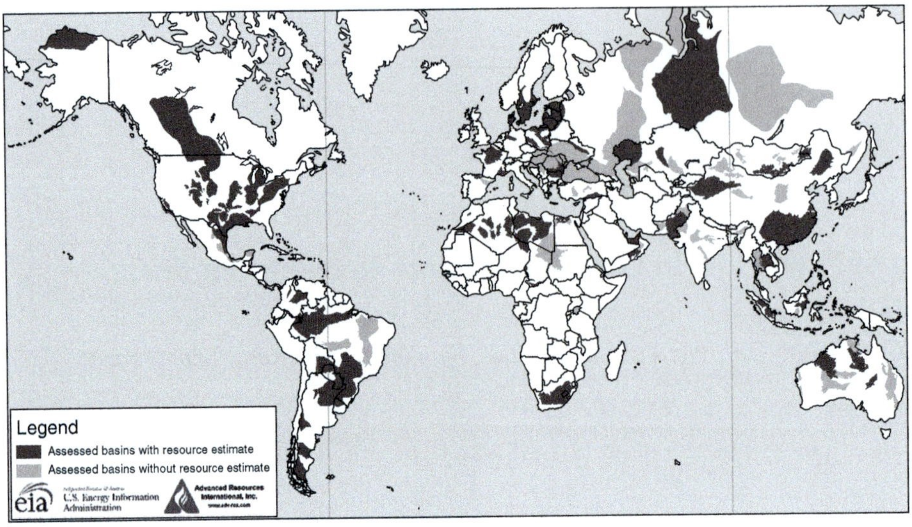

Figure 13.1 Shale oil and gas resources in the world. *Source:* US Energy Information Administration (2015)/Public Domain/U.S. Energy Information Administration.

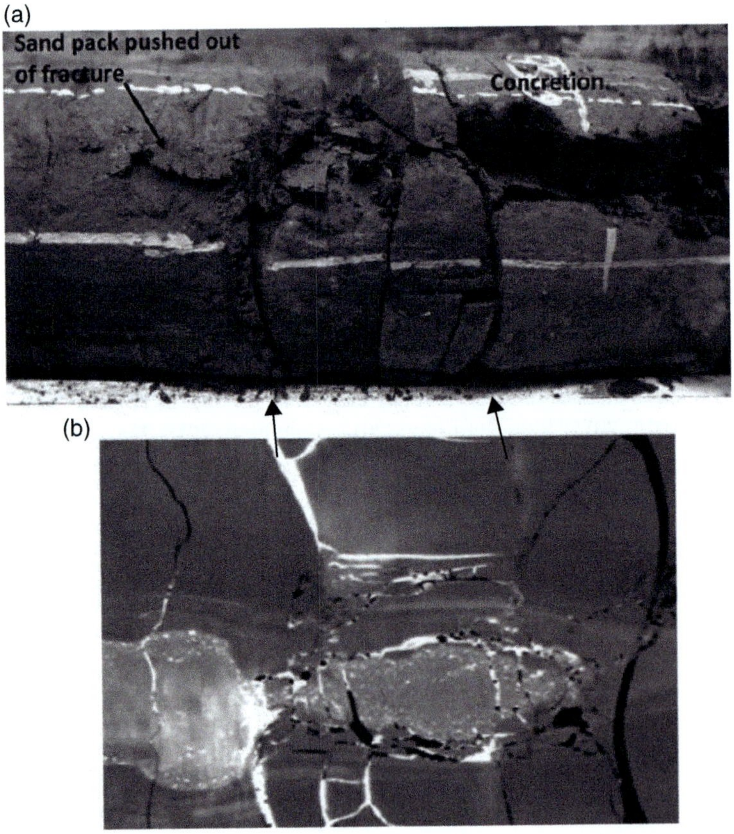

Figure 13.2 A complex fracture network observed in a fractured shale sample. (a) Core observation and (b) μ-CT scan. *Source:* Gale et al. (2018)/reproduced with permission from American Association of Petroleum Geologists.

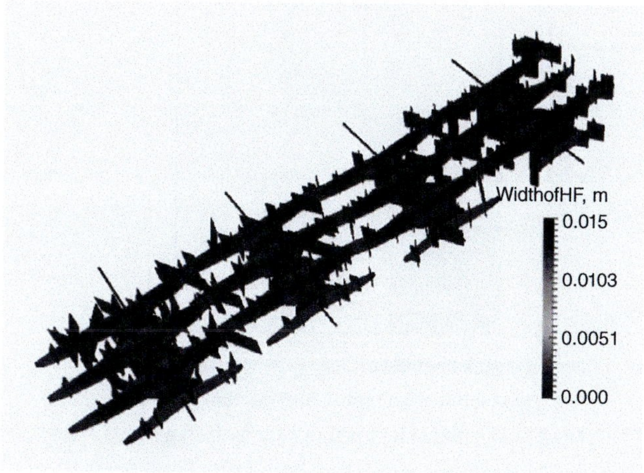

Figure 13.3 An example of complex fracture network in multiple wells, which was simulated by a complex fracture propagation model.

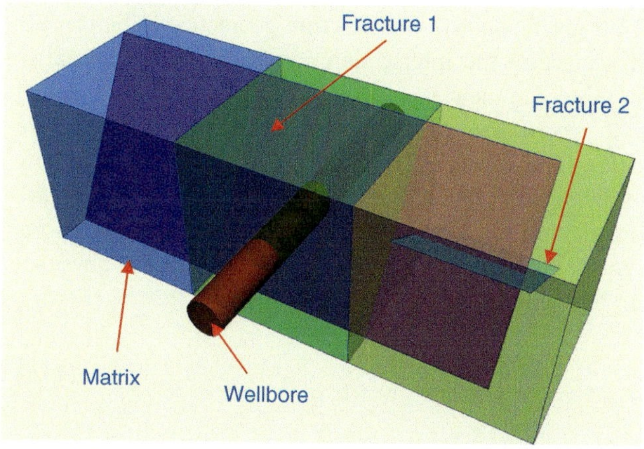

Figure 13.4 Definition and demonstration of EDFM's special non-neighboring connection concepts through physical domains. *Source:* Xu et al. (2017)/Society of Petroleum Engineers.

hydraulic fractures and cannot model natural fractures explicitly and accurately (Karimi-Fard et al. 2004; Bourbiaux 2010; Moinfar et al. 2011). LGR and unstructured girds can model every single fracture explicitly. However, in real-field applications, it is still limited because of the high complexity in gridding, especially around fracture intersections, and high computational cost (Du et al. 2017; Fumagalli et al. 2017). Hence, the development of a new fracture modeling technology is necessary.

13.3 Embedded Discrete Fracture Model

In this chapter, a novel numerical fracture modeling method, named the embedded discrete fracture model (EDFM), which merges the strengths of both dual-continuum models and discrete fracture models, is introduced. The EDFM method does not compromise the original

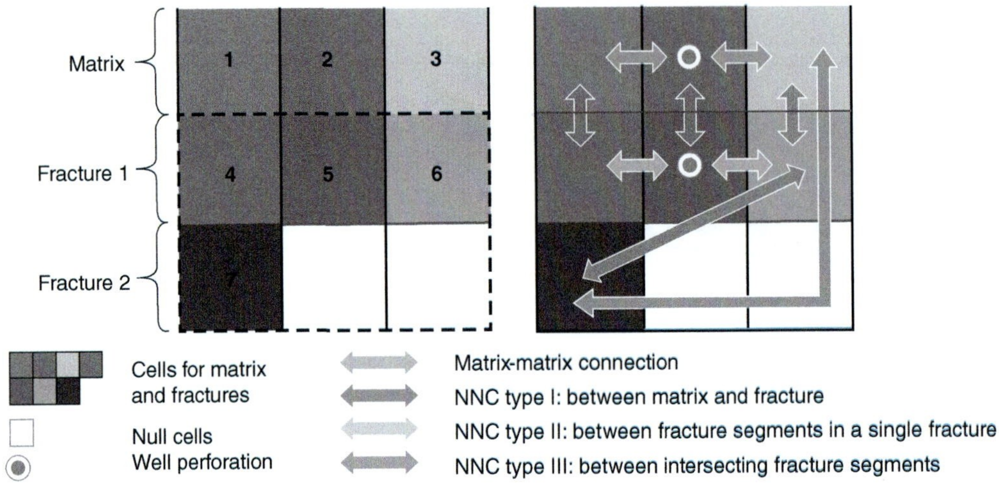

Figure 13.5 Definition and demonstration of EDFM's special non-neighboring connection concepts through computational domains. *Source:* Xu et al. (2017)/Society of Petroleum Engineers.

gridding scheme of the matrix block, even given sophisticated fracture geometries. The discretization of any fracture is performed by detecting the intersection conditions with the matrix block boundaries. Xu et al. (2017) introduced a new concept of special connections called "non-neighboring connections" (NNCs) to record flow information between any in-situ fracture to other reservoir/fracture medium. The flow information is a measure of how easily fluid can flow with respect to the effective length of flowing path. Four categories of special connections are established, namely matrix-fracture, intra-fracture segments, inter-fracture, and fracture-wellbore. The basic governing equations for these connections are listed in Eqs. (13.1)–(13.4), and a physical/numerical corresponding illustration can be found in Figures 13.4 and 13.5:

$$T_1 = \frac{A_e K_{me}}{d_e} \tag{13.1}$$

where T_1 is transmissibility of the first type of special connection (matrix-fracture), A_e is the effective flowing area, K_e is the effective matrix permeability, d_e is the effective flowing path normal distance.

$$T_2 = \frac{K_{fe} A_{fs}}{d_a + d_b} \tag{13.2}$$

where T_2 is transmissibility of the second type of special connection (intra-fracture segments), K_{fe} is the equivalent fracture permeability, A_{fs} is the common surface area of the two adjacent fracture segments a and b, and d_a and d_b are the flowing path distances from segment a and segment b to the connection, respectively.

$$T_3 = \frac{L K_{fx} K_{fy} w_{fx} w_{fy}}{K_{fx} w_{fx} d_{fy} + K_{fy} w_{fy} d_{fx}} \tag{13.3}$$

where T_3 is transmissibility of the third type of special connection (inter-fracture), L is the length of the intersection line, K_{fx} and K_{fy} are the fracture permeability of fracture x and y, respectively, w_{fx}

and w_{fy} are the fracture aperture of fracture x and y, respectively, and d_{fx} and d_{fy} are the equivalent normal distance from fracture x and y to the intersection.

$$T_4 = \frac{2\pi K_{fe} w_{fe}}{\ln\left(\frac{0.14\sqrt{L_{fs}^2 + H_{fs}^2}}{r_w}\right)} \tag{13.4}$$

where T_4 is the well index of the fourth type of special connection (fracture-wellbore), K_{fe} is the effective fracture permeability, w_{fe} is the effective fracture aperture, L_{fs} and H_{fs} are the length and height of intersecting fracture segment, and r_w is the wellbore radius. This equation is adapted and modified from the Peaceman formulation (Peaceman 1983; Moinfar et al. 2013).

Xu et al. (2017) have also extensively tested the EDFM method through multiple complex fracture configurations as well as single-flowing phase simulations and validated against the traditional LGR method. The computational costs and the accuracy of EDFM were proven to have a superior balance when compared to the LGR. Miao et al. (2018) have also validated the EDFM method with multi-flowing phase simulations, further confirming the robust accuracy of the EDFM.

Since the EDFM methods rely solely on the connection through boundary concept, it is very advantageous over the LGR or unstructured gridding method. These traditional methods rely rather on the refinement/re-griding of the grid blocks in the near-fracture regions. These concepts are inefficient in both input (model setup) and output (long simulation running time). Furthermore, they face extreme limitations when the matrix medium is represented through complicated geometry (such as corner point gridding) and when the fracture boundary geometry is intricate and 3D configured (natural fracture with dip angles/non-planar fractures, etc.). More importantly, the computational cost of the EDFM only increases linearly with increasing number of fractures; but such costs associated with LGR/unstructured gridding method would increase exponentially. The profound power of EDFM was illustrated in several recent studies (Moinfar et al. 2014; Panfili et al. 2015; Du et al. 2017).

Another significance of the EDFM method is its non-intrusiveness, which means that it can realize the NNC concepts without accessing the source codes when coupled with a reservoir simulator. The EDFM has been successfully applied within several in-house and commercial reservoir simulators, which have NNC functionality. With more than 10 years of technical development, EDFM technology has appeared and assisted in multiple applications within various fields in the petroleum industry. Xu and Sepehrnoori (2019) have extended the EDFM method to the corner point model. To list a few of such applications, sophisticated and dynamic behavior of fracture mechanism can be extensively researched using the EDFM (Tavassoli et al. 2018; Xu et al. 2019). Investigations of well interference and stress interference can be easily performed using the EDFM (Yu et al. 2017; Pei et al. 2020). Novel fishbone technology can also be readily captured by the EDFM method (Cavalcante Filho et al. 2015). Traditional/new fracture diagnostic tools, such as microseismic and distributed sensing, can also be cross-validated (Shakiba and Sepehrnoori 2015; Sun et al. 2020). Through various studies, the EDFM approach has been shown to be a very robust tool for comprehensive fracture analysis.

In this chapter, the superior EDFM technology is utilized to study the effect of sophisticatedly orientated natural fractures on horizontal well performance. In addition, an innovative fracture connectivity detection tool formulated by Leines-Artieda (2020) is also used to help explore the relationship between natural fracture connectivity with well performance. Although many works regarding natural fractures have been carried out, most of them focus on 2D-oriented and simplified natural fractures with limited numbers. Moreover, nobody systematically researched different

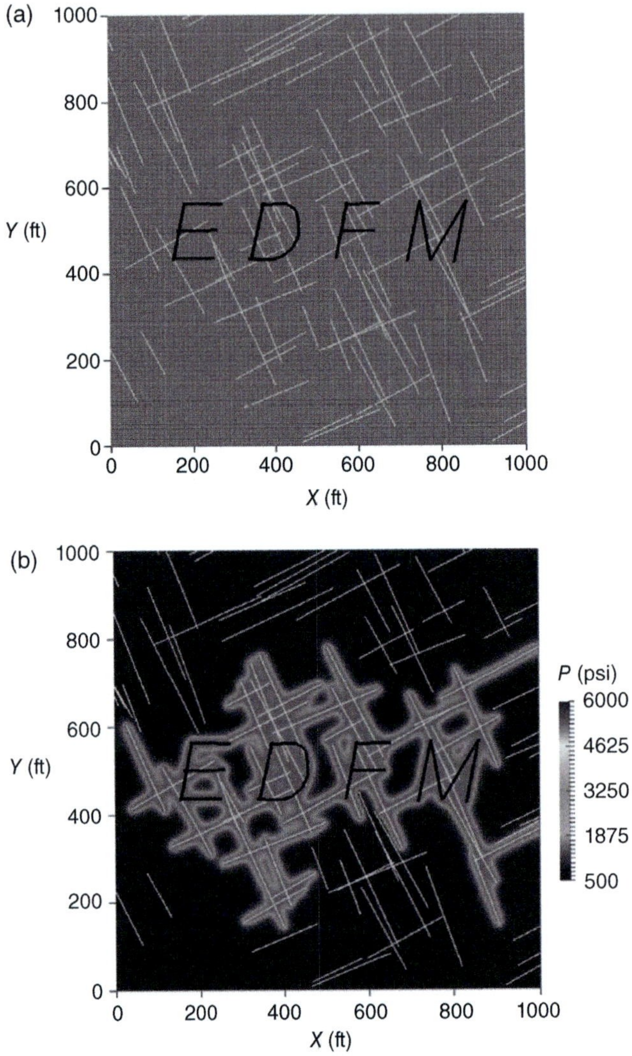

Figure 13.6 Two examples of the application of the EDFM method for easy and efficient modeling of complex fracture patterns using structured grids. (a) Fracture configuration (Yu et al. 2019) and (b) pressure drainage distribution. *Source:* Yu et al. (2019)/Unconventional Resources Technology Conference.

aspects of natural fractures (such as dip, azimuth, length, numbers, etc.). The work of Yu et al. (2019), as shown in Figure 13.6, has initiated the goal to investigate these details of natural fracture network systematically and holistically on the well performance.

13.4 EDFM Verification

Before utilizing EDFM to study the effect of natural fractures on the well performance, a verification of EDFM with traditional numerical fracture modeling method, i.e. LGR, is carried out. The reservoir model considered is based on petrophysical data provided for a real field, with vertical porosity heterogeneity and varying thickness. The model size is 2500 m in x direction, 2400 m in y direction, and 17.4 m in z direction. Number of grid blocks in each of these directions are 125, 120, and 4, respectively. The depth of the reservoir top is situated around 3350 m. The reservoir pressure is 85 mega-Pascal (MPa), and reservoir temperature is 150 °C. Matrix permeability is 50 nano-Darcy (nD), and initial water saturation is set to be equal to residual water saturation, i.e. 0.3. Specific gas gravity is 0.677, and water compressibility is 6.18×10^{-7} kPa^{-1}. The Langmuir isotherm gas adsorption is also modeled by considering a Langmuir volume of 0.002 m^3/kg, a Langmuir inverse pressure of 0.00029 kPa^{-1}, and a bulk density of 3560 kg/m^3 for shale. A comprehensive list of reservoir properties is tabulated in Table 13.1.

There is a single horizontal well with 1980 m depth, 100 hydraulic fractures, and a cluster spacing of 20 m in the verification model. For the fracture geometry, uniform half-lengths of 70 m, uniform heights of 5 m, and apertures of 0.1 m are designed. Fracture conductivity and fracture water saturation are 10 milliDarcies-times-meter (mD-m) and 0.3, respectively. A well constraint of constant minimum bottom-hole pressure of 20 MPa is set up. The fracture closure coefficient, which modifies the equivalent fracture conductivity as a function of pressure, is 0.0494. The model of the fracture closure phenomenon is given by Eq. (13.5)

$$k_f' = k_f^i \exp\left[-\alpha\left(P^i - P'\right)\right] \tag{13.5}$$

Table 13.1 Key reservoir inputs of all studied models in our work.

Parameter		**Value**	**Unit**
Reservoir size		$2500 \times 2400 \times 17.4$	meter (m)
Reservoir depth		3350	m
Reservoir thickness	Layer 1	5.90	m
	Layer 2	5.00	m
	Layer 3	1.60	m
	Layer 4	4.90	m
Reservoir pressure		85	MPa
Reservoir temperature		150	°C
Matrix permeability		50	nD
Reservoir porosity	Layer 1	5.99	%
	Layer 2	6.02	%
	Layer 3	5.53	%
	Layer 4	4.38	%
Initial water saturation		0.3	—
Specific gas gravity		0.677	—
Water compressibility		6.18×10^{-7}	kPa^{-1}
Bulk density of shale		3560	kg/m^3
Langmuir adsorption volume		0.002	m^3/kg
Langmuir adsorption inverse pressure		0.00029	kPa^{-1}

Table 13.2 Key fracture parameters of the EDFM verification model.

Parameter	Value	Unit
Fracture height	5	m
Fracture half-length	70	m
Fracture spacing	20	m
Fracture conductivity	10	mD-m
Fracture width	0.1	m
Fracture water saturation	0.3	—
Fracture closure coefficient	0.0494	—
Minimum bottom hole pressure	20	MPa

where k_f' is the reduced fracture permeability at arbitrary reservoir pressure P', k_f^i is the initial fracture permeability, α is the fracture closure coefficient, and P^i is the initial reservoir pressure. A complete table of fracture properties for the verification model is shown in Table 13.2.

The compaction curve associated with the 0.0494 fracture closure coefficient is given in Figure 13.7. A severe compaction is designed to mimic the aggravated fracture closure phenomenon typically observed in this field. The relative permeability curves for water phase and gas phase are plotted in Figure 13.8. The system is gas (hydrocarbon)-wet, which is typically observed in shale reservoirs. Figure 13.9 shows the 2D top view and 3D view for the verification model. The well is situated in the middle of the reservoir.

Figure 13.10 shows a comparison of simulation results using the traditional LGR and novel EDFM methods. Both gas flow rate and cumulative gas production plots show that the differences between the two methods are negligible meaning that the EDFM can be accurately and reliably used to further study the effect of natural fractures on the well performance. We should point out that only four processing cores were employed in the verification simulation and the rest of all studies. The CPU time for the EDFM takes about 32 seconds, and 35 seconds for the LGR

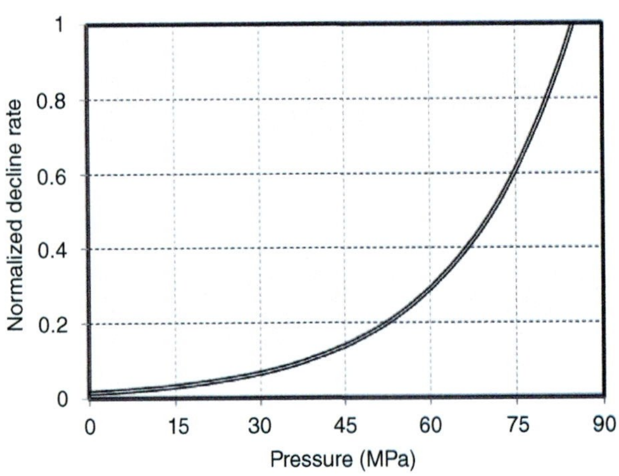

Figure 13.7 Pressure dependent permeability curve in reservoir simulation.

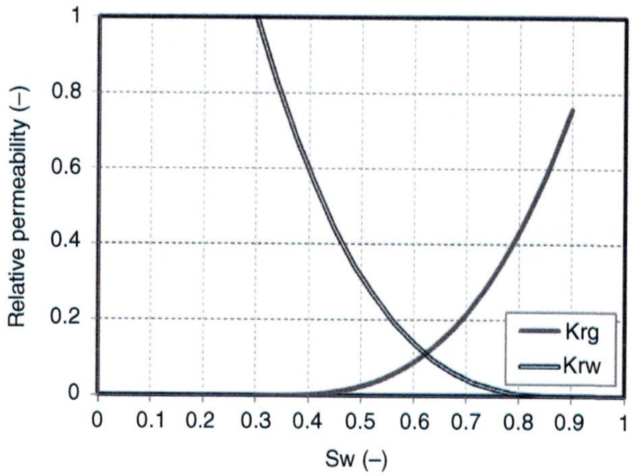

Figure 13.8 Relative permeability curves used in this verification study.

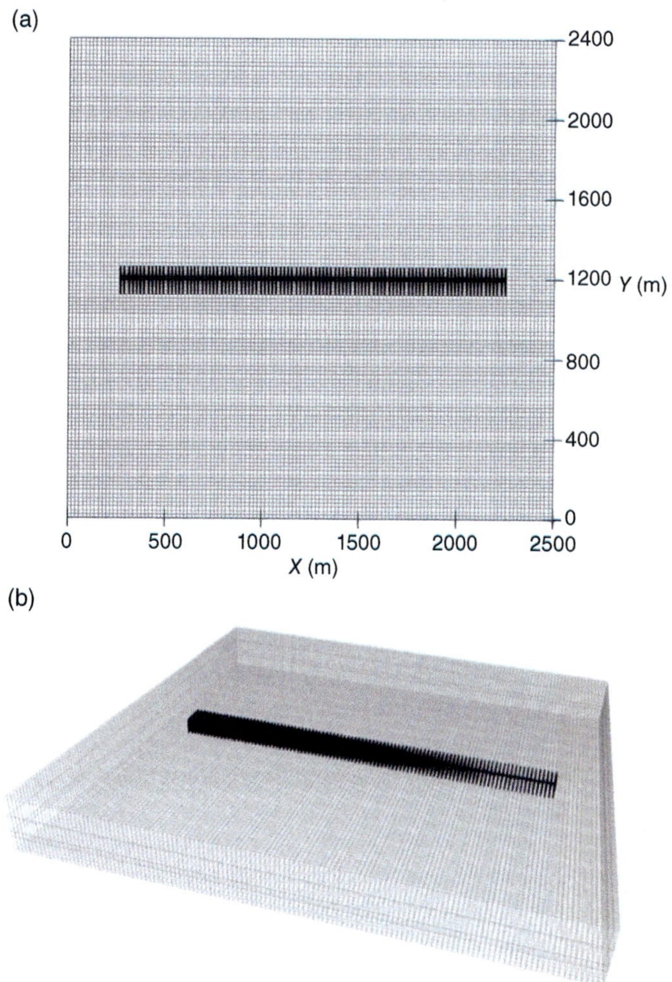

Figure 13.9 Validation model with a horizontal well and 100 hydraulic fractures and different layer thicknesses and porosities. (a) 2D view and (b) 3D view.

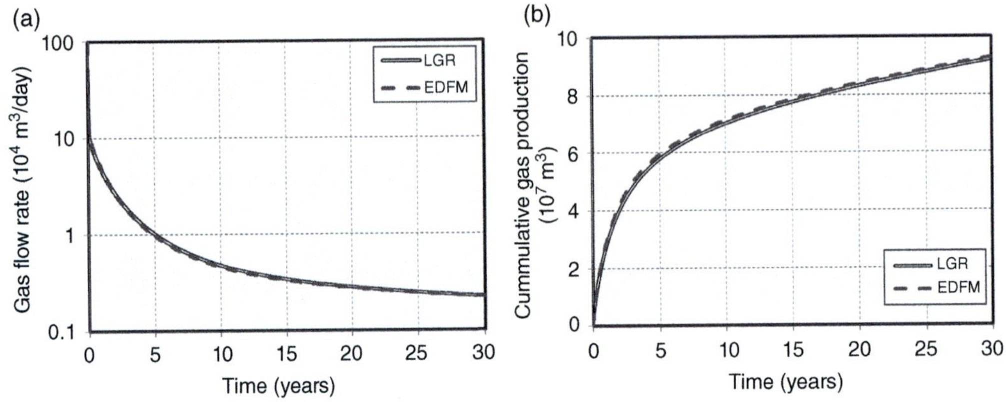

Figure 13.10 Comparing the well productivity via the EDFM method with those by the LGR approach. (a) Gas flow rate and (b) cumulative gas production.

method. Although the difference is small, the advantage of computational efficiency would be more pronounced as natural fractures are included, and the EDFM does not require cumbersome re-gridding work as required by LGR, which could be highly difficult for complex natural fracture configurations. A more complex well constraint would also magnify the efficiency of the EDFM.

13.5 Well Performance Study – Base Case

Once the verification procedure is completed, a base case scenario for studying the effect of natural fractures on well performance is set up. The basic reservoir properties are identical to those given in Table 13.1. However, multiple horizontal wells are placed in the reservoir model. In addition, the fracture geometrical properties are drastically different from the verification model. Figures 13.11 and 13.12 show the 2D top view, a zoomed-in view, and the 3D view of the base case model. The half-length and height of each of the clusters are different, but the geometries of hydraulic fractures along one whole well are copied to the two other wells.

Statistical plots of individual cluster half-length and height are shown in Figure 13.13. As can be observed in this figure, the half-length within stages changes drastically. This is because such complex fracture geometries are linked to realistic injection rates and sand concentration observed in the field. Moreover, the uneven distribution of fracturing fluid in each cluster within a single stage causes the fracture to grow differentially, which is called the "heel bias effect." However, the height stays mostly uniform because the fracture is difficult to grow vertically due to gravity. This geometry is mainly controlled by the stress difference between bedding layers and thus is constrained by layer thicknesses.

Next, 20-year production-enhanced ultimate recovery (EUR) forecast for the base case reservoir and fracture model is performed. The well is constrained by flowing bottom-hole pressure trends typically seen from the field response, shown in Figure 13.14. The bottom-hole pressure decreased from initial reservoir pressure of 85 to 20 MPa in the first three and half years, then it reached a constant value afterward. The relative permeability curves and proppant compaction curve, which simulates the fracture closure phenomenon, are the same as those used in the verification model.

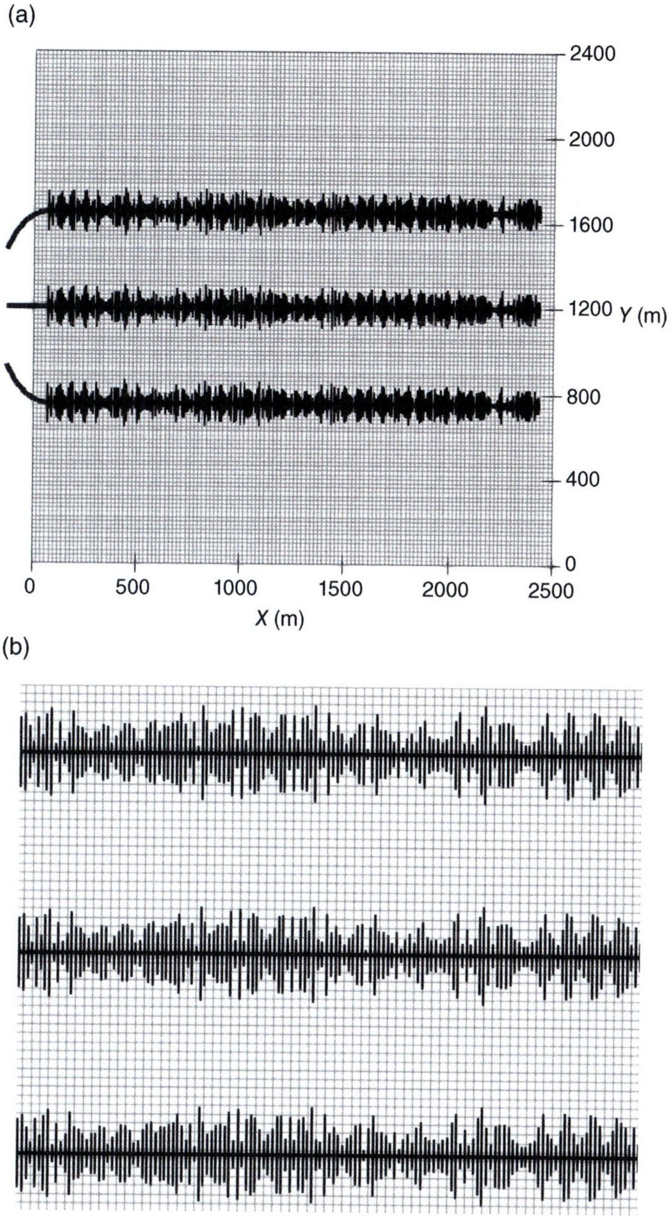

Figure 13.11 Base case model with three horizontal wells and complex hydraulic fracture geometries (2D view). (a) 2D view and (b) magnified view of some hydraulic fractures.

The additional EDFM fracture cells are 10,560 and the CPU time is 141 seconds. Figure 13.15 shows the 2D pressure distribution and 3D drainage volume with hydraulic fractures after 20 years of production. It can be seen that pressure distribution is heterogeneous both horizontally and vertically due to variation of hydraulic fracture geometries. Furthermore, the extent of pressure drawdown and the drainage volume is limited laterally without natural fractures. We should note that the presence of natural fractures might extend the drawdown and drainage volume significantly.

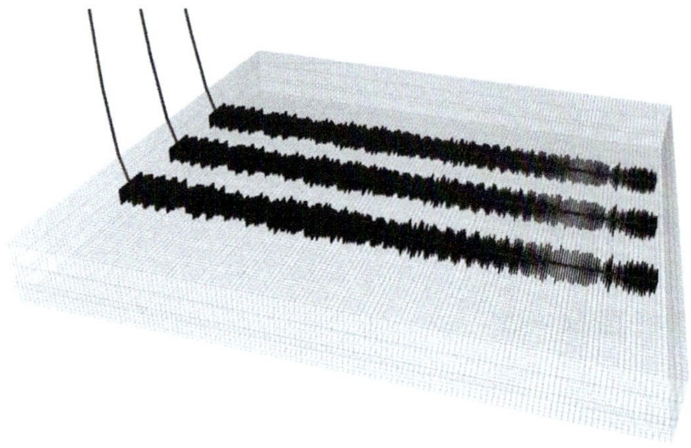

Figure 13.12 Base case model with three horizontal wells and complex hydraulic fracture geometries (3D view).

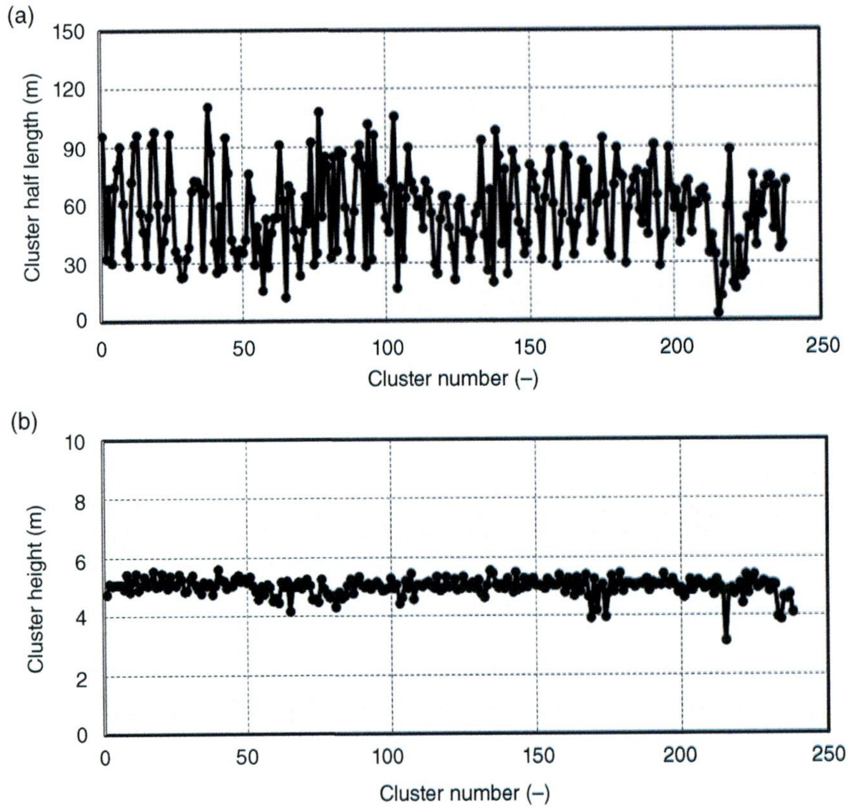

Figure 13.13 Statistical plots of complex hydraulic fracture geometries. (a) Hydraulic fracture half-lengths for different clusters along the well and (b) hydraulic fracture heights for different clusters along the well.

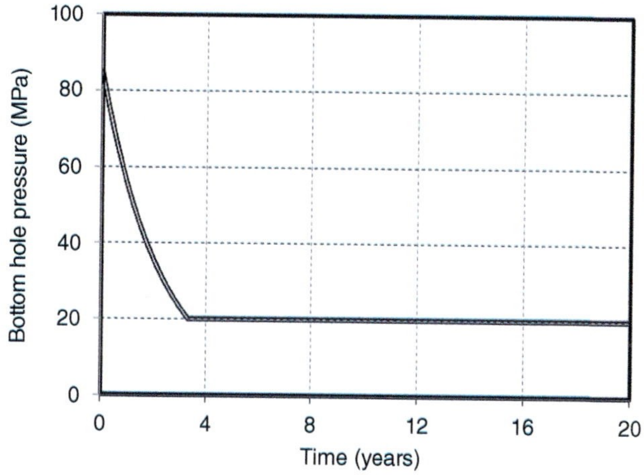

Figure 13.14 Wellbore flowing bottom hole pressure.

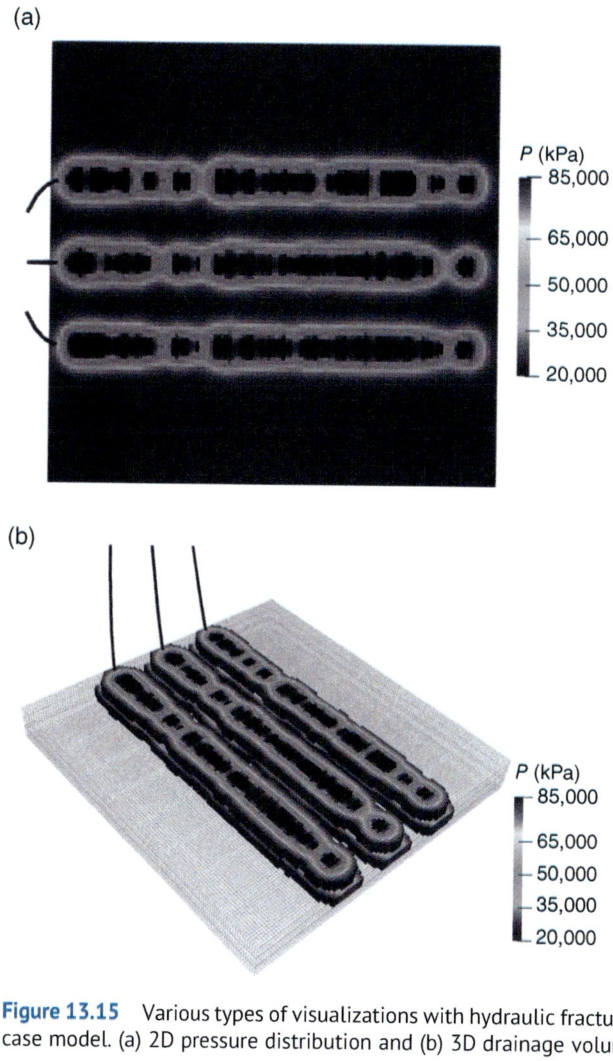

Figure 13.15 Various types of visualizations with hydraulic fractures after 20 years of production for the base case model. (a) 2D pressure distribution and (b) 3D drainage volume.

13.6 Effect of Natural Fracture Connectivity on Well Performance

First, the influence of natural fracture connectivity is analyzed. The connectivity is defined by the number of natural fractures intersecting with hydraulic fractures. Parameters controlling natural fracture discrete fracture network (DFN), such as azimuth, total number, length, set number, and dip angle, would all affect connectivity in various degrees. The analysis is initiated with the study of natural fracture azimuth.

13.6.1 Effect of Natural Fracture Azimuth

The effect of two different azimuths of natural fractures on well performance is examined. The natural fracture azimuth is defined as the angle relative to the horizontal well direction (along the *x*-axis). 2D and 3D views of natural fracture distribution with azimuth of 45°–55° and 90° are shown in Figures 13.16 and 13.17, respectively. The natural fracture total number, aperture, length, height, dip angle, and conductivity are fixed at 2500, 0.01 m, 50 m, 5–10 m, 80°–90°, and 0.1 mD-m, respectively. The additional EDFM fracture cells for natural fracture azimuth of 45°–55° are 40,320, and the CPU time is 355 seconds. The additional EDFM fracture cells for natural fracture azimuth of 90° are 33,120, and the CPU time is 298 seconds.

Comparisons of gas flow rate and cumulative gas production between two different natural fracture azimuths are shown in Figure 13.18. It can be seen that natural fractures play a positive effect on gas flow rate and well production compared to the base case without natural fractures. It should also be noted that there are some inflection points in the gas flow rate curves, which are caused by the flowing bottom hole pressure reaching the minimum constraint. In addition, we find that the gas recovery after 20 years of production increases by 69% and 50% corresponding to azimuth of 45°–55° and 90°, respectively. In other words, natural fractures that are not parallel with hydraulic fractures can boost well productivity. The numbers of intersections between natural and hydraulic fractures are 1236 and 45, respectively. Therefore, the azimuth is a highly sensitive parameter affecting the intersection between natural and hydraulic fractures. Hence, with increasing intersections between natural and hydraulic fractures, there are more connected fracture areas, resulting in improved well performance.

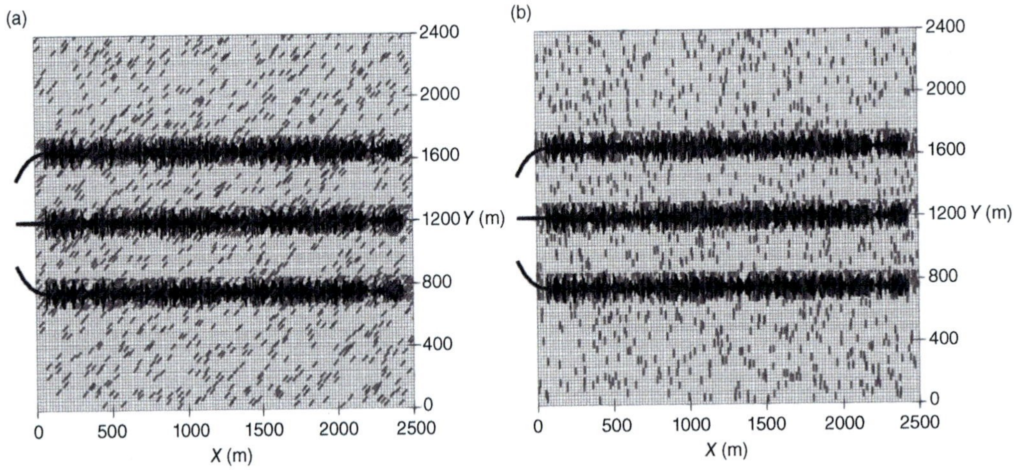

Figure 13.16 Three horizontal wells with two different natural fracture azimuths (2D view). (a) Azimuth of 45°–55° and (b) Azimuth of 90°.

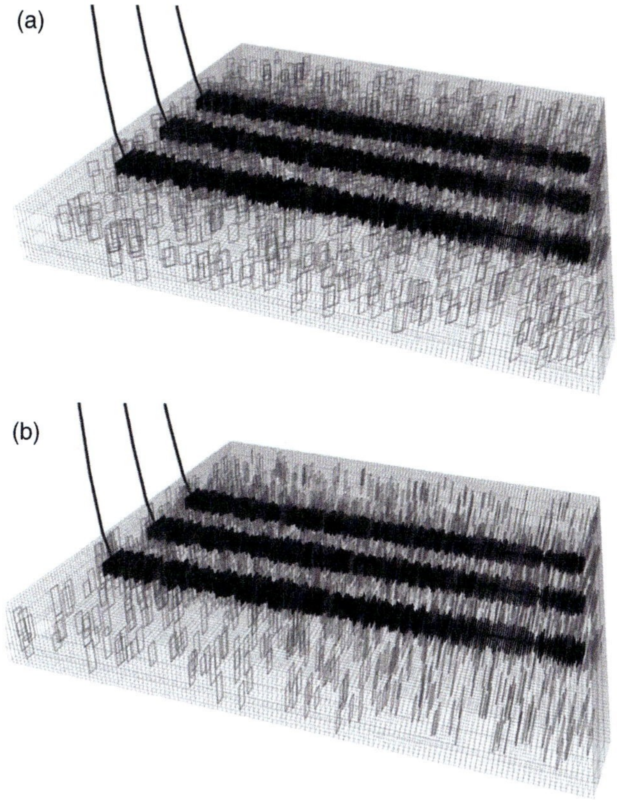

Figure 13.17 Three horizontal wells with two different natural fracture azimuths (3D view). (a) Azimuth of 45°–55° and (b) Azimuth of 90°.

Comparisons of pressure distribution and drainage volume, which are shown in Figures 13.19 and 13.20, can further explain the effect of natural fracture azimuth on shale gas production. With more intersections between natural and hydraulic fractures, the drainage area of azimuth of 45°–55° becomes larger than that of azimuth of 90°. Due to the limitation of natural fracture length, the drainage area is enlarged mainly near the wellbore region.

13.6.2 Effect of Number of Natural Fractures

Based on the case study with azimuth of 45°–55° in Section 13.6.1, the effect of number of natural fractures on well performance is investigated. The new number of natural fractures is 5000; the 2D and 3D view of natural fracture distribution are shown in Figure 13.21. The natural fracture aperture, length, height, azimuth, dip angle, and conductivity are 0.01 m, 50 m, 5–10 m, 45°–55°, 80°–90°, and 0.1 mD-m, respectively. The additional EDFM fracture cells for 5000 natural fractures are 70,560, and the CPU time is 520 seconds.

Figure 13.22 compares the gas flow rate and cumulative gas production for different numbers of natural fractures. It is obvious that more natural fractures can extract more gas from the shale reservoir. In contrast with the base case, the gas recovery after 20 years of production increases by 69% and 117% corresponding to the number of natural fractures 2500 and 5000, respectively. The reason behind is that more natural fractures lead to more intersections between natural and hydraulic fractures. In this case, as the number of natural fractures increases from 2500 to 5000, the number of

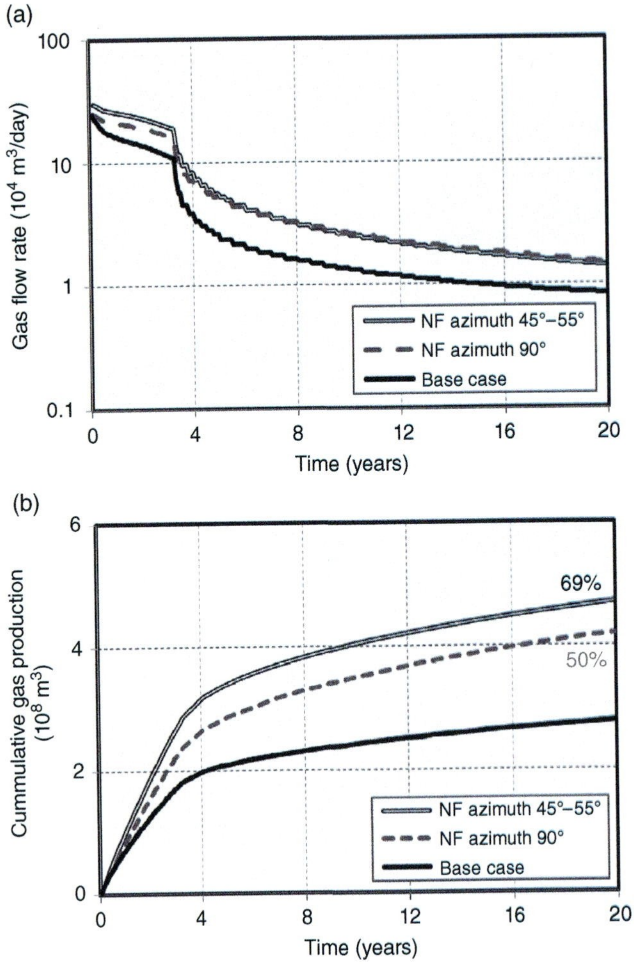

Figure 13.18 Comparisons of well productivity between two scenarios of natural fracture azimuth. (a) Gas flow rate and (b) Cumulative gas production.

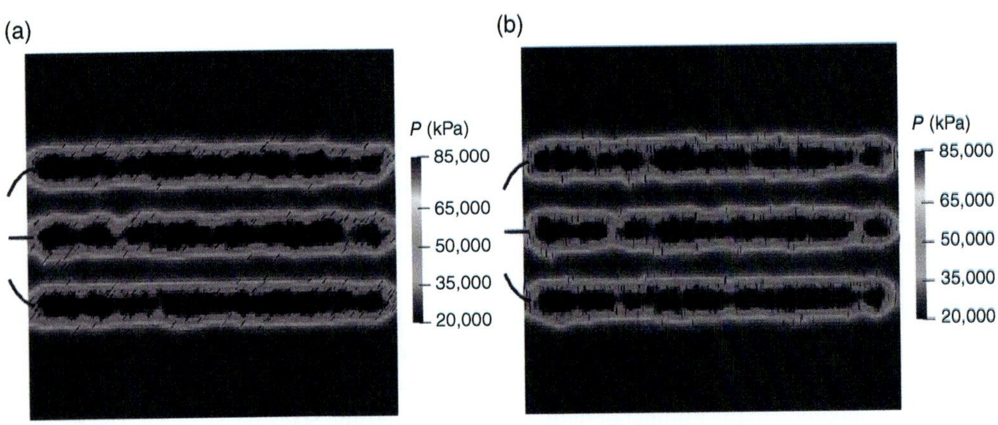

Figure 13.19 Comparisons of pressure distribution between two different natural fracture azimuths after 20 years of production. (a) Azimuth of 45°–55° and (b) Azimuth of 90°.

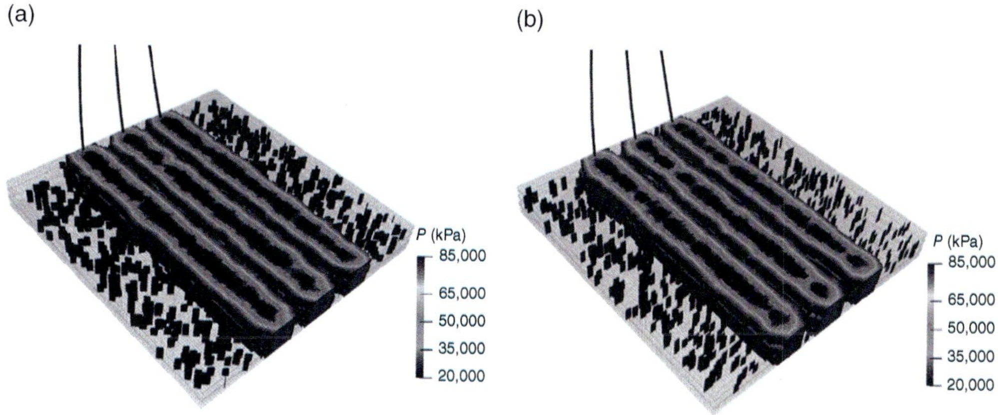

Figure 13.20 Comparisons of drainage volume between two different natural fracture azimuths after 20 years of production. (a) Azimuth of 45° – 55° and (b) Azimuth of 90°.

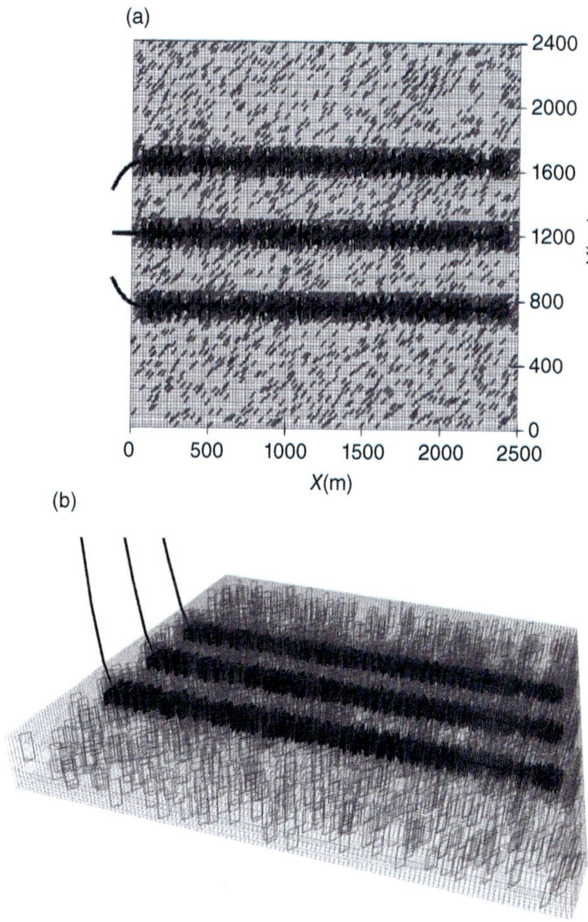

Figure 13.21 Three horizontal wells with 5000 natural fractures. (a) 2D view and (b) 3D view.

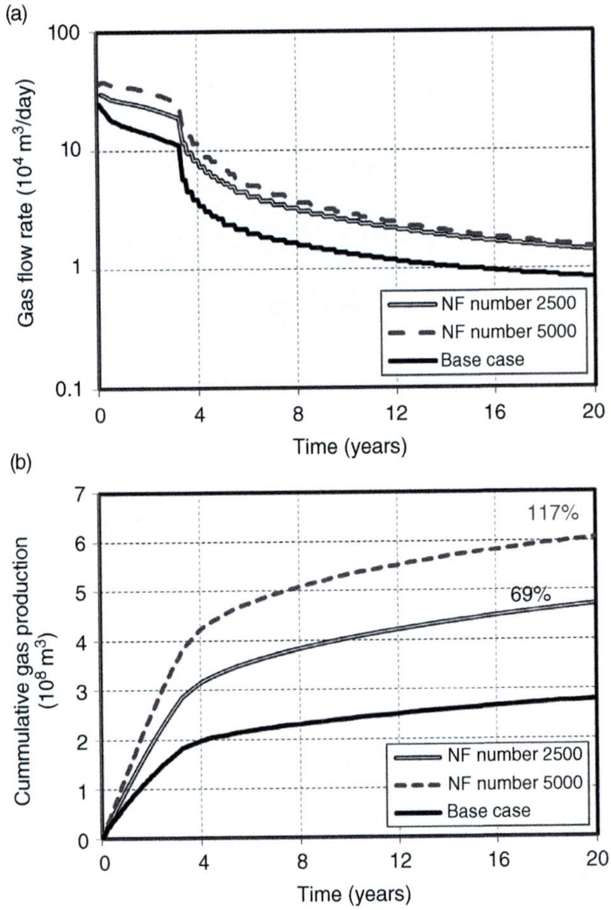

Figure 13.22 Comparisons of well productivity between two scenarios of total natural fracture number. (a) Gas flow rate and (b) cumulative gas production.

intersecting fractures increases from 1236 to 2535. Therefore, more natural fractures would enlarge the connected area with hydraulic fractures resulting in higher gas recovery.

The 2D pressure distribution and 3D drainage volume with 5000 natural fractures at different times i.e. 3 years, 10 years, and 20 years are shown in Figure 13.23. Compared to the case with 2500 natural fractures shown in Figures 13.19a and 13.20a, it is clear that increasing the number of natural fractures mainly increases the drainage area near the wellbore region instead of the far field region. Because of the limitation of the length of natural fractures, the gas in the far field region could not be reached by the intersecting fractures. Therefore, there is still a considerable amount of gas in the far-field region, and additional efforts should be made to further exploit the far-field region resource by well placement optimization. In the next section, the effect of natural fracture length on well performance is investigated, especially on the far-field region.

13.6.3 Effect of Natural Fracture Length

As indicated in Section 13.6.2, more natural fractures mainly help to extract gas in the near wellbore region due to the limitation of natural fracture length, and there still remains a large amount of

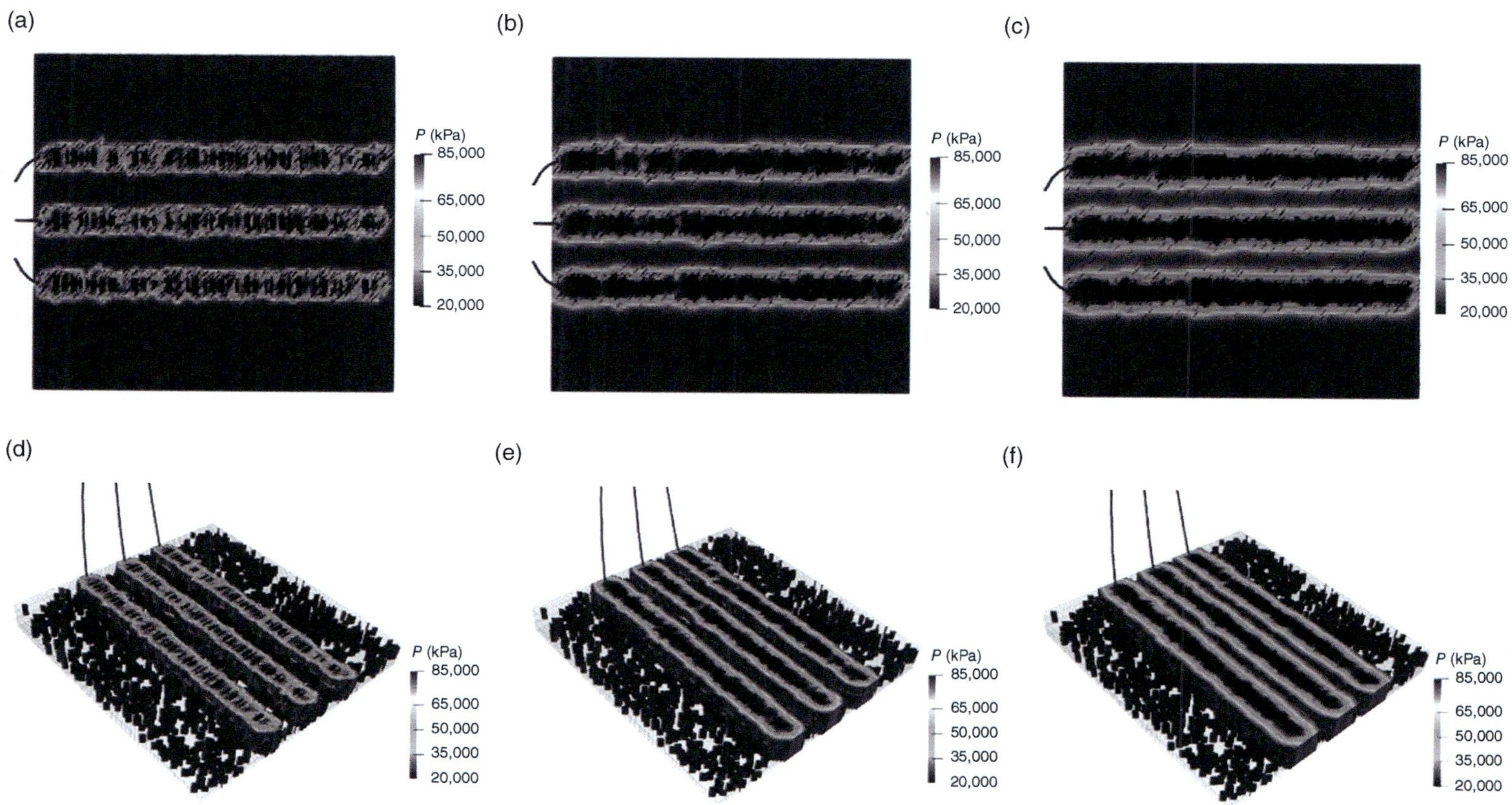

Figure 13.23 Different types of visualizations with 5000 natural fractures at different times. (a) 2D pressure distribution after 3 years of production, (b) 2D pressure distribution after 10 years of production, (c) 2D pressure distribution after 20 years of production, (d) 3D drainage volume after 3 years of production, (e) 3D drainage volume after 10 years of production, and (f) 3D drainage volume after 20 years of production.

remaining gas in the far-field region. In this section, the effect of natural fracture length on well performance from the far-field region is investigated. Based on the case study in Section 13.6.1 with azimuth of 45°–55° and natural fracture length of 50 m, the length of natural fractures is extended to 100 m; the 2D and 3D view of natural fracture distribution are shown in Figure 13.24. The natural fracture number, aperture, height, dip angle, and conductivity are 2500, 0.01 m, 5–10 m, 80°–90°, and 0.1 mD-m, respectively. The additional EDFM fracture cells for natural fracture length of 100 m are 64 800, and the CPU time is 547 seconds.

Figure 13.25 shows the simulated gas flow rate and cumulative gas production for different lengths of natural fractures. As can be seen, more gas is extracted with increasing the natural fracture length. Compared with the base case, the gas recovery after 20 years of production increases by 69% and 96% corresponding to the length of natural fractures 50 m and 100 m, respectively. The reason for this is quite obvious: the longer natural fractures lead to more intersections between natural and hydraulic fractures. It can be explained with the help of 2D pressure distribution and 3D drainage volume at different times i.e. 3 years, 10 years, and 20 years shown in Figure 13.26. The production at the early stage is mainly dominated by the perforated hydraulic fractures, while the natural fractures play a key role at the later stage, and well interference can be seen after 10 years of

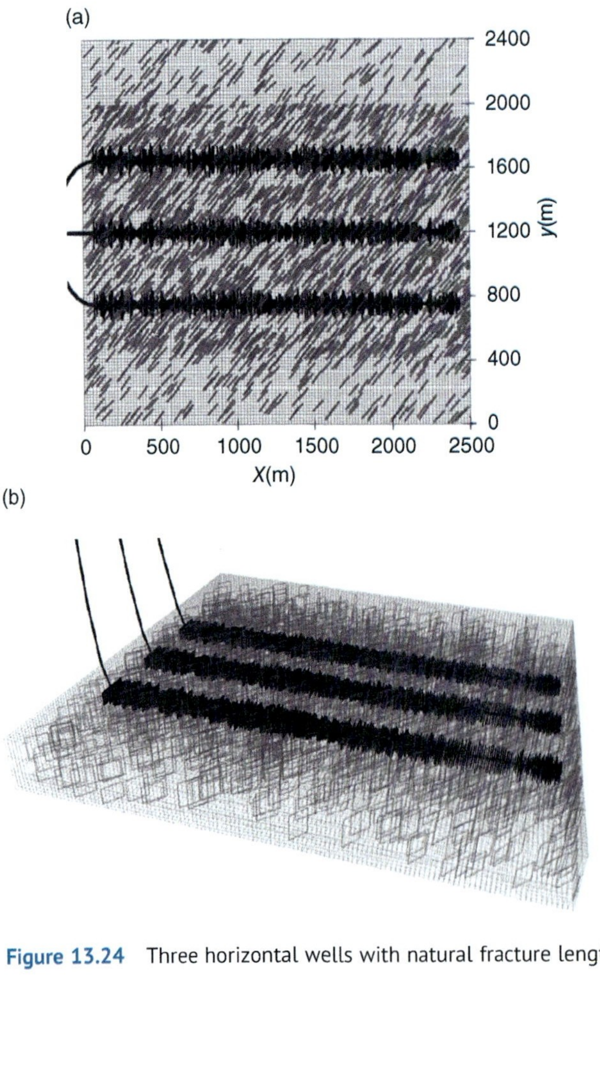

Figure 13.24 Three horizontal wells with natural fracture length of 100 m. (a) 2D view and (b) 3D view.

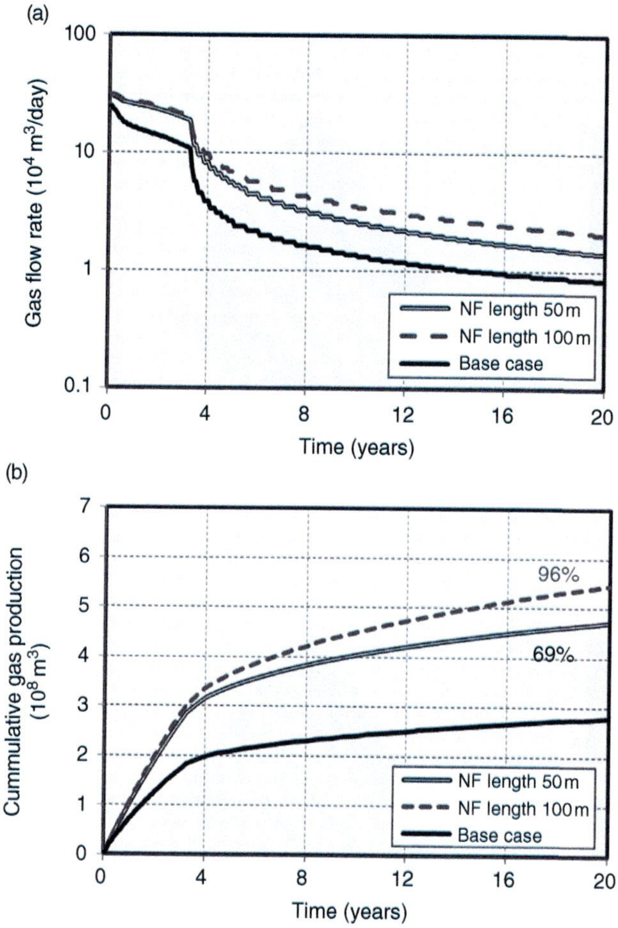

Figure 13.25 Comparisons of well productivity for two scenarios of natural fracture length. (a) Gas flow rate and (b) cumulative gas production.

production. Compared to the case with natural fracture length of 50 m shown in Figures 13.18a and 13.19a, the extended length of natural fractures is indicated to increase the drainage area and volume beyond the wellbore region. Hence, the longer natural fractures would not only increase the intersections between natural and hydraulic fractures in the near wellbore region but also increase the drainage area and volume beyond the wellbore region. To a certain degree, proper well interference may be beneficial to gas recovery, which needs to be further studied.

13.6.4 Effect of Number of Sets of Natural Fractures

Based on the case study with natural fracture length of 100 m in Section 13.6.3, 2500 natural fractures are divided into two sets, one set with azimuth of 45°–55° (shown in green) and the other one with azimuth of 135°–145° (shown in yellow) to examine the effect of different sets of natural fractures on the well performance. Each set has 1250 natural fractures perpendicular to each other. Figure 13.27 shows the 2D and 3D views of the natural fracture distribution. The natural fracture number, aperture, length, height, dip angle, and conductivity are 2500, 0.01 m, 100 m, 5–10 m,

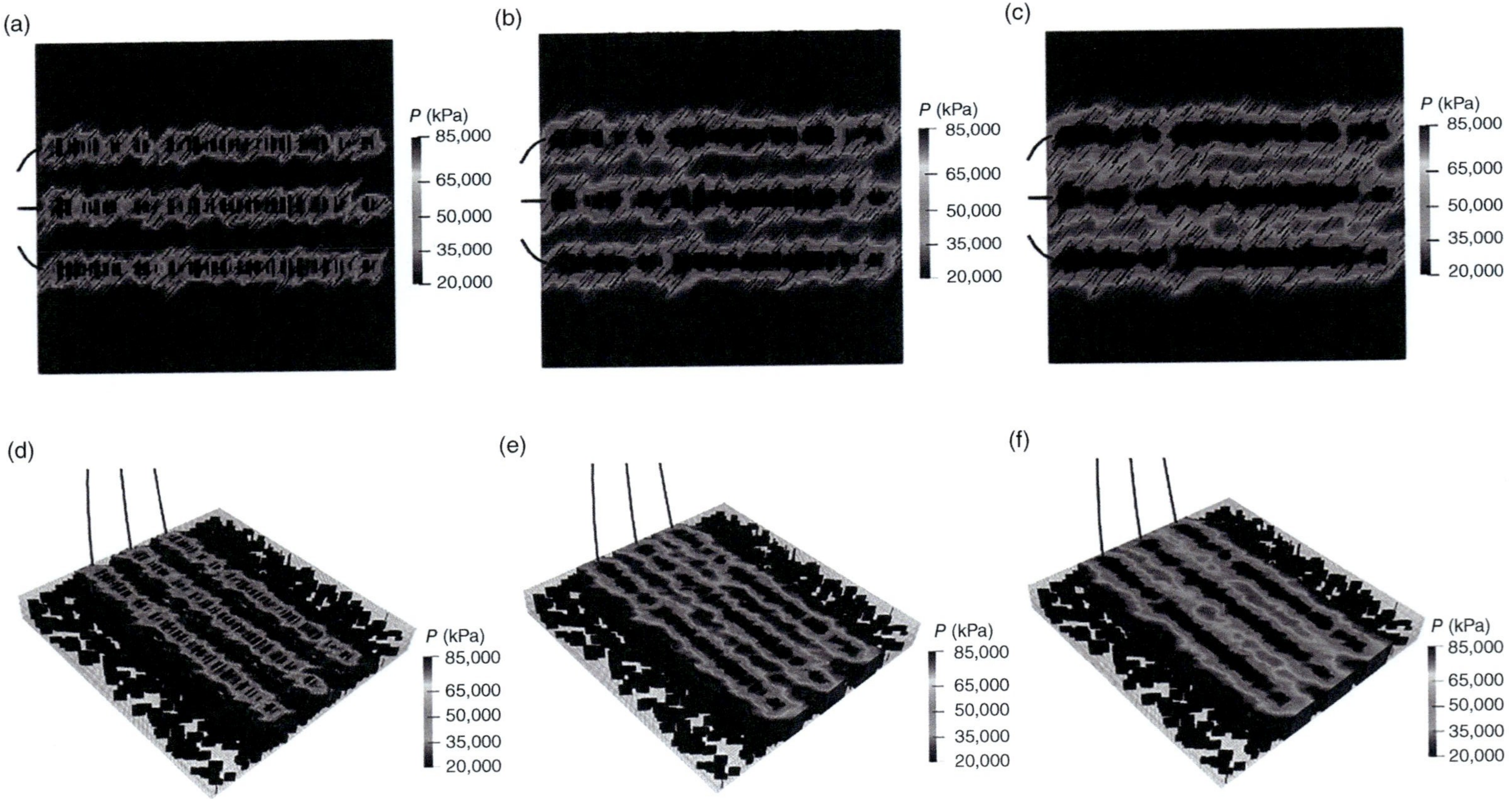

Figure 13.26 Different types of visualizations with natural fracture lengths of 100 m at different times. (a) 2D pressure distribution after 3 years of production, (b) 2D pressure distribution after 10 years of production, (c) 2D pressure distribution after 20 years of production, (d) 3D drainage volume after 3 years of production, (e) 3D drainage volume after 10 years of production, and (f) 3D drainage volume after 20 years of production.

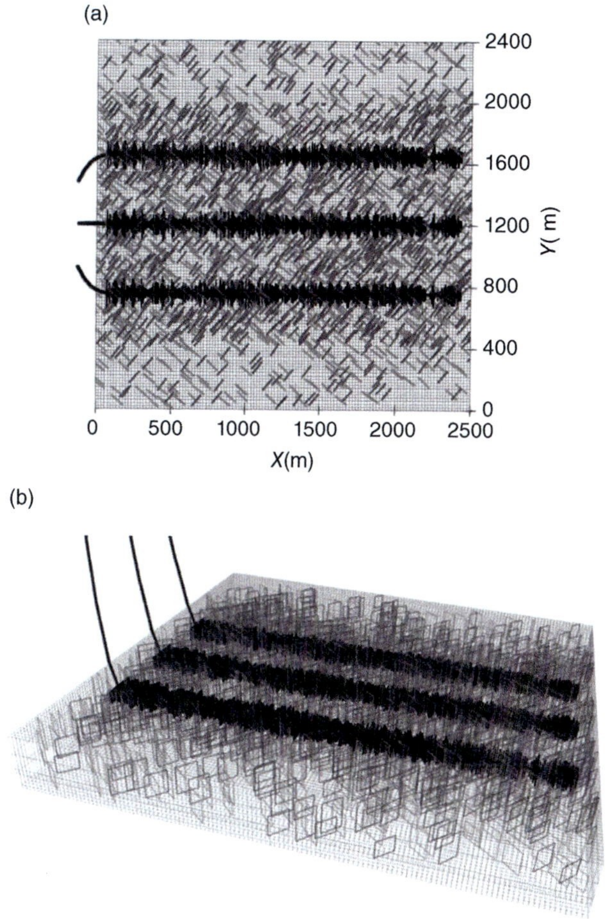

Figure 13.27 Three horizontal wells with two sets of natural fractures. (a) 2D view and (b) 3D view.

80°–90°, and 0.1 mD-m, respectively. The additional EDFM fracture cells for the two sets of natural fractures are the same as that of one set of natural fractures, which are 64,800, and the CPU time is 550 seconds.

Comparisons of the gas flow rate and cumulative gas production for different sets of natural fractures are shown in Figure 13.28. Results show that two sets of natural fractures have higher gas production than the case with only one set of natural fractures. Compared to the base case, the gas recovery after 20 years of production increases by 96 and 121% from one set of natural fractures to two sets of natural fractures. This is because two sets of natural fractures have more intersecting fractures with hydraulic fractures than one set of natural fractures, despite the fact that the total number of natural fractures are the same. The intersecting fracture number is 783 and 1672, respectively, for one set of natural fractures and two sets of natural fractures. Once again, as the number of intersecting fractures increases, there would be more connected fracture areas resulting in more production.

The 2D pressure distribution and 3D drainage volume with two sets of natural fractures are displayed in Figure 13.29. Compared to the case with only one set of natural fractures as given in Figure 13.26, the increase of the set of natural fractures can increase the drainage area and volume

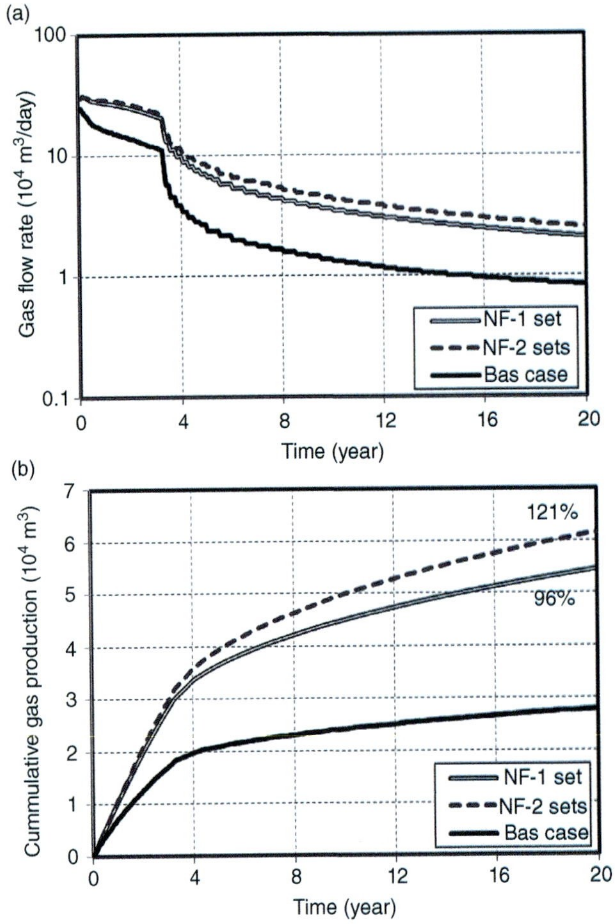

Figure 13.28 Comparisons of well productivity between two scenarios of natural fracture set number. (a) Gas flow rate and (b) cumulative gas production.

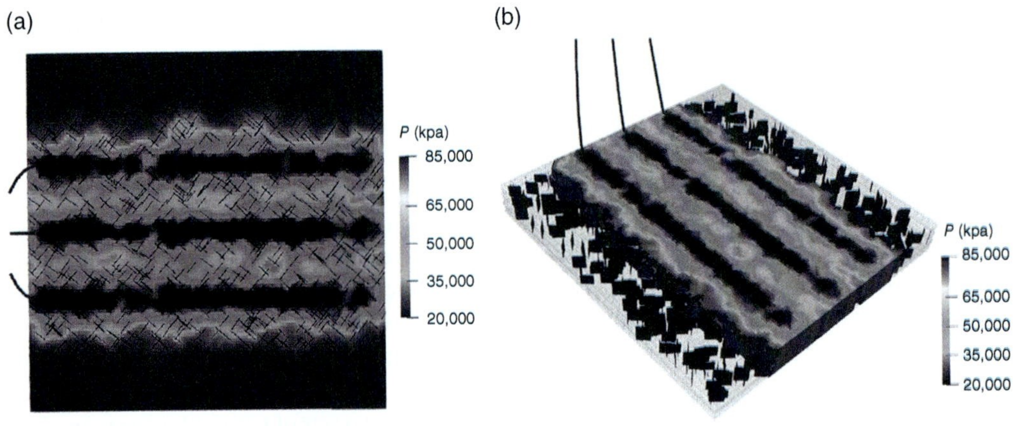

Figure 13.29 Different types of visualizations with two sets of natural fractures. (a) 2D pressure distribution and (b) 3D drainage volume.

in both the near wellbore and the far-field regions. Therefore, the more sophistically oriented natural fractures would lead to higher gas recovery.

13.6.5 Effect of Natural Fracture Dip Angle

The effect of the dip angle of natural fractures on well performance is investigated in this section. Based on the case study with natural fracture length of 100 m in Section 13.6.3, the dip angles are modified from 80°–90° to 10° and 45°. The 2D and 3D views of natural fracture distribution are provided in Figures 13.30 and 13.31, respectively. The magnified view of natural fractures with three different kinds of dip angles is shown in Figure 13.32. It should be noted that the natural fracture dip angle is defined as the angle relative to the horizontal plane (the X–Y face). The natural fracture number, aperture, length, height, azimuth, and conductivity are 2500, 0.01 m, 100 m, 5–10 m, 45°–55°, and 0.1 mD-m, respectively. The additional EDFM fracture cells for natural fracture dip angles of 10° and 45° are, respectively, 40, 320, and 58,560, and the CPU time is 337 and 493 seconds.

Comparisons of the gas flow rate and cumulative gas production for three different kinds of natural fracture dip angles are shown in Figure 13.33. It can be seen that the gas recovery after 20 years of production increases by 18%, 80%, and 96% corresponding to the dip angle of 10°, 45°, and 80–90°, respectively. The gas production of lower dip angle natural fractures (10°) is far less than that of the higher dip angle natural fractures (80°–90°). However, the gas production of middle dip angle natural fractures (45°) is close to the higher dip angle natural fractures (80°–90°). The reason for this is the decreased intersecting fracture number between the lower dip angle natural fractures and hydraulic fractures. The intersecting fracture number is 319, 703, and 783 for natural fracture dip angle of 10°, 45°, and 80–90°, respectively. Figures 13.34 and 13.35 display the pressure distribution and drainage volume of natural fractures with dip angles 10° and 45° after 20 years of production. It can be clearly seen that the drainage area and volume of lower dip angle case are less than those of the middle dip angle case due to the less intersecting fractures between hydraulic and natural fractures. Comparing Figures 13.34b and 13.35b with Figure 13.26 indicates that middle dip angle (45°) natural fractures almost have the same drainage area and volume as of the high dip

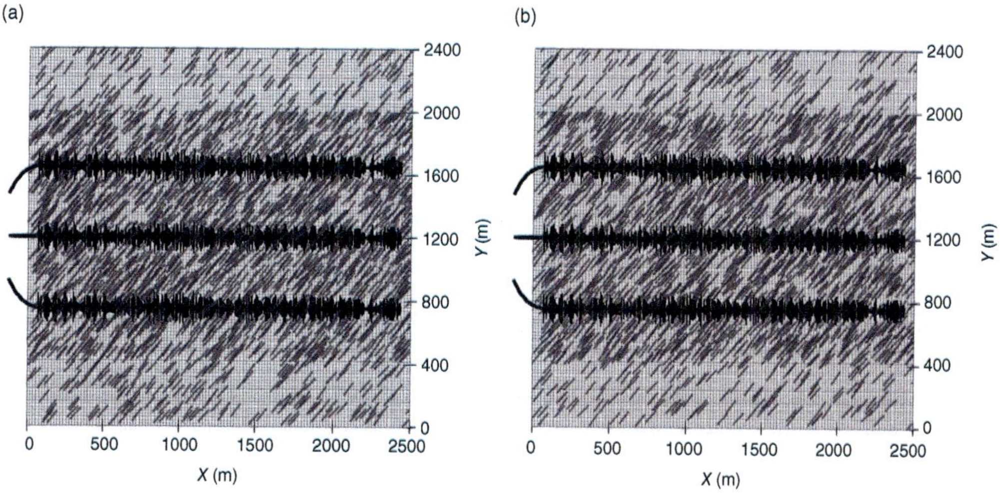

Figure 13.30 Three horizontal wells with two different natural fracture dip angles (2D view). (a) Dip angle of 10° and (b) dip angle of 45°.

(a)

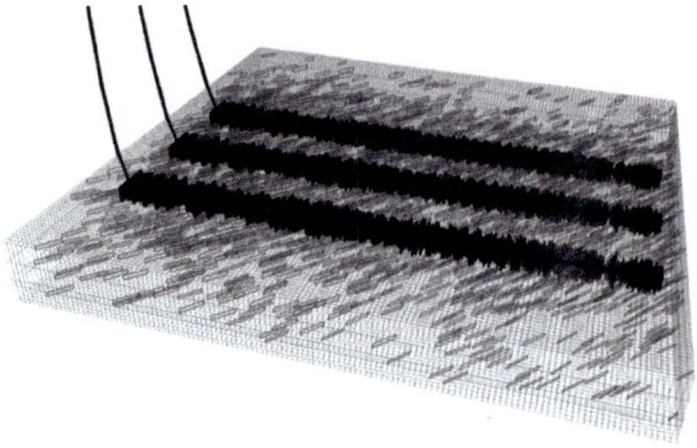

(b)

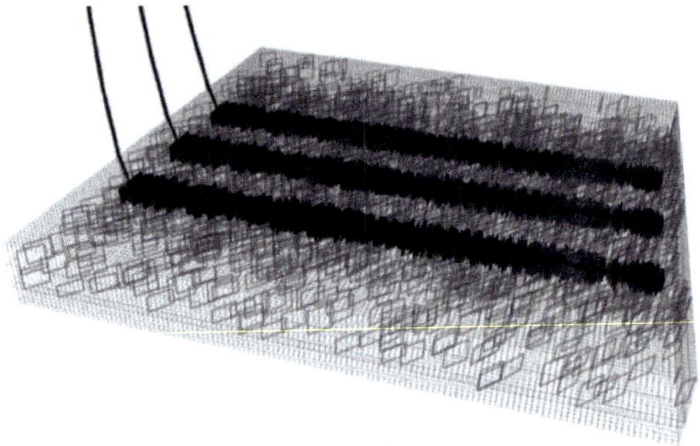

Figure 13.31 Three horizontal wells with two different natural fracture dip angles (3D view). (a) Dip angle of 10° and (b) dip angle of 45°.

angle case since the intersecting fractures of these two cases are close. Therefore, the dip angle also shares the same mechanism for increasing the gas recovery with the effect of azimuth, number of natural fractures, and number of sets of natural fractures on well performance, i.e. with the increasing intersecting fractures, there are more connected fracture area resulting in more production.

13.7 Effect of Natural Fracture Conductivity on Well Performance

In addition to the natural fracture connectivity, the natural fracture conductivity is also a parameter that plays a key role in the well performance of shale reservoirs. Based on the case study with azimuth of 45°–55° in Section 13.6.1, the fracture conductivity values are considered to be 0.001, 0.01, 1 , and 10 mD-m. The 2D and 3D views of natural fracture distribution are already given in Figures 13.16a and 13.17a. The natural fracture number, aperture, length, height, azimuth, and dip angle are 2500, 0.01 m, 50 m, 5–10 m, 45°–55°, and 80°–90°, respectively.

Figure 13.36 compares the gas flow rate and cumulative gas production for different conductivities. Compared to the base case, the gas recovery after 20 years of production increases by 42%, 54%, 63%, 69%, and 70% corresponding to conductivities 0.001 mD-m, 0.01 mD-m, 0.1 mD-m, 1 mD-m, and 10 mD-m, respectively. It is noticeable that gas production increases with increasing natural

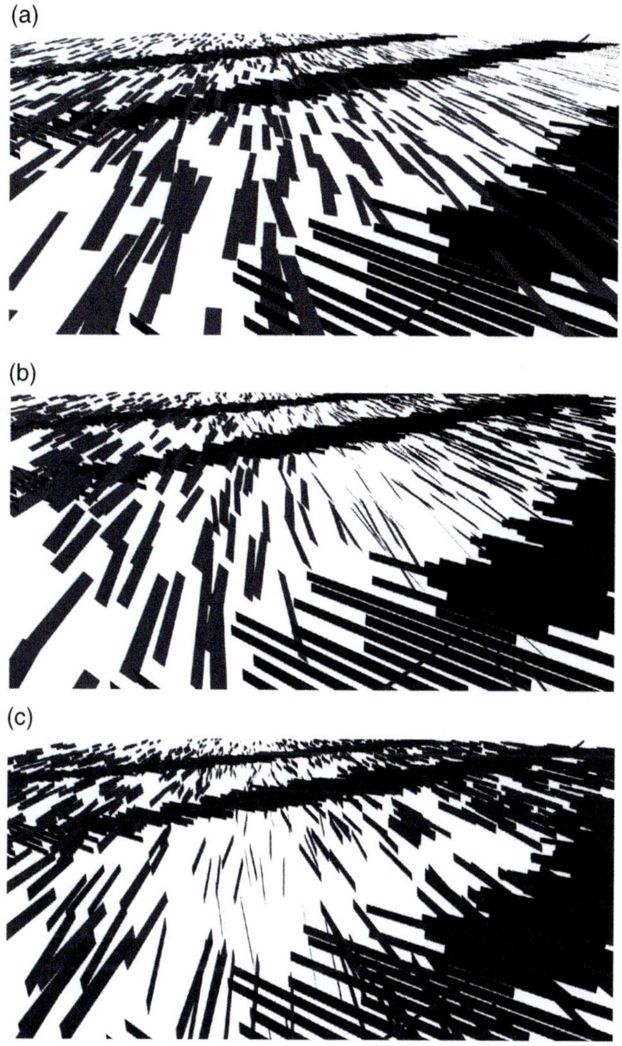

Figure 13.32 Magnified view of natural fractures with three different kinds of dip angles (red lines represent well bores; blue planes represent hydraulic fractures; green planes represent natural fractures). (a) Dip angle of 10°, (b) dip angle of 45°, and (c) dip angle of 80–90°.

fracture conductivity. However, when the conductivity is larger than 1 mD-m, the increase in conductivity has no influence on the well performance, which means it has reached the infinite conductivity.

Figures 13.37 and 13.38 further compare the 2D pressure distribution and 3D drainage volume of two different natural fracture conductivities after 20 years of production corresponding to 0.001 mD-m and 1 mD-m, respectively. Comparatively, the increase of conductivity undoubtedly increases the drainage area and drainage volume in the near wellbore region.

The sensitivity studies in Sections 13.6 and 13.7 have shown the capability and flexibility of the EDFM for complex natural fracture simulations. The number of cells including matrix cells, EDFM fracture cells, and total cells for each case and the corresponding CPU time are summarized in Table 13.3. In the DPDK model, the number of fracture cells are always equal to the number of matrix cells. Nevertheless, the EDFM creates less fracture cells than the DPDK model in some cases,

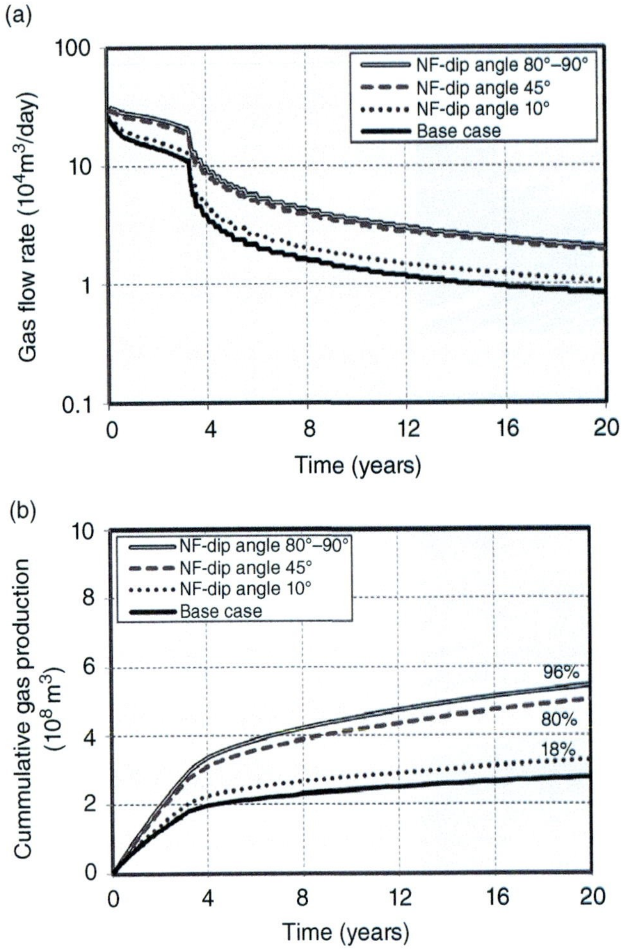

Figure 13.33 Comparisons of well productivity with different natural fracture dip angles. (a) Gas flow rate and (b) cumulative gas production.

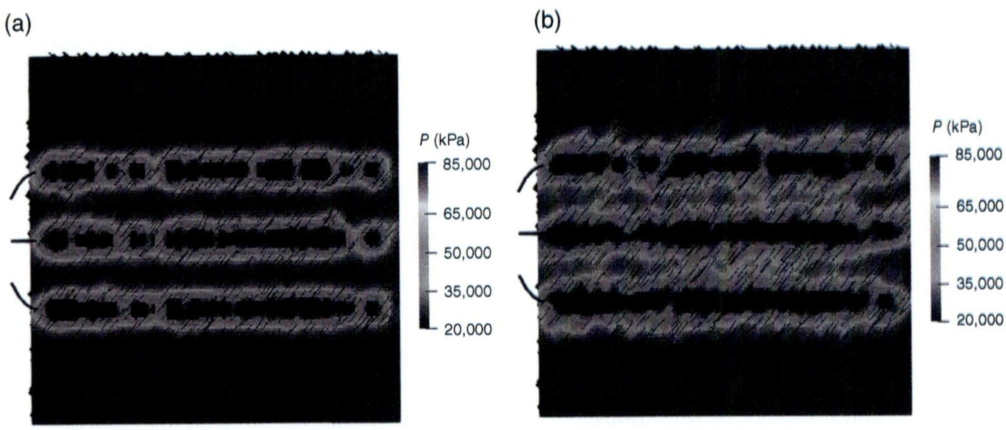

Figure 13.34 Pressure distribution with two different natural fracture dip angles after 20 years of production. (a) Dip angle of 10° and (b) dip angle of 45°.

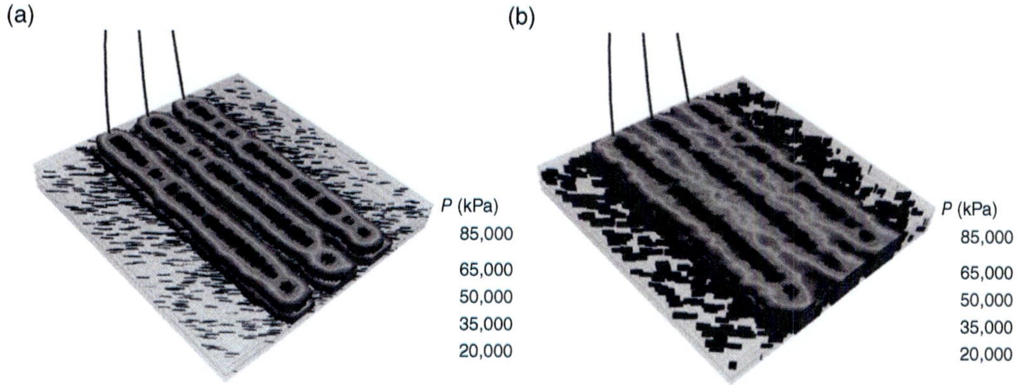

Figure 13.35 Drainage volume with two different natural fracture dip angles after 20 years of production. (a) Dip angle of 10° and (b) dip angle of 45°.

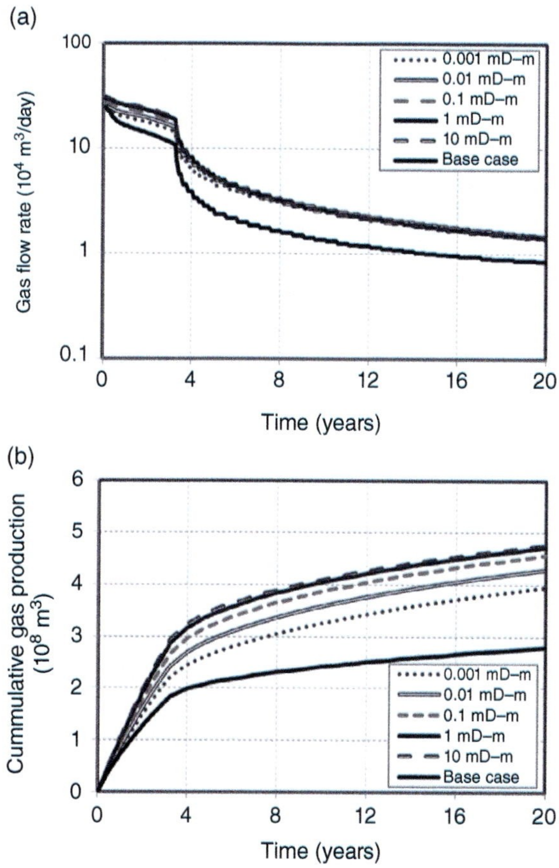

Figure 13.36 Comparisons of well productivity among six different conductivity scenarios of natural fractures. (a) Gas flow rate and (b) cumulative gas production.

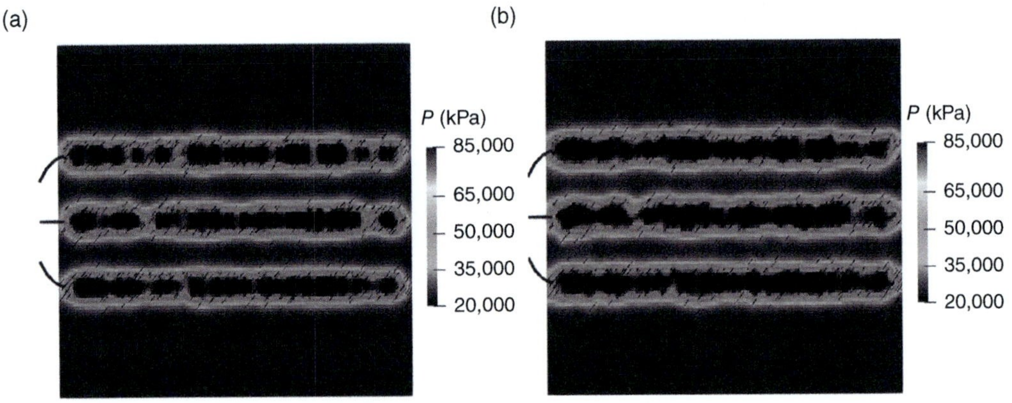

Figure 13.37 Comparisons of pressure distribution with two natural fracture conductivities after 20 years of production. (a) Natural fracture conductivity of 0.001 mD-m and (b) natural fracture conductivity of 1 mD-m.

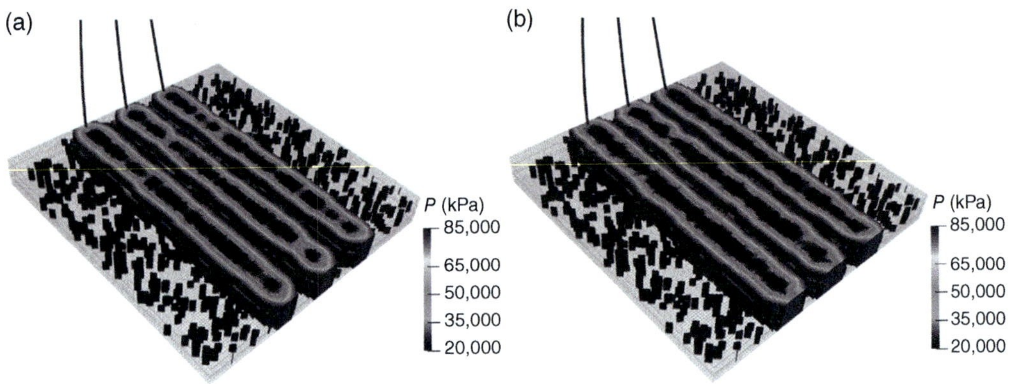

Figure 13.38 Comparisons of drainage volume with two natural fracture conductivities after 20 years of production. (a) Natural fracture conductivity of 0.001 mD-m and (b) natural fracture conductivity of 1 mD-m.

Table 13.3 The number of grid blocks information and simulation time for each case.

Case name	Matrix cells	EDFM fracture cells	Total cells	CPU time (seconds)
13.5 Base case	60,000	10,560	70,560	141
13.6.1 Natural fracture azimuth 45–55°	60,000	40,320	100,320	355
13.6.1 Natural fracture azimuth 90°	60,000	33,120	93,120	298
13.6.2 Natural fracture number 5000	60,000	70,560	130,560	520
13.6.3 Natural fracture length 100 m	60,000	64,800	124,800	547
13.6.5 Natural fracture dip angle 10°	60,000	40,320	100,320	337
13.6.5 Natural fracture dip angle 45°	60,000	58,560	118,560	493

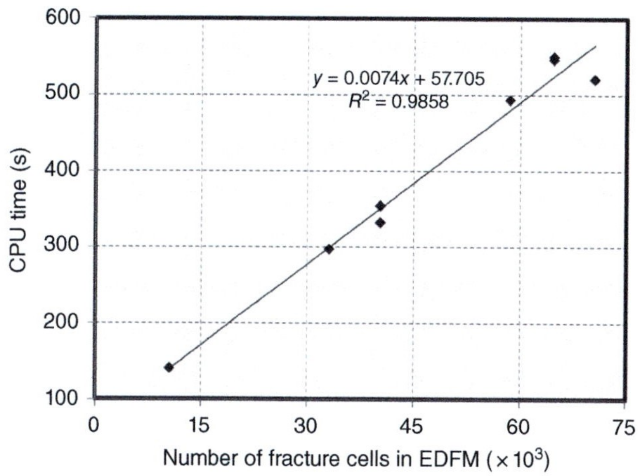

Figure 13.39 Relationship between simulation CPU time and the number of fracture cells in EDFM.

such as the base case explained in Section 13.5, natural fracture azimuth cases explained in Section 13.6.1, and natural fracture dip angle cases explained in Section 13.6.5. In other cases, such as the 5000 natural fractures case explained in Section 13.6.2, and 100 m natural fractures length case described in Section 13.6.3, the EDFM creates more fracture cells than the DPDK model. It is also found that there exists a linear relationship between the additional EDFM fracture cells and the CPU time, illustrated in Figure 13.39.

13.8 Conclusions

In this study, the method of EDFM was briefly introduced and benchmarked against the traditional LGR method, and the effects of complex 3D natural fractures on long-term well performance were extensively analyzed. Sensitivity on key geometry-controlling parameters, such as natural fracture DFN azimuth, total number, length, set number, and dip angle were designed, and flow-controlling parameter such as conductivity was also examined. On the basis of the presented case comparisons, the following conclusions can be drawn:

1) The accuracy of the EDFM is well benchmarked with the traditional LGR method and is proven reliable with a cumulative gas production percentage difference of 0.95% at 30 years. The EDFM can easily deal with complex natural and hydraulic fractures without redefining the gridding schemes of the reservoir with a complex algorithm. This advantage showcases the EDFM's superior efficiency to fulfill multiple instances of sensitivity studies.
2) The difference of connected fractures between azimuth of 45°–55° and 90° is 27.47-fold. Doubling the number of natural fractures from 2500 to 5000 increased the connected fractures number by 2.05-fold. Doubling the set of natural fractures from a single set to two sets increased the connected fractures number by 2.14-fold. Modifying natural fractures from high dip angles (80°–90°) to low dip angles (10°) decreased the connected fractures number by 0.4 times. Therefore, the azimuth is the most sensitive parameter controlling the number of connected fractures in this study.

3) The change of azimuth angle from 90° to 45°–55° increased production by 12.7%. Increasing the total natural fracture number from 2500 to 5000 improved production by 28.4%. Extending the lengths of natural fractures from 50 to 100 m enlarged production by 16.0%. Doubling the set number of natural fractures raised production by 8.2%. Reducing dip angle from 80°–90° to 10° and 45° decreased production by 40.0% and 7.8%, respectively. Therefore, the total natural fracture number and dip angle are the two most sensitive parameters to improve well performance.
4) Based on the drainage volume visualization results, it is concluded that the intensity of natural fractures around the near-wellbore region and in the inter-wellbore region is the contributing factor to the development of drainage area/volumes and enhancement of well productivity. This implies that the degrees of overall connectivity may not directly impact well performance but rather the regional connectivity (near/inter-wellbore connectivity).
5) As the conductivity of natural fractures reached approximately 1 mD-m, or one-tenth of the conductivity of hydraulic fracture (10 mD-m), the infinite conductivity situation was reached meaning further increments of conductivity would not affect well productivity.
6) The accurate and reliable EDFM have proven its practicality in analyzing complex fracture geometries and DFN network, and seamless coupling between numerical fracture modeling and numerical flow modeling. This application of EDFM unlocks the possibility of further investigate more in-depth fracture-flow mechanism in the future.
7) The number of additional EDFM fracture cells can be either more or less than the additional fracture cells produced by DPDK model due to different degrees of fracture complexity. However, the CPU time only increases linearly with the number of fracture cells for EDFM method, given the total matrix cell number and other matrix properties are kept the same.

References

Bourbiaux, B. (2010). Fractured reservoir simulation: a challenging and rewarding issue. *Oil & Gas Science and Technology - Revue d'IFP Energies nouvelles* 65 (2): 227–238.

Cavalcante Filho, J.S.A., Xu, Y., and Sepehrnoori, K. (2015). Modeling fishbones using the embedded discrete fracture model formulation: sensitivity analysis and history matching. *SPE Annual Technical Conference and Exhibition* (28–30 September 2015), Houston, Texas, USA.

Cipolla, C.L., Lolon, E., Erdle, J.C., and Rubin, B. (2009). Reservoir modeling in shale-gas reservoirs. SPE Eastern Regional Meeting (23–25 September 2009), Charleston, West Virginia, USA.

Du, S., Liang, B., and Yuanbo, L. (2017). Field study: embedded discrete fracture modeling with artificial intelligence in Permian basin for shale formation. *SPE Annual Technical Conference and Exhibition* (9–11 October 2017), San Antonio, Texas, USA.

Fisher, M.K. and Warpinski, N.R. (2012). Hydraulic-fracture-height growth: real data. *SPE Production & Operations* 27 (1): 8–19.

Fisher, M.K., Heinze, J.R., Harris, C.D. et al. (2004). Optimizing horizontal completion techniques in the Barnett shale using microseismic fracture mapping. *SPE Annual Technical Conference and Exhibition* (26–29 September 2004), Houston, Texas, USA.

Fumagalli, A., Zonca, S., and Formaggia, L. (2017). Advances in computation of local problems for flow-based upscaling in fractured reservoirs. *Mathematics and Computers in Simulation* 137: 299–324.

Gale, J.F.W., Reed, R.M., and Holder, J. (2007). Natural fractures in the Barnett shale and their importance for hydraulic fracture treatments. *AAPG Bulletin* 91 (4): 603–622.

Gale, J.F.W., Elliott, S.J., Laubach, S.E. (2018). Hydraulic fractures in core from stimulated reservoirs: core fracture description of HFTS slant core, Midland Basin, West Texas. *SPE/AAPG/SEG Unconventional Resources Technology Conference* (23–25 July), Houston, Texas.

Karimi-Fard, M., Durlofsky, L.J., and Aziz, K. (2004). An efficient discrete-fracture model applicable for general-purpose reservoir simulators. *SPE Journal* 9 (2): 227–236.

Leines-Artieda, J.A. (2020). Discrete fracture network modeling and simulation using EDFM. Master thesis. The University of Texas at Austin.

Maity, D. (2018). Microseismicity analysis for HFTS pad and correlation with completion parameters. *SPE/AAPG/SEG Unconventional Resources Technology Conference* (23–25 July), Houston, Texas.

Maxwell, S.C., Urbancic, T.I., Steinsberger, N., and Zinno, R. (2002). Microseismic imaging of hydraulic fracture complexity in the Barnett shale. *SPE Annual Technical Conference and Exhibition* (29 September–2 October 2002), San Antonio, Texas, USA.

Miao, J., Yu, W., Xia, Z. et al. (2018). An easy and fast EDFM method for production simulation in shale reservoirs with complex fracture geometry. *2nd International Discrete Fracture Network Engineering Conference* (20–22 June 2018), Seattle, Washington, USA.

Moinfar, A., Narr, W., Hui, M.-H. et al. (2011). Comparison of discrete-fracture and dual-permeability models for multiphase flow in naturally fractured reservoirs. *SPE Reservoir Simulation Symposium* (21–23 February 2011), The Woodlands, Texas, USA.

Moinfar, A., Varavei, A., Sepehrnoori, K., and Johns, R.T. (2013). Development of a coupled dual continuum and discrete fracture model for the simulation of unconventional reservoirs. *SPE Reservoir Simulation Symposium* (18–20 February 2013), The Woodlands, Texas, USA.

Moinfar, A., Varavei, A., Sepehrnoori, K. et al. (2014). Development of an efficient embedded discrete fracture model for 3D compositional reservoir simulation in fractured reservoirs. *SPE Journal* 19 (2): 289–303.

Olson, J.E. (2008). Multi-fracture propagation modeling: applications to hydraulic fracturing in shales and tight gas sands. *42nd U.S. Rock Mechanics Symposium* (29 June–2 July 2008), San Francisco, California, USA.

Panfili, P., Colin, R., Cominelli, A. et al. (2015). Efficient and effective field scale simulation of hydraulic fractured wells: methodology and application. *SPE Reservoir Characterisation and Simulation Conference and Exhibition* (14–16 September 2015), Abu Dhabi, UAE.

Peaceman, D.W. (1983). Interpretation of well-block pressures in numerical reservoir simulation with nonsquare grid blocks and anisotropic permeability. *SPE Journal* 23 (3): 531–543.

Pei, Y., Yu, W., and Sepehrnoori, K. (2020). Investigation of vertical fracture complexity induced stress interference in multilayer shale gas reservoirs with complex natural fractures. *SPE Annual Technical Conference and Exhibition, Virtual* (26–29 October 2020).

Raterman, K.T., Farrell, H.E., Mora, O.S., et al. (2017). Sampling a stimulated rock volume: an Eagle Ford Example. *SPE/AAPG/SEG Unconventional Resources Technology Conference* (24–26 July), Austin, Texas.

Shakiba, M. and Sepehrnoori, K. (2015). Using embedded discrete fracture model (EDFM) and microseismic monitoring data to characterize the complex hydraulic fracture networks. *SPE Annual Technical Conference and Exhibition* (28–30 September 2015), Houston, Texas, USA.

Sun, H., Yu, W., and Sepehrnoori, K. (2020). Integrated fracture characterization and flow profiling with distributed temperature sensing. *SPE Annual Technical Conference and Exhibition, Virtual* (26–29 October 2020).

Tavassoli, S., Xu, Y., and Sepehrnoori, K. (2018). Modeling fault reactivation using embedded discrete fracture method. *SPE Annual Technical Conference and Exhibition* (24–26 September 2018), Dallas, Texas, USA.

U.S. Energy Information Administration (2015). World shale resource assessments. https://www.eia.gov/analysis/studies/worldshalegas ().

U.S. Energy Information Administration (2019). EIA adds new play production data to shale gas and tight oil reports. Press release (15 February).

Vermylen, J. and Zoback, M.D. (2011). Hydraulic fracturing, microseismic magnitudes, and stress evolution in the Barnett shale, Texas, USA. *SPE Hydraulic Fracturing Technology Conference* (24–26 January), The Woodlands, Texas.

Wang, Y. (2015). A hybrid dual-continuum discrete fracture modeling approach for numerical simulation of production from unconventional plays. *SPE Annual Technical Conference and Exhibition* (28–30 September 2015), Houston, Texas, USA.

Warpinski, N.R., Mayerhofer, M.J., Vincent, M.C. et al. (2009). Stimulating unconventional reservoirs: maximizing network growth while optimizing fracture conductivity. *Journal of Canadian Petroleum Technology* 48 (10): 39–51.

Warren, J.E. and Root, P.J. (1963). The behavior of naturally fractured reservoirs. *SPE Journal* 3 (3): 245–255.

Weng, X., Kresse, O., Cohen, C.E. et al. (2011). Modeling of hydraulic-fracture-network propagation in a naturally fractured formation. *SPE Production & Operations* 26 (4): 368–380.

Wu, K. and Olson, J.E. (2014). Mechanics analysis of interaction between hydraulic and natural fractures in shale reservoirs. *SPE/AAPG/SEG Unconventional Resources Technology Conference* (25–27 August 2014), Denver, Colorado, USA.

Wu, K. and Olson, J.E. (2015). Simultaneous multifracture treatments: fully coupled fluid flow and fracture mechanics for horizontal wells. *SPE Journal* 20 (2): 337–346.

Xu, Y. and Sepehrnoori, K. (2019). Development of an embedded discrete fracture model for field-scale reservoir simulation with complex corner-point grids. *SPE Journal* 24 (4): 1552–1575.

Xu, Y., Cavalcante Filho, J.S.A., Yu, W. et al. (2017). Discrete-fracture modeling of complex hydraulic-fracture geometries in reservoir simulators. *SPE Reservoir Evaluation & Engineering* 20 (2): 403–422.

Xu, Y., Yu, W., and Sepehrnoori, K. (2019). Modeling dynamic behaviors of complex fractures in conventional reservoir simulators. *SPE Reservoir Evaluation & Engineering* 22 (3): 1110–1130.

Yu, W., Xu, Y., Weijermars, R. et al. (2017). A numerical model for simulating pressure response of well interference and well performance in tight oil reservoirs with complex–fracture geometries using the fast embedded–discrete–fracture–model method. *SPE Reservoir Evaluation & Engineering* 21 (2): 489–502.

Yu, W., Wu, K., Zuo, L. et al. (2019). Embedded discrete fracture model assisted study of gas transport mechanisms and drainage area for fractured shale gas reservoirs. *Unconventional Resources Technology Conference* (22–24 July 2019), Denver, Colorado, USA.

14

A Closed-Form Relationship for Production Rate in Stress-Sensitive Unconventional Reservoirs

Hui-Hai Liu, Huangye Chen, and Yanhui Han

Aramco Americas: Aramco Research Center–Houston, Houston, TX, USA

14.1 Introduction

During hydrocarbon production from unconventional reservoirs, the largest resistance to hydrocarbon flow occurs in rock matrix that, at the same time, has the largest storage capacity for hydrocarbon. This is mainly because the matrix has small pore sizes and thus extremely low permeability on the order of 10–100 nD, while both natural and hydraulic fractures have significantly higher permeabilities. Consequently, the hydrocarbon production rate, to a large extent, is controlled by matrix permeability among several other factors (e.g. Carratu et al. 2014; Liu et al. 2016).

Reservoir matrix permeability can be measured or estimated in laboratory or field. The commonly used laboratory approach is the pressure pulse decay method that estimates permeability from a core plug (with a typical size being approximately 1 in.) by analyzing pressure data generated by transient gas flow through the plug. The laboratory results show that shale matrix permeability is a strong function of both stress and pore pressure because of the shale stress sensitivity and impacts of Knudsen diffusion and slippage flow (Sone and Zoback 2013). The upscaling of permeability obtained from laboratory measurements is generally needed for large-scale reservoir simulations because of the subsurface heterogeneity (Liu et al. 2018).

Rate transient analysis (RTA) is widely used to estimate field-scale reservoir permeability using early production data. RTA employs closed-form relationships between the production rate, time, and related reservoir flow parameters including matrix permeability (that is derived from analytical solutions to fluid flow in reservoirs) to fit the production data. Although numerical modeling is more flexible, and thus can avoid certain approximations, RTA is almost exclusively based on analytical solutions because of its ability to diagnose different flow regimes and provide useful physical insight into parameter estimation, as well as its ease of implementation in practical applications.

The majority of analytical solutions used in the RTA are for linear flow equations in which permeability is constant, as reviewed by Bello (2009). This is because analytical solutions to linear fluid flow equations can be relatively easily derived using the Laplace transform and other mathematical schemes. There is a consensus in the unconventional-reservoir engineering community that the early production process in an unconventional reservoir is mainly characterized by the so-called linear flow regime in which hydrocarbon production is controlled by one-dimensional flow from

Physics of Fluid Flow and Transport in Unconventional Reservoir Rocks, First Edition.
Edited by Behzad Ghanbarian, Feng Liang, and Hui-Hai Liu.

the reservoir rock to hydraulic fractures. Flow can be considered as occurring in an infinitely long domain because the penetration depth from the hydraulic fracture, owing to the production process, is much smaller than the hydraulic fracture spacing. It is important to distinguish between linear flow equation and linear flow regime herein. The linear flow regime can correspond to both linear and nonlinear governing equations for fluid flow process in it. In this study, we focus on the linear flow regime.

In a stress-sensitive unconventional reservoir, pore pressure is a function of both time and location as a result of pore pressure evolution during production. If the total stress in a reservoir is approximated by the overburden that is a constant, the effective stress, or the total stress minus pore pressure, can be fully characterized by pore pressure variations. In this case, reservoir permeability would not be constant, but a strong function of time and location because pore pressure is a function of such factors. Accordingly, the fluid flow equation becomes highly nonlinear. The RTA, based on analytical solutions to linear flow equations in which permeability is treated as a constant parameter, is not applicable any longer.

Effort has been made to develop RTA methods for nonlinear flow equations within the linear flow regime. Two general approaches have been proposed. The first approach is to linearize the nonlinear flow equation using transformed variables such as pseudo pressure and pseudo time (e.g. Nobakht and Clarkson 2012; Tobatabaie et al. 2017). After linearization, the governing equation for flow with transformed variables still contains some coefficients that depend on pore pressure. As an approximation, the coefficients are evaluated at a typical pore pressure, such as the average pore pressure in the region of influence resulting from production, and then treated as constants in developing analytical solutions (e.g. Nobakht and Clarkson 2012; Tobatabaie et al. 2017). In other words, this approach ignores the fact that the pore pressure-dependent coefficients are functions of location.

The second approach directly uses the closed-form relationships among production rate, pseudo pressure, production time, and reservoir parameters that are obtained from analytical solutions to the linear flow equation corresponding to constant permeability. Then, the effects of the nonlinearity are accounted for by introducing a correction factor into the above closed-form relationships. The correction factor was developed empirically based on numerical simulation results (Ibrahim and Wattenbarger 2006) or using exact solutions to the nonlinear flow equation for some specific form of pressure-dependent coefficients in the equation (Qanbari and Clarkson 2013). Both approaches are important steps to addressing the nonlinearity effects and have been shown to be useful in practical applications (Ibrahim and Wattenbarger 2006; Nobakht and Clarkson 2012; Clarkson et al. 2013; Qanbari and Clarkson 2013; Tobatabaie et al. 2017; Johnson and Jamiolahmady 2018). While considerable progress has been made in developing the RTA methodologies based on solutions to the nonlinear flow equation, it is desirable to further investigate such an important issue.

This chapter presents a new closed-form relationship among the production rate, time, and reservoir properties under both constant and variable pressure drawdown conditions in production well for stress-sensitive unconventional reservoirs. The relationship of this kind is critical for the RTA because it can be used for estimating reservoir permeability with production rate as a function of time. We develop such a closed-form relationship based on an exact integral solution to the one-dimensional nonlinear flow equation and an approximation of the ratio of fluid flux to that at the interface between the reservoir and hydraulic fracture. The relationship is compared with numerical experimental results for validation purposes. The equivalent state approximation (ESA) approach is then used to extend the analytical solution of production rate from the constant pressure drawdown condition to the more general variable drawdown condition.

14.2 Production Rate as a Function of Time in the Linear Flow Regime Under the Constant Pressure Drawdown Condition

It is well documented in the literature that in the linear flow regime with a constant pressure drawdown, the cumulative production rate is proportional to the square root of production time for the linear flow equation with a constant permeability (e.g. Bello 2009). In fact, for stress-sensitive unconventional reservoirs that correspond to nonlinear flow equations, the above relationship between the production rate and time still holds (e.g. Qanbari and Clarkson 2013). For completeness, this section provides a detailed derivation of the relationship for nonlinear flow equations. It also serves as the starting point for developing the relationship among the production rate, time, and reservoir properties in the next section.

As previously indicated, in the linear flow regime, production is controlled by one-dimensional flow from the rock matrix to hydraulic fractures. Thus, this study focuses on solving the one-dimensional flow equation under appropriate boundary and initial conditions. The governing (mass balance) equation for one-dimensional fluid flow is given by

$$\frac{\partial(\rho\varphi)}{\partial t} = \frac{\partial}{\partial x}\left(\frac{k\rho}{\mu}\frac{\partial p}{\partial x}\right) \tag{14.1}$$

where t is the time, x is the spatial coordinate with zero value at the interface between rock matrix and hydraulic fracture, k is the permeability that is a function of pressure, μ, ρ, and p are the fluid viscosity, density, and pressure, respectively, and φ is the porosity.

Letting

$$m = \rho\varphi \tag{14.2}$$

and considering that m is only a function of pore pressure under the isothermal conditions, we can rewrite Eq. (14.1) as

$$\frac{\partial m}{\partial t} = \frac{\partial}{\partial x}\left(D\frac{\partial m}{\partial x}\right) \tag{14.3}$$

where D is so-called "diffusivity" and defined as follows:

$$D = \frac{k\rho}{\mu}\frac{\mathrm{d}p}{\mathrm{d}m} \tag{14.4}$$

Since D is a function of m (or p), Eq. (14.3) is nonlinear. If D is constant, Eq. (14.3) becomes linear. In the linear flow regime, the flow can be considered to occur in an infinitely long domain, and Eq. (14.3) is subject to the following boundary and initial conditions:

$$\begin{aligned} m(x,t) &= m_i \ (x \geq 0, t = 0) \\ m(x,t) &= m_0 \ (x = 0, t > 0) \\ m(x,t) &= m_i \ (x \to \infty, t > 0) \end{aligned} \tag{14.5}$$

Using the Boltzmann transformation

$$\lambda = xt^{-\frac{1}{2}} \tag{14.6}$$

we can transform Eq. (14.5) into

$$m(\lambda) = m_i \ (\lambda \to \infty)$$

$$m(\lambda) = m_0 \ (\lambda = 0) \tag{14.7}$$

and Eq. (14.3) into

$$-\frac{\lambda}{2}\frac{\mathrm{d}m}{\mathrm{d}\lambda} = \frac{\mathrm{d}}{\mathrm{d}\lambda}\left[D\frac{\mathrm{d}m}{\mathrm{d}\lambda}\right] \tag{14.8}$$

Note that Eq. (14.8) can be much more easily solved than Eq. (14.1), although they are equivalent. This is because Eq. (14.8) is an ordinary differential equation with λ as the only independent variable.

Based on the fluid mass balance, we can determine the cumulative fluid flow rate $Q(t)$ getting out of the reservoir (at $x = 0$) as:

$$Q(t) = A_\mathrm{r}\int_0^\infty (m - m_i)\mathrm{d}x = A_\mathrm{r}(m - m_i)x\,|_0^\infty - A_\mathrm{r}\int_{m_0}^{m_i} x\mathrm{d}m = A_\mathrm{r}\int_{m_i}^{m_0} x\mathrm{d}m \tag{14.9}$$

where A_r is the surface area of hydraulic fracture. In the above equation, the integration by parts is used. Note that the term $A_\mathrm{r}(m - m_i)\,x$ is equal to zero for both $x = 0$ and $x = \infty$ (or $m = m_i$). Combining Eqs. (14.6 and 14.9) gives

$$Q(t) = At^{1/2} \tag{14.10}$$

where A is a constant and given by

$$A = A_\mathrm{r}\int_{m_i}^{m_0} \lambda \mathrm{d}m \tag{14.11}$$

As indicated by Eqs. (14.10 and 14.11), the cumulative flow rate is still proportional to the square root of time in the linear flow regime with a constant pressure drawdown even when the flow equation is highly nonlinear. Eq. (14.10) can be used for fitting the early production data in order to estimate the parameter A. As long as the relationship between A and permeability is known, we can determine the field-scale permeability for the reservoir under consideration.

14.3 An Approximate Relationship Between Parameter *A* and Stress-Dependent Permeability

This section focuses on the development of the relationship between the parameter A in Eq. (14.10) and the stress-dependent permeability by modifying the method of McWhorter and Sunada (1990) used for developing an integral solution to one-dimensional two-phase flow. Mathematically, there is a considerable similarity between the governing equation for two-phase flow in rigid porous media and the governing equations for single-phase fluid flow in stress (or pore pressure)-sensitive porous media. This work is devoted to the single-phase fluid flow although the obtained results can be extended to multiphase fluid flow as well.

To derive the relationship between the parameter A and stress-dependent permeability, we first follow McWhorter and Sunada (1990) to solve the nonlinear flow equation for a flux ratio, F, defined by

$$F = \frac{q}{q_0} \tag{14.12}$$

where $q(x, t)$ is the fluid mass flux (or flow rate per unit cross-sectional area) and q_0 is the flux at $x = 0$ (or the interface between hydraulic fracture and matrix) determined from Eq. (14.10) as follows:

$$q_0 = \frac{1}{A_r}\frac{dQ}{dt} = \frac{1}{2}at^{-\frac{1}{2}} \tag{14.13}$$

where

$$a = \frac{A}{A_r} \tag{14.14}$$

The flow equation, Eq. (14.3), can be rewritten as

$$\frac{\partial m}{\partial t} = -\frac{\partial q}{\partial x} \tag{14.15}$$

Using $q = Fq_0 = \dfrac{1}{2}aFt^{-1/2}$ and Eq. (14.6), we can convert Eq. (14.15) to an ordinary differential equation as follows:

$$a\frac{dF}{dm} = \lambda \tag{14.16}$$

In the derivation of the above equation, the derivatives with respect to x and t in Eq. (14.15) need to be converted to the derivatives with respect to λ. Based on Darcy's law, the mass flux q can be expressed as

$$q = -\frac{k\rho}{\mu}\frac{\partial p}{\partial x} = -D\frac{\partial m}{\partial x} \tag{14.17}$$

Recall that D was defined in Eq. (14.4). Inserting $q = Fq_0 = \dfrac{1}{2}aFt^{-1/2}$ and Eq. (14.6) into Eq. (14.17) yields

$$\frac{d\lambda}{dm} = -\frac{2D}{aF} \tag{14.18}$$

Combining Eqs. (14.16 and 14.18) and eliminating λ result in

$$\frac{d^2F}{dm^2} = -\frac{2D}{a^2F} \tag{14.19}$$

Then, we solve the above equation for F. Two direct integrations of Eq. (14.19) give

$$F = -G(m) + C_1m + C_2 \tag{14.20}$$

where C_1 and C_2 are constants and $G(m)$ is given by

$$G(m) = \frac{2}{a^2}\int_m^{m_0}\int_\xi^{m_0}\frac{D}{F}d\beta d\xi \tag{14.21}$$

Using integration by parts, we can rewrite Eq. (14.21) as (see Appendix 14.A)

$$G(m) = \frac{2}{a^2}\int_m^{m_0}\frac{(\xi - m)D(\xi)}{F(\xi)}d\xi \tag{14.22}$$

By the definition of F (Eq. (14.12)) and noting that $q = 0$ at $x \to \infty$, we have

$$F = 1\ (x = 0 \text{ or } m = m_0) \tag{14.23a}$$

$$F = 0\,(x \to \infty \quad \text{or} \quad m = m_i) \tag{14.23b}$$

From these boundary conditions, the constants in Eq. (14.20) can be determined as

$$C_1 = \frac{1 - G(m_i)}{m_0 - m_i} \tag{14.24}$$

$$C_2 = G(m_i) - \frac{(1 - G(m_i))}{m_0 - m_i} \tag{14.25}$$

Next, Eq. (14.20) can be rewritten as

$$F = -\frac{2}{a^2}\int_m^{m_0} \frac{(\xi - m)D(\xi)}{F(\xi)}\mathrm{d}\xi + \frac{1 - G(m_i)}{m_0 - m_i}m + G(m_i) - \frac{(1 - G(m_i))}{m_0 - m_i} \tag{14.26}$$

It is the exact integral solution to the nonlinear flow Eq. (14.20). In order to get a unique solution to the above equation, we need to know the relationship between F and a.

Equation (14.16) leads to

$$\frac{\mathrm{d}F}{\mathrm{d}m} = 0\,(\lambda = 0 \text{ or } m = m_0) \tag{14.27}$$

Applying this condition to Eq. (14.26) and noting that $(\mathrm{d}G/\mathrm{d}m) = 0$ for $m = m_0$ (from Eq. 14.21), we obtain (from Eq. 14.26) that $G(m_i) = 1$, which, based on Eq. (14.22), is equivalent to

$$a^2 = 2\int_{m_i}^{m_0} \frac{(m - m_i)D}{F}\mathrm{d}m \tag{14.28}$$

This changes Eq. (14.26) to

$$F = \frac{2}{a^2}\int_{m_i}^{m} \frac{(\xi - m)D(\xi)}{F(\xi)}\mathrm{d}\xi \tag{14.29}$$

F can be uniquely determined from Eqs. (14.28 and 14.29) for a given m via an iteration method. For initial guesses of a and $F(m)$, Eq. (14.29) is solved for F as a function of m. With the obtained $F(m)$, an updated a is determined from Eq. (14.28). After that, the updated a and $F(m)$ are used in Eq. (14.29) to determine the new F. The above iteration procedure continues until F does not change significantly between two adjacent iterations for a given m.

After $F(m)$ is obtained, the relationship between parameter A and pore pressure-dependent permeability (included in D) will be given by Eq. (14.28). Inserting Eqs. (14.2 and 14.14) into Eq. (14.28) yields

$$A = A_r\left[2\int_{p_i}^{p_0} \frac{(p - p_i)C_t k\rho}{F\mu}\mathrm{d}p\right]^{\frac{1}{2}} \tag{14.30}$$

where p_O is the pressure at $x = 0$ (or the interface between hydraulic fracture and matrix), p_i is the initial pore pressure in the reservoir, and C_t is defined by

$$C_t = \frac{m - m_i}{p - p_i} \tag{14.31}$$

Equation (14.30) is equivalent to:

$$A = A_r(p_i - p_0)\left[2\int_0^1 \frac{\tilde{p}C_t k\rho}{F\mu}\mathrm{d}\tilde{p}\right]^{\frac{1}{2}} \tag{14.32}$$

in which

$$\tilde{p} = \frac{p - p_i}{p_0 - p_i} \tag{14.33}$$

The application of Eqs. (14.30 or 14.32) is not convenient in practice because F needs to be determined through iterations. This work will construct an approximate relation between F and $\tilde{p}$ such that the iterations are not needed. The constructed relationship needs to satisfy the conditions defined in Eqs. (14.23 and 14.27) that still hold when m is replaced by $\tilde{p}$. These conditions are $F = 0$ for $\tilde{p} = 0$ and $F = 1$ for $\tilde{p} = 1$. The simplest function satisfying those conditions is

$$F = \tilde{p}(2 - \tilde{p}) \tag{14.34}$$

Obviously, there are many other mathematical functions satisfying the conditions defined in Eqs. (14.23 and 14.27). We chose Eq. (14.34) mainly because it is simple and behaves well mathematically. The accuracy of this approximation will be evaluated in the next section.

Inserting Eq. (14.34) into Eq. (14.32) gives the following closed-form relationship between A and stress or pore-pressure dependent permeability:

$$A = A_r (p_i - p_0) \left[2 \int_0^1 \frac{C_t k \rho}{(2 - \tilde{p})\mu} \, \mathrm{d}\tilde{p} \right]^{\frac{1}{2}} \tag{14.35}$$

The integration in the above equation can be interpreted as a weighting average of $C_t k\rho/\mu$ for the pressure range between the bottom-hole pressure in a production well (p_0) and the initial reservoir pore pressure (p_i). A large $\tilde{p}$ (corresponding to a location close to the hydraulic fracture surface) will give a relatively large $1/(2 - \tilde{p})$ and therefore a relatively large weight. In other words, $C_t k\rho/\mu$ (especially permeability k that is generally more pore-pressure sensitive than any of C_t, ρ, and μ) from locations near the hydraulic fracture surface plays a more important role in determining A than that at a location away from the fracture surface. This is expected because the production rate is more sensitive to reservoir permeability near the hydraulic fracture. Finally, it is important to emphasize that Eq. (14.35) is determined based on Eq. (14.13) obtained for a constant pressure drawdown. Thus, Eq. (14.35) is applicable to that case only.

14.4 Evaluation of the Relationship Between Parameter *A* and Stress-Dependent Permeability

This section is devoted to evaluating Eq. (14.35). To do so, we first compare Eq. (14.35) with an exact result obtained when the one-dimensional flow equation is linear. Note that Eq. (14.35) was developed for the nonlinear flow equation including the linear equation as its special case. Thus, Eq. (14.35) is applicable to the linear flow equation as well.

When C_t and $k\rho/\mu$ are constants, the flow equation that is linear and under the constant pressure drawdown condition has the exact solution to the fluid flux at the interface between hydraulic fracture and rock matrix as (Crank 1975; Bello 2009).

$$q = \frac{1}{\sqrt{\pi}} (p_i - p_0) \left(\frac{C_t k \rho}{\mu} \right)^{\frac{1}{2}} t^{-\frac{1}{2}} \tag{14.36}$$

Comparing Eq. (14.36) with Eq. (14.13) yields

$$A = \frac{2}{\sqrt{\pi}} A_r (p_i - p_0) \left(\frac{C_t k \rho}{\mu} \right)^{\frac{1}{2}} \tag{14.37}$$

For constant C_t and $k\rho/\mu$, we can also obtain A from Eq. (14.35) as

$$A = \sqrt{2 \ln 2} A_r (p_i - p_0) \left(\frac{C_t k \rho}{\mu} \right)^{\frac{1}{2}} \tag{14.38}$$

The ratio of A value obtained from Eq. (14.35) to that obtained from Eq. (14.36) is $(\sqrt{\pi \ln 2}/2) \approx 1.04$. Thus Eq. (14.38) is a very good approximation of the exact solution given in Eq. (14.37). To ensure that Eq. (14.35) reduces to Eq. (14.37) when the flow equation becomes linear, we multiply Eq. (14.35) by a correction factor close to one $((2/\pi \ln 2)^{1/2})$ such that the modified Eq. (14.35) becomes

$$A = \frac{2}{\sqrt{\pi \ln 2}} A_r (p_i - p_0) \left[\int_0^1 \frac{C_t k \rho}{(2 - \tilde{p})\mu} \mathrm{d}\tilde{p} \right]^{1/2} \tag{14.39}$$

Equation (14.39) is our final relationship between A and the related parameters.

While we have demonstrated that Eq. (14.39) (or Eq. (14.35)) is practically identical to the exact solution to the fluid flux for the linear flow equation, we next examine the validity of Eq. (14.39) for the more general and nonlinear flow condition. However, a general and closed-form analytical solution to the nonlinear flow equation does not exist, which is the major motivation for developing the approximate relationship given in Eq. (14.39). Numerical modeling is then used for examining the validation of Eq. (14.39) in the rest of this section.

The implicit finite difference method is used to numerically solve the one-dimensional nonlinear partial differential equation for fluid flow (Eq. 14.1). The one-dimensional flow process is schematically demonstrated in Figure 14.1. While the details of the numerical model were published previously (Liu et al. 2019), we briefly discuss it here. The boundary and initial conditions used in the numerical simulations are consistent with those used in the development of

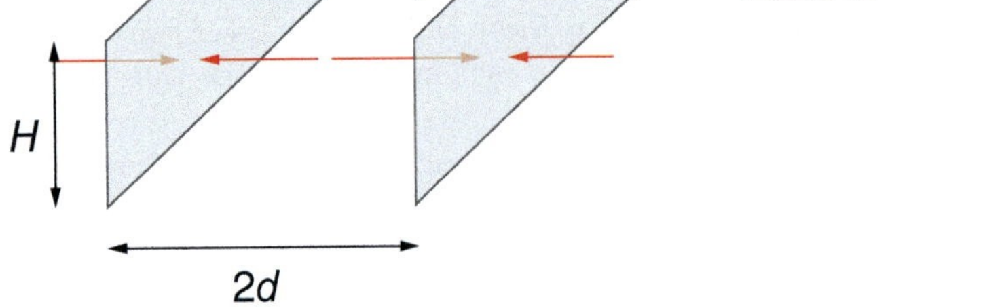

Figure 14.1 Schematic one-dimensional flow from reservoir to hydraulic fractures. H and L are the height and half-length of hydraulic fracture, respectively, and d is the half-spacing of hydraulic fractures.

Eq. (14.39). In other words, initial pore pressure (p_i) is uniform along the rock column, and the pore pressure at $x = 0$ changes to p_0 and does not vary with time. The length of the one-dimensional rock column (l) is large such that the pore pressure at $x = l$ does not change during the simulation, which ensures that the simulated flow is in the linear flow regime. In this study, we consider natural gas to be pure CH_4, which will not impact our evaluation results that depend on the degree of nonlinearity of the flow equation rather than specific fluids. The gas compressibility factor is calculated with the Peng-Robinson equation (Starling and Savidge 1992). The gas viscosity can then be calculated as a function of gas density and temperature (Selim and Sloan 1989). The numerical code was validated by comparing its simulation results with an analytical solution to the one-dimensional diffusion problem; they were found to be essentially identical (Liu et al. 2019). Note that for the gas flow equation with constant fluid and rock properties, the equation is mathematically equivalent to the well-known mass diffusion equation with a constant diffusion coefficient that has analytical solutions.

For stress-sensitive unconventional reservoirs, the reservoir permeability is a strong function of pore pressure. Following previous studies (e.g. Bhandari et al. 2015; Zheng et al. 2015; Liu 2017), we express the pore pressure dependence of permeability in an exponential function form

$$k = k_i \exp[\alpha(p - p_i)] \tag{14.40}$$

where k_i is the reservoir permeability under the initial reservoir pressure and α is the pressure-sensitivity factor for permeability. A larger α value corresponds to a higher degree of permeability sensitivity due to the change in pore pressure. From Eq. (14.40), α is expressed as

$$\alpha = \frac{1}{k}\frac{dk}{dp} \tag{14.41}$$

Thus, the physical meaning of the pressure-sensitive factor α can be interpreted as the relative change in permeability for a unit change in pore pressure. Our numerical simulations focus on relatively large pore pressures and thus the effects of diffusion/slippage flow on permeability can be ignored (Liu et al. 2019). Furthermore, the fundamentals regarding the exponential relationship between permeability and pore pressure (or effective stress) can be found in Liu (2017).

The numerical simulations were performed for $p_i = 8000$ psi and 4000 psi, respectively, and for different pressure drawdowns at the inlet of the one-dimensional flow domain. The temperature in the domain was 190 °F, and permeability was 100 nD. The values for pressure-sensitivity factor (α) for permeability ranged from 4.0×10^{-5}/psi to 1.0×10^{-3}/psi. This covers the typical α values for shale reservoirs that are generally on the order of 1.0×10^{-4}/psi (e.g. Bhandari et al. 2015). The simulated cumulative gas mass production rates are fitted to obtain A values. In the simulations, we set the cross-sectional area (A_r) to be equal to one.

As previously stated, the key assumption in the derivation of the relationship for parameter A (Eq. (14.39)) is the connection between flux ratio (F) and normalized pressure ($\tilde{p}$) given in Eq. (14.34). Figure 14.2 presents the comparison of the simulated flux ratio F with the analytical one (Eq. (14.34)) against the normalized pressure $\tilde{p}$ for $p_i = 8000$ psi, pressure drawdown $p_i - p_0 = 1000$ psi, and $\alpha = 4 \times 10^{-4}$/psi. As can be seen, they match fairly well. Figure 14.3 compares the simulated cumulative production rate with the analytical one as a function of time for the same case shown in Figure 14.2. The analytical curve determined from the closed-form relationship (Eq. (14.39)) lays on top of the simulation curve, indicating that Eq. (14.39) gives very accurate results. However, Eq. (14.34) exhibits certain errors as demonstrated in Figure 14.2. This is largely because Eq. (14.34) is in the integral form that is not very sensitive to the small error in $F(\tilde{p})$.

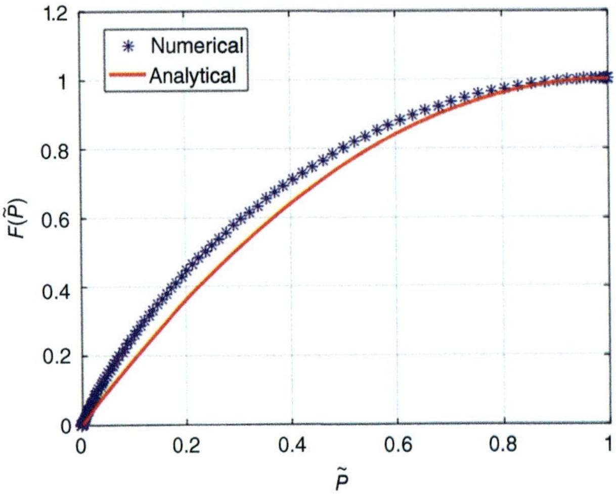

Figure 14.2 Comparing the simulated and analytical flux ratio F against the normalized pressure $\tilde{p}$ for p_i = 8000 psi, pressure drawdown $p_i - p_0$ = 1000 psi, and $\alpha = 4 \times 10^{-4}$/psi.

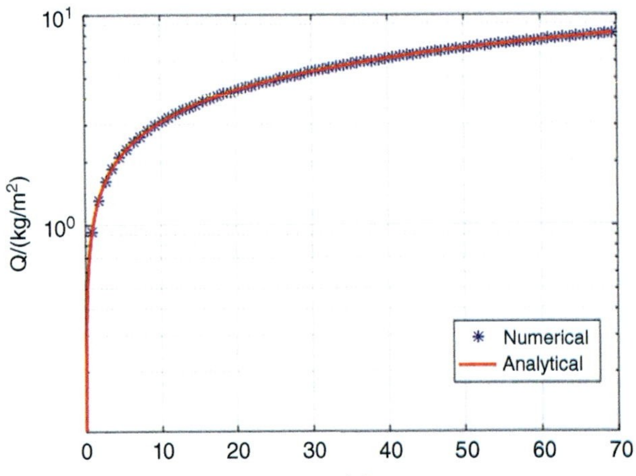

Figure 14.3 Comparing the simulated and analytical cumulative production rate against time for p_i = 8000 psi, pressure drawdown $p_i - p_0$ = 1000 psi, and $\alpha = 4 \times 10^{-4}$/psi.

To further investigate the accuracy of the closed-form relationship given in Eq. (14.39), we define the relative error of parameter A determined from Eq. (14.39) by

$$\delta = \frac{A - A_{\text{num}}}{A_{\text{num}}} \tag{14.42}$$

where A_{num} is the A value determined from the simulations and considered to be the true value here. Tables 14.1 and 14.2 show the relative errors determined for the parameter A for simulation cases where $p_i = 8000$ psi and 4000 psi, respectively, and for different pressure drawdowns and pore pressure sensitivity factors. Clearly, a greater pressure drawdown and/or a greater pore pressure

Table 14.1 The relative error[a] determined for the parameter *A* estimated by Eq. 14.39 for p_i = 8000 psi and different pressure drawdowns and pore pressure sensitivity factors.

$p_i - p_0$ (psi)	δ (%)	M_n
	$\alpha = 4.0 \times 10^{-5}$/psi	
10	0.00	6.98×10^{-5}
1000	−0.04	0.007
3000	−0.09	0.028
5000	−0.11	0.076
7000	−0.60	0.225
	$\alpha = 1.0 \times 10^{-4}$/psi	
10	0.00	4.00×10^{-4}
1000	−0.001	0.041
3000	−0.13	0.121
5000	0.26	0.213
7000	2.80	0.356
	$\alpha = 4.0 \times 10^{-4}$/psi	
10	0.00	0.002
1000	1.20	0.188
3000	4.62	0.444
5000	9.64	0.603
	$\alpha = 1.0 \times 10^{-3}$/psi	
1000	4.30	0.400

[a] The relative error of *A* is defined in Eq. (14.42), and nonlinearity indicator M_n is defined in Eq. (14.43).

sensitivity factor results in a greater relative error, as a result of enhancing the degree of nonlinearity of the flow Eq. (14.3).

We use the following parameter M_n (Emancipator and Kroll 1993) to characterize the nonlinearity of flow equation in Tables 14.1 and 14.2:

$$M_n = \left\{ \int_0^1 \left[\frac{D}{D_i} - 1 \right]^2 d\tilde{p} \right\}^{\frac{1}{2}} \tag{14.43}$$

where D is given in Eq. (14.4) and D_i is the D value under the initial reservoir conditions. Note that the nonlinearity of the flow Eq. (14.3) results from D only. For a constant D (corresponding to a linear problem), M_n is zero. A greater M_n corresponds to a higher degree of nonlinearity and thus to a greater relative error in the estimation of parameter A from Eq. (14.39) that gives accurate results for linear problems.

Figure 14.4 shows the relative error as a function of M_n from the data given in Tables 14.1 and 14.2. If we consider $\delta \leq 5\%$ to be acceptable for practical applications, then based on Figure 14.4, Eq. (14.39) should be valid under the condition:

$$M_n \leq 0.45 \tag{14.44}$$

Table 14.2 The relative error[a] determined for the parameter A estimated by Eq. (14.39) for p_i = 4000 psi and different pressure drawdowns and pore pressure sensitivity factors.

$p_i - p_0$ (psi)	δ (%)	M_n
	$\alpha = 4.0 \times 10^{-5}$/psi	
10	0.00	4.0×10^{-4}
1000	0.10	0.058
2000	0.87	0.153
3000	3.20	0.296
	$\alpha = 1.0 \times 10^{-4}$/psi	
10	0.00	4.0×10^{-4}
1000	0.36	0.090
2000	1.40	0.205
3000	4.16	0.354
	$\alpha = 4.0 \times 10^{-4}$/psi	
1000	1.71	0.227
2000	4.62	0.410
3000	9.80	0.560
	$\alpha = 1.0 \times 10^{-3}$/psi	
1000	4.94	0.427

[a] The relative error of A is defined in Eq. (14.42), and nonlinearity indicator M_n is defined in Eq. (14.43).

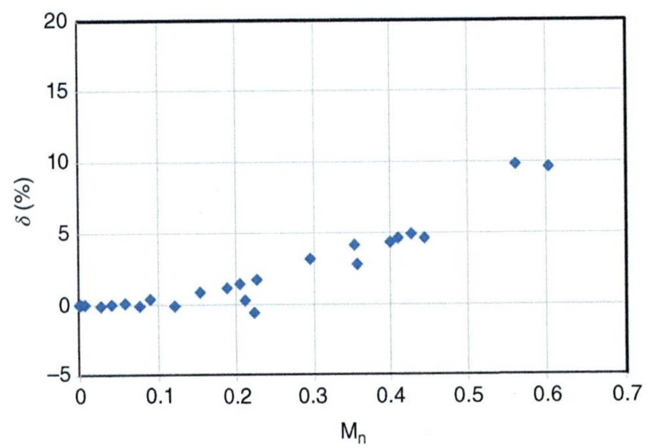

Figure 14.4 The relative error determined for the parameter A estimated by Eq. (14.39) as a function of M_n. The data are from Tables 14.1 and 14.2.

When M_n is beyond the above range, the iteration method, discussed in the above section, should be used to accurately determine the value of A. Since our focus here is on the closed-form relationship between the parameter A and stress-dependent permeability, further discussion of the iteration method is beyond the scope of this chapter.

14.5 Equivalent State Approximation for the Variable Pressure Drawdown Conditions

All the above discussions were relevant to the case where the pressure drawdown is constant. In practice, the pressure drawdown often varies with time. In this case, the relation of the production rate for the constant pressure drawdown cannot be directly applied and needs to be modified. Liu et al. (2022) proposed a principle called ESA (equivalent state approximation) to develop approximate solutions to fluid flow in porous media with variable boundary conditions from those with constant boundary conditions.

ESA includes three steps. Firstly, reformulate the solution for the constant boundary conditions such that it is expressed without explicitly using time as a variable. Secondly, assume the mathematical form of the reformulated solution to be applicable to variable boundary conditions as well. Thirdly, acquire the final solution by incorporating the variable boundary condition into the result obtained from the second step. It is important to note that the practical usefulness of the ESA approach has already been independently demonstrated in several different areas, although the ESA has not been clearly and systematically defined in those studies. For example, Warrick et al. (2005) reported that a solution to water infiltration into soils with variable water ponding depth on the soil surface can be adequately approximated by the solution with a constant ponding depth for the same cumulative infiltration depth of water. Blasingame et al. (1991) demonstrated that the production rate as a function of time for the variable drawdown can be well approximated by that with a constant pressure drawdown as long as the time in the latter case is represented with so-called materials balance time or the ratio of cumulative production to the production rate. Both results can be easily obtained by the ESA principle. Next, we demonstrate how to use ESA to obtain an approximate relation between production rate and time for a variable pressure drawdown case.

To apply ESA, we first need to reformulate the result for the constant pressure drawdown. Eq. (14.13) can be rewritten as:

$$q_0 A_r = \frac{dQ}{dt} = \frac{1}{2} A t^{-1/2} \tag{14.45}$$

Eliminating time by combining Eqs. (14.15 and 14.45) yields

$$Q q_0 A_r = \frac{1}{2} A^2 \tag{14.46}$$

Clearly, the above equation does not explicitly include time t as a variable. According to the ESA, the equation, although derived for the constant pressure drawdown case, can approximately hold for the variable pressure drawdown case. While A is still expressed by Eq. (14.39) in the latter case, it is not constant any longer, but a function of the bottom hole pressure in the production well, p_0, which is a temporal variable.

Finally, we use Eq. (14.46) to derive the production rate as a function of time by incorporating the variable pressure drawdown. Inserting $q_0 A_r = (dQ/dt)$ into Eq. (14.46) gives

$$\frac{dQ^2}{dt} = A^2 \tag{14.47}$$

Then, we obtain the cumulative production as

$$Q = \left(\int_0^t A^2 \mathrm{d}t\right)^{1/2} \tag{14.48}$$

Liu et al. (2022) demonstrated that the above equation gives acceptable results for some problems of practical interest, while more investigations are still needed to determine the general validity range of the ESA.

14.6 Discussions

For stress-sensitive unconventional reservoirs, the conventional RTA needs to be modified or improved, because in these reservoirs, as previously discussed, the flow equation may become highly nonlinear, while the conventional RTA is mainly based on solutions to linear flow equations. Several efforts have been made to deal with this challenging issue (e.g. Ibrahim and Wattenbarger 2006; Nobakht and Clarkson 2012; Clarkson et al. 2013; Qanbari and Clarkson 2013; Tobatabaie et al. 2017; Johnson and Jamiolahmady 2018). The major difference between our method and the previous ones is that we directly solve the nonlinear flow equation.

The key assumption to derive the relationship for the parameter A (Eq. (14.39)) is the connection between the flux ratio (F) and normalized pressure ($\tilde{p}$). Eq. (14.34) likely gives the simplest form of the relationship between F and $\tilde{p}$. There is considerable room to improve the current work with different ways to deal with this assumption. Firstly, it is possible that other mathematical forms of the relationship may give more accurate results, while the current treatment gives satisfactory results under certain conditions. Secondly, a correction factor may also be introduced in Eq. (14.39) for $M_n > 0.45$ to extend the applicable range of the current work. As demonstrated in Figure 14.4, the accuracy of Eq. (14.39) is mainly determined by the parameter M_n that is a measure of the nonlinearity. Thus, the correction factor should be fully characterized by M_n for practical applications. Thirdly, we can also develop an efficient iteration method to solve Eqs. (14.28 and 14.29) for parameter A such that we can accurately estimate A without using any assumption regarding the relationship between F and $\tilde{p}$. However, the closed-form relationships are always preferred because they are relatively easy to use and can provide useful physical insights. We will leave all these options to future studies.

Our closed-form relationship for the production rate in stress-sensitive unconventional reservoirs (Eqs. 14.10 and 14.39) can be easily used with the production data from the linear flow regime to estimate reservoir permeability under ambient conditions (k_i in Eq. 14.40) as long as we know the stress sensitivity parameter α. Liu et al. (2018) theoretically demonstrated that the large-scale α for a heterogeneous reservoir is simply the average of its small-scale (or core scale) values. It is worth noting that applying the conventional RTA to a stress-sensitive reservoir would underestimate the reservoir permeability under the ambient condition (k_i), as explained below. The conventional RTA assumes a constant permeability during production and, thus, estimates an effective permeability that is a weighted average of permeability in the reservoir zone influenced by the production. Since permeability values within the zone are always smaller than the permeability under the ambient condition owing to the stress effect, the latter is always larger than the estimated effective permeability.

The closed-form relationship for production rate in stress-sensitive unconventional reservoirs (Eqs. 14.10 and 14.39) was developed for single-phase flow and can be extended to multiphase flow conditions by ignoring the capillary pressures between different phases. This is because the governing partial differential equation of multiphase flow in terms of total mass for all the phases is analogous to Eq. (14.3). In this case, both m and D in Eq. (14.3) need to include contributions from all the phases. If assuming no changes in the composition of different phases, we then can approximately consider m and D to be dependent on pore pressure only. Thus, the mathematical derivation procedure discussed in this work can still apply. More details of this extension can be found in the literature (e.g. Saleh and Stewart 1992; Heidari and Gerami 2011; Behmanesh et al. 2017).

14.7 Concluding Remarks

In this study, we developed a relationship linking the production rate to reservoir and fluid properties as well as time for the one-dimensional flow process in a stress-sensitive reservoir. This relationship is the key element of the RTA for estimating field-scale permeability based on early production data from an unconventional reservoir. Our approach provides a new method to incorporate the effects of stress sensitivity of permeability into the RTA framework.

We presented an exact integral solution to the one-dimensional and nonlinear flow equation for single-phase fluid flow in the linear flow regime. An approximate and closed-form relationship for the production was then obtained by introducing an approximate relationship between the fluid flux ratio and a normalized pressure to the integral solution. The closed-form relationship was demonstrated to return results very close to the exact results when the flow equation becomes linear. Numerical simulations are performed to further validate our relationship for different initial reservoir pressures, pressure drawdowns, and pressure sensitivity factors for permeability. Satisfactory comparisons between analytical and numerical results are achieved in most cases under consideration. The validity range of the developed relationship among production rate, reservoir and fluid properties, and time is also provided in terms of a measure of the nonlinearity. Room to further improve this work has been discussed. The ESA approach is used to extend the analytical solution of production rate from the constant pressure drawdown condition to the variable drawdown condition.

Nomenclature

A = a variable defined in Eq. (14.10), $MT^{-1/2}$
a = a variable defined in Eq. (14.14), $MT^{-1/2}L^{-2}$
A_{num} = A value determined from numerical experiments, $MT^{-1/2}$
A_r = fracture surface area, L^2
C_t = compressibility defined in Eq. (14.31), $L^{-2}T^{-2}$
D = a variable defined in Eq. (14.4), L^2T^{-1}
F = flux ratio defined in Eq. (14.12) (dimensionless)
k = permeability, L^2
k_i = permeability at initial reservoir pore pressure, L^2

m = apparent gas density defined in Eq. (14.2), ML^{-3}
M_n = dimensionless parameter defined in Eq. (14.43)
p = fluid pressure, FL^{-2}
p_i = initial reservoir pore pressure, FL^{-2}
p_0 = pressure in production well, FL^{-2}
$\tilde{p}$ = dimensionless pressure defined in Eq. (14.33)
q = mass flux $ML^{-2}T^{-1}$
Q = accumulative mass flow rate M
t = time, T
x = location, L
α = stress sensitive parameter defined in Eq. (14.41), L^2F^{-1}
δ = relative error for estimating parameter A (dimensionless)
μ = viscosity, $MT^{-1}L^{-1}$
ρ = gas density, ML^{-3}
ϕ = porosity
λ = Boltzmann transformation variable, $LT^{-1/2}$
ξ = dummy variable defined in Eq. (14.29), ML^{-3}

Subscript

i = initial reservoir condition
0 = fluid property at production well

Appendix 14.A Derivation of Eq. (14.22) with Integration by Parts

Equation (14.21) can be rewritten as

$$G(m) = \frac{2}{a^2}\int_m^{m_0} Y(\xi)d\xi \tag{14.A.1}$$

where

$$Y(\xi) = \int_\xi^{m_0} \frac{D}{F}d\beta \tag{14.A.2}$$

Note that $Y(\xi)$ has the following properties:

$$Y(m_0) = 0 \tag{14.A.3}$$

and

$$\frac{dY}{d\xi} = -\frac{D}{F} \tag{14.A.4}$$

Using the integration by parts, Eq. (14.A.1) becomes

$$\frac{a^2}{2}G(m) = m_0Y(m_0) - mY(m) - \int_m^{m_0} \xi dY \tag{14.A.5}$$

Inserting Eqs. (14.A.2–14.A.4) into Eq. (14.A.5) yields

$$\frac{a^2}{2}G(m) = -m\int_m^{m_0}\frac{D}{F}\mathrm{d}\beta + \int_m^{m_0}\xi\frac{D}{F}\mathrm{d}\xi \tag{14.A.6}$$

Rearranging Eq. (14.A.6) gives

$$G(m) = \frac{2}{a^2}\int_m^{m_0}\frac{(\xi - m)D(\xi)}{F(\xi)}\mathrm{d}\xi \tag{14.22}$$

References

Behmanesh, H., Hamdi, H., and Clarkson, C.R. (2017). Production data analysis of gas condensate reservoirs using two-phase viscosity and two-phase compressibility. *Journal of Natural Gas Science and Engineering* 47 (*Suppl. C*): 47–58.

Bello, R.O. (2009). Rate transient analysis in shale gas reservoirs with transient linear behavior. Ph.D dissertation. Texas A&M University.

Bhandari, A.R., Flemings, P.B., Polito, P.J. et al. (2015). Anisotropy and stress dependence of permeability in the Barnett Shale. *Transport in Porous Media* 108: 393–411. https://doi.org/10.1007/s11242-015-0482-0.

Blasingame, T.A., McCray, T.L., and Lee, W.J. (1991). Decline curve analysis for variable pressure drop/variable flowrate systems. *SPE Gas Technology Symposium*, Houston.

Carratu, J.C., Ozkan, S., and Ozkan, E. (2014). Sensitivities of fractured horizontal wells in shale gas plays to technical and economic parameters. *SPE Western North American and Rocky Mountain Joint Meeting (SPE-169562)*, Denver, Colorado, (17–18 April).

Clarkson, C.R., Qanbari, F., Nobakht, M., and Heffner, L.E. (2013). Incorporating geomechanical and dynamic hydraulic fracture property changes into rate-transient analysis: example from the haynesville shale. *SPE Reservoir Evaluation & Engineering* 16: 303–316.

Crank, J. (1975). *The Mathematics of Diffusion*. Oxford, U.K.: Clarendon Press.

Emancipator, K. and Kroll, M.H. (1993). A quantitative measure of nonlinearity. *Clinical Chemistry* 39 (5): 766–772.

Heidari, S.M. and Gerami, S. (2011). A new model for modern production decline analysis of gas-condensate reservoirs. *Journal of Canadian Petroleum Technology* 50 (7): 14–23.

Ibrahim, M.H. and Wattenbarger, R.A. (2006). Analysis of rate dependence in transient linear flow in tight gas wells. *Abu Dhabi International Petroleum Exhibition and Conference (SPE 100836)*, Abu Dhabi, UAE. (5–8 November 2006).

Johnson, C. and Jamiolahmady, M. (2018). Production data analysis using type curves in tight gas condensate reservoirs: impact of pressure-dependent permeability. *The 80th EAGE Conference and Exhibition (SPE 190832)*, Copenhagen, Denmark (11–14 June).

Liu, H.H. (2017). *Fluid Flow in the Subsurface: History, Generalization and Applications of Physical Laws*. Switzerland: Springer.

Liu, H.H., Ranjith, P.G., Georgi, D.T., and Lai, B.T. (2016). Some key technical issues in modelling of gas transport process in shales: a review. *Geomechanics and Geophysics for Geo-Energy and Geo-Resources* 2 (4): 231–243.

Liu, H.H., Chen, H., Han, Y.H. et al. (2018). On the relationship between effective permeability and stress for unconventional rocks: analytical estimates from laboratory measurements. *Journal of Natural Gas Science and Engineering* 56: 408–413.

Liu, H.H., Chen, H., and An, C. (2019). A criterion for evaluating the effect of shale-matrix dual-continuum flow on gas production. *Geomechanics and Geophysics for Geo-Energy and Geo-Resources* 5 (1): 87–102.

Liu, H.H., Zhang, J., Feng, L. et al. (2022). Is water saturation a fracability indicator for organic-rich, yet low-clay content, tight carbonate source rock reservoirs? *Canadian Journal of Chemical Engineering* 100 (6): 1202–1213.

McWhorter, D.B. and Sunada, D.K. (1990). Exact integral solution for two-phase flow. *Water Resources Research* 28 (3): 399–413.

Nobakht, M. and Clarkson, C.R. (2012). A new analytical method for analyzing linear flow in tight/shale gas reservoirs: constant-flowing-pressure boundary condition. *SPE Reservoir Evaluation & Engineering* 15: 370–384.

Qanbari, F. and Clarkson, C.R. (2013). Analysis of transient linear flow in stress sensitive formations. *SPE Reservoir Evaluation & Engineering* 14: 98–104.

Saleh, A.M. and Stewart, G. (1992). Interpretation of gas condensate well tests with field examples. *The 67th Annual Technical Conference and Exhibition of the Society of Petroleum Engineers (SPE 24719)*, Washington, DC (4–7 October 1992).

Selim, M.S. and Sloan, E.D. (1989). Heat and mass transfer during the dissociation of hydrates in porous media. *AICHE Journal* 35 (6): 1049–1052.

Sone, H. and Zoback, D. (2013). Mechanical properties of shale-gas reservoir rocks – part 1: static and dynamic properties and anisotropy. *Geophysics* 75 (5): D381–D392.

Starling, K.E. and Savidge, J.L. (1992). Compressibility factors for natural gas and other related hydrocarbon gases. American Gas Association (AGA) Transmission Measurement Committee Report No. 8, *American Petroleum Institute (API) MPMS*, Chapter 14.2, Second Edition.

Tobatabaie, S.H., Pooladi-Darvish, M., Mattar, L., and Tavallali, M. (2017). Analytical modelling of linear flow in pressure-sensitive formations. *SPE Reservoir Evaluation & Engineering* 20 (1): 215–227.

Warrick, A.W., Zerihun, D., Sanchez, C.A., and Furman, A. (2005). Infiltration under variable ponding depths of water. *Journal of Irrigation and Drainage Engineering* 131: 358–363. https://doi.org/10.1061/(ASCE)0733-9437(2005)131:4(358).

Zheng, J., Zheng, L., Liu, H.H., and Ju, Y. (2015). On the relationships among permeability, porosity, and effective stress for low-permeability sedimentary rock. *International Journal of Rock Mechanics and Mining Sciences* 78: 304–318.

15

Sweet Spot Identification in Unconventional Shale Reservoirs

Rabah Mesdour, Mustafa Basri, Cenk Temizel, and Nayif Jama

Saudi Aramco, Dhahran, Saudi Arabia

15.1 Introduction

The oil and gas industry has explored different ways to produce oil and gas from shales for several decades (Glaser et al. 2013). However, the production became economic only when the horizontal drilling and multistage hydraulic fracturing were applied, especially after the successful development of the Barnett and Haynesville shales in the United States. In addition to the available technologies of hydraulic fracturing and horizontal-well drilling, the identification of so-called sweet spot has played an important role in the successful development. The sweet spot referrers to locations of shale formations that are favorable for economic production.

There are many countries where unconventional resources have not been characterized and developed mainly because of the lack of technology and infrastructure that allow for exploration and characterization in the known territory. The first step to develop such resources is to learn from the success in other countries by case studies and adapt them to each situation. Following this premise, it is pertinent to develop a practical method to identify sweet spots in shale formations with available data over there (Liu and Wang 2016). Thus, this chapter will briefly review the current available methods for sweet spot identification.

The generic term "shale" refers to a wide range of fine-grained, low-permeability lithologies, including fine-grained carbonates, clay-rich shales, and mudstones (Zou et al. 2012). The identification of sweet spot relies on the reliable characterization of shale properties related to hydrocarbon in place, hydrocarbon flow process, and the easiness of hydraulic fracturing. They include thickness, thermal maturity, lithology, porosity, natural fractures, formation depth, rock traceability, kerogen content, permeability, hydrocarbon fluid properties, and rock mechanical properties (Kinley et al. 2008; Glaser et al. 2013). A number of researchers have developed methods for identifying the sweet spot based on some of the above-mentioned data types obtained from different sources such as seismic survey and well logs. Sweet spot identification is essentially a part of the reservoir characterization, although the sweet spot may be further refined with more data available during the field development process after the characterization.

The objective of this chapter is to review the currently available methods of sweet spot identification in unconventional or shale reservoirs and to identify the related challenges and gaps to guide further research activities in the area.

Physics of Fluid Flow and Transport in Unconventional Reservoir Rocks, First Edition.
Edited by Behzad Ghanbarian, Feng Liang, and Hui-Hai Liu.

15.2 Reservoir Characterization

This section briefly discusses the reservoir characterization that plays a vital role in oil-field development, reserves estimation, crude oil production, reservoir development, and management, while more information on the reservoir characterization and the associated activities generally conducted in the oil and gas industry is available in the literature (e.g. Chopra and Michelena 2011; Varga and Wellmann 2016; Al-Amri et al. 2017; Sun et al. 2015; Abdel-Fattah et al. 2020). As previously indicated, sweet spot identification is essentially a part of the reservoir characterization. Sweet spot locations are crucial for placing the wells on the oil and gas fields.

Reservoir characterization involves estimating and predicting reservoir properties such as rock permeability, porosity, water saturation, identification of lithofacies, reservoir pressure, and temperature (Chaki et al. 2015). A shale formation can be separated into individual layers that have different minerals (such as kaolinite, quartz, and dolomite) and organic material compositions (Glaser et al. 2013). The mineral and organic compositions for each layer depend on the history of geologic conditions through time in the area (Neal et al. 1993; Aplin et al. 1999). Consequently, the composition differences among the layers result in different reservoir properties mentioned above.

During the reservoir characterization, data is collected from different sources, including core analysis (or laboratory measurement of rock properties with core samples), wireline logging from wells, and seismic survey. Core data is used for calibrating the well-log data and relatively accurate understanding of the sedimentary process/history. Well-log data provides petrophysical parameter values that are used for a quantitative lithofacies modeling to discover subsurface lithofacies (Loucks and Ruppel 2007; Jia et al. 2016). The seismic data can be used for three-dimensional facies modeling that maps lithofacies and improves the detection, characterization, and development strategy for shale resources (Woo et al. 2020). Logging and seismic data can be combined to identify the sweet spots, as will be discussed later. Geological analysis, structural interpretation, logging evaluation, and attribute calculation constitute the prerequisites for the successful prediction of multiple shale oil and gas sweet spots. Seismic attributes mainly include amplitude, frequency, phase, and attenuation; during the reservoir evaluation, hundreds of attributes are generally available.

15.3 Sweet Spot Identification

The sweet spot identification is a complex process that can vary significantly among different shale reservoirs because each shale is unique. Giles and Tennant (2014) further classified sweet spots into simple ones, determined by a single factor, and composite sweet spots, determined by multiple factors. Herein, we are more interested in the latter because they are more common in practical applications.

In general, a sweet spot, a three-dimensional volume of a reservoir, should have (i) enough hydrocarbon in place for recovery, (ii) relative easiness for the gas and/or oil recovery (e.g. relatively high permeability) after the stimulation, and (iii) relative easiness for reservoir stimulation or hydraulic fracturing. Some economic factors, such as the cost for transportation, may also need to be considered (Giles and Tennant 2014). Consequently, sweet spot identification includes quantifying lithological formation properties and depositional characteristics such as total organic carbon (TOC), maturity, hydrocarbon saturation, and porosity (Glaser et al. 2013; Tandon and Heidari 2018).

Sweet spots generally correspond to high brittle mineral content, high organic matter and gas contents, and large shale thickness (Liu and Wang 2016).

Clearly, there is not a universal method applicable to all shale formations. Thus, this chapter attempts to provide general guidance for sweet spot identification with different data types and indicators by discussing several typical methods available in the literature, although overlapping exists in these methods.

15.3.1 The Method Based on Organic, Rock and Mechanical Qualities

Aldrich and Seidle (2018) proposed a method for sweet spot identification based on so-called organic quality (OQ), rock quality (RQ), and mechanical quality (MQ), as shown in Figure 15.1. A sweet spot is then identified as reservoir location where all these three Qs are high and correspond to the area where the three circles intersect (with high OQ, RQ, and MQ, respectively) in Figure 15.1.

The OQ is used to characterize the ability for a shale reservoir to generate hydrocarbons. The parameters for OQ include the TOC, maturity, kerogen type, and storage capacity. TOC represents hydrocarbon-generation potential in a source rock. Shales are defined as organic-rich if their TOC is greater than 2% by weight (Boyer et al. 2006). The preservation richness of organic matter depends on the relative rates of its production, dilution, and destruction (happening in bacterial consumption) (Bohacs et al. 2012). According to Aldrich and Seidle (2018), the sweet spot may have TOC from 3 to 12% by weight (Table 15.1). Kerogen is the main source of organic matter and its type and maturity determine the fluid types (e.g. wet gas and dry gas). The storage capacity for shale gas includes capacity to store free gas in pore space and adsorbed gas on the kerogen surface. Sweet spots with kerogen-rich locations generally contain large volumes of adsorbed gas and are stored on kerogen surfaces (Boyer et al. 2006; Pan et al. 2016). However, the adsorbed gas cannot be easily produced unless the bottom hole pressure is very low because only at very low pore pressures a significant amount of absorbed gas can be released from the kerogen surface and become part of free gas in pore space during the production.

The RQ characterizes the ability for a shale reservoir to store and transport hydrocarbon. The parameters for RQ mainly include formation thickness, porosity, and permeability. For a production well, a larger formation thickness is expected to correspond to a larger amount of hydrocarbon in place and likely to larger hydraulic fracture height and thus a potentially larger fracture surface area for the production. According to Aldrich and Seidle (2018), the thickness of the producing zone larger than 100 ft is preferred for identifying the sweet spots (Table 15.1). High matrix porosity

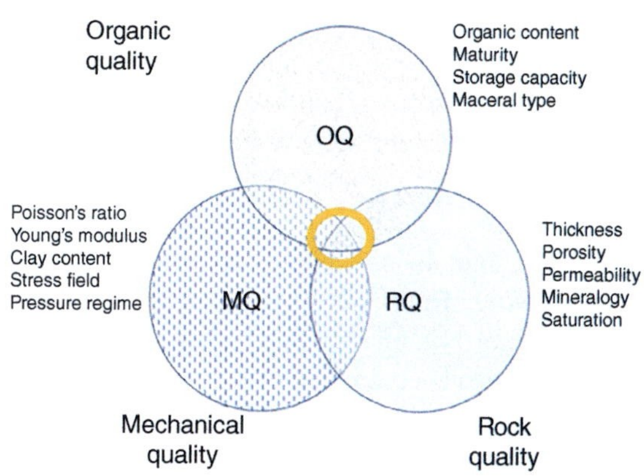

Figure 15.1 An illustration for identifying the sweet spot based on OQ, RQ, and MQ. *Source:* Adapted from Aldrich and Seidle (2018).

Table 15.1 Guidelines for sweet spot identification with OQ, RQ, and MQ.

Quality indicator	Parameter	Range for commercial shales
OQ	Organic content	>3% and <12% of correct macerals
	Maturity (Vro, Tmax, CAI)	Wet gas or dry gas windows
	Storage capacity (Langmuir Isotherm, BCF/ac-ft, etc.)	Variable by shale and thickness
RQ	Thickness	>100′ and bounded by Frac
	Porosity	>8% – can be much higher
	Permeability	>0.001 mD – not firm rule
MQ	Clay content	<40%
	Poisson's ratio	<0.5
	Young's modulus	$>5 \times 10^6$
	Pressure	Generally – overpressure is positive

Source: Aldrich and Seidle (2018)/Society of Petroleum Engineering.

contains a significant amount of free hydrocarbons and allows for a rapid payout of a horizontal well (Glaser et al. 2013). Aldrich and Seidle (2018) proposed that for sweet spots the porosity should be larger than 8% (Table 15.1). During production, the permeability determines how easily the hydrocarbon can be produced from the reservoir. A larger reservoir permeability generally gives a larger production rate when all the other conditions (such as bottom hole pressure) remain the same. Also, permeability is positively correlated with porosity for a given shale reservoir. Aldrich and Seidle (2018) suggested that the sweet spot permeability should be 1 μD (or 0.001 mD) or higher. The suggested permeability value seems to be on the high side, while Aldrich and Seidle (2018) acknowledged that their suggested values for different parameters can only be served as guidelines, not hard rules. We believe that 100 nD may be a more realistic permeability cutoff for the sweet spots based on our experience. Again, this is not a hard rule either. As we previously indicated, every shale is unique in many ways.

MQ is an indicator for reservoir stimulation with hydraulic fracturing. The major purpose of hydraulic fracturing is to enhance the reservoir contact area by artificially inducing fracture surfaces in the reservoir such that the hydrocarbon can be economically produced. Aldrich and Seidle (2018) indicated that the MQ parameters may include clay content, Poisson's ratio, Young's modulus, and pressure regime (Table 15.1), with the first three ones being related to so-called brittleness index (BI). Formations with high brittleness are more susceptible to fractures. During hydraulic fracturing process, complex fracture networks that are preferable for hydrocarbon production tend to form for brittle rocks because complex networks give relatively large fracture areas. Different BI formulations have been proposed and recently reviewed by Mews et al. (2019), Jin et al. (2015), and Holt et al. (2015). A larger fraction of the brittle minerals (generally including silicate minerals, carbonate minerals, and pyrite) corresponds to a larger BI. Ductile minerals include clay and organic matter. Thus, Aldrich and Seidle (2018) suggested that sweet spots should have clay contents less than 40%, and the TOC is less than 12% by weight (Table 15.1). The Young's modulus is the ratio of tensile or compressive stress to the corresponding elastic strain and often served as an indicator of rock stiffness. The Poisson's ratio is defined as the ratio of transverse to longitudinal strains under uniaxial stress condition. As indicated by Mews et al. (2019), a higher BI corresponds to a lower Poisson's ratio and higher Young's modulus, which has a more significant impact on BI than Poisson's ratio. Consequently, Aldrich and Seidle (2018) reported that as guidelines, sweet spots may have Poisson's ratio less than 0.2 and Young's modulus larger than 5.0×10^6 psi (Table 15.1). It is also preferred for a sweet spot to be in the over-pressure regime in which pore pressure is larger than the hydrostatic pressure. This is because in the over-pressure regime, the effective stress, defined as overburden minus pore pressure multiplied by Biot coefficient (that is a positive constant between zero and one), is relatively small and the tensile failure (or hydraulic fracturing) easily occurs.

It is important to note that MQ in Table 15.1 focuses on parameters related to BI. Several recent studies indicate that BI alone cannot sufficiently characterize the fracability (or the easiness of hydraulic fracturing), and some reservoirs with high BI values are still difficult to fracture (Daniel et al. 2011; Bai 2016). In addition to brittleness, the fracability is related to rock structures (e.g. mechanical weakness and natural fractures at different scales), fracture toughness, and in-situ stresses as well. As previously indicated, these additional factors clearly determine how easily a rock can be fractured in an unconventional reservoir.

To determine MQ and its spatial distributions in practical applications, it is desirable to have a geomechanical model for the reservoir, including the depth profiles of vertical stress, maximum horizontal stress, minimum horizontal stress, pore pressure, Young's modulus, Poisson's ratio, and fracture toughness, and the direction of the maximum or the minimum horizontal stress. All the mechanical properties mentioned above, if available, should be included for determining MQ because they may all contribute to the fracability. As an example, a workflow for geomechanical modeling in a hydraulic fracturing study is presented by Rahman and Gui (2016). They also showed that the mechanical properties, such as Young's modulus, pore pressure, and in-situ stress, can vary considerably with location.

15.3.2 Methods Based on Geological and Engineering Sweet Spots

A framework to identify sweet spots based on OQ, RQ and MQ was discussed in Section 15.3.1. Comprehensive data sets from the reservoir characterization are needed for determining these quality indicators. During the exploration phase of the unconventional resource development, well logging and seismic data are obtained. Thus, some frameworks for sweet spot identification are developed mainly using well logging and seismic data. One typical example is based on so-called geological sweet spot and engineering sweet spot, which has been widely used in China (e.g. Chen et al. 2017; He et al. 2020). The geological sweet spot essentially is the zone with high OQ and RQ. The engineering sweet spot is the zone with high MQ that allows for the generation of desirable hydraulic fracture networks. Final sweet spots are areas that are both geological sweet spot and engineering sweet spot.

Chen et al. (2017) reported a workflow to identify the sweet spots for the lower Silurian Longmaxi formation of Sichuan basin in China using interpretations of well logging and seismic data (including gamma rays, density, P-wave, and S-wave logs) as shown in Figure 15.2.

Rock physics analysis aims at locating shale gas sweet spots and defining shale petrophysical properties using logging data (Figure 15.2). Chen et al. (2017) used reservoir thickness, TOC content and gas content to determine the geological sweet spot. The Longmaxi shale geological sweet spot is defined to be zones with the thickness larger than 30 m, TOC content above 3%, and gas content above 2.5 m^3/t. The reservoir thickness and porosity were estimated with the inverted P-velocity and the relationships derived from rock physics analysis. The TOC content was determined based on the density data, and gas content was calculated with multivariate regression method.

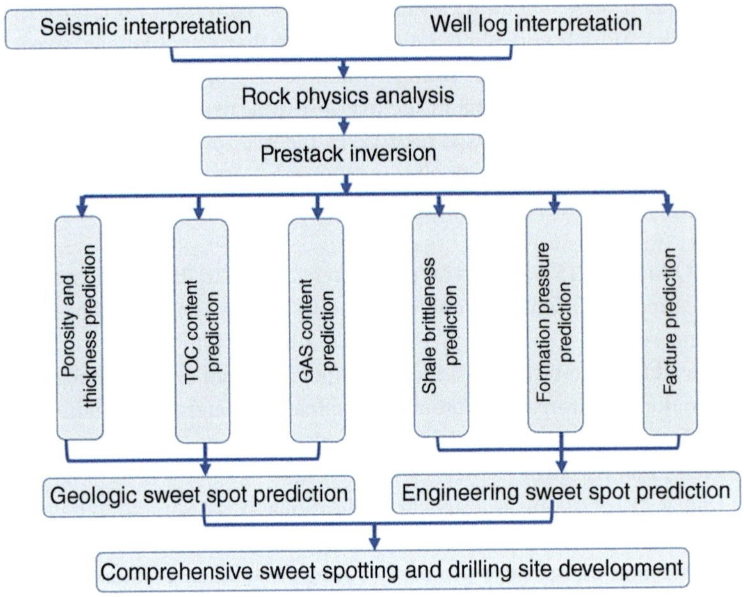

Figure 15.2 Sweet spots prediction workflow using logging and 3D seismic data. *Source:* Chen et al. (2017)/ with permission of Sage Publications.

The detailed procedures to determine these parameters for the geological sweet spot are available in Chen et al. (2017).

Chen et al. (2017) also used high rock brittleness, formation pore pressure, and abundance of natural fracture to define the engineering sweet spot. The brittleness was estimated with rock Young's modulus and Poisson's ratio that were derived from prestack simultaneous inversion on seismic data (Figure 15.2). The formation pressure, the driven force for hydrocarbon migration and accumulation, was calculated with Fillippone (1982) method that assumes overpressure to cause the interval velocity drop. They used curvature attributes derived from the seismic data to characterize the abundance of natural fracture. Larger curvatures may correspond to more fractures because bending stress tends to generate fractures.

To test the usefulness of their workflow, Chen et al. (2017) compared gas production rates from six wells within the identified sweet spots and those from six wells within regions that are either geological sweet spot or engineering sweet spots, but not both. The former rates, on average, are about six times higher than the latter rates, convincingly demonstrating the successful application of the workflow to the reservoir under consideration. Thus, the workflow of Chen et al. (2017) may be used as a guidance to identifying sweet spots in other areas as well, although as previously indicated, there is not a universal method applicable to different shale formations.

Most recently, He et al. (2020) further extended the method of sweet spot identification based on the concepts of geological sweet spot and engineering sweet spot, when they identified the sweet spots in the basin-margin transition zone of southeastern Chongqing that is also located in Sichuan basin. Their method includes 13 key indicators and 11 auxiliary indicators (Table 15.2). In addition to geological sweet spot and engineering sweet spot, the so-called economic sweet spot (that will be discussed later) is also considered. The final sweet spot is the area that is geological sweet spot, engineering sweet spot, and economic sweet spot as well.

Table 15.2 Parameters comparison of shale gas of different areas in the Sichuan Basin, southwest China.

Parameters				Sweet spot
Geological sweet spot	Resource evaluation	Key parameters	High-quality shale thickness (m)	>30
			organic matter abundance (%)	>3
			High-quality shale distribution area (km^2)	>100
			Resource abundance (10^8 m^3/km^2)	>8
		Auxiliary parameters	Thermal evolution degree (%)	2.0–3.0
	Preservation conditions	Key parameters	Formation pressure coefficient	>1.1
			Porosity (%)	>4
			Gas content (m^3/t)	>4
		Auxiliary parameters	Distance from the denudation boundary (km)	>4
			Faults	Underdeveloped and small scale
			Distance from open fault (km)	>1.5
			Pore size (nm)	Mainly >10 nm
			Specific surface area (m^2/g)	>20
Engineering sweet spot	Tectonic stress field	Key parameters	Tectonic style	Anticline, slope, syncline with wide and gentle fold
			Burial depth (m)	2000–3800
			Bedding fracture	Developed
			Microfracture	Developed
			in-situ stress (Mpa)	<80
			Stress difference coefficient	<0.2
			Silicon content (%)	>50
			Curvature	Medium
		Auxiliary parameters	Posson's ratio	<0.25
			Clay content (%)	<30
			Brittleness index (%)	>50
			Top/floor	Good stress barrier
Economic sweet spot	Drilling condition	Key parameters	Surface condition	Abundant flat and wide lands with convenient traffic
		Auxiliary parameters	Water source	Well-developed rivers and lakes

Source: Adapted from He et al. (2020).

For identifying the geological sweet spot, He et al. (2020) divided the indicators into two groups related to resource evaluation and shale gas preservation conditions, respectively (Table 15.2). The major parameters for the resource evaluation include shale thickness, organic matter abundance, shale distribution, and resource abundance, while the thermal evolution or maturity (measured with vitrinite reflectance value) is considered as an auxiliary indicator. A geological sweet spot should have relatively high values of these parameters. The major parameters for the preservation conditions were selected by considering the tectonic evolution and transformation as well as the shale gas accumulation mechanism and conditions. They include formation pore pressure, porosity, and gas content, while the auxiliary parameters are the distance to the denudation boundary, fault conditions, distance to the open fault, specific surface area of the shale, and pore size. A geological sweet spot should not be close to open faults and the denudation boundaries and have relatively high values for other parameters associated with the preservation conditions. The specific ranges of parameter values for the geological sweet spot are given in Table 15.2. Some of these ranges are similar to those reported in Chen et al. (2017).

He et al. (2020) selected parameters for the engineering sweet spot by considering the tectonic stress field and rock mechanical properties that are relevant to the existence of natural fractures and generation of hydraulic fractures. The major parameters include the tectonic style, burial depth, stress conditions, and curvature while the silicon content, clay content, BI, and Poisson's ratio were the auxiliary parameters (Table 15.2). As previously indicated, the curvature can serve as an index for the density of natural fracture that generally enhances the shale gas production. The stress difference is defined as the difference between the maximum and minimum principal stresses. The small stress difference allows for generating more complex hydraulic fracture networks, which is beneficial to the hydrocarbon production. It is particular of interest that the in-situ stress conditions were considered by He et al. (2020), although they did not mention the details regarding how these conditions were determined in their study. As previously indicated in Section 15.3.1, the conditions are related to the fracability. All the auxiliary parameters are related to rock brittleness. The specific ranges of parameter values and descriptions of nonnumerical parameters for the engineering sweet spot are given in Table 15.2.

He et al. (2020) also proposed parameters for the so-called economic sweet spot for the reservoir under their consideration, including surface conditions and water resources. An economic spot should have a flat surface with open terrains that are convenient for transportation. It also should be close to rivers or lakes with sufficient water supply for drilling and hydraulic fracturing. The major purpose for He et al. (2020) to propose the concept of economic sweet spot was to minimize the pre-drilling engineering costs. However, the parameters of the economic sweet spot could be significantly different among different reservoirs.

The method to identify sweet spot with the concepts of geological and engineering sweet spots and the alike have been widely used in Chinese unconventional community. We discussed the studies by Chen et al. (2017) and He et al. (2020) herein as examples including detailed lists of the related parameters and value ranges of them (or descriptions of nonnumerical parameters) for sweet spots. We also refer the readers to the related references cited in these two studies for more applications of the method.

15.3.3 Methods Based on Other Quality Indicators

Sections 15.3.1 and 15.3.2 present two systematic approaches to identify sweet spots based on OQ, RQ, and MQ, and the concepts of geologic and engineering sweet spots, respectively. Since the importance of sweet spot identification and the varieties of shale formations in terms of geological conditions and data availability, many methods using different quality indicators or indices are available in the literature, although all the methods have the same goal to identify reservoir zones or locations that have high hydrocarbon in place and relative easiness for hydrocarbon flow (or recovery) and reservoir stimulation. In this section, we briefly review some selected methods that are based on other quality indicators or indices than those discussed in Sections 15.3.1 and 15.3.2, while the similarities between them and those previously discussed may exist.

Partially motivated by the sweet spot identification, Gamero-Diaz et al. (2013) introduced a mineralogy-based mudstone classification system called sCore that is based on a ternary diagram with clay content, QFM (quartz, feldspar, and mica), and carbonate content as the three corner points. Sixteen mudstone classes can be classified by sCore diagram. The system allows for determining correlations between reservoir mineralogy and parameters associated with reservoir quality (RQ) and completion quality (CQ). The definition of the RQ in sCore is not exactly the same as that in Table 15.1 and its parameters include rock porosity, TOC, matrix permeability, and hydrocarbon saturation. The CQ, similar to previously discussed MQ and engineering sweet spot, is used for predicting the successful reservoir stimulation through hydraulic fracturing. Its parameters include minimum fracture closure stress and rock brittleness. Gamero-Diaz et al. (2013) found that there is a good correlation between mineralogy and CQ in most US shale plays, but the correlation between mineralogy and RQ is not as strong. Nevertheless, sCore allows for identifing lithofacies with the best RQ and CQ, or sweet spots, using minimal processing and no interpretation of the mineralogy data (including those from logs, cores, and cuttings) and provides a common metric for the comparison of such reservoirs worldwide.

Rojas et al. (2016) proposed a method to identify sweet spots for cases in which data are available only for determining rock brittleness and TOC as RQ indicators. One may argue that brittleness and TOC are the major parameters for MQ and RQ, respectively, as discussed in Section 15.3.1. Rojas et al. (2016) provided workflows to estimate the brittleness based on either mineralogy data or rock mechanical property data and indicated that fracturing fluid also plays an important role in determining the sweet spot. They did not provide a detailed criterion in terms of the brittleness and TOC to determine the sweet spots, but gave parameter values of sweet spots from several formations that may serve as references for quickly and roughly identifying the sweet spots with limited information (Table 15.3).

Glaser et al. (2013) reported case studies to identify sweet spots using specific seismic attributes only as quality indicators. For a Woodford Shale in the Arkoma basin of southeast Oklahoma, United States, it was found that a seismic frequency attribute at certain frequencies corresponded to high gas production areas. It was interpreted that the frequency anomalies likely represent areas

Table 15.3 Identified sweet spots and optimum fracture fluid (modified from Rojas et al. (2016)).

Formation	Sweet spot (ft)	BI average (%)	TOC average (wt%)	Optimum fracture fluid
Barnett	7,180–7,390	61.6	3.77	Slickwater
Marcellus	5,694–5,700	71.09	4.57	Slickwater
Geverick Member	13,933–13,950	58	5.51	Hybrid
Posidonia	5,485–5,555	25.5	2.27	Crosslinked
Qiongzhusi	8,740–8,760	67.1	2.35	Slickwater

Source: Rojas et al. (2016)/Society of Petroleum Engineering.

with increased porosity and microfracture density, and thus resulted in high production. The similar observations were also made in some locations in the Delaware basin, United States. In a study area of Bakken Formation, it was found that the seismic anisotropy was closely related to the first 90 days of production from the wells in the area. A larger degree of anisotropy corresponds to a better production. One speculation is that a large degree of anisotropy generally results from highly layered organic matter with well-connected pores and/or the existence of horizontal fractures in reservoirs. Both help the enhancement of hydrocarbon flow and thus the production. Since the seismic data is generally available for the reservoir characterization, the results from the above case studies are practically useful. However, the relations between the sweet spots and the discussed seismic attributes are empirical and likely reservoir specific. More studies on limitations of the use of these seismic attributes and potential generalization to different cases are required with data from different reservoirs.

Avanzini et al. (2016) proposed a method to identify sweet spots using so-called static RQ and dynamic RQ as quality indicators. The former is characterized by lithology and fluid content data and the latter reflects geomechanical behavior associated with hydraulic fracturing. Essentially, they are similar to RQ and CQ used by Gamero-Diaz et al. (2013), respectively, although the parameters used for defining the related RQ indicators are somewhat different in Avanzini et al. (2016) and Gamero-Diaz et al. (2013). Petrophysical data analysis was performed from core measurements and well-log data and geomechanical data analysis was done based on well-log data (Figure 15.3). Cluster analysis was employed to determine geomechanical facies and petrophysical facies. The sweep spot was then identified with those facies with both favorite static RQ and dynamic RQ.

In many cases, the data available for identifying sweet spots from a reservoir is purely from core measurements and well logs (e.g. Avanzini et al. 2016). Since sweet spots are three-dimensional (3D), the 3D distributions of rock properties are needed for sweet spot identification. To deal with this issue, Ismail et al. (2020) recently proposed a color transformation overlay method. It consists of

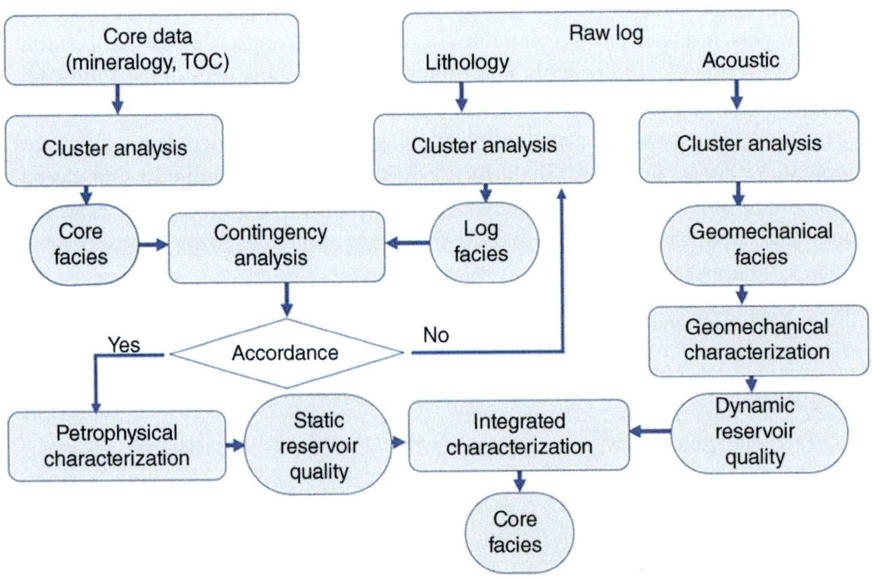

Figure 15.3 Workflow for the productive sweet spot. *Source:* Avanzini et al. (2016)/Society of Exploration Geophysicists.

three steps. The first is to determine distributions of reservoir petrophysical properties and mechanical properties along wells where the well logs data are available. The second step is to determine the 3D distributions of the reservoir properties based on geo-statistics methods. Different color maps are used for representing these distributions. The third step is to determine sweet spots by identifying overlapping zones of favorite reservoir properties with so-called color transformation overlay. With the currently available computational and visualization capabilities, the method is practically useful when the parameter value ranges of the associated quality indicators used for defining the sweet spots are given. However, the method is largely dependent on geo-statistics methods for 3D interpolations, while the uncertainty associated with the geo-statistics methods can be significant when the number (or spatial coverage) of the data points is limited. Thus, the usefulness of the method needs to be further evaluated especially when multi-dimensional seismic data is available.

15.3.4 Methods Based on Data Mining and Machine Learning

In Sections 15.3.1–15.3.3, we reviewed the typical methods for sweet spot identification while keeping in mind that these methodologies may serve as useful guidelines for readers in their practical applications. All these methods are based on different RQ indicators (e.g. OQ, RQ, and MQ), with each indicator being characterized by a group of parameters, although the differences among these indicators, under different names, may not be significant in reality. Then the sweet spots are determined by empirically defining the ranges of the related parameters. These rule-based methods, however, have one major limitation that they all ignore the complex interactions among these parameters. Take TOC as an example. TOC is an important parameter for RQ and can be important for MQ as well for shales with small clay contents, but not for shales with large clay contents. This is because the rock brittleness can be determined by TOC in the former case but mainly by the clay content in the latter case. Therefore, there are different TOC ranges for sweet spots in the two cases.

One method to potentially overcome the above limitation is machine learning and/or data mining method that has become increasingly popular in the oil and gas industry because of their capabilities in efficiently re-cognizing hidden patterns in extremely complex, nonlinear, and multivariate data. Sweet spot identification is a supervised learning problem. For a machine learning and/or data mining method, parameters used for the model input are called features (such as seismic attributes), and entities to match during the model training are called target(s) (such as sweet spot or its properties). After model training, the trained model (or algorithm) is then used to predict sweet spots based on feature values in a reservoir of interest. As will be demonstrated later in this section, the challenge in applying machine learning and/or data mining method is how to appropriately define the target.

Hauge and Hermansen (2017) reported a study on the use of machine learning methods for sweet spot identification. They used and compared several machine learning methods for their study, including logistic regression, *k*-nearest neighbor, support vector machine, and random forest. Note that the focus of this chapter is on the methodology of sweet spot identification. We refer readers to elsewhere (such as textbooks for machine learning) for the details of machine learning methods. Hauge and Hermansen (2017) used seismic attributes as features, including P-wave velocity, S-wave velocity, density, acoustic impedance, 4D residuals of prestack time migration, and average magnitude of reflectivity. They defined the sweet spot with certain ranges of gamma ray and TOC and trained the machine learning models to match the observed gamma ray and TOC data. The trained model was then used for identifying the sweet spots. However, their definition of sweet spot in terms of gamma ray and TOC only is arguable for different practical applications.

Tahmasebi et al. (2017) presented a machine-learning approach to estimate rock TOC and brittleness with well-logs, X-ray diffraction data and core measurements. However, they did not touch on how to identify sweet spots with TOC and brittleness values, while they are obviously related to sweet spot identification. Tandon (2019) discussed integration of machine learning into sweet spot identification mainly based on numerical simulation results. Using large number of numerical simulations, Tandon (2019) developed a machine-learning model to map the rock mechanical properties (including shear modulus, Poisson's ratio, fracture coefficient of friction, maximum horizontal stress, fracture toughness, fracture closure stress, shear dilation angle, and initial hydraulic fracture aperture) as features to the simulated stimulated reservoir volume (SRV) as target. The SRV was used to define sweet spots that correspond to zones with large SRVs. The usefulness of the method, however, needs to be further investigated because it is difficult, if not impossible, to get spatial distributions of those mechanical properties in practice.

The machine learning and/or data mining method works best when the production data are available from certain number of wells in the reservoir under investigation. This is because the target (or sweet spot) can be easily and reliably defined with the production data. Furthermore, completion parameters and their impacts can also be considered. There are many studies in the literature on the production prediction with machine learning and/or data mining methods for shale reservoirs (Liu et al. 2021). These studies can be viewed as part of identification and/or refinement of sweet spots. Herein we only discuss several selected ones as examples. Ketineni et al. (2015) reported an approach to identify sweet spots using artificial neural network (ANN) model for field development planning. They used various seismic attributes and well logs (including gamma ray, neutron porosity, total porosity, bulk density, and deep resistivity) as features and the production as the target. With the trained machine learning model, they were able to identify 100 sweet spots for potentially drilling production wells.

Luo et al. (2018) discussed production prediction using machine learning in Bakken shale. They selected approximately 2000 horizontal wells from Mid Bakken Formation with detailed completion records and initial production rates during 2013 and 2014 and targeted on the prediction of first-year production. The eight features (or model input variables) include the total thickness of the formation, the average volume of proppant per stage, formation depth, porosity, number of stages, the average volume of fracturing fluid per stage, average stage length, and water saturation. The trained ANN model correctly predicts 61% of the variance in well performance on the wells that were not involved in model training. The study also indicates that the average volume of proppant per stage is the most important completion feature impacting the well production.

Liang and Zhao (2019) presented a machine-learning study on production from Eagle Ford shale formation. The study was based on production, petrophysical, and completion data from 4000 wells and had three major components. Firstly, a multivariate linear regression model was created to predict the overall EUR (estimated ultimate recovery). Secondly, the spatial autocorrelation analysis was carried out to identify whether spatial variables could affect the accuracy of the multivariate regression model. Thirdly, the Random Forest Regression models were trained to examine their reliability in predicting EUR with spatially autocorrelated data. The trained model correctly predicts about 60% of the variance in EUR from the wells. The key variables to predict Oil EUR are PVT window, well depth, X coordinate, Y coordinate, lateral length of horizontal well, resources density (the product of hydrocarbon saturation, porosity, and formation thickness), formation thickness of lower Eagle Ford, and hydrocarbon saturation. The top 10 key predictors for gas EUR are PVT window, Y coordinate, vitrinite reflectance equivalent, well depth, gross thickness of shale formation, API, pressure gradient, the product of hydrocarbon saturation, porosity and formation thickness, X coordinate, and TOC.

Crnkovic-Friis and Erlandson (2015) used geology data only to estimate EUR and GOR (gas oil ratio) for a region in Eagle Ford shale formation. The geological parameters include formation thickness, porosity, bulk density, vitrinite reflectance, water saturation, TOC, and brittleness. Data from over 800 wells were used for training and testing their deep-learning model. They also divided the study region into three parts. Data from one part were used for training, while the data from the other two parts were used for validation and testing, respectively. The mean absolute percentage error for predicting the EUR is about 30% and shown to be smaller than the type-curve-average method.

Bowie (2018) reported the use of an ANN and multiple linear regression for optimizing well performance from Duvernay formation around Fox Creek, Alberta, Canada. The Duvernay is an upper Devonian source rock in the Western Canadian Sedimentary basin. The Duvernay ranges in thermal maturity from dry gas to black oil. Data from 262 wells were used for developing the ANN model. There were 21 model input variables (features) that are almost all related to well completion except the initial condensate yield; the geological and petrophysical data were not available for the model development. The developed machine learning model seems to have relatively high predictive power, correctly predicting 78% of the variance in well performance on 52 wells that were not involved in model training. This implies that the performance for the studied wells there are largely driven by well completion parameters for the formation. Key conclusions from the study include that the total proppant amount is a key driver of well performance, and the more expensive fracturing procedures, such as using ceramic or resin-coated proppant, do not provide significant benefit to the well performance for Duvernay formation. The study by Wang and Chen (2016) also indicates that the average proppant placed per stage is the most important parameter contributing to the well production from Duvernay formation.

15.4 Discussion

Section 15.3 presents commonly used methods for sweet spot identification. In this section, we discuss the related challenges and gaps to guide further research activities in the area of sweet spot identification.

Because of its complexity, sweet spot identification has been largely relied on empirical approaches. Consequently, it is important to demonstrate how these approaches actually work in terms of the observed production rates from wells located in the identified sweet spots when the rate data is available, such as the study of Chen et al. (2017). However, studies on the evaluation of the sweet spots identified in the reservoir characterization stage in terms of the production data available in the field development stage are relatively rare in the literature. More studies on such evaluation activities will provide significant insight into the advantages and limitations of different methodologies for sweet spot identification.

It is well known that natural fractures play an important role in determining well productions that generally prefer more natural fractures corresponding to larger reservoir permeability and contact areas. However, the natural fractures have not been considered in some methods for sweet spot identification (e.g. Aldrich and Seidle 2018 or have been considered only qualitatively in other methods (e.g. Chen et al. 2017). The key technical challenge is how to characterize natural fractures reliably and accurately by interpreting and analyzing core measurements, well logs, and especially the seismic data. More studies to address this challenge and develop ways to quantitatively incorporate information on natural fractures into sweet spot identification are desirable.

Most methods for sweet spot identification use rock brittleness to characterize the fracability, or the easiness of hydraulic fracturing. As previously indicated, it is not sufficient to use rock brittleness alone for characterizing the fracability, and some reservoirs with high BI values are still difficult to fracture (Daniel et al. 2011; Bai 2016). In addition to brittleness, the fracability is related to rock structures, fracture toughness, and in-situ stresses as well. The future studies on sweet spot identification need to consider more factors impacting the fracability in addition to the brittleness.

While most methods for sweet spot identification are rule-based, machine learning methods provide new ways for the identification and can overcome the limitation of the rule-based methods that are not able to consider the nonlinear interactions among different parameters used to calculate RQ indicators. They, however, generally need the well production data as their targets. For the newly developed unconventional resources, the production data is limited. Liu et al. (2021) call the prediction of well production for this case as a gray-box problem with incomplete understanding of physics and certain amount of data. They also indicated that the most effective way to deal with the gray-box problem is to develop and employ gray-box approaches that combine the strengths of physics-based models and machine learning that is used for dealing with certain components of the prediction where physical understanding is poor or difficult. More studies on the gray-box approaches for sweet spot identification, including case studies, are highly desirable.

15.5 Conclusion

The success of developing unconventional resources is largely determined by identifying the so-called sweet spots, or locations of shale formations that are favorable for economic production. Sweet spot identification is an important part of reservoir characterization and seeks reservoir zones that have enough hydrocarbon in place for recovery, relative easiness for the gas and/or oil recovery (e.g. relatively high permeability) after the stimulation, and relative effectiveness for reservoir stimulation or hydraulic fracturing. Some economic factors, such as the cost for transportation, may also need to be considered for sweet spot identification.

While we acknowledge that there is not a universal method applicable to different shale formations, this chapter intends to provide a state-of-art review on the currently available methods used for sweet spot identification, including the method based on organic, rock, and mechanical qualities, the method based on geological and engineering sweet spots, the method based on other quality indicators, and the method based on data mining and machine learning. Efforts are made to present parameter-value ranges (when available) used to define the sweet spots in these methods such that the chapter can serve as a guidance for future improvement and applications of these methods.

To motivate future researches on sweet spot identification, this chapter also discusses the related challenges and gaps, including the needs of more studies on the evaluation of the sweet spots identified in the reservoir characterization stage in terms of the production data available in the field development stage, the technical challenge for characterizing natural fractures reliably and accurately by interpreting and analyzing core measurements, well logs and especially the seismic data, consideration of more factors impacting the fracability in addition to the brittleness, and the further development of gray-box approaches that combine the strengths of physics-based models and machine learning for well performance production.

References

Abdel-Fattah, M.I., Pigott, J.D., and El-Sadek, M.S. (2020). Integrated seismic attributes and stochastic inversion for reservoir characterization: insights from Wadi field (NE Abu-Gharadig Basin, Egypt). *Journal of African Earth Sciences* 161 (October 2019): 103661. https://doi.org/10.1016/j.jafrearsci.2019.103661.

Al-Amri, M., Mahmoud, M., Elkatatny, S. et al. (2017). Integrated petrophysical and reservoir characterization workflow to enhance permeability and water saturation prediction. *Journal of African Earth Sciences* 131: 105–116. https://doi.org/10.1016/j.jafrearsci.2017.04.014.

Aldrich, Jeffrey B, and John P Seidle. 2018. ""Sweet spot" identification and optimization in unconventional reservoirs." *Search and Discovery* 80644. https://doi.org/10.13140/2.1.1580.8960.

Aplin, A.C., Fleet, A.J., and Macquaker, J.H.S. (1999). Muds and mudstones: physical and fluid-flow properties. *Geological Society Special Publication* 158: 1–8. https://doi.org/10.1144/gsl.sp.1999.158.01.01.

Avanzini, A., Balossino, P., Brignoli, M. et al. (2016). Lithologic and geomechanical facies classification for sweet spot identification in gas shale reservoir. *Interpretation* 4 (3): SL21–31. https://doi.org/10.1190/int-2015-0199.1.

Bai, M. (2016). Why are brittleness and fracability not equivalent in designing hydraulic fracturing in tight shale gas reservoirs. *Petroleum* 2016 (2): 1–19.

Bohacs, K.M., Carroll, A.R., Mankiewicz, P.J. et al. (2012). Production, destruction, and dilution - the many paths to source-rock development. *Society for Sedimentary Geology (SEPM)* 82: 61–101.

Bowie, B. (2018). Machine learning applied to optimize Duvernay well performance. Paper presented at the SPE Canada Unconventional Resources Conference, Calgary, Alberta, Canada (13 March 2018). Society of Petroleum Engineers. https://doi.org/10.2118/189823-MS.

Boyer, C., Kieschnick, J., Suarez-Rivera, R. et al. (2006). Producing gas from its source. *Oilfield Review* 18 (3): 36–49.

Chaki, S., Routray, A., Mohanty, W.K., and Jenamani, M. (2015). A novel multiclass SVM based framework to classify lithology from well logs: a real-world application. *12th IEEE India International Conference (INDICON)* (17–20 December 2015), pp. 1–5. New Delhi, India. IEEE.

Chen, S., Zhao, W., Ouyang, Y. et al. (2017). Prediction of sweet spots in shale reservoir based on geophysical well logging and 3D seismic data: a case study of lower Silurian Longmaxi formation in W4 block, Sichuan Basin, China. *Energy Exploration & Exploitation* 35 (2): 147–171. https://doi.org/10.1177/0144598716679961.

Chopra, S. and Michelena, R.J. (2011). Introduction to this special section: reservoir characterization. *The Leading Edge* January: 35–37.

Crnkovic-Friis, L. and Erlandson, M. (2015). Geology driven EUR prediction using deep learning. Paper presented at the SPE Annual Technical Conference and Exhibition, Houston, Texas, USA (28 September 2015). Society of Petroleum Engineers. https://doi.org/10.2118/174799-MS.

Daniel, M.J., Ronald, J.H., and Tim, E.R. (2011). A comparative study of the Mississippian Barnett shale, Fort Worth Basin, and Devonian Marcellus shale, Appalachian Basin. *AAPG Bulletin* 91: 475–499.

Varga, Miguel De, and J Florian Wellmann 2016. "Structural geologic modeling as an inference problem: a Bayesian perspective Interpretation" 4 (3): 1–16. https://doi.org/10.1190/INT-2015-0188.1.

Fillippone, W.R. (1982). Estimation of formation parameters and the prediction of overpressures from seismic data. *SEG Technical Program Expanded Abstracts* 1 (1): 482–483.

Gamero-Diaz, H., Miller, C., and Lewis, R. (2013). Score: a mineralogy based classification scheme for organic mudstones. *Proceedings - SPE Annual Technical Conference and Exhibition* 3 (October): 2465–2481. https://doi.org/10.2118/166284-ms.

Giles, M.R. and Tennant, S.H. (2014). Sweet spots: what are they, where are they, how are they created and are they important anyway?. Paper presented at the SPE/EAGE European Unconventional Resources Conference and Exhibition, Vienna, Austria (February 2014). SPE-167760-MS.

Glaser, K.S., Miller, C.K., Johnson, G.M. et al. (2013). Seeking the sweet spot: reservoir and completion quality in organic shales. *Oilfield Review* 25 (4): 16–29.

Hauge, V.L. and Hermansen, G.H. (2017). Machine learning methods for sweet spot detection: a case study. *Quantitative Geology and Geostatistics* 19: 573–588. https://doi.org/10.1007/978-3-319-46819-8_38.

He, X., Zhang, P., He, G. et al. (2020). Evaluation of sweet spots and horizontal-well-design technology for shale gas in the basin-margin transition zone of Southeastern Chongqing, S.W. China. *Energy Geoscience* 1 (3–4): 134–146. https://doi.org/10.1016/j.engeos.2020.06.004.

Holt, R.M., Fjær, E., Stenebraten, J.F., and Nes, O.M. (2015). Brittleness of shales: relevance to borehole collapse and hydraulic fracturing. *Journal of Petroleum Science and Engineering* 131: 200–209.

Ismail, Atif, Arshad Raza, Raoof Gholami, and Reza Rezaee. 2020. "Reservoir characterization for sweet spot detection using color transformation overlay scheme" Journal of Petroleum Exploration and Production Technology 10 (6): 2313–34. https://doi.org/10.1007/s13202-020-00913-5.

Jia, C., Zheng, M., and Zhang, Y. (2016). Some key issues on the unconventional petroleum systems. *Petroleum Research* 1 (2): 113–122. https://doi.org/10.1016/S2096-2495(17)30036-4.

Jin, X., Shah, S.N., Roegiers, J.C., and Zhang, B. (2015). An integrated petrophysics and geomechanics approach for fracability evaluation in shale reservoirs. *SPE Journal* 2015: 518.

Ketineni, S.P., Ertekin, T., Anbarci, K., and Sneed, T. (2015). Structuring an integrative approach for field development planning using artificial intelligence and its application to an offshore oilfield. *Proceedings - SPE Annual Technical Conference and Exhibition*, pp. 2040–2060 (January 2015). https://doi.org/10.2118/174871-ms.

Kinley, T.J., Cook, L.W., Breyer, J.A. et al. (2008). Hydrocarbon potential of the Barnett shale (Mississippian), Delaware Basin, West Texas and Southeastern New Mexico. *AAPG Bulletin* 92 (8): 967–991.

Liang, Y. and Zhao, P. (2019). A Machine learning analysis based on big data for eagle ford shale formation. Paper presented at the SPE Annual Technical Conference and Exhibition, Calgary, Alberta, Canada (23 September 2019). Society of Petroleum Engineers. https://doi.org/10.2118/196158-MS.

Liu, N. and Wang, G. (2016). Shale gas sweet spot identification and precise geo-steering drilling in Weiyuan Block of Sichuan Basin, S.W. China. *Petroleum Exploration and Development* 43 (6): 1067–1075. https://doi.org/10.1016/S1876-3804(16)30124-0.

Liu, H.H., Zhang, J., Liang, F. et al. (2021). Incorporation of physics into machine learning for production prediction from unconventional reservoirs: a brief review of the gray-box approach. *SPE Reservoir Evaluation & Engineering* 24 (4): 847–858.

Loucks, R.G. and Ruppel, S.C. (2007). Mississippian Barnett shale: lithofacies and depositional setting of a deep-water shale-gas succession in the Fort Worth Basin, Texas. *American Association of Petroleum Geologists Bulletin* 91 (4): 579–601. https://doi.org/10.1306/11020606059.

Luo, G., Tian, Y., Bychina, M., and Ehlig-Economides, C. (2018). Production optimization using machine learning in Bakken Shale. *Unconventional Resources Technology Conference* (9 August). https://doi.org/10.15530/URTEC-2018-2902505.

Mews, K.S., Alhubail, M.M., and Barati, R.G. (2019). A review of brittleness index correlations for unconventional tight and ultra-tight reservoirs. *Geosciences* 9: 319.

Neal, J., Risch, D., and Vail, P. (1993). Sequence stratigraphy – a global theory for local success. *Oilfield Review* 5 (1): 51–62.

Pan, R., Gong, Q., Yan, J., and Jin, J. (2016). Elements and gas enrichment laws of sweet spots in shale gas reservoir: a case study of the Longmaxi Fm in Changning block, Sichuan Basin. *Natural Gas Industry B* 3 (3): 195–201. https://doi.org/10.1016/j.ngib.2016.05.003.

Rahman, K. and Gui, F. (2016). Geomechanical sweet spot identification in unconventional resources development. *Society of Petroleum Engineers - SPE Asia Pacific Oil and Gas Conference and Exhibition 2016*. https://doi.org/10.2118/182247-ms.

Rojas, L.F., Peña, Y.Q., and Calderon Carrillo, Z.H. (2016). Brittleness analysis: a methodology to identify sweet spots in shale gas reservoirs. Paper (SPE-180955-MS) presented at the SPE Argentina Exploration and Production of Unconventional Resources Symposium held in Buenos Aires, Argentina (1–3 June 2016).

Sun, T., Merletti, G., Patel, H. et al. (2015). Advanced petrophysical, geological, geophysical and geomechanical reservoir characterization – key to the successful implementation of a geo-engineered completion optimization program in the eagle ford shale. *American Association of Petroleum Geologists* 694–703. https://doi.org/10.15530/urtec-2015-2152246.

Tahmasebi, P., Javadpour, F., and Sahimi, M. (2017). Data mining and machine learning for identifying sweet spots in shale reservoirs. *Expert Systems with Applications* 88: 435–447. https://doi.org/10.1016/j.eswa.2017.07.015.

Tandon, S. (2019). Integrating machine learning in identifying sweet spots in unconventional formations. *SPE Western Regional Meeting Proceedings* 2019: https://doi.org/10.2118/195344-ms.

Tandon, S. and Heidari, Z. (2018). Effect of internal magnetic-field gradients on nuclear-magnetic-resonance measurements and nuclear-magnetic-resonance-based pore-network characterization. *SPE Reservoir Evaluation & Engineering* 21 (3): 609–625.

Wang, S. and Chen, S. (2016). A comprehensive evaluation of well completion and production performance in Bakken shale using data-driven approaches. Paper presented at the SPE Asia Pacific Hydraulic Fracturing Conference (24–26 August 2016). Society of Petroleum Engineers. https://doi.org/10.2118/181803-MS.

Woo, J., Rhee, C.W., Jeoung, T.J., and Jang, S. (2020). Application of multi-resolution graph-based clustering for electrofacies prediction: a case study from the Horn River shale, British Columbia, Canada. *Geosciences Journal* 24 (5): 507–518. https://doi.org/10.1007/s12303-019-0046-3.

Zou, C., Zhu, R., Liu, K. et al. (2012). Tight gas sandstone reservoirs in China: characteristics and recognition criteria. *Journal of Petroleum Science and Engineering* 88–89: 82–91. https://doi.org/10.1016/j.petrol.2012.02.001.

Index

Physics of Fluid Flow and Transport in Unconventional Reservoir Rocks, First Edition.
Edited by Behzad Ghanbarian, Feng Liang, and Hui-Hai Liu.

Printed and bound by CPI Group (UK) Ltd, Croydon, CR0 4YY

17/09/2026

14964314-0004